EUROPEAN MINERALOGICAL UNION
NOTES IN MINERALOGY

Commissioning Editor: Roberta Oberti

Volume 20

THE CONTRIBUTION OF MINERALOGY TO CULTURAL HERITAGE

UNIVERSITY TEXTBOOK

Edited by

GILBERTO ARTIOLI and ROBERTA OBERTI

Mineralogical Society

Published by the European Mineralogical Union and the Mineralogical Society of Great Britain & Ireland, London, 2019

The publication of this textbook is supported by the European Mineralogical Union

Editors of this Volume

Gilberto Artioli, Department of Geosciences, Università di Padova, Padova, Italy
Roberta Oberti, CNR, Pavia, Italy

Managing Editor and Indexer: Kevin Murphy, London

Front cover design: Michel H. Guay

On the front cover: Egyptian faience was coloured by Cu to reproduce turquoise minerals, or by Co to reproduce lapis lazuli. Here a specimen of lapis lazuli from the classical localities in the Badakshan Province, Afghanistan is related to a faience cat from Egypt, 26th Dynasty (cat image courtesy of Sadigh Gallery Ancient Art Inc., New York).

ISSN: 1417 2917
ISBN: 978-0903056-61-8

Published by the European Mineralogical Union and the Mineralogical Society of Great Britain & Ireland (12, Baylis Mews, Amyand Park Road, Twickenham TW1 3HQ, UK)

The EMU Notes in Mineralogy Series

Published volumes

Volume	*Year*	*Editors*	*Title*
1	1997	S. Merlino	*Modular aspects of minerals*
2	2000	D.J. Vaughan R. Wogelius	*Environmental mineralogy*
3	2001	C.A. Geiger	*Solid solutions in silicate and oxide systems*
4	2002	C. Gramaccioli	*Energy modelling in minerals*
5	2003	D.A. Carswell R. Compagnoni	*Ultrahigh pressure metamorphism*
6	2004	A. Beran E. Libowitzky	*Spectroscopic methods in mineralogy*
7	2005	R. Miletich	*Mineral behaviour at extreme conditions*
8	2010	F.E. Brenker G. Jordan	*Nanoscopic approaches in earth and planetary sciences*
9	2010	G. Christidis	*Advances in the characterization of industrial minerals*
10	2010	M. Prieto	*Ion-partitioning in ambient-temperature aqueous systems*
11	2011	M.F. Brigatti A. Mottana	*Layered mineral structures and their application in advanced technologies*
12	2012	J. Dubessy M.-C. Caumon F. Rull	*Applications of Raman Spectroscopy to earth sciences and cultural heritage*
13	2013	D.J. Vaughan R.A. Wogelius	*Environmental mineralogy II*
14	2013	F. Nieto K.J.T. Livi	*Minerals at the nanoscale*
15	2015	M.R. Lee H. Leroux	*Planetary mineralogy*
16	2017	W. Heinrich R. Abart	*Mineral reaction kinetics: Microstructures, textures, chemical and isotopic signatures*
17	2017	I.A.M. Ahmed K.A. Hudson-Edwards	*Redox reactive minerals: properties, reactions and applications in clean technologies*
18	2017	A.F. Gualtieri	*Mineral fibres: crystal chemistry, chemical-physical properties, biological interaction and toxicity*
19	2017	J. Plášil J. Majzlan S. Krivovichev	*Mineralogical crystallography*
20	2019	G. Artioli R. Oberti	*The contribution of mineralogy to cultural heritage*

Copies of the EMU Notes (volumes 1–7) are distributed in Europe by the larger member societies of the European Mineralogical Union:

Società Italiana di Mineralogia e Petrologia:
www.socminpet.it

Mineralogical Society of Great Britain & Ireland:
www.minersoc.org

Société Française de Minéralogie et de Cristallographie:
www.sfmc-fr.org

and by the Secretary of the EMU, Prof. Juraj Majzlan:
juraj.majzlan@uni-jena.de

in America by the Mineralogical Society of America:
www.minsocam.org

Institutional orders as well as individual requests from outside Europe and America should be sent to the Secretary of EMU, Prof. Juraj Majzlan:
juraj.majzlan@uni-jena.de

For volumes 8 onwards, the Mineralogical Society of Great Britain acts as co-publisher and copies may be ordered from www.minersoc.org

or from:
Mineralogical Society
12 Baylis Mews,
Amyand Park Road,
Twickenham TW1 3HQ
UK

E-mail: admin@minersoc.org
Tel. + 44 (0)20 8891 6600
Fax: + 44 (0)20 8891 6599

Contents

Preface

Mineralogy and its companion discipline, crystallography, are fundamental sciences, which from their original observations of natural minerals and their practical applications of economic interest (metal resources) and aesthetic appeal (gems) slowly became core parts of the physical and chemical developments of science in the period following the industrial revolution. Clashing with their fundamental importance in scientific education, mineralogy and crystallography are now encountering an identity crisis in academic curricula, and the net effect is an overall and diffuse rapid shrinking of competences concerning natural materials and their transformation products. All basic and applied sciences are suffering enormously from this general attitude of academic myopia. Some of these issues are discussed in the first chapter of this volume, where the editors argue that mineralogy is a truly interdisciplinary fundamental science as it lies in a very special position between the Earth sciences and materials science. The competent mineralogist should possess a profound perception of the complexity of natural materials, he/she should have the necessary knowledge of the ancient and recent geological and physicochemical processes acting on them and on the artifacts produced by human activities, and he/she should master most of the methods and techniques useful for investigating our common heritage.

The chapters contributed to the volume recognize the important and diverse contributions of mineralogy to the valorization, characterization, interpretation and conservation of cultural heritage. The book focuses on examples of materials and methodological issues rather than technical/analytical details. We have attempted to deal with the cultural heritage materials in chronological order of their technological developments, to relate them to past human activities, and to highlight unresolved problems in need of investigation.

The chapters collected in the present volume, of necessity, show only a selection of examples of the mineralogical applications to cultural heritage. Natural minerals and rocks [silica (chapter 2), obsidian (chapter 11), lithics (chapter 10), gemstones (chapters 8 and 9), pigments (chapter 7)] used by man in antiquity for various purposes are described. Insights into transformation products provide fundamental information on pre-historic and historic man-made materials [glass (chapter 3), mortars (chapter 4), metals (chapter 5), ceramics (chapter 6)]. An exception to this materials-ground approach is chapter 12, which illustrates some state-of-the-art methodological developments based on synchrotron radiation.

The editors originally planned to cover other key topics of mineralogical research related to human heritage, *e.g.* the important advances in our knowledge of apatite and apatite-composites, related to bio-materials such as bones and teeth. Lack of time precluded several authors from contributing. The reader will hopefully forgive the volume limitations and value the large amount of information collected by the authors. The authors are thanked for their efforts and competence; as editors we are greatly indebted for their knowledge and patience.

Gilberto Artioli and Roberta Oberti
Padova and Pavia, July 2019

EMU Notes in Mineralogy, Vol. 20 (2019), Chapter 1, 1–11

Introduction: The role of modern mineralogy in cultural heritage studies

Gilberto ARTIOLI[1] and Roberta OBERTI[2]

[1]*Department of Geosciences and CIRCe Centre, Università di Padova, Via Gradenigo 6, 35131 Padova, Italy*
e-mail: gilberto.artioli@unipd.it
[2]*CNR-Istituto di Geoscienze e Georisorse, sede di Pavia, Via Ferrata 1, 27100 Pavia, Italy*
e-mail: oberti@crystal.unipv.it

This short introduction aims to rethink the role of modern mineralogy and highlights the diverse and important contributions that it may provide in the study of materials and processes relevant to cultural heritage. It is argued that mineralogy lies in a very special position between Earth and materials sciences and that mineralogists have a profound perception of the structural and chemical complexity of natural materials. They possess knowledge of both the ancient and recent geological and physicochemical processes which produced the raw materials used by humans, and of the analogue processes used to transform them into artefacts. It is thus highly appropriate that a volume in the EMU series acknowledges some of the recent contributions of mineralogy to the investigation of human history, art and technology.

1. Traditional *vs.* modern mineralogy

Mineralogy and crystallography are historically closely related disciplines, and unfortunately both have experienced a serious decrease in their appeal as basic formative disciplines in the most recent university curricula. Many courses have recently been cancelled or merged into general courses on geomaterials. This is not the right place to discuss such worrying trends, which are consequences of a simplistic perception of science by society and, in turn, by the recent generations of students. It is a fact, however, that mineralogy and crystallography are perceived as 'old' disciplines, with little to offer for future technologies and societal needs. During the International Year of Crystallography in 2014 (www.iycr2014.org), the International Union of Crystallography (IUCr) did revitalize the concepts and fascination of crystals and minerals for a much larger audience. The many events worldwide showed clearly that far from being a 'crystallized' technique, the basics concepts and the recent advances in crystallography are still fundamental to any investigation of the solid state and thus still have a significant impact on all aspects of human life.

DOI: 10.1180/EMU-notes.20.1

Crystallography no longer exists, in practice, as an independent discipline in most university curricula. This is unfortunate if we think that throughout most of the 20th century many laboratories of physics and chemistry actively developed crystallographic techniques and applications. The long list of Nobel prizes (www.iucr.org/people/nobel-prize) awarded to scientific achievements directly related to, or involving the use of, crystallographic methods and techniques is an impressive demonstration of the impact of crystallography. Also, a few decades ago entire Departments were dedicated to crystallography, such as the Department of Crystallography in Pittsburgh, created by George Jeffery. A few still remain, albeit devoted mostly to biological studies, such as at CSIC, Madrid, or the University of Madras, among others. The Italian CNR still has an Institute of Crystallography based in Bari, and the French CNRS has a 'Laboratoire de cristallographie et sciences des matériaux' based in Caen. However, the small amount of teaching in crystallography and diffraction left in academies is commonly disguised as 'techniques for structural investigations', 'techniques for materials science', or embedded in the basic courses of mineralogy inserted in the curricula of geosciences.

The many mutual interactions between mineralogy, crystallography and the geosciences are very interesting. A brief recapitulation may help to make clear the present and future state of the research in the field.

Crystallography has had illustrious pioneers in Vannoccio Biringuccio (*De la pirotechnia*, 1540), Kepler (*On the six-cornered snowflake*, 1611), Robert Hook (*Micrographia*, 1665) and Steno (*De solido intra solidum naturaliter contento*, 1669). This shows the constant fascination with which crystals were held in scientific minds, besides of course being valued since prehistory for their intrinsic beauty as gems and precious stones (Rapp, 2009). The developments of crystallography and mineralogy were inextricably linked. Metals and ore mining have always driven mineralogical knowledge, and after the Middle Ages the discipline started to find practical encoding (Georgius Agricola, *De re metallica*, 1556). The availability of ores and the need for metal, especially silver, in the late 15th and early 16th centuries led to the need for mineralogical expertise, strict control of mining and the establishing of state mining schools.

Saxon mining officials played a huge role in the development of mineralogy and geology for about two hundred years (Laudan, 1987; Schneer, 1995). Agricola lived in the booming mining towns of Joachimstahl and Chemnitz. Roesler (*Speculum metallurgie*, 1700), Zimmerman (*Obersächsische Bergakademie*, 1740), von Oppel (*Anleitung zur Markscheidekunst*, 1749), Baumer (*Geographia et hydrographia subterranea*, 1779), von Trebra (*Erfahrungen vom Innern der Gebirge*, 1785; Fig. 1), Hoffman (*Dissertatio de Matricibus Metallorum*, 1738) were all connected extensively with mining and the mine services in the Saxony area.

Much of what we now consider geological knowledge was at that time included in disciplines defined by mineralogy and cosmogony (Fig. 2). At the end of the 18th century, Abraham Gottlob Werner (1749–1817), who is said to have originally proposed the chronological succession of rocks, defined mineralogy "*as being made up of three major subdivisions that, taken together, closely approximate the scope of*

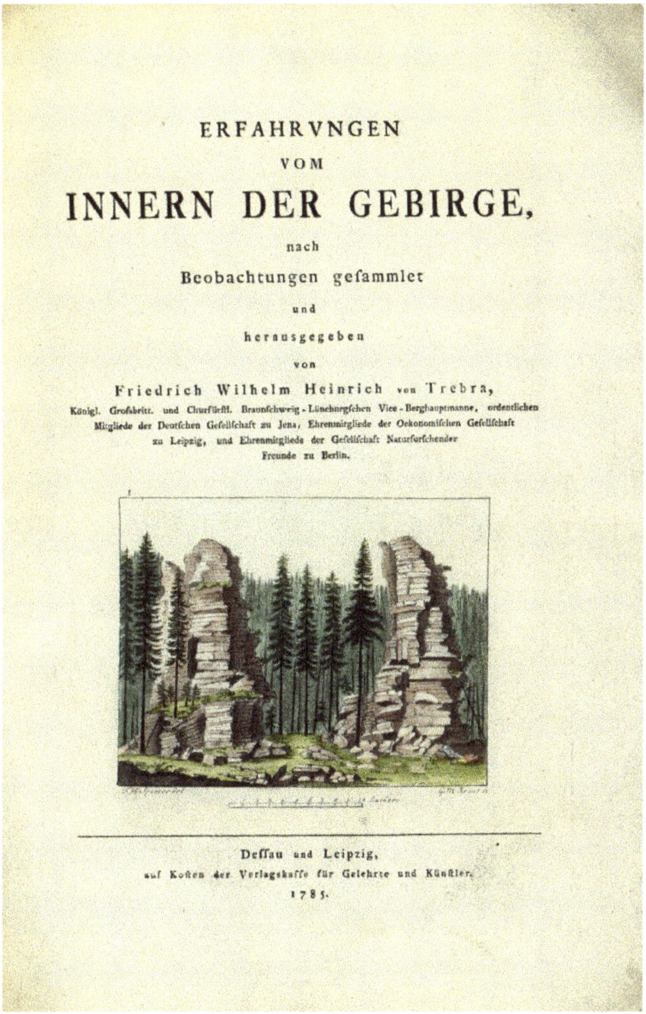

ERFAHRVNGEN

VOM

INNERN DER GEBIRGE,

nach

Beobachtungen geſammlet

und

herausgegeben

von

Friedrich Wilhelm Heinrich von Trebra,

Königl. Großbritt. und Churfürſtl. Braunſchweig-Lüneburgſchen Vice-Berghauptmanns, ordentlichen Mitgliede der Deutſchen Geſellſchaft zu Jena, Ehrenmitgliede der Oekonomiſchen Geſellſchaft zu Leipzig, und Ehrenmitgliede der Geſellſchaft Naturforſchender Freunde zu Berlin.

Deſſau und Leipzig,

auf Koſten der Verlagskaſſe für Gelehrte und Künſtler.

1785.

Figure 1. Cover of a mineralogical volume by Friedrich Wilhelm Heinrich von Trebra (1740–1819), Saxon Oberberghauptmann and a friend of Johann Wolfgang von Goethe. Author of *Erfahrungen vom Innern der Gebirge* (Essays from the interior of mountains, 1785).

modern geology: oryctognosy (the identification and classification of minerals), mineral geography (the distribution of rocks and minerals), and geognosy (the formation and history of rocks and minerals)" (Laudan, 1987, p. 21).

In addition to mining and metallurgy, in the eighteenth century the glass and porcelain industry also fostered greatly the study of mineralogy and geology in Europe. Chemical mineralogy and indeed much of inorganic chemistry were to stem from these premises.

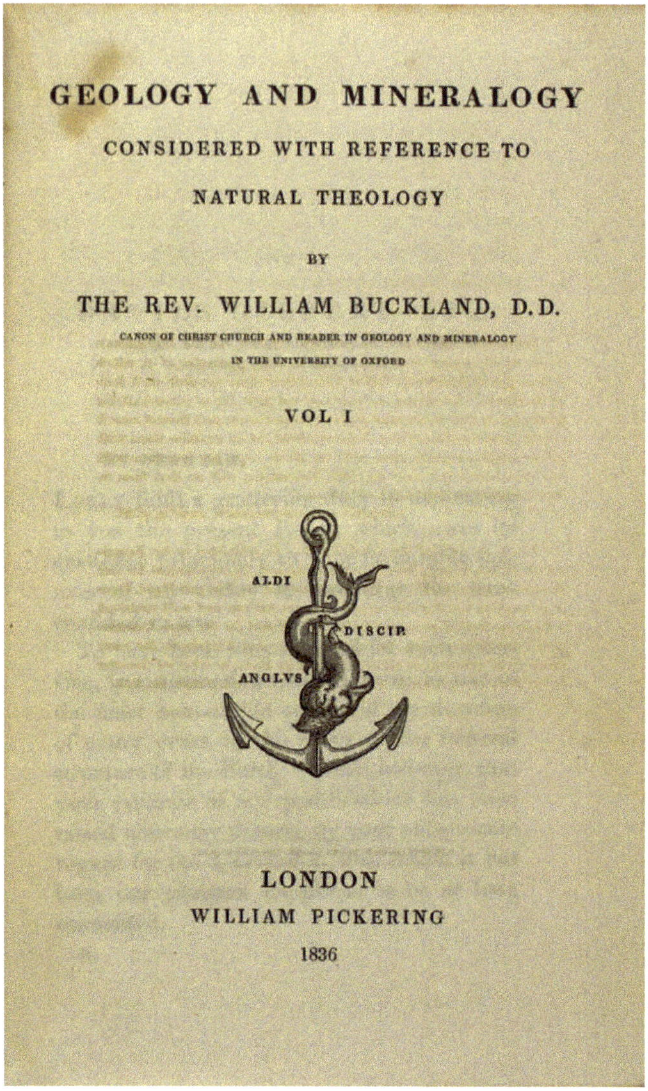

GEOLOGY AND MINERALOGY

CONSIDERED WITH REFERENCE TO

NATURAL THEOLOGY

BY

THE REV. WILLIAM BUCKLAND, D.D.

CANON OF CHRIST CHURCH AND READER IN GEOLOGY AND MINERALOGY

IN THE UNIVERSITY OF OXFORD

VOL I

ALDI DISCIP. ANGLVS

LONDON

WILLIAM PICKERING

1836

Figure 2. The cover of the work by William Buckland, in three volumes (1836).

A couple of centuries ago the standing of mineralogy was therefore quite different from the academic niche in which it tends to be confined today. This queen of the geosciences is not really abdicating from her role, however; we hope that the present situation is just another short-lived example of academic myopia. Mineralogy, and crystallography as well, are alive and kicking, and still in a position to contribute to science and society. As discussed above, mineralogy has always been linked with geology, petrology and geochemistry, but has benefited significantly from having integrated the methods and techniques of modern crystallography and advanced materials science. Not only because crystallography has deciphered the structure and the crystal-chemistry of minerals and their solid solutions, but also because it allowed understanding and modelling of the high-*T* behaviour of minerals and inorganic materials, with an impressive impact on our knowledge of both the planets and material sciences.

Arguably, this knowledge makes mineralogy ideally suited and equipped to meet the demanding challenges posed by archaeometric analysis and conservation problems (Berthoud and Cleuziou, 1981; Chiari, 2000; Maggetti, 2006; Artioli, 2010a). These are its most important assets: (1) Mineralogy (and crystallography) have strong historical and theoretical roots in fundamental physics and chemistry; hence they should not be considered to be naturalistic or phenomenological sciences. (2) Mineralogy deals with all the relevant materials: natural minerals and their transformations by man, in short the materials surrounding us in everyday life. (3) Mineralogy investigates the distribution and properties of minerals in the Earth's crust as well as their availability as resources. Geochemistry, petrology and the survey of mineral deposits stemmed from the long tradition of mineralogical investigation of ore sources. (4) Mineralogy is well acquainted with the complexity of natural systems, commonly composed of a large number of chemical elements and several crystal-chemical phases. Therefore, it is technically equipped to face the methodological and experimental challenges involved in the analysis of complex materials, and in the interpretation of the processes acting upon them (Chiari, 2000; Artioli, 2010b). (5) Modern mineralogy routinely uses the full range of crystallographic techniques to explore the solid state. In many instances it is at the forefront of the analysis of complex materials, *e.g.* at non-ambient conditions. Therefore, mineralogy is well equipped to face intricate analytical problems related to production processes, weathering and conservation.

Mineralogy and crystallography clearly have the core knowledge and expertise that are fundamental to investigating issues in the field of cultural heritage (CH).

The boundaries of cultural heritage investigations cannot be defined easily (Chiari, 2000). They are commonly related to the field of archaeology (archaeometry, archaeological sciences) and to the field of conservation (conservation science, technical art history, authentication). The two areas are currently well distinguished in academia, in professional curricula and in their theoretical and methodological traditions. However, all those disciplines are concerned with the investigation of materials produced by past human activities. From an analytical point of view, artistic and archaeological materials should be investigated using the same scientific techniques and methods, although the primary questions to be answered and the objectives of the research may be quite different. As Chiari (2000) clearly and rightly stated, the material sampled during restoration interventions could well be studied from the archaeometric point of view, and archaeometric investigations are, of course, well suited to providing the scientific basis for proper conservation approaches.

The various participants in CH studies have very different backgrounds, sometimes very specialized, so that it is often difficult to overcome cultural gaps. It is therefore mandatory that scientists, archaeologists, art historians and conservators learn how to interact efficiently when dealing with the cultural heritage (Fig. 3). Indeed, the hyper-specialization of modern academia does not favour interaction. Sometimes the language codes and semantic values of the terms used in the different disciplines are so different that a discussion itself is nearly impossible or, at best, very difficult.

Pollard and Bray (2007) recall the general perception of scientists, who often feel that it is easier to teach the fundamentals of archaeology to a student in physics or chemistry (or mineralogy, may we add), in order to bring him or her to a sufficient level of understanding of the problems to be confronted at the experimental level, than to teach physics and chemistry to an archaeologist. Of course we can quote alternative experiences, where excellent archaeology students started to follow with great success scientific courses in order to reach a deep understanding of the materials investigated, whereas science-based students never quite grasped the anthropological, archaeological or social consequences of their scientific data. Nonetheless, the lesson to be kept in mind is that cultural heritage is a truly interdisciplinary field. Patience, willingness to listen to other people's points of view, and the flexibility to change long-trusted assumptions and protocols are crucial for a fruitful collaboration.

Scientific hyper-specialization has often been welcomed as the sign of increased competence and a sound way to progress. However, the amount of information required by each specific discipline is so vast, that it can hardly be grasped competently by single researchers. Nowadays, computers, databases and networking can greatly facilitate knowledge acquisition, retrieval and exchange. Unfortunately, disciplines perceived as devoted to natural sciences and classification are going out of fashion because some may think that their knowledge, if needed, can be retrieved easily from databases.

This is not the right way to go. What we need in CH studies is an appropriate training in critical investigation and a deep understanding of how to relate the data observed to the problems at hand. The real problem facing cultural heritage is not 'mineralogy or not mineralogy' .... The real problem is to find a person who: (1) is able to deal with natural and man-made materials; (2) is well trained in terms of characterization

Figure 3. Appropriate positive and intense interaction between all players involved in the analysis and management of cultural heritage is mandatory.

techniques; (3) knows the basics of statistics and mathematical treatment of data; and (4) is able to translate complex problems into analytical strategies and to interpret the experimental observations correctly in order to reconstruct the whole picture.

All scientific disciplines can contribute to solving cultural heritage issues, provided that the problems and questions are well posed. The real problem is to avoid the syndrome of the "friendly scientist" (Chiari, 2000), who thinks that he or she can solve any problem by using the fistful of techniques he or she has mastered (*i.e.* the "scientist without parachute", according to the metaphor of Pollard and Bray, 2007). In public lectures, we often quote the honest archaeologist handing a piece of pottery to a mineralogist and petrologist friend (then getting XRD-XRF-OM data back), or to a chemist friend (getting FTIR-Raman data back) or to a physicist friend (getting TEM-XPS data back). Of course each friend will use the techniques with which he is best acquainted, those that are readily available in his laboratory, or the cheapest/easiest ones to apply. If the "friendly scientist" is not an expert in the problem under investigation, the answer will be trivial irrespective of the technique adopted and in most cases it will offer poor information (Fig. 4).

In order to solve a complex problem properly, the competent scientist should be able to select the most appropriate techniques, optimize sampling, methodologies, and measurement protocols according to the material and the questions being asked, and interpret the results rigorously and stringently. To fulfil these conditions, the 'modern mineralogist' should be acquainted with mineralogy and the traditional experimental techniques, such as X-ray diffraction and optical microscopy. However, he/she should also extend his/her mineralogical background to industrial products and synthetic

Figure 4. Poor technical choice is bound to provide a trivial answer (from Gary Larson's *Far Side*, retrieved from www.pinterest.ca/pin/321585229621926143/).

systems, including phase transformations and the physical properties of manufactured materials (Rapp, 2009). Many of these topics are commonly treated in the area of applied mineralogy or applied petrology. An updated knowledge of modern physical and chemical techniques for the investigation of matter (microscopy, imaging, spectroscopy) is also fundamental to improve the characterization of CH materials (Janssens and Van Grieken, 2004). Thefore, the solid crystallographic and crystal-chemical backgrounds of mineralogists are commonly an advantage. Knowledge of advanced techniques should include the proficient use of large-scale facilities, such as those providing brilliant electromagnetic radiation (synchrotrons, free-electron lasers) and neutron beams (nuclear reactors, spallation sources) (Kockelmann *et al.*, 2006; Creagh and Bradley, 2007; Bertrand *et al.*, 2012; Mannes *et al.*, 2015). These are commonly treated in courses of solid-state physics and analytical chemistry. Finally, a solid background in data treatment and statistics is also crucial nowadays.

The conscientious mineralogist approaching this field must also be aware of the fact that the material to be investigated is only the interface between scientific analysis and a more general approach to cultural heritage problems. To avoid producing trivial answers (as in Fig. 4), the material investigated needs to be inserted into a "big picture" (Chiari, 2000) and used to answer well defined questions. The work which follows must be conducted with continuous and intense confrontation with the humanistic collaborators, and must produce reciprocal understanding and mutual improvements. To sum up, a multidisciplinary investigation ought to share: (a) common targets, (b) common language, and (c) reciprocal respect and intellectual equilibrium.

There is a constantly increasing attendance of scientists at archaeological and conservation workshops and conferences, but unfortunately we cannot say the same for the attendance of archaeologists and conservators at scientific meetings. Even the attendences at International Symposia of Archaeometry (www.ims.demokritos.gr/ISA/), which ought to be natural fora for innovative applications and fruitful discussions between archaeologists and scientists, are weighted heavily towards scientists.

2. The materials of cultural heritage

The ultimate aim of the scientific investigation of CH artefacts is to relate them to human life and history. Most people visiting an archaeological museum or an art exhibition enjoy a more complete experience if the object is not only valued for its aesthetic appeal, but is also immersed in its context, so that its origin, practical use and anthropological/social meaning can be appreciated readily.

We can consider this process as the insertion of a given artefact into the time-line of its thermodynamic evolution (*e.g.* fig. 1 in Artioli and Angelini, 2011).

Any change in the physical state of the material, starting from production of the specimen from the raw geological/mineralogical sources through manufacturing, diffusion through trade and use, and ending with the deployment in the archaeological record. Even the archaeological excavation itself causes sudden changes to the thermodynamic equilibrium of the material. Chemical-physical reactions are therefore triggered in an attempt to bring the material back to the ground state. Conservation treatments cannot stop these thermodynamically driven processes, they can hope only to slow down their kinetics and extend the lifetime of the object.

Any information provided by scientific investigation will increase our knowledge of the object. Also, scholar competence will be expanded, the general context of production, use and distribution of the material/object will be better defined, and as a consequence public awareness and acceptance will be expanded as well.

In CH studies, various participants often have a very different perception of the materials. Most scientists generally have a good perception of the physicochemical aspects of the material at the atomic and molecular scale (nano- and micro-scale). Some also have a correct vision of the structure and texture of the material at the meso-scale. Some others only have a macroscopic perception of the material. The choice of the spatial scale of investigation is a critical point in understanding the object as a whole. Bear in mind that the various analytical techniques operate at very different space-time-energy scales. Appropriate analytical strategies should follow.

It is not appropriate to recall here all the issues involved in the investigation of CH materials. In many cases the parameters controlling the type and mode of the experimental investigations to be designed are critically linked to the following choices:

- Whether the measurements can be carried out in specialized laboratories (Laboratory work), either by sampling the object or by moving it to the laboratory, or the measurements are to be carried out in the field (Field work, *in situ* measurements), *e.g.* during archaeological excavations or when non-movable objects such as large architectural components or precious museum items are to be analysed.
- The qualitative *vs.* the quantitative character of the results. Very often the non-invasive character of the investigation as well as the limited time/funding available for the analysis imply that the quantitative information obtained is limited.
- The quantity of sample that can be extracted for the analysis may vary from none to a substantial amount. This imposes severe constraints on the techniques that can be adopted for the investigation. The quantity of sample investigated often determines whether or not the results are representative.

Nowadays, the rapid development of portable instrumentation for diffraction, spectroscopy and imaging has created enormous expectations in the CH community. Many curators, museum managers, archaeologists and even some scientists are increasingly convinced that a sufficient amount of information can be obtained quickly with reasonably cheap instruments, operated by specialists or even by non-specialists. This is of course true in many fields, where portable instruments can indeed provide appropriate data. However, it is important (albeit difficult) to convey the concept that measurements can be done on the same material/object in many different ways but with very different precision and accuracy. For example, measuring the chemical composition of a given material in the field with a portable X-ray fluorescence spectrometer (pXRF), or using the same instrument in the laboratory, or using conceptually different probes (such as mass spectrometry or optical emission spectroscopy) is expected to provide different results. As scientists, we should be aware of this issue but this is a concept not always shared by all counterparts. As an example, recently the authors made a proposal for micro-sampling metal objects for lead isotope analysis and provenance investigations which was rejected by the (in)competent authorities, because their view was that the analysis should be done by non-invasive pXRF analysis. Should we write in each proposal that XRF cannot measure isotopic ratios because of the physical nature of the signal?

To sum up, the study of CH materials is strongly interdisciplinary, and must be carried out by competent people in a complex network of collaboration. Indeed the scientific investigation of cultural heritage is a very privileged area where different disciplines may interact efficiently and evolve jointly. Because mineralogists broadly have the correct time-space-energy perception of materials, they ought to be given a relevant role in the team.

The chapters collected in the present volume, of necessity, show only a selection of examples of the mineralogical applications to CH. Natural minerals and rocks (silica, obsidian, lithics, gems, pigments) used by man in antiquity for different purposes are described, as well as their transformation products to provide fundamental man-made materials (glass, mortars, metals, ceramics).

A recent comprehensive survey (López Varela, 2018) encompasses a wider variety of other important mineral materials, such as those produced by normal or pathological bio-mineralization (Mann, 2001). Iron oxides produced by bacterial activity, amber resulting from the diagenesis of plant resins, carbonates bio-synthesized by corals and shells are all examples of biologically produced materials that have been employed by humans in the past for tools and decorations, or transformed into usable compounds, though they are not fully reviewed in the present volume.

Apatite-related minerals (Harlov and Rakovan, 2015) present in bones and teeth of vertebrates are especially important because of their implication for medicine, forensic science, anthropology, palaeontology and archaeology (White and Folkens, 2005). The mineralogical investigation of the mechanisms of hydroxylapatite crystallization and the detailed understanding of the transformations occurring after the death of the organism (Hedges, 2002; Kendall *et al.*, 2018) represent yet another example of the relevant contribution of mineralogy to modern CH. Recent advances in the mineralogical characterization of modern and fossil bones include the spectroscopic assessment of the microstructural evolution of carbonated hydroxylapatite during diagenesis, burial and cremation (Keenan, 2016; Dal Sasso *et al.*, 2018a,b; Mamede *et al.*, 2018). Notwithstanding our efforts, these studies could not be included in the present volume.

References

Artioli, G. (2010a) Mineralogy and cultural heritage. *Chimia*, **64**, 712–715.

Artioli, G. (2010b) *Scientific Methods and the Cultural Heritage*. Oxford University Press, Oxford, UK.

Artioli, G. and Angelini, I. (2011) Mineralogy and archaeometry: Fatal attraction. *European Journal of Mineralogy*, **23**, 849–855.

Berthoud, Th. and Cleuziou, S. (1981) Archéométrie et archéologie: les leçons d'une expérience. In: *Colloque Français d'Archéométrie 3 (Valbonne, 5–7 ottobre 1981), Valbonne*, 29–42.

Bertrand, L., Robinet, L., Thoury, M., Janssens, K., Cohen, S.X. and Schöder, S. (2012) Cultural heritage and archaeology materials studied by synchrotron spectroscopy and imaging. *Applied Physics A*, **106**, 377–396.

Chiari, G. (2000) Mineralogy and cultural heritage. Pp. 351–381 in: *Environmental Mineralogy* (D.J. Vaughan and R. Wogelius, editors). European Mineralogical Union Notes in Mineralogy, **2**. Eötvös Universit Press, Budapest..

Creagh, D.C. and Bradley, D. (2007) *Physical Techniques in the Study of Art, Archaeology and Cultural Heritage*, Volume **2**. Elsevier, Amsterdam.

Dal Sasso, G., Angelini, I., Maritan, L. and Artioli, G. (2018a) Raman hyperspectral imaging as an effective and highly informative tool to study the diagenetic alteration of fossil bones. *Talanta*, **179**, 167–176.

Dal Sasso, G., Asscher, Y., Angelini, I., Nodari, L. and Artioli, G. (2018b) A universal curve of apatite crystallinity for the assessment of bone integrity and preservation. *Scientific Reports*, **8**, 12025.

Harlov, D.E. and Rakovan, J.F. (editors) (2015) Apatite: A mineral for all seasons. *Elements*, **11**, June issue.

Hedges, R.E. (2002) Bone diagenesis: an overview of processes. *Archaeometry*, **44**, 319–328.

Janssens, K. and Van Grieken, R. (editors) (2004) *Non-destructive Micro Analysis of Cultural Heritage Materials* Comprehensive Analytical Chemistry, Volume **42**. Elsevier, Amsterdam.

Keenan, S.W. (2016) From bone to fossil: A review of the diagenesis of bioapatite. *American Mineralogist*, **101**, 1943–1951.

Kendall, C., Eriksen, A.M.H., Kontopoulos, I., Collins, M.J. and Turner-Walker, G. (2018) Diagenesis of archaeological bone and tooth. *Palaeogeography, Palaeoclimatology, Palaeoecology*, **491**, 21–37.

Kockelmann, W., Chapon, L.C., Engels, R., Schelten, J., Neelmeijer, C., Walcha, H.M., Artioli, G., Shalev, S., Perelli-Cippo, E., Tardocchi, M. and Gorini, G. (2006) Neutrons in cultural heritage research. *Journal of Neutron Research*, **14**, 37–42.

Laudan, R. (1987) *From Mineralogy to Geology: The Foundations of a Science, 1650–1830*. University of Chicago Press, Chicago, Illinois, USA.

López Varela, S.L. (editor) (2018) *The Encyclopedia of Archaeological Sciences*. Wiley-Blackwell, Hoboken, New Jersey, USA.

Maggetti, M. (2006) Archaeometry: Quo vadis? Pp. 1–8 in: *Geomaterials in Cultural Heritage* (M. Maggetti and B. Messiga, editors). Special Publications, **257**. Geological Society, London.

Mamede, A.P., Gonçalves, D., Marques, M.P.M. and Batista de Carvalho, L.A. (2018) Burned bones tell their own stories: A review of methodological approaches to assess heat-induced diagenesis. *Applied Spectroscopy Reviews*, **53**, 603–635.

Mann, S. (2001) *Biomineralization: Principles and Concepts in Bioinorganic Materials Chemistry*. Vol. **5**. Oxford University Press, Oxford, USA.

Mannes, D., Schmid, F., Frey, J., Schmidt-Ott, K. and Lehmann, E. (2015) Combined neutron and X-ray imaging for non-invasive investigations of cultural heritage objects. *Physics Procedia*, **69**, 653–660.

Pollard, A.M. and Bray, P. (2007) A bicycle made for two? The integration of scientific techniques into archaeological interpretation. *Annual Review of Anthropology*, **36**, 245–259.

Rapp, G. (2009) *Archaeomineralogy*. Springer Science and Business Media.

Schneer, C.J. (1995) Origins of mineralogy: the age of Agricola. *European Journal of Mineralogy*, **7**, 721–734.

White, T.D. and Folkens, P.A. (2005) *The Human Bone Manual*. Elsevier, Amsterdam.

EMU Notes in Mineralogy, Vol. 20 (2019), Chapter 2, 13–85

Variations on the silica theme: Classification and provenance from Pliny to current supplies

ELISABETTA GLIOZZO

Department of Earth, Environmental and Physical Sciences, University of Siena, Italy
e-mail: gliozzo@unisi.it

Over recent decades, numerous studies have highlighted the importance of opal, chalcedony and quartz varieties, chiefly in volcanic, but also in metamorphic and sedimentary environments. The focus is to define accurately their structures, composition and properties, as well as to identify the factors controlling the formation and the ageing of different forms of silica. In the field of archaeological sciences efficient discriminants are the bases from which the origin and provenance of materials may be traced. Substantial efforts were made in the attempt to combine geochemical, mineralogical, petrographic and geological features with archaeological and archaeometric information. However the results show that data integration is complicated, and several unanswered questions remain. On the one hand, archaeological research has focused on technological and ethnographic aspects, mainly concerning use-wear and heat-treatment studies. Mineralogical characterization has often been limited to the identification of the material, frequently by Raman microspectroscopy alone. On the other hand, the Earth sciences have provided basic mineralogical, crystal-chemical and geological knowledge, but failed to provide a systematic data collection of sources and their geochemistry. As a consequence, large gaps persist in the identification of archaeological opals, chalcedonies and quartz varieties, and in the geographic mapping of possible sources.

In this context, the present review aims to summarize the current academic debate on such issues, possibly to encourage further work in the field. After a brief introduction to terminology, the structure of opals, their colours and properties are discussed, followed by an introduction to silica dissolution/precipitation and opal-formation processes.

The next section reviews the information available on use of opals and provenance from historical sources, mainly Pliny the Elder, followed by a short list of ancient and modern opal supply areas, together with a (necessarily incomplete) summary of the geological and geochemical information.

The discussion then encompasses chalcedony, agate and chalcedony varieties (carnelian, sard, onyx, sardonyx, chrysoprase, Cr-chalcedony, 'gem silica' or 'chrysocolla chalcedony' and heliotrope), following the same scheme as was adopted for opals. Terminology, distinguishing features, formation conditions, information derived from Pliny's books, past and current supply areas and, finally, archaeometric provenance issues are addressed for each type of material. As for chalcedony, a comprehensive note on moganite has been included.

The next section focuses on chert, flint and jasper. Given the large amount of materials available on this topic, the present review must necessarily be considered introductory and partial. The discussion aims to provide useful indications on how to distinguish chert from flint and chert from jasper; secondly, the information provided by Pliny and the archaeometric state of the art on these materials is reviewed.

DOI: 10.1180/EMU-notes.20.2

The last section examines quartz varieties: hyaline quartz (rock crystal), milky quartz, smoky quartz, rose and pink quartz, amethyst, citrine, prasiolite and blue quartz. An exhaustive mineralogical discussion on quartz is beyond the scope of this review; conversely a review of the historical information is provided, together with a brief list of major supply areas, a summary of the archaeometric studies performed on these materials, as well as an indication of the geological literature which can be used proficiently for provenance studies.

1. Classifying opal, chalcedony and quartz varieties

"Nec est aliud difficilius quam discernere haec genera" (Pliny N.H. 37, 98)
(Nothing is harder than the attempt to distinguish the varieties of this stone)

The terminology usually adopted for the characterization of natural silica-based materials may be a potential source of confusion, as it combines well established but somewhat vague terms (*e.g.* carnelian, sard), technical terms (*e.g.* chalcedony and quartzine) and commercial terms (*e.g.* fire opal).

A wider knowledge of these materials (*e.g.* of factors controlling the formation of different forms of silica) would be needed in order to provide a systematic review of their nomenclature. Table 1 illustrates schematically how to distinguish opal from chalcedony and quartz varieties, using broad distinctions and without taking into account the possible variations due to secondary processes, such as recrystallization.

Among others, the terms "silica sinter" and "jasperoid" were not included in the table as their use is confined generally to geological studies on hydrothermal alteration. Silica sinter indicates those deposits which formed by precipitation from aqueous solutions and where silica prevails over all other possible components (travertine represents its carbonate equivalent). Conversely, the name jasperoid has been applied to "epigenetic siliceous bodies formed largely by replacement, thus excluding cherts that may have formed in lime muds either syngenetically and diagenetically" (Lovering, 1962, on the characteristics and distribution of jasperoids in American carbonates). A different case is presented by aventurine which is not a mineral in the strict sense. This is generally described as a quartzitic rock, coloured green by fuchsite.

2. Opal-A$_N$, Opal-A$_G$, Opal-CT, Opal-C

2.1. An introduction to their structures, colours and properties

The distinction between opal-A$_N$ and opal-A$_G$ was introduced by Langer and Flörke (1974), based on small-angle X-ray and neutron scattering experiments.

The network-like opal-A$_N$ (rarely used as a gem) is characterized by a water-containing glass-like structure (Flörke *et al.*, 1973), indicated by diffuse scattering of X-rays and neutrons at small angles. Previously called hyalite, opal-A$_N$ is far less widespread than the opal-A$_G$ variety and is typically botryoidal (generally forming a millimetric crust on volcanic rocks), colourless and shows strain-induced birefringence under the polarizing microscope (Flörke *et al.*, 1985; Graetsch, 1994; Gaillou *et al.*, 2008a).

Table 1. Basic criteria for the distinction between opal, chalcedony and quartz varieties. Grain size can be fixed arbitrarily as follows: macro- (crystals visible to the naked eye, >30 μm), micro-crystalline (<30 μm) and cryptocrystalline (<1 μm). Prev. = prevalent; min.= minor contents; ev. = eventual.

	Macrogroup	Growth form	Phases	Varieties	Main features
High solubility Low structural order	OPAL	Amorphous	–	Opal-A_N	Prev. hyalite. glass-like structure.
			–	Opal-A_G	Close-packed arrangement of homometric spherical particles of amorphous silica (150–350 nm), with a long-range order not exceeding ~9 Å
		Amorphous and microcrystalline	Cristobalite and tridymite	Opal-CT	Prev. lussatite. Crystalline or rather "para-crystalline". Ordered and disordered inter-stratifications of cristobalite and tridymite layers. Silica nanograins with different degrees of organization. Includes "fire opal".
			Mostly cristobalitic. Presence/absence of tridymite controversial	Opal-C	Prev. lussatine. Higher degree of crystallinity and lower disorder in the crystal structure with respect to opal-CT.
	Chalcedony (prev.), opal (ev.), quartz macrocrystalline (last filling)	Amorphous to macrocrystalline	Chalcedony + moganite/quartzine (+ opal) (+quartz)	Agate	Colours: red, orange, brown, yellowish, purple, white, grey, green and, more rarely, blue. Must be banded. Possible causes of banding: (a) alternating phases of activity and rest in the crystallization process, (b) Fe oxide 'self-purification' process during crystallization, (c) Al^{3+} and Fe^{3+} substitutions for smaller Si^{4+} and (d) textural and structural changes.
	CHALCEDONY	Fibrous microcrystalline and nanocrystalline (prev.), amorphous (ev.)	Chalcedony + moganite/quartzine	Carnelian	Orange variety
				Sard	Dark orange variety
				Onyx	Black and white variety
				Sardonyx	Brown and white variety
				Chrysoprase	Apple green variety (Ni-based chromophore)
				Cr-chalcedony	Emerald green variety (Cr-based chromophore)
				Heliotrope	Green variety dotted with small, hematite red, blood-like spots
	Quartz (prev.), chalcedony (min.)	Granular microcrystalline (prev.), fibrous microcrystalline (min.)	Chalcedony + moganite/quartzine + quartz	Chert	Used in geological papers. Broad term indicating large bodies of white or lightly coloured rocks. Mainly in "soft" carbonates.
				Flint	Mainly used in archaeological papers. Darker variety of chert. Nodules in chalk and marl.
Low solubility High structural order	QUARTZ	Macrocrystalline granular	Quartz	Rock crystal	Transparent variety
				Milky quartz	Milky appearance due to fluid inclusions.
				Smoky quartz	Smoky appearance due to aluminium (Al^{3+}) impurities substituting silicon (Si^{4+})
				Rose quartz	Rose variety. Colour due to Fe^{2+}–Ti^{4+} intervalence change transfer in dumortierite
				Pink quartz	Pink variety. Colour due to Al-O^--P hole centres induced by ionizing radiation.
				Amethyst	Purple variety. Colour due to (mainly) trivalent Fe replacing Si in its tetrahedral sites
				Citrine	Rare yellow variety. Colour due to trivalent iron (Fe^{3+}) and colour centres often associated with Al impurities
				Prasiolite	Green variety (≠ from prase). Radiation-induced colour variant of quartz (iron chromophore).
				Blue quartz	Rare blue variety. Colour due to Rayleigh scattering by submicron-sized inclusions.

In the gel-like opal-A_G variety (generally synonymous with opal-A), the basic structure comprises a close-packed arrangement of homometric spherical particles of amorphous silica (150–350 nm), with long-range order not exceeding ~9 Å (Graetsch, 1994; Rossmann, 1994).

The presence of variable water contents (1–8 wt.%) characterizes both species (further details in Langer and Flörke, 1974), reducing their refractive index and density. Mostly molecular, water is lost largely at 200°C and is almost absent at 400°C (Jones and Segnit, 1971; Langer and Flörke, 1974). In opal-A_G water fills the interstices between the spherical silica particles and the scattering intensities imply interparticle interferences superimposed on diffuse scattering (Langer and Flörke, 1974; Graetsch and Ibel, 1997). Both the common ("potch") and the precious opal are mainly included in this variety, the distinction being in the absence or presence of the play-of-colour.

In precious opals, the play-of-colour is caused by diffraction of light by the randomly faulted, cubic- or hexagonally-packed spheres of amorphous silica (150–350 nm); schematically, the greater the diameter of the particles, the greater the lattice parameter and the grating which establish the range of colours diffracted by the opal (Sanders, 1964, 1968; Sanders and Darragh, 1971; Rossman, 1994 with references therein). Small spheres will give rise to blue and violet colours, while large spheres will be responsible for red and orange colours. In sedimentary environments, gravitational settling has generally been considered as the main factor controlling the formation of ordered lattices of spheres in the fractures and pores of the host rocks (Darragh *et al.*, 1966; Rey, 2003; Pecover, 2007; Gaillou *et al.*, 2008a, Pewkliang *et al.*, 2008), however, Liesegang and Milke (2018) oppose this reconstruction claiming that "gravitational settling/ordering is the exception rather than the rule".

In common opals, the lack of play-of-colour is caused mostly by the non-uniformity, inappropriate packing and dimensions of silica sphere (Sanders and Darragh, 1971) but also by the complete filling of the interstices by silica cement (Graetsch, 1994). Gaillou *et al.* (2008a) noted that: (1) diffraction was inhibited by a variation of only 5% in sphere diameters; (2) imperfectly shaped spheres occurred relatively commonly; (3) the minimum (~140 nm) dimensions range of the sphere in order to diffract light was still applicable, while doubts remained about the maximum value of ~650 nm; and (4) the effect of several compaction features was significant.

Within the aforementioned dimensional range of the spheres, the development of a non-uniform array seems to be controlled mainly by the deposition temperature (see Section 2.3.) and the silica concentration; in fact, a high concentration of spheres has limited freedom of movement and makes regular stacking nearly impossible (Filin *et al.*, 2003). In this regard, it is worth noting that the absence of voids seems to represent a characteristic feature of Australian opals, in contrast to the Californian ones from the Lead Pipe Springs and Last Chance Canyon deposits (Sanders and Dietz, 1967).

Unlike opal-A_G and opal-A_N, opal-CT is crystalline or rather "paracrystalline" (Smith, 1998). Previously known as lussatite (*i.e.* a variety of fibrous cristobalite defined in Mallard, 1890), opal-CT consists of silica nanograins (12–32 nm in size in Rice and Elzea, 1993, average of ~25 nm in Gaillou *et al.*, 2008a) with four different degrees of organization: (1) random; (2) arranged in fibres; (3) arranged in platelets; or (4) arranged in lepispheres (Gaillou *et al.*, 2008a). The random arrangement (type 1) characterizes most "fire" opals (*i.e.* transparent and bright orange) which, characteristically, have nanograin diameters ranging ~10–~50 nm. The fibrous arrangement (type 2) is present in opaque pink opals, which possess fibres with diameters of ~20–25 nm and lengths of ~4 μm (see also the Mexican pink opal-CT associated with palygorskite in Fritsch *et al.*, 2004). Type 3 has nanograins arranged as platelets (with blades corresponding to the {111} or {001} forms of cristobalite and tridymite, respectively). They are typically ~25 nm thick and ~300 nm wide but their thickness seems to vary depending on their provenance (Graetsch, 1994; Gaillou *et al.*, 2008a). The lepispheres which characterize type 4 are almost equidimensional platelets intersecting at a nearly common centre, thus assuming a sphere-like shape. The lepispheres vary in diameter from ~250 to ~1000 nm and are generally embedded in a silica nanograin matrix. Gaillou *et al.* (2008a) found all the four types in Mexican opals, but only types 1 and 3 were present in opals from Australia, Ethiopia and Turkey; type 1 CT opals were reported from Brazil, Kazakhstan (an interesting review on precious and decorative non-metallic minerals from east Kazakhstan is in Chernenko and Chlachula, 2017) and Tanzania; type 2 opals from Peru; and type 3 from France, Madagascar and Venezuela.

As demonstrated by numerous authors (*e.g.* Mizutani, 1970, 1977; Jones and Segnit, 1971; Flörke *et al.*, 1975; Williams and Crerar, 1985; Elzea *et al.*, 1994; Graetsch, 1994; Guthrie *et al.*, 1995; Cady *et al.*, 1996; Nagase and Akizuki, 1997), opal-CT is characterized structurally by both ordered and disordered interstratifications of cristobalite and tridymite layers; the proportion of the two phases being largely cristobalitic (Jones and Segnit, 1971; Rice *et al.*, 1995) or largely tridymitic (Mitchell, 1967; Mitchell and Tufts, 1973; Wilson *et al.*, 1974; Elzea and Rice, 1996) or nearly equal (Flörke *et al.*, 1991; Graetsch, 1994; Guthrie *et al.*, 1995). Continuous attempts to model the structure of opal-CT have resulted in several authors proposing to distinguish a new tridymitic end-member – "opal-T" – from opal-CT, alongside the better-known opal-C (Eversull and Ferrell, 2008). Wilson (2014) proposed defining opal-CT as a "paracrystalline substance in which a disordered matrix predominates over the crystalline domains, as well as the proposition that these crystalline domains are largely tridymitic where opal-CT yields an X-ray diffraction (XRD) maximum near 4.10 Å".

With respect to opal-CT, opal-C (previously known as lussatine) is characterized by a higher degree of crystallinity and a lower disorder in the crystal structure (Elzea *et al.*, 1994). As is the case for opal-CT, the presence/absence of tridymite is still so controversial that some authors did not find it (Rice *et al.*, 1995), others indicated a very limited incidence of tridymite stackings (Jones and Segnit, 1971), yet others constrained the tridymite percentage to 20–30% with respect to cristobalite (Graetsch, 1994).

The content of molecular water present in the pores may be a discriminating characteristic, being ~1–3 wt.% in opal-C while ranging between 3 and 10 wt.% in opal-CT (Graetsch *et al.*, 1985, 1987; Flörke *et al.*, 1991; Cady *et al.*, 1996).

Using XRD, opal-CT and opal-C can be referred to mistakenly as α-cristobalite (Elzea *et al.*, 1994, Önal *et al.*, 2007) and, for this reason, Elzea *et al.* (1994) proposed observation of the (101) reflection. In opal-C, this appears intense, centred at near 4.0 Å and with a peak-width ranging from 0.222 to 0.453 Å. Conversely, in opal-CT, the same reflection would appear less intense, centred at near 4.07–4.10 Å, and with a peak-width ranging from 0.506 to 0.883 Å. The efficacy of heat treatments has been tested further by numerous authors, based on the simple concept that the (101) peak will shift for hydrated phases such as opals while it will not change position for α-cristobalite (Elzea *et al.*, 1994; Elzea and Rice, 1996; Hand *et al.*, 1998; Kahraman *et al.*, 2005; Önal *et al.*, 2007). Phosphoric acid (H_3PO_4) digestion was applied by Miles (1994), Kahraman *et al.* (2005) and Önal *et al.* (2007) in order to sharpen the (101) peak and to dissolve non-silica phases in bentonites and kaolins.

Lastly, as for colours, it is well established that impurities or mineral inclusions are responsible for the various hues such as the blue colour of Peruvian blue opals (Cu silicate; Rossmann, 1994; Gaillou *et al.*, 2008b), the yellow to orange, red or brown colour of many opals such as the "fire opals" (nanoscale inclusions of Fe-oxyhydroxides in Gaillou *et al.*, 2008b; Fe^{3+} in Rossmann, 1994) or a few Ethiopian opals (see *e.g.* Johnson *et al.*, 1996; Muyal, 2017; the latter-named for dyed specimens), and the opaque orange colour of Austrian "forcherite" opals (realgar; see Gaillou *et al.*, 2008b with references therein). The wide range of opal colours has led to the creation of a plethora of colloquial and trade names such as the aforementioned fire opals and forcherite (yellow, with realgar and orpiment), as well as the hydrophane (translucent in water), the milk opal and the honey opal (Duffin, 2006).

2.2. Silica dissolution and precipitation

Silica dissolution and precipitation control the formation of opal and of all silica varieties; therefore, this topic is referred to here but concerns quartz and chalcedony varieties too.

The main processes which can lead to large concentrations of dissolved silica are: (1) weathering of silicate minerals, which is generally considered to be the most effective process in terms of quantity levels; (2) dissolution of amorphous silica, which can be observed in certain sedimentary environments; (3) dissolution of quartz, which is less effective than the other processes, given the low solubility of quartz; and (4) hydrothermal alteration of Si-rich rocks, which produces large concentrations of silica but is confined within retricted geographical limits. Therefore, the most likely places where dissolved silica can concentrate are thermal springs and all those geological formations where silicate minerals are present and, as a consequence, weatherable.

As for hydrothermal fluids, there are many factors that can enhance, reduce or inhibit the deposition of amorphous silica from silica supersaturated solutions: (1) rapid cooling, generally associated with decompressional boiling; (2) pH changes, *e.g.* due to the mixing of different waters; (3) presence of dissolved salts; and (4) interaction between fluids and volcanic glass (Fournier, 1985). The deposition of amorphous silica is favoured by near-neutral pH and high salinity (Morey *et al.*, 1961; Makrides *et al.*, 1980; Crerar *et al.*, 1981; Weres *et al.*, 1982). In weathering environments, the rates of dissolution and precipitation of silica are slow and the system implicates "a complex of spatial and temporal conditions" (Thiry *et al.*, 2014).

Dissolution and leaching, followed by evaporation and concentration, are the basic processes but they can, however, be affected by multiple concomitant events. In this regard, the activity of microbes have been considered further, given their capacity to weather clay minerals and feldspars through the excretion of acids and enzymes (see *e.g.* Banfield *et al.*, 1999; Chen *et al.*, 2000).

In the oceans, the amounts of dissolved silica are extremely small (of the order of a few tenths of a ppm); in fact, silica is extracted and fixed by diatoms, radiolarians and sponges to form part or all of their skeletons but it cannot be precipitated inorganically. Similar amounts are present in rivers and groundwaters (12/18 ppm SiO_2; see White *et al.*, 1963; Garrels and Christ, 1965).

Neutral to slightly alkaline (*i.e.* basic) waters produce thick silica sinters while highly acidic waters (pH <3) form silica deposits; in the latter case, the presence of H^+ ions inhibits silica polymerization (Rothbaum *et al.*, 1979; Weres *et al.*, 1982). Acidic waters, from outer rocks or internal oxidation, may attack silicate minerals (*e.g.* feldspars) and cause the release of silica into the solution and the consequent supersaturation of the solution itself. The latter case may be neutralized and precipitate amorphous silica. It is possible, although infrequent, that highly alkaline waters may also cause high rates of dissolved silica in hydrothermal waters, due to their ability to favour the dissociation of silicic acid (Fournier, 1985).

As for the solubility of silica phases, it has been established that amorphous silica is more soluble than all other silica varieties and, similarly to all the other silica phases, its solubility is influenced by temperature, pressure, pH, type of dissolved constituents in the solution (*e.g.* salts, organic compounds, *etc.*), particle size, structure and properties of the silica–water interface (see Dove and Rimstidt, 1994 for a comprehensive review of this topic).

To give a few examples relating to this topic, the solubility of amorphous silica in pure water at 100°C is equal to that of quartz at ~225°C (Fournier, 1985). With pH <9, the solubility is invariably ~120 mg/L SiO_2, while it increases strongly at pH >9 (Krauskopf, 1956). In the presence of most dissolved salts, below ~300°C, the solubility decreases slightly, while it increases significantly in the presence of Na_2SO_4 at all measured temperatures (Marshall, 1980; Marshall and Warakomski, 1980; Chen and Marshall, 1982; Marshall and Chen, 1982a,b; Fournier and Marshall, 1983). This latter phenomenon has been correlated with the formation of silica-sulfate complexes (Marshall and Chen, 1982b; Fournier and Marshall, 1983).

Finally, the presence of organic compounds (Bennet, 1991) may further increase the solubility. Thiry *et al.* (2014), for instance, demonstrated that, in weathering environments, the dissolution of silica occurs mainly at the contact between organic compounds and organic-rich soils, while the precipitation of silica originates from pore waters.

2.3. Formation sequence

The debate on opal formation is part of a wider-ranging research topic regarding the diagenetic sequence which may develop from the highly soluble, non-crystalline opal-A to the crystallization of less soluble, highly structurally ordered quartz:

$$\text{opal-A} \rightarrow \text{opal-CT} \rightarrow \text{opal-C} \rightarrow \text{chalcedony} \rightarrow \text{quartz}$$

The procedures, constraints and timeframes in which this sequence proceeds have been the subject of numerous studies (*e.g.* Carr and Fyfe, 1958; Fyfe and McKay, 1962; Ernst and Calvert, 1969; Mizutani, 1970; Mukhopahyay and Chanda 1972; Murata and Nakata, 1974; Bettermann and Liebau, 1975; Flörke *et al.*, 1975, 1991; Murata and Larson, 1975; Murata and Randall, 1975; Kastner *et al.*, 1977; Murata *et al.*, 1977; Kano and Taguchi, 1982; Williams *et al.*, 1985; Jansen and Van der Gaast, 1988; Graetsch, 1994; Cady *et al.*, 1996; Lynne and Campbell, 2004). It is currently accepted that both the complete sequence (mostly in deep-sea deposits) and the direct formation of late phases (*i.e.* all except for opal-A_G) may happen, depending on formation environment (*i.e.* time, pH, salinity, solubility, saturation, temperature, presence of dissolved Mg, *etc.*). Authors do not agree about which factor should be the most determinant over all the others but it seems unlikely that a single factor may drive the entire sequence from the beginning to its end.

The degree of silica saturation in the solution is of fundamental importance in order to determine which type of silica polymorph will precipitate (Flörke *et al.*, 1982; Heaney and Davis, 1995). In fact, the greater the number of nuclei and crystal defects, the smaller the dimensions and degree of crystallinity of the silica variety that forms; on the other hand, the smaller the amount of silica in the solution, and the smaller the number of nuclei and crystal defects, the greater the dimensions and degree of crystallinity of the silica variety formed (Lee, 2007 with references therein). In this respect, the influence of temperature must not be underestimated. Low temperatures (~45°C) – common in sedimentary environments – favour opal-A_G formation (see *e.g.* Rondeau *et al.*, 2004), while higher temperatures (around 170°C) are necessary for opal-CT to deposit (*e.g.* Spencer *et al.*, 1992). Muldarisnur and Marlow (2017) have investigated opal formation from suspensions, while others (*e.g.* Carr and Fyfe, 1958; Rice *et al.*, 1995) have demonstrated experimentally that opal-A transforms to opal-CT/C during heating to temperatures of >1000°C. Moreover, these data may further explain why opal-CT/C does not seem to have an opal-Ag precursor in numerous silica deposits of volcanic origin (see *e.g.* Graetsch *et al.*, 1994).

Lastly, Gaillou *et al.*, (2008b) observed that gem opal-CT was always found in volcanic environments while opal-A could have been found in both sedimentary (the most widespread) and volcanic environments. The study conducted by Gaillou *et al.* (2008b) showed how further geochemical investigations may distinguish between sedimentary (Ba >110 ppm, Eu and Ce anomalies) and volcanic opals (Ba <110 ppm, no Eu or Ce anomaly) and hence reconstruct the conditions of formation. In this regard, Chauviré *et al.* (2017a) have recently proposed a new method (near infrared signature) for the assessment of the geological conditions of opal formation.

2.4. Pliny on opals: description and provenance

Pliny's *Naturalis Historia* (1st century AD) reported on the then current state of mineralogy, petrography, geology and numerous branches of Natural Sciences. Due to its encyclopaedic nature, *Naturalis Historia* represents an ideal and indispensable starting point for Mediterranean Archaeometry.

Based on Pliny (*N.H.* 37, 80-84), only emeralds precede opals (*opalus*). Pliny (*N.H.* 37, 80) specifies that they display the "fires of the *carbunculus*" (*i.e.* carnelian or garnet), "the bright purple of the amethyst" and "the sea-green tint of the emerald", all combined together. However, still according to Pliny (*N.H.* 37, 83), the gemstone may also have shortcomings such as (1) a colour similar to that of the plant called *heliotropium* (*i.e.* intense purple if corresponding to the actual turnsole in the borage family, *Boraginaceae*; or green and red by analogy with the heliotope gemstone), or (2) a colour similar to that of rock crystal or even to that of ice, or (3) the presence of salt-like specks (or of salt efflorescence) and stains.

Pliny (*N.H.* 37, 81-82) further tells the story of the senator *Nonius* who was outlawed by Mark Antony for the sake of a precious opal, valued at 2,000,000 *sestertii*. It is possible to contextualize this information considering that, during the period of Caesar and Augustus, a senator had to have at least one million sesterces of annual income (Eck, 2000).

Pliny teaches us a method to recognise fakes and adds that opal was easily counterfeited in glass (*N.H.* 37, 83). The method would have consisted of observing the reflection of light through the stone orientated towards the sun. If the stone was fake, it would have shown a single colour; in contrast, the light would have passed through the real stone, shedding a bright light on fingers and showing different colours changing in the different parts of the stone.

The adoption of this method, along with the list of the defects that Pliny assigns to opal, seems to indicate that the name *opalus* corresponded to different types of stones. This is not surprising and may be confirmed by the mention of the "*paederos*" gemstone which, in Pliny's opinion, was the Greek term for opal while, according to others, it was a separate variety, corresponding to the Indian "*sangenon*" (*N.H.* 37, 84).

Pliny himself provides a controversial description of the "*paederos*" gemstone: initially as a sky-blue and purple opal, never showing the green tints of emerald (*N.H.* 37, 84), but some chapters later as a precious gem, characterized by a rock-crystal-like transparency, a sky-green tint, a glint of purple and the shining of golden wine (*N.H.* 37, 129-130). In this respect, it is interesting to note that today the "*Paederota Bonarota*" is a perennial herb characterized by blue-lilac blooms, which was named after the Plinian opals, forgetting perhaps that Dioscorides (40-90 AD) used this term for a type of *achantus* (*De materia medica* 4-74).

Greek mythology and Roman and Medieval superstition testify further to the worldwide popularity of opals. In Greek mythology, Zeus wept tears of joy which turned into opals when he won against the Titans. Centuries later, the Romans considered the opal capable of attracting large fortunes (similarly to the rainbow). The connection between opal and eyes (*ophtalmis lapis or Lapis optallion; LAPID.24*) was a Late Antique and medieval must (*ORPH.L.* 282-284; Kostov, 2008) and a commonplace of modern crystal therapy. No wonder then that opal was considered a "miracle" and the "Queen of gems" (Shakespeare, The Twelfth Night).

As for *opalus* provenance, Pliny indicates India as the sole producer (*N.H.* 37, 80) and, centuries later, Isidore of Seville (*Etimologiae*, 16) reports the same information ("*Nomen habet ex patria; sola enim eum parturit India*") . Conversely, the supply areas listed for the *paederos* gemstone are more numerous (*N.H.* 37, 84, 130): Egypt (for a type of *paederos* called "*syenites*" or "*tenites*"), Arabia, *Pontus* (gems of lower quality; southern coast of the Black Sea, in Turkey), *Galatia* (*i.e.* central Turkey), Thasos (Greece), Cyprus and Thrace. In this instance, it is worth noting that these sites are more or less the same as those indicated for amethyst. The localization of the toponyms mentioned by Pliny is shown in Fig. 1.

2.5. Past and current supply areas

It is reasonable to suppose that the Carpathian (or "Hungarian") opal sources were the main opal suppliers in Roman times (*e.g.* Frondel, 1962; Caucia *et al.*, 2013). The Slovakian mines were located between the districts of Prešov, Košicem, Červenica and Dubnik (Fig. 2). Considered the oldest mine in the world for opal mining, the site of Červenica (Slovakia) has been restored recently and operates in the field of Geotourism (Klamár *et al.*, 2017). Eckert (1997) also mentioned the town of Opalbanya, but he merely said that it must have been close to Dubnik.

Today, the major opal fields are located in Australia (Andamooka and Coober Pedy, New South Wales; Lambina and Mintabie, Queensland), Brazil, Canada (British Columbia), Ethiopia (Johnson *et al.*, 1996; Rondeau *et al.*, 2010, 2012; Chauviré *et al.*, 2017b; Muyal, 2017), Honduras, Mexico and the USA (Opal Butte in Oregon, Idaho, Nevada and Utah; for details of uraniferous opals from Spor Mountain and the Thomas

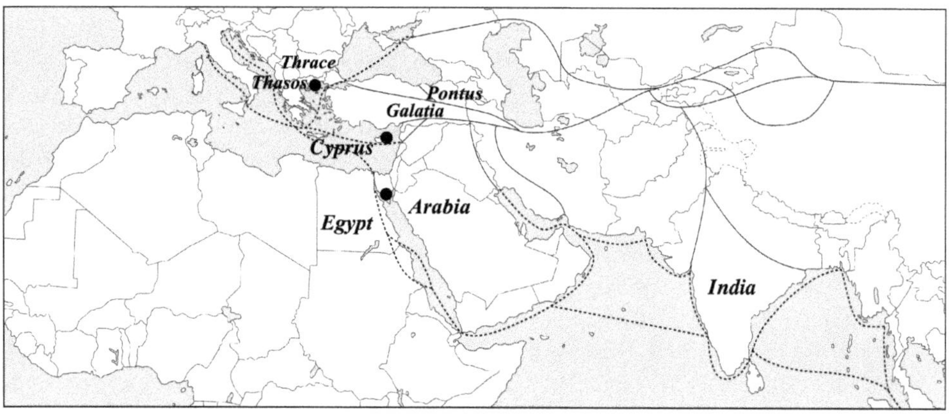

Figure 1. From the *Naturalis Historia* by Pliny the Elder: the distribution of *opalus* and *paederos*. The silk road and the principal maritime routes are shown.

Figure 2. Opal mines in Slovakia probably exploited during Roman times.

Range, Utah see Schindler *et al.*, 2017). Around the Mediterranean, sporadic occurrences have been registered in almost all countries but relevant deposits seem to be found in Bulgaria, Hungary, Czech Republic and Slovakia only (see also Sodo *et al.*, 2016). For a descriptive list of all modern mining sites and occurrences see Eckert (1997).

2.6. Opal provenance in archaeometry and information gained from the geological literature

In archaeometry, the provenance issue does not seem to concern opals, in fact specific publications are completely lacking. In this context, the Neutron Activation Analysis (NAA) investigations of 25 archaeological opals from Troy (Oregon, USA) represent an isolated case (Hess, 1996; Cummings *et al.*, 1989).

Papers available from the geological literature, by comparison, are slightly more numerous but still very few if compared to the worldwide occurrence of opals. Excluding materials from Japan as well as oceanic materials, geochemical analyses are available for at least 262 exemplars (see Appendix 1). Further spectroscopic investigations have been performed by Sodo *et al.* (2016). A further 50 samples of orange, yellow, honey, green, blue and pink opals and 18 samples of colourless opals from recognized fields in Australia, Peru, Mexico and USA must be added to the list (McOrist and Smallwood, 1995).

3. Chalcedony

3.1. Terminology and distinguishing features of chalcedony

For a long time the literature ignored the presence of moganite, or identified it with the different name of lutecite. This shortcoming meant omitting the effects of a determining variable, in relation to the physical and chemical properties of the material, as well as to its mode of formation and stability (*e.g.* solubility, geothermometers based on the concentration of dissolved silica, *etc.*).

Studying the evaporites of the Paris Basin, Michel-Levy and Munier-Chalmas (1892) introduced a 'variety' of length-slow chalcedony (*i.e.* with crystallites parallel to the *c* axis) in order to describe a silica phase intergrown with length-fast (*i.e.* crystallites stacked perpendicular to the *c* axis) chalcedony. This classification was further elaborated by Cady *et al.* (1998), who proposed to distinguish four microfibrous varieties: (1) chalcedony, with a parallel extinction and a length-fast elongation; (2) pseudochalcedony, with extinction subparallel to the *a* axis and a length-fast elongation; (3) quartzine, with a parallel extinction and a length-slow elongation; and (4) lutecite, with extinction subparallel to the *c* axis and a length-slow elongation. More precisely, the term lutecite was used when the *c* axis was at an angle of ~30′ with respect to the elongation direction of the fibres (Michel-Lery and Munier-Chalmas, 1892). This distinction has not taken root, while the distinction between length-fast and length-slow chalcedony (alternatively termed quartzine) is still in use. As far as lutecite is concerned, Heaney and Post (1992) and Heaney (1995) proposed to identify this length-slow chalcedony with moganite, although accepting the existence of length-slow quartz, or rather quartzine.

Nowadays, there is overall agreement in relation to the fact that the principal characteristic of chalcedony is its microcrystalline grain size and fibrous habit. According to these two features, chalcedony has been considered either a quartz variety with a different petrographic texture or a disordered mixture of microcrystalline fibrous quartz and moganite. Given that quartz and moganite are two polymorphs and thus two distinct phases, chalcedony would not properly be a mineralogical phase but rather a mineralogical assemblage. In the absence of moganite, the application of this definition may become critical if we do not consider that microcrystalline fibrous quartz is structurally different from ideal macrocrystalline α-quartz (Heaney *et al.*, 1994).

With respect to α-quartz, the main distinguishing features of chalcedony are represented by its lower refractive index and density. These properties are due to the numerous grain boundaries and the presence of (on average) ~1–2 wt.% water; the latter being both molecular water $(H_2O)_{mol}$ (entrapped mainly in fluid inclusions and open pores) and chemically bound hydroxyl $(H_2O)_{SiOH}$ (*i.e.* silanole), concentrated mainly within grain boundaries and defect sites (Flörke *et al.*, 1982; Miehe *et al.*, 1984; Schmidt *et al.*, 2013). The solubility of chalcedony is greater than that of quartz and less than that of cristobalite and amorphous silica (Pelto, 1956; Fournier and Rowe, 1966; Fournier, 1977, 1985; Walther and Helgeson, 1977). This feature is strictly dependent on the quartz:moganite ratio, *i.e.* the larger the moganite content, the greater the solubility of chalcedony (Heaney and Post, 1992).

In practice, the use of optical microscopy makes it possible to distinguish the length-fast phase, which is chalcedony *s.s.*, from the length-slow phase. The latter can be examined further by means of XRD or Raman spectroscopy, in order to distinguish length-slow moganite (previously lutecite) from length-slow quartz, *i.e.* quartzine.

3.2. Formation conditions of chalcedony

In sedimentary and low-temperature hydrothermal environments, near-surface conditions are generally described as the most favourable ones for chalcedony formation (see *e.g.* White and Corwin, 1961; Cady *et al.*, 1998).

There is no overall agreement regarding the temperature of deposition of chalcedony. Taking into account results from laboratory syntheses and from natural outcrops, the temperature of deposition ranges from <50°C to 300°C. Most laboratory syntheses are performed currently at temperatures >150°C but this generally reflects the voluntary choice of decreasing the processing time (see Heaney, 1993 with examples and references therein). In natural materials, the temperature range varies between 25°C and 180°C, although generally being <100°C (Fallick *et al.*, 1985; Harris, 1989; Saunders, 1990; Heaney, 1993). However, decompression and boiling conditions may lead to extreme supersaturation of silica with respect to quartz that, in turn, may induce the deposition of chalcedony at temperatures >200°C (Fournier, 1985). With respect to pressure, Carr and Fyfe (1958) demonstrated how pressure constrains chalcedony formation to shallow depths.

Regarding precipitation of chalcedony, both direct precipitation from solution and transformation from an amorphous and viscous silica gel must be considered. The latter case was supported experimentally by laboratory syntheses – which obtained chalcedony using gels and opals (Oehler, 1976; Kastner, 1980) – and models (Wang and Merino, 1990). Some authors suggested recognizing a gel precursor with the occurrence of the following features: dehydration cracks, slump structures, preferential silica deposition on the bottoms of cavities rather than on the sides and tops (Fournier, 1985), bulbous voids often disrupting the wall-banding, fortification structures (Moxon, 1991) and crenulated shapes of some agates (Garlick and Jones, 1990).

A complete resizing of these observations was proposed by Heaney (1993) who (1) provided alternative explanations (*e.g.* pseudomorphism), (2) showed the rarity of certain features (*e.g.* liesegang banding) and (3) demonstrated the partiality of several results obtained through laboratory experiments (*e.g.* silica gels and non-silica solids). Heaney (1993) concluded that chalcedony precipitates from a non-viscous mother fluid and indicated the fossiliferous Formation of Hillsborough Bay (Tampa, Florida) as direct evidence for this interpretation. In this regard, it is worth underlining that an "opal phase" has been observed in non-fibrous microcrystalline quartz (Graetsch *et al.*, 1985) and its transformation into chalcedony seems to involve the dissolution of cristobalite instead of a solid–solid reaction (Mizutani, 1966; Stein and Kirkpatrick, 1976; Williams and Crerar, 1985).

From a diagenetic point of view, the most interesting conclusion reached by Heaney (1993) regards the role of the solution chemistry in deciding which polymorph will form: under conditions of equal pressure and temperature, chalcedony requires smaller concentrations of silica than cristobalite and larger than quartz.

The role of pH proved to be of critical importance in determining the spherulitic texture of chalcedony. Numerous studies demonstrated that the linearity of siloxane (Si–O–Si) chains is a function of low-alkaline pH and that Na^+ ions might inhibit branch formation, decreasing the number of bridging oxygens (Prassas *et al.*, 1982; Schaefer and Keefer, 1984; Schaefer, 1989).

At high temperatures (250–600°C), chalcedony gradually loses silanole while molecular water is synthesized according to the reaction formulated by Schmidt *et al.* (2011, 2012b): $Si{-}OH + HO{-}Si \rightarrow Si{-}O{-}Si + H_2O$, further involving the closing of open pores. In length-slow chalcedony, the temperature of silanole loss is slightly lower (200°C). In moganite, a temperature of ~100°C seems to be supported by experimental data (Schmidt *et al.*, 2013), however, the experiments are never performed with pure moganite specimens (*i.e.* always moganite mixed with chalcedony). Moreover, unlike chalcedony, the water/silanole loss of moganite appears constant and regular, with no major events.

3.3. Moganite

The International Mineralogical Association's Commission on New Minerals and Mineral Names (CNMMN) approved moganite as a mineral species in 1999. It was recognized in the 70s by Flörke *et al.* (1976, 1984) and structurally characterized from the 80s (Miehe *et al.*, 1984). This phase derives its name from the Mogán formation on the island of Gran Canaria (Canary Islands); the formation consists of a sequence of pyroclastic and lava flows.

Moganite has monoclinic symmetry (Miehe *et al.*, 1988) with space group *I*2/*a* (Miehe and Graetsch, 1992). Its structure is described generally as alternate stacking of layers of right- and left-handed quartz, with a periodic twinning according to the Brazil-twin law at the unit-cell scale (Miehe and Graetsch, 1992; Graetsch *et al.*, 1994). A reversible transition from monoclinic α-moganite to orthorhombic β-moganite has been observed at temperatures > ~573 K, combined with the symmetry increasing from *I*2/*a* to *Imab* (Heaney and Post, 2001; Heaney *et al.*, 2007).

The water content is both molecular and silanole (Graetsch *et al.*, 1994). It can reach up to 3 wt.%, thus exceeding the chalcedony content of 1–2 wt.% (Graetsch *et al.*, 1985; Miehe and Graetsch, 1992). Moreover, both the silanole/molecular water ratio and the concentration of silanole groups are higher in moganite than in chalcedony, while molecular water contents are comparable or slightly lower (Flörke *et al.*, 1984; Miehe and Graetsch, 1992; Graetsch *et al.*, 1994; Schmidt *et al.*, 2013).

As for moganite solubility, Gislason *et al.* (1997) provided the value of 44 mg/kg (quartz solubility is 6 mg/kg), further determining that the dissolution rate of moganite is 7.4 times faster than that of quartz at pH 3.5 and 25°C. The instability of moganite is regarded as being the most reliable explanation for its gradual decrease with the age and/or the alteration of the hosts, until its complete "disappearance".

As for the occurrence moganite, Heaney and Post (1992) demonstrated that this is a common mineral, being present in chalcedony, agate, chert and flint with percentages that range between 5 and 20 wt.% on average. The same authors detailed this estimation providing the following values: between <5 wt.% and >20 wt.% in both agate and chalcedony; between <5 wt.% and >75 wt.% in chert; and from 13 to 17 wt.% in flint. Heaney (1995) further distinguished evaporitic from non-evaporitic silica, respectively including 20–75 wt.% or 5–15 wt.% of moganite. Herdianita *et al.* (2000a,b) and Rodgers and Cressey (2001) found moganite levels of <13 vol.% in microcrystalline quartz-bearing sinters (between 20,000 and 200,000 y old) from the Taupo and the Coromandel Volcanic Zones of New Zealand. Some years later, Moxon and Rios (2004) indicated a value of 14 wt.% as the maximum moganite content in their agates.

Among archaeometric studies, Roman age gemstones from Vigna Barberini in Rome (Gliozzo *et al.*, 2011) showed variable moganite contents: 19–23 wt.% in two colourless chalcedonies, 0–35 wt.% in 13 orange carnelians, 12–14 wt.% in an onyx, 15–27 wt.% in three green Cr-bearing chalcedonies. Similarly, moganite contents ranged between 4 and 53 wt.% or between 6 and 16 wt.%, respectively, in carnelians and chert found in the Fezzan desert (Gliozzo *et al.*, 2014).

The general consideration that can be drawn from these data is that the wide variability of moganite contents are not visibly related to the variety (*i.e.* chalcedony, agate, chert, *etc.*) but rather to geological origin (host types and geographical location).

Another important aspect of moganite occurrence is that this phase can form in various environmental conditions and that its thermodynamic stability field is still uncertain. It can be argued that moganite is a strictly low-pressure crustal mineral, which can form in low-temperature evaporitic brines as well as in silica-rich hot waters (Heaney *et al.*, 2007). Heaney (1995) indicated that moganite can be used as a "valuable indicator for vanished evaporites", based on the frequent association of chalcedony with evaporitic regimes (*e.g.* Folk and Pittman, 1971). Moganite occurrence in the presence of hydrothermally altered basalt veins (Flörke *et al.*, 1982; Heaney and Post, 1992; Parthasarathy *et al.*, 2001), as well as with volcanic activity in general (Moxon and Ríos, 2004) has been shown repeatedly. Moreover, Parthasarathy *et al.* (2001) described moganite stability up to the laumontite zone of zeolite facies. The identification of moganite has also been the research topic in the Martian studies performed by Hardgrove and Rogers (2013).

As for dating, the first studies fixed ~100 Ma as being the maximum age of the host-rocks (*i.e.* Cretaceous Period; Heaney and Post, 1992). Rodgers and Cressey (2001) never found moganite in sinters younger than 20,000 y BP while it was present in rocks more than 40,000 and less than 200,000 y BP (Herdianita *et al.*, 2000a,b; Rodgers and Cressey, 2001). These authors further specified that, within this wide chronological range, moganite was rare or absent in most Tertiary sinters (*i.e.* from Paleocene to Pliocene). Moxon and Rios (2004) and Moxon and Reed (2006) showed that traces of moganite were present in agates older than 400 Ma. This limit in the Devonian period has been taken as the reference chronology now in use. Indeed, it is worth underlining that the chronologies provided by several authors are those of the host-rocks and there are some cases in which moganite may be secondary and, as a consequence, later.

3.4. Diagenetic transformations and selective growth of moganite

Heaney *et al.* (2007) described moganite as a "metastable transitory phase during silica diagenesis". Heaney (1995) further proposed a sequence of transformations with moganite as a follower of magadiite and a precursor of quartzine which, in turn, would have been a precursor of quartz:

$$\text{Magadiite} \rightarrow \text{(Kenyaite)} \rightarrow \text{Moganite} \rightarrow \text{Quartzine} \rightarrow \text{Quartz}$$

Many scholars have recognized moganite as the typical intermediate phase in the diagenesis of geothermal opal-A to quartz (Rodgers and Cressey, 2001; Rodgers and

Hampton, 2003; Hampton *et al.*, 2004). Herdianita *et al.*, (2000b) indicated the duration times for both the Opal-A → Opal-CT and/or Opal-C transformation (10,000–50,000 years; shorter in the presence of other components) and the quartz crystallization (20,000–50,000 years); however, the placement of moganite (detected in aged samples) within the sequence was not specified. Conversely, moganite (found in 20,000–200,000 years old sinters) and quartz were regarded as the resulting products of the maturation process of an opaline precursor by Rodgers and Cressey (2001):

$$\text{Opal-C} \rightarrow \text{Quartz} + \text{Moganite}$$

A few perplexities arise in relation to moganite transformation into chalcedony postulated by Moxon and Ríos (2004) and Moxon and Reed (2006).

As for selective growth, the reasons why moganite forms rather than quartz are not yet completely understood. Heaney (1995) provided two possible explanations: (1) structural affinities between moganite and its precursor (*i.e.* magadiite in that instance); and (2) the chemistry of the parent fluid. In the second case, Heaney and Post (1992) and Heaney (1995) suggested that alkaline and/or sulfatic fluids, possibly combined with ferric iron, were likely to control moganite precipitation. In this regard, Heany (1994, with related reference therein) showed that the transformation of Gran Canaria moganite to quartz was strongly promoted by ferric iron.

In summary, the diagenetic transformation of moganite depends on time, temperature and weathering and it progresses to the end-transformation into quartz, regardless of whether it crystallized directly or from an opaline precursor.

3.5. Moganite and quartz identification and quantification: a tool for the provenance issue in Archaeometry

The structural relationships between moganite and quartz have long created difficulties in discriminating these two phases by XRD; in fact, many of the major peaks of moganite overlie those of quartz (Fig. 3). However, the present state of the art is able to distinguish moganite from quartz and to quantify the respective percentages in weight by combining XRD with Rietveld refinement.

For quartz:moganite ratio determination, Raman spectroscopy is used increasingly in view of its ease and rapidity. Kigma and Hemley (1994) demonstrated that the moganite pattern is easily distinguishable from that of quartz, based on its intense peaks centred at 501, 220 and 129 cm^{-1} and less intense bands in the ranges 250–480, 650–850 and 1040–1210 cm^{-1} (Fig. 4a).

Götze *et al.* (1998) made a step forward demonstrating that the intensity ratios of the main symmetrical stretching-bending vibrations of α-quartz (465 cm^{-1}) and moganite (502 cm^{-1}) were indicative of the respective weight percentages of the phases. They further established that Raman measurements were not affected by the non-orientation or the non-preparation of the samples and that the resulting values for moganite contents were systematically higher than those obtained by XRD. Given that Raman spectroscopy was focused on an area 2.5 μm in diameter while powdered samples were used for XRD investigations, this shift was imputed to the different time of observation

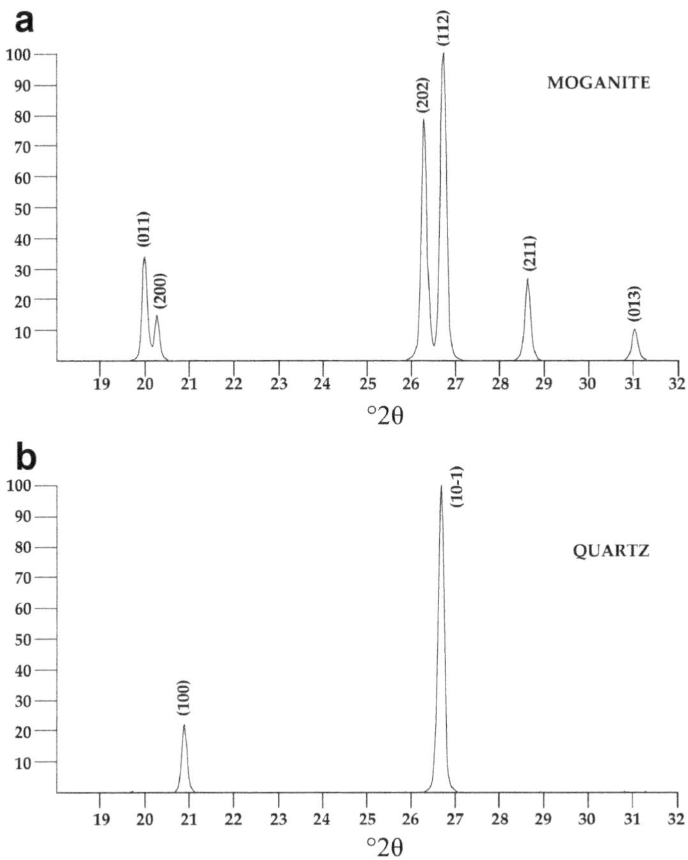

Figure 3. X-ray diffraction patterns (Cu-$K\alpha$) of moganite and quartz (°2θ scale) from (a) Miehe and Graetsch (1992) and (b) Smith and Alexander (1963).

of these two techniques. Raman results were considered more significant than those obtained by XRD, the latter having greater difficulties in quantifying poorly crystalline phases. Schmidt *et al.* (2014) warned of spectral interferences resulting from silanol bands which can lead to an overestimation of moganite content. For this reason, they recommended heat-treatment in order to reduce the interferences.

Gliozzo *et al.* (2014) applied both XRD and Raman spectroscopy to the investigation of a collection of carnelians and cherts. The intercalibration was extremely satisfactory, clearly indicating that when experiments and data analysis are performed carefully both techniques yield reasonable assessment of the mineral-phase quantities.

Moreover, the utility of (thermal) infrared spectroscopy for the quantitative determination of quartz and moganite has also been demonstrated (Graetsch, 1994, *e.g.* fig. 9; Hardgrove and Rogers, 2013).

In geology, the quantification of the quartz:moganite ratio is becoming increasingly common (see above cited literature as well as Pop *et al.*, 2004) while it is still infrequent

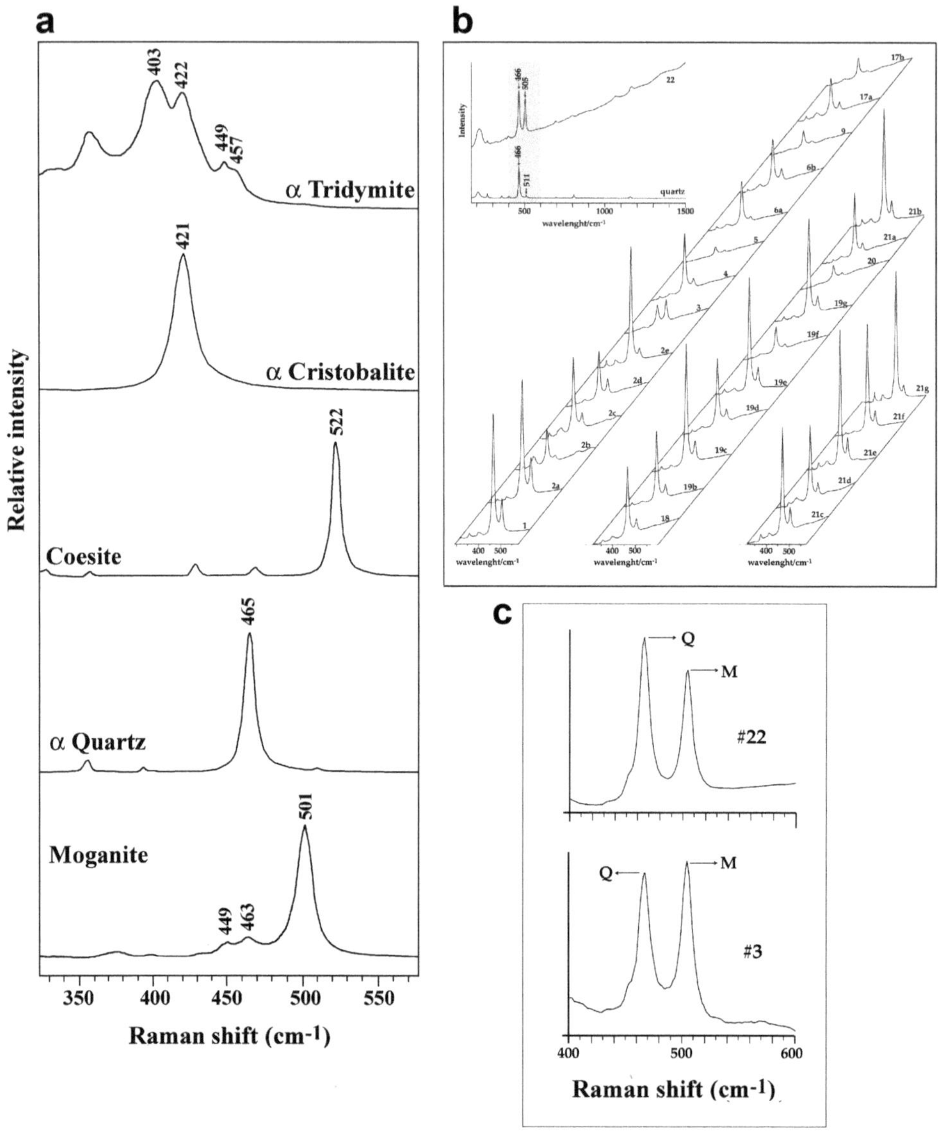

Figure 4. Raman spectroscopy. (a) Raman spectra of tridymite, cristobalite, coesite, quartz and moganite after Kigma and Hemley (1994). (b–c) The Raman spectra of 30 samples of carnelian and chert from the Fezzan desert; (c) the detailed view of two spectra where quartz and moganite contents are nearly equivalent.

in archaeometry, although it may represent a powerful tool for discrimination and, as a consequence, for provenance studies. There is very little information available on the topic (Pretola, 2001; Gliozzo *et al.*, 2011, 2014; Capél Ferron *et al.*, 2014; Gołyźniak *et al.*, 2016; Patrizi *et al.*, 2016; Prinsloo *et al.*, 2017; Wadley *et al.*, 2017; note that in some papers terminology and interpretations are not accurate) but this specific investigation, combined with bulk chemical data on major-, minor- and trace-element contents can support reasonable hypotheses on gemstone provenance. For instance, the results obtained on a collection of 16 carnelians and 15 cherts from the Fezzan desert by means of LA-ICP-MS, XRD and Raman spectroscopy (Fig. 4b–c) ruled out their importation from India and defined a reliable reference group for these materials, which were likely derived from a unique geological source (Gliozzo *et al.*, 2014).

Conversely, the exclusive application of Raman spectroscopy, especially when used merely for a phase determination, leads to such a partial characterization of the materials – even when non-destructive techniques are the only practicable ones – that it is no longer acceptable, except for a preliminary study aimed at research planning. A rapid overview of the principal applications of Raman spectroscopy in the field of archeo-gemmology is provided by Bersani and Lottici (2010, with references therein) while a review on the application of Raman spectroscopy in the wider field of gemmology is provided by Kiefert *et al.* (2012) and Kiefert and Karampelas (2011).

4. Agate

Agate is constituted mostly by chalcedony but it can also contain opal and macrocrystalline quartz as the final filling. It is characterized by multiple colours, typically distributed with a banded pattern. Many varieties are currently distinguished for commercial use based on their macroscopic characteristics (*e.g.* fire agate and Thunderegg; Fig. 5) and colour (*e.g.* blue lace agate) or area of origin (*e.g.* moss agate/mocha stone/Mokkastein from the ancient supply area of Mocha in Yemen, see Frondel, 1962 for further details on supply areas; Botswana agate, Laguna agate, *etc.*).

It can show the full range of colours (white, yellow, red, green and light blue) and, as clearly described by Graetsch (1994), can be of two types: wall-lining (concentric banding) or horizontally banded (Fig. 6). The former shows fibres radiating towards the centre of the geode, forming characteristic "parabolic fiber bundles"; conversely, in the horizontally layered agates, each layer may show "closely packed radiating spherulites" (Fig. 7).

The colours of agate are red, orange, brown, yellowish, purple, white, grey, green and, more rarely, blue. Agate with particularly intense colours are frequently dyed (*e.g.* Fig. 8), mainly by immersing a porous specimen in a coloured solution (*e.g.* cobalt nitrate for intense blue agate); however, many different kind of treatments are currently in use (see *e.g.* Scholz *et al.*, 2012 and Sindelar *et al.*, 2016, with references therein).

Given the enormous variety of rocks that goes under the name of agate, it is clear that the term does not indicate a well defined group of materials. Two main features should be required in order to classify a specimen as an agate: (1) it must be made primarily of chalcedony, or rather of fibrous quartz and variable amounts of moganite, and (2) it must show a banded pattern, otherwise it should be classified as chalcedony. As for feature 1, however, it is worth noting that an age- and weathering-dependent phase such as moganite can be absent.

Figure 5. A Thunderegg from Germany (Photo courtesy of Richard Nass).

Similarly to chalcedony, the origin of agate formation and its growth mechanism have long been debated without finding a single, widely accepted, model. Although Nacken provided the first reliable synthesis of chalcedony with a geode-like structure in 1948 (Nacken, 1948), the fact that nobody has yet been able to reproduce agate in a laboratory does not help.

It is difficult to generalize whether the entire sequence from amorphous silica to chalcedony (*e.g.* Moxon *et al.*, 2013) or the direct precipitation of chalcedony is involved in agate formation. These processes are closely tied to the surrounding conditions and there are still many unresolved problems in silica transportation, deposition and crystallization assessment. For instance, the self-organizational crystallization of agate postulated by Merino *et al.* (1995) and Wang and Merino (1995) involved a chemically controlled (*e.g.* silica saturation) self-organization of a non-crystalline precursor. Otherwise, in the model proposed by Heaney in 1993 (see also Frondel, 1978), a screw dislocation along the $[11\bar{2}0]$ direction should activate a spiral growth mechanism, which induces short-chain linear polymers to join *via* bridging silica monomers. Heaney and Davis (1995) explained banding in Iris agate

Figure 6. A specimen of agate from South Africa (author's personal photo collection).

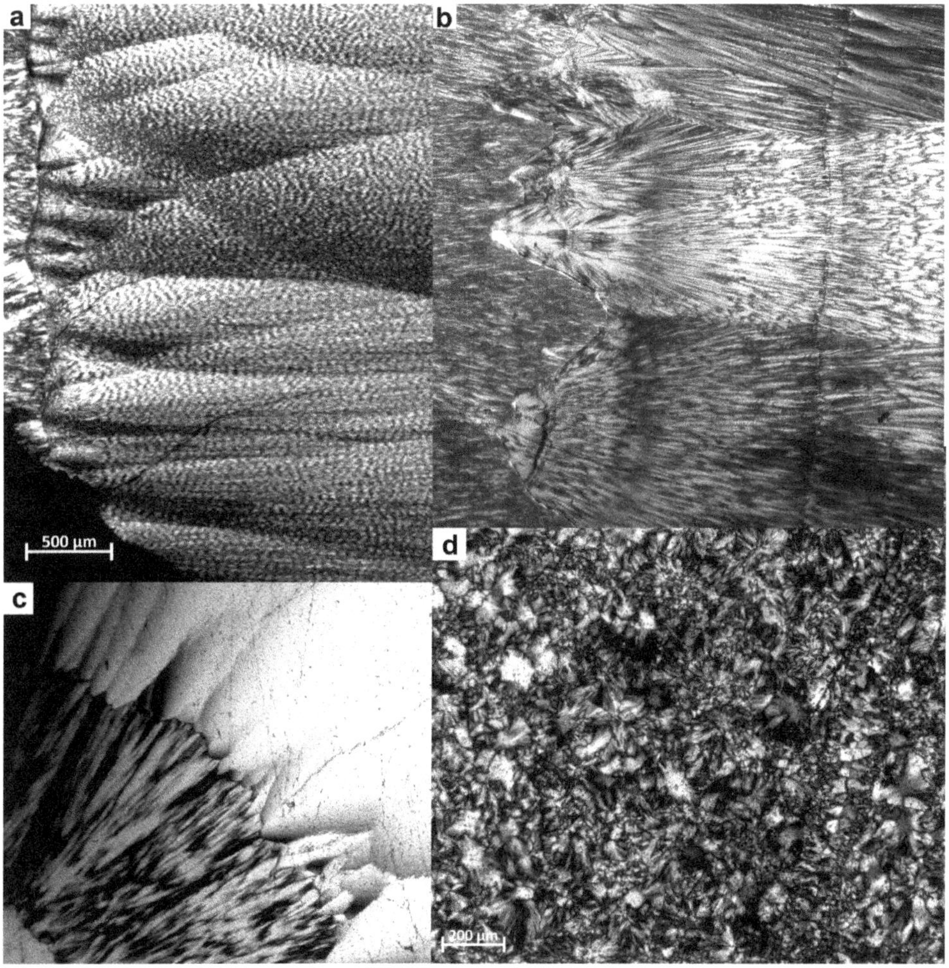

Figure 7. Agate at the optical microscopy scale (500 µm scale in a; 200 µm scale in b–d). (a–b) Length-fast fan-shaped chalcedony in a wall-lining agate. (c) Length-slow quartzine/moganite between length-fast chalcedony; and (d) zebraic chalcedony in a horizontally banded agate (author's personal photo collection).

with Ostwald-Liesegang crystallization cycles from silica-rich fluids. The Landmesser (1995, 1998) model proposed the following sequence: amorphous silica → cristobalite = tridymite → cristobalite → (±calcedony + moganite) → granular quartz. Lastly, Götze *et al.* (2016) investigated agates from Permian volcanic rocks of the Sub-Erzgebirge basin (Saxony, Germany) and explained their formation by "crystallization *via* an amorphous silica precursor under non-equilibrium conditions".

The banding phenomenon may represent alternating phases of activity and rest in the crystallization process. It can be influenced by the Fe oxide 'self-purification' process

Figure 8. A treated (dyed) agate.

during crystallization (Götze *et al.*, 2009), and/or by Al^{3+} and Fe^{3+} substitutions for smaller Si^{4+} (Wang and Merino, 1990, 1995) and/or by textural and structural changes (fine-grained highly defective chalcedony and moganite/coarse-grained low-defect quartz; see Heaney and Davis, 1995; Götze *et al.*, 1998). French *et al.* (2003) postulated that bands form "as a result of discrete influxes of silica-rich fluid" and recognized a constant sequence within individual bands ("chalcedony-A (with amorphous silica and nano-crystalline quartz) → chalcedony-MQ (with microcrystalline quartz) → quartz") which is proposed as "a viable model for the growth of chalcedony in agates". Furthermore, horizontal banding is generally assumed to be gravity-controlled (Moxon and Reed, 2006).

One of the few points on which there seems to exist a broader agreement concerns the formation temperature, which is estimated to be <100°C (Heaney, 1993) in basic igneous rocks (Fallick *et al.*, 1985; Saunders, 1990); however, the higher temperatures (>400°C) proposed for agates in volcanic environments (Moxon, 1996 with references therein) have been supported experimentally by chalcedony reproductions.

As for the estimation of the age of agate formation, Moxon (2017) investigated mobile and total water content of 60 agates (from 23 to 2717 Ma) in relation to host-rock age and it turned out that water contents showed age-related trends in agates up to ~60 Ma. Furthermore, it was established that Brazilian (135 Ma) and New Zealand (89 Ma) agates formed much later than their host-rock.

Lastly, the "unusually high uranium content" in the agates from Germany, Poland, Mexico and USA have been investigated by Götze *et al.* (2015).

4.1. Pliny on agate: description and provenance

Pliny was the first to list the many terms used for describing agate (*N.H.* 37, 139): "*iaspachates*" (jasper-agate), "*cerachates*" (wax-agate), "*smaragdachates*" (emerald-agate), "*haemachates*" (bloodagate), "*leucachates*" (white agate), "*dendrachates*" (dendritic agate = moss/mocha agate), "*antachates*" (not translatable with certainty, text maybe corrupted), "*corallachates*" (coral agate, "sprinkled with golden particles"), and "*sacra*" (in reference to its power to counteract the bites of spiders and scorpions).

Three paragraphs (*N.H.* 37, 136-138) were dedicated to the Iris agate (= moss agate). According to the author, this variety (a) corresponded to rock crystal – the reason why it was also called '*radicem crystalli*" (root of crystal), (b) had hexagonal faces, like the rock crystal, (c) scattered the light shining upon it, and (d) projected lighting at the same time to the objects around (*N.H.* 37, 137). Also according to Pliny and his sources, there would also be another rainbow stone in Persia called '*Iritis*' and other gems of similar appearance but black and white in colour, called '*leros*' (*N.H.* 37, 138).

Pliny further supplemented his description of agates with information gathered by the science of the Magi, although the latter were frequently attacked in tones of scorn and contempt (*N.H.* 28, 6-7, 228-229: 30, 17-19, 95). Judged extravagant, frivolous and patently irrational, they were often defined as false and impostors but their beliefs about agates were still reported. In fact, Pliny (*N.H.* 37, 142) described the ability of agates to keep scorpions away (when the gemstone resembled a lion's skin), to avert storms and waterspouts, to stop the course of rivers (when linked to the hairs of the mane of a lion), or even to make athletes invincible (a particular type of monochrome agate). In addition, Pliny reported the test to run in order to verify that the stones were actually effective: the stone should have been immersed for a least two hours in a pot of boiling oil, together with unspecified pigments. At the end of the test, all the pigments would have to have acquired a vermilion colour.

As for provenance, Pliny reported that agate was discovered first in Sicily, near the homonymous river. The most likely identification of this ancient river *Achates* is with the current River Acate-Dirillo. This river comes from the Iblei mountains, runs by the village of Acate, flows near Gela and along its course of 54 km assumes different names, among which is Acate. A further identification with the modern "Carabi (Canatello)" river, "not far from Giuliana" was proposed by Eichholz (1962) but it is difficult to find a correspondence between these toponyms and the actual geography.

Still, according to what Pliny says, Sicily was only one of the many supply areas of this gemstone which enjoyed great esteem in the past but no longer had a great value in his time (*N.H.* 37, 139). Agates with the appearance of rivers, woods and livestock (see for instance the *murrina vasa* in Del Bufalo, 2016) were arriving from India and were used to make *paterae* (*i.e.* libation bowls), figurines, horse-trappings and small mortars used by pharmacists. Moreover, it was said that, among other things, agate possessed an important property to dispel thirst by holding it in the mouth (*N.H.* 37, 140).

The "coral agate" was characterized by golden particles like those of lapis lazuli and was used extensively in Crete; however, the Plinian text does not clarify whether Crete was also the place of extraction (*N.H.* 37, 139); moreover, given the wide use of agate for the seals in Minoan Crete (Walters, 1897; Boardman, 1970; Spier, 1992), it is likely that this information resulted from a suggestion.

The Phrygian (modern Turkey), Egyptian (Thebes) and Cypriot agates were also mentioned, specifying that the first contained no green hues, the second did not show red and white veins, while the third had transparent glassy portions (*N.H.* 37, 141). Other supply areas were the Trachis region (near Mount Oeta, place of Heracles' death), Mount Parnassus, Lesbos, Rhodes and the Messenia district (southwestern Peloponnese); the last-named providing agates with flower-like patterns (*N.H.* 37, 141). As for the provenance of Iris agate, Pliny reported that it was excavated on an island in the Red Sea, 60 Roman miles (~88 km) from the city of Berenice (*N.H.* 37, 136). It is perhaps no coincidence that, at approximately the same distance (82 km as the crow flies), there is the volcanic island of Zabargad (Saint John's island), which was the supply area of the best peridot (Wilson, 1976).

The localization of the toponyms mentioned by Pliny is shown in Fig. 9.

4.2. Past and current supply areas

Agate occurrence is generally related to volcanic environments, especially to basaltic, andesitic, dacitic and rhyolitic hosts (Götze *et al.*, 2009); conversely, agates are rarer in sedimentary and metamorphic hosts (Clark, 2002; Moxon and Reed, 2006; Moxon *et al.*, 2006, 2007).

In Europe, one of the larger deposits was that at Idar-Oberstein (Germany; Fig. 9), where agate was extracted and worked on an industrial scale until the 19th century (Frazier *et al.*, 1998–1999). When the deposit was almost completely exhausted, agate was imported from the Brazil mines of Rio Grande do Sul which were discovered by Idar-Oberstein migrants and shipped as ballast. The exploitation of the Idar-Oberstein deposit in an earlier period has not yet been demonstrated, although it is very likely. The area is still the object of amateur research and the seat of the important Edelstein museum where agates from all over the world are collected (Fig. 10).

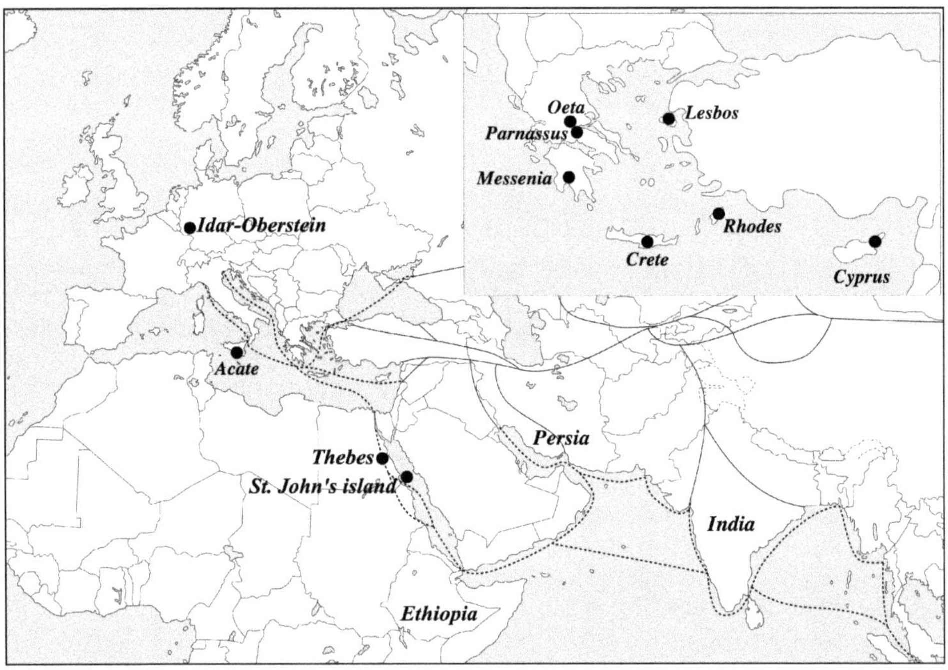

Figure 9. From the *Naturalis Historia* by Pliny the Elder: the distribution of *achates*. The silk road and the principal maritime routes are shown, as well as the localization of Idar-Oberstein.

Figure 10. Some of the marvellous agates preserved at the Idar-Oberstein Museum. The upper two are from Argentina while the other is from Freisen, a locality close to the town (author's personal photo collection).

The major supply sources for the modern market are Brazil (see Scholz *et al.*, 2012 for further details), Madagascar, Argentina, Uruguay, Mexico, Poland (for those from Płóczki Górne see Dumańska-Słowik *et al.*, 2017), Botswana, India (Deccan), Australia and the United States (Oregon, Arizona, Montana, Wyoming, South Dakota and Michigan).

Recent interest has arisen also in relation to both Moroccan agates (Natkaniec-Nowak, 2016) and Serbian deposits. The latter are poorly known but rich in opal, carnelian, bluish chalcedony, chrysoprase, jasper, rock crystal, amethyst and many other precious hard stones (Miladinović *et al.*, 2016).

Lastly, it is perhaps worth remembering that coral and fossilized wood are frequently termed as "agatized" when silicified.

4.3. Agate provenance in archaeometry and information gained from the geological literature

Excluding chert and flint, agate has received the widest attention in archaeometric studies. In fact, a total of 183 samples (including carnelians), found in Afghanistan, Cambodia, India, Iran, Italy, Lybia, South Africa and Thailand were investigated geochemically. Among these studies, that performed by Law *et al.* (2012) is of outstanding importance because it demonstrated how different sources could be differentiated, providing reasonable discrimination for the three agate deposits in India as well as with respect to eastern Iranian and Thai agates (Fig. 11). Although lacking in archaeometric analyses, the study conducted by Kenoyer *et al.* (1994) is of further interest as it includes the recognition of numerous ancient agate working sites in India.

The number of investigated geological materials is smaller but with a wider geographic distribution. Excluding materials from Japan as well as oceanic materials, geochemical analyses are available for at least 171 examples from: Brazil, Canada, the Czech Republic, Germany, India, Iran, Lesotho, Mexico, Namibia, Poland, Thailand, Turkey, the United Kingdom and the USA (see Appendix 2).

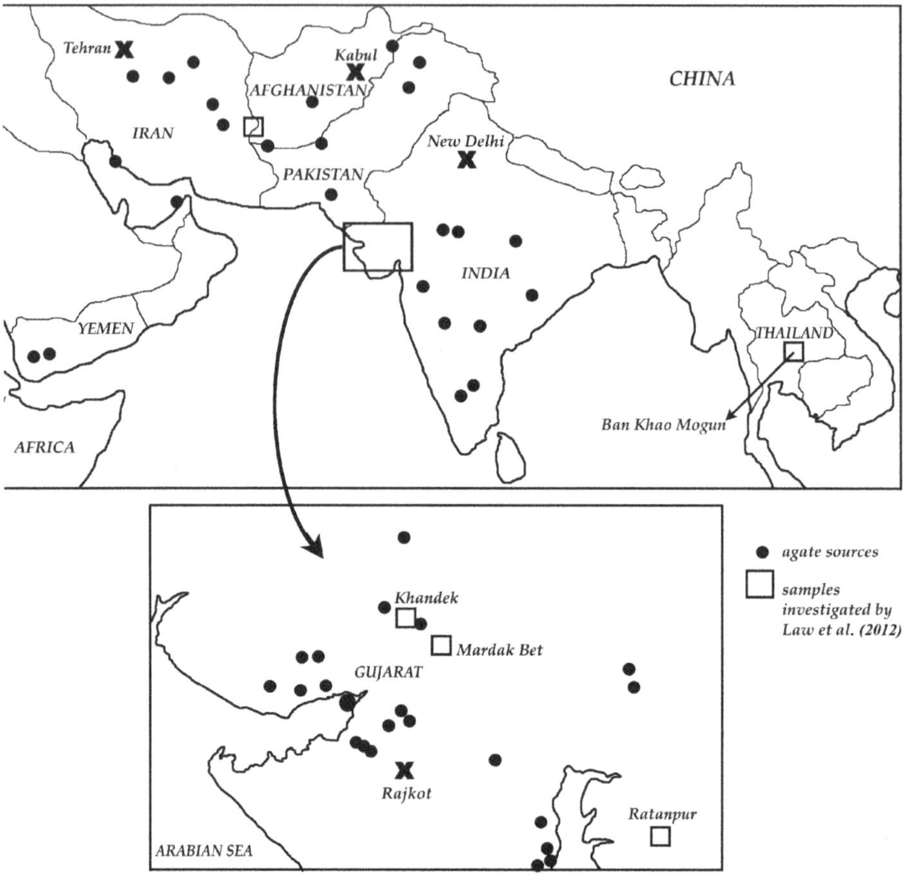

Figure 11. Agate – source distribution in Yemen, Iran, Afghanistan, Pakistan, India and Thailand from Law *et al.* (2012). The localization of samples investigated by Law *et al.* (2012).

Given that quantifications are provided for agate and carnelian together, please also see Section 5.4. on "Carnelian provenance in archaeometry and the information gained from geological literature".

5. Chalcedony varieties

Similarly to the scheme adopted previously for agate, the following sections will provide information about:

(1) the fundamental characteristics of chalcedony varieties from a mineralogical point of view, in order to establish the basic criteria to distinguish them;

(2) the information passed on by Pliny, especially with regard to the origin of these gemstones, in order to assess the knowledge and beliefs about these minerals and the supply areas known at the time;

(3) today's main extractive areas and deposits, in order to gather additional information on the availability of these gemstones:

(4) the archaeometric studies performed on archaeological gemstones, in order to obtain an insight into the current state of the art on these topics.

5.1. Red and orange chalcedony: carnelian and sard

The term carnelian (*corniola* in Italian, *cornaline* in French, *karneol* in German) is applied to the orange-red translucent variety (Fig. 12) while the term 'sard' (*sarda* in Italian, *sardoine* in French, σάρδιον in ancient Greek) is used for the dark-orange to brown varieties. These chalcedony varieties consist of a mixture of microcrystalline quartz and moganite and owe their colour to small hydrous iron oxide inclusions (Fritsch and Rossman, 1988).

Carnelian can be obtained by heat-treating agate, in order to enhance the colouring effect of hematite. The same result can be obtained using an iron nitrate solution or other appropriate dyes. Usually, dyed carnelians show a striped pattern against the light, while untreated carnelian has a cloudy appearance.

Both carnelians and sards have always functioned as decoration and status markers. The Egyptian civilization used carnelian for the manufacture of scarabs and beetle-shaped amulets, symbols of resurrection and the human soul's endurance. In more recent times, this tradition was passed down by Phoenicians, Greeks, Romans and Etruscans, as well as by the first Christian communities, for its ideological connection with the resurrection. During Roman times, carnelian and sard were used mainly to make engraved gems (*N.H.* 37, 105) for luxury items, with an apotropaic finality or lucky charms. Hot wax does not stick to these gemstones and this property has determined their further use as seal rings since the Assyrian civilization.

5.2. Pliny on carnelian, sard, *carbunculus* and *charchedonia*: description and provenance

As frequently happens with Plinian text, the description of carnelians and sards suggests that different types of gems were grouped under one name. It seems, in fact,

Figure 12. Carnelian and quartz from the territory of Idar-Oberstein, Germany (Personal photo collection).

that the same material was unknowingly named in four different ways: carnelian, sard, *carbunculus* and *charchedon(ia)*.

According to Pliny, the colour and the aspect of carnelians and sards were used in order to distinguish their provenance (*N.H.* 37, 105-107). Greek and Egyptian sards were characterized by being backed with gold foil, while those from India were of three different varieties: a red variety, a '*pionia*' variety (so-called for its greasy lustre) and a third variety backed with silver foil. The honey- and the earthenware-coloured varieties were so despised by Pliny that he does not even say where they came from.

Pliny asserted that sard was first discovered at *Sardis*, identified as Σάρδεις in Lydia (modern Sart) and not as Σαρδώ (modern Sardegna) as had been hypothesized. Onsite investigations have never been performed and modern researchers presume that the word *Sardis* has nothing to do with that town. In fact, the term should derive from "sered", a Persian word meaning yellowish-red.

However, Pliny added that the most valuable specimens were found near Babylon; moreover, he mentioned several other sources such as Paros and Assos (modern Behramkal, Turkey), India, Arabia, Epirus and Egypt, as well as exhausted sources in Persia (*N.H.* 37, 105-106).

According to Eichholz (1962), India should be the true source of those quoted by Pliny as being from Persia and Babylon. However, the summary traced by Charpentier *et al.* (2017) indicates that many other possibilities were absolutely reliable. In fact, the first productions of carnelian and agate beads emerged in Baluchistan (eight millennia BCE) and in northern Mesopotamia (a millennium later; Inizan, 1995) and became frequent in lower Mesopotamia (Ubaid culture, fifth millennium). Given that geological sources have not been identified along the Arabian coast, Inizan (1995, 1999) suggested that Mesopotamian beads were traded by the coastal Neolithic societies of the Gulf while Crawford (1998) proposed an Iranian origin. During the Middle Neolithic period, hard-stone beads made their appearance in Arabia (Inizan, 2007) while the earlier occurrences of carnelian beads in southern Arabia date back to the second millennium BCE (Inizan, 2007; Crassard, 2008). Furthermore, between the sixth and the fourth millennia, carnelian and agate from the Jebel al-Ma'taradh (Emirate of Ra's al-Khaimah) were surely exploited to make beads (Charpentier *et al.*, 2017).

In this regard, it is worth discussing the eventual correspondence of carnelians/sards with the incombustible *carbunculus* gemstone cited by Pliny (*N.H.* 37, 92-97, 103-104), but more frequently referred to as a pyrope garnet (see for instance http://www.mindat.org/gm/28904).

Pliny stated clearly that there were several types of *carbunculi*. The first variety from India was flawed, with a parched lustre and rather opaque but it allowed the creation of large vessels and showed brighter colours than the second variety.

The latter is described twice in *Liber* 37th. The first mention appears in chapters 92-97 and presents these *carbunculi* as gemstones linked to the Garamantes people, who were also called *Charchedonios* (*i.e.* Carthaginian), in honour of the greatness of Carthage. In fact, according to *Eusebius*, the Tyrians or Phoenicians *Charchedon* and

Zorus founded Carthage thirty years before the destruction of Troy. In Rome, Garamantian carbuncles were known as "Carthaginian stones" possibly because Roman traders initially accessed these gemstones *via* Carthaginian traders (Gliozzo *et al.*, 2014). Still in this section of his book, Pliny lingered on the description of this variety, also reporting the opinion of his ancient colleagues *Callistratus* and *Archelaus*. The information gathered described the Garamantic *carbunculi* as small gems, with an overall dark appearance, tending to a violet hue in the shade or to a fire-red hue in the open air. They also sparkled if directed towards the sun, and were able to melt wax in the dark when used as signets. The further notation on the asterism of male *carbunculi* can be interpreted readily as implying that among the red stones called *carbunculi*, rubies were also included. The second mention appears in chapter 104 where Pliny insists on the African origin of *charchedonia*. He specified that these stones originated in the territory of the Nasamones and skeptically added that local people thought that they were formed by rains of divine origin. Similarly, Strabo indicated the Saharan mountains as the land from where the "Carthaginian stones" were brought (*Geography*, 17.3.11; 17.3.19), thus supporting the Libyan origin of the *charchedon*.

If, on the one hand, the ancient literary sources seem to indicate decisively the Libyan Sahara as a supply area for the *charchedon*, on the other hand we have no tools to further limit the possible areas of extraction. The mention of *Mons Gyri* as the source location in the territory of the Garamantes (Pliny, *NH* 5.37) does not find direct correspondence with modern toponymy. In this regard, the volcanic area of the Jabal Bin Ghanima (in the basalt province of al-Haruj), the Tibesti mountains or the plateau/mountains to the north of the Wadi ash-Shati may be potential source areas (Gliozzo *et al.*, 2014).

Another interesting aspect seems to be the confirmation that the Garamantes were particularly active in the gemstone market, not only for *carbunculus* but also for *sandastros* (also called Garamantic stone; see Pliny *N.H.* 37, 100-102). On reasonable grounds, therefore, this information should be interpreted as a very important provenance indication.

Going back to the description of *carbunculus* varieties provided by Pliny, the third one was of Ethiopian origin and characterized by a greasy appearance without any lustre (*N.H.* 37, 92).

The fourth variety was named *Alabanda* and was a darker and rougher variety found at *Orthosia* but treated at *Alabanda* (both sites were located in *Caria*, *i.e.* western Anatolia). The correspondence with almandine – which takes its name from the city of Alabanda – is absolutely clear in this case.

The fifth variety – the *lychnis* – was also supplied in Caria and the neighbouring regions (*N.H.* 37, 103). It was characterized by lamps lit with kindling wood and it was sometimes named 'mild carbuncle' when resembling the so-called 'Flower of Jove' (*i.e. Lychnis flos-jovis*, a flowering plant with pink flowers of the *Caryophyllaceaea* family; please note that in the same family of the *Caryophyllaceaea* is also included the red *Lychnis Chalcedonica*).

Among the varieties of *carbunculi* not affected by fire, Pliny mentioned that from *Miletus*, while *Theophrastos* quoted *Orchomenos* in Arcadia (dark variety used for mirror-making), Chios, Troezen (in Argolis), Corinth and Marseille as additional areas of discovery. *Lucius Cornelius Bocchus* (another Lusitanian colleague of Pliny) further wrote about how they were extracted with great difficulty near Lisbon (*N.H.* 37, 96-97).

Among other ancient authors, further information about these gems has been handed down by *Teophrastos* and the Greek epigrammatist poet *Posidippus* (Pella, *ca.* 310–240 BCE). The former distinguished a female and a male variety which could correspond to the distinction between carnelian and sard (*Lap.* 8, 23, 30). Pliny faithfully reported this belief (*N.H.* 37, 106), adding further that male and female gemstones could be distinguished based on their brighter or more opaque lustre, respectively, and, among the former, the 'stones of Syrtis' (*syrtitas*) were particularly appreciated for their red (core) and purple-amethyst (edges) bichromy (*N.H.* 37, 93).

This gender aspect was of particular importance especially when using crystals for therapeutic reasons; in fact, it was believed that their differential use had specific properties, in close analogy with the concept of yin and yang in Chinese philosophy.

Posidippus wrote about a very bright Sard, of at least a 66 cm side (if squared) or diameter (if circular), that excelled Indian rubies and bore the engraved image of Darius and his chariot (I 36- II 2; see the comments made by Casanova, 2004).

Lastly, Pliny provides further information on: (1) treatments in vinegar (14 days immersion) to make the gemstone brighter and (2) numerous counterfeits in glass which could be exposed through the use of grindstones and scales.

The locations of the toponyms mentioned by Pliny are shown in Fig. 13.

5.3. Past and current supply areas

The Indian deposits were the largest and their exploitation probably predates the Harappan period (see *e.g.* Allchin, 1979; Kenoyer *et al.*, 1994). They were not the only ones, however, because traces of processing have been found in Indonesia, West Java (Garut, Bungbulang mountains), East Sumatra (Kalienda, Lamphung region) and central Thailand (Lopburi area) (Bellina, 2003). In the last-named areas (Fig. 14), production centres and bead distribution between the last centuries BCE and the first millennium CE were studied by Bellina (2003).This author claimed that multiple Asian sources must have been exploited, therefore, an exclusive import from India would have been unlikely.

Recently, research into Fezzan (Libya) carnelians and cherts demonstrated for the first time that carnelian could have been of Saharan provenance (Gliozzo *et al.*, 2014). Similarly, the study conducted by De Waele and Haerinck (2006), helped to characterize numerous etched beads from northeast and southeast Arabia, in relation to a wide carnelian trade, possibly involving numerous source areas in the Deccan Plateau, Iran (Ratanpur in the Rajpipla, Shahr-i Sokhta in the Helmand Basin and on the Bushehr peninsula; see Whitehouse, 1975), Arabian Peninsula (al-Ghail, ~40 km south of Ras al-Khaimah, UAE), Yemen and western Saudi Arabia (references in De Waele and Haerinck, 2006).

Modern supply was concentrated mainly in India, before being flanked and partially supplanted by Brazil, USA, Sri Lanka and Uruguay. Among the numerous supply areas

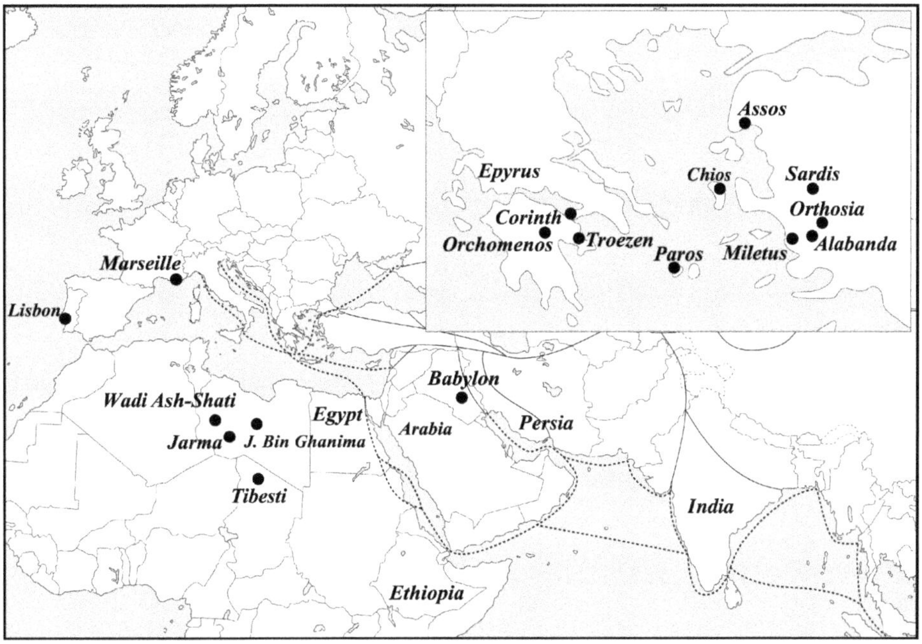

Figure 13. From the *Naturalis Historia* by Pliny the Elder: the distribution of carnelians. The silk road and the principal maritime routes are shown.

in the United States, Frondel (1962) mentioned Specimen Ridge and Amethyst Mountain in the Yellowstone National Part (Wyoming), Yellowstone River (Montana), Chehalis (Washington), Colorado Springs (Colorado), Fairburn (South Davota), Isle Royale (Michigan), Flint Ridge (Ohio) and other unspecified localities in Oregon.

5.4. Carnelian provenance in archaeometry and information gained from the geological literature

The trade in carnelian beads has been studied for a long time with the objective of tracing the interactions between different cultures and geographic areas (see Kenoyer, 2008; Kenoyer *et al.*, 2013 with references therein). Use-wear analyses and technological (/ethnographic) studies have invaded the archaeometric literature (*e.g.* Chevalier *et al.*, 1982; Gorelick and Gwinnett, 1990; Inizan and Tixier, 2000; Roux, 2000; Rosenfeld and Dvorachek, 2003; Kenoyer, 2005; Groman-Yaroslavski and Bar-Yosef Mayer, 2015), while provenance studies were virtually neglected. A few geochemical studies are currently available on archaeological carnelians from Cambodia, India, Nigeria, Sri Lanka, Thailand, Italy and Libya:

- Iron Age carnelian and agate beads from Cambodia and Thailand, investigated by Carter and Dussubieux (2016);
- Carnelians from modern mines and ancient manufacturing sites in India (Gujarat), Nigeria and Libya, investigated by Insoll *et al.* (2004), using UV-LA-ICP-MS;

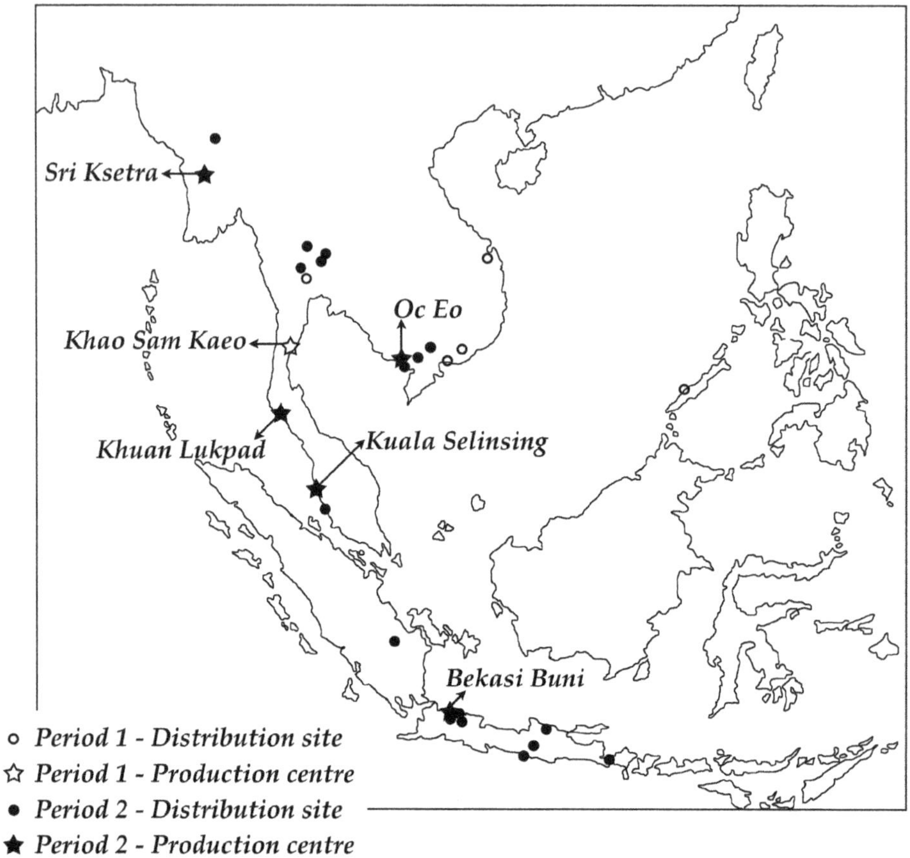

Figure 14. Agate and carnelian – Production centres and bead distribution between the last centuries BCE (Period I) and the first millennium CE (Period II) reconstructed by Bellina (2003).

- Carnelians from several sites in India, Sri Lanka and Thailand, investigated by Theunissen *et al.*, (2000), using PIXE/PIGME;
- Republican to 7th century AD carnelians from the archaeological site of Crypta Balbi (Rome, Italy), investigated by Andreozzi *et al.* (1996);
- 1st to 2nd century AD carnelians (11 samples) from the archaeological site of Vigna Barberini (Rome, Italy), investigated by Gliozzo *et al.* (2011), using XRD and PIXE;
- 4th century BCE to 4th century AD carnelians (16 samples) from the Fezzan region (Libya), investigated by Gliozzo *et al.* (2014), using XRD, Raman microspectroscopy and LA-ICP-MS.

Other studies only supply information about the discovery of carnelians, provide a very preliminary characterization of a few samples (Behar *et al.*, 2016) or include a detailed mineralogical characterization, but without geochemical measurements (*e.g.* Anghel *et al.*, 2016, among others cited above).

Geological literature on carnelians is also extremely rare. Two carnelians from Coşava (Banat, Romania) were investigated using LA-ICP-MS by Schmidt *et al.* (2012a). Geological sources for agate and carnelian from Thailand, Iran and India have been investigated by Carter and Dussubieux (2016).

Lastly, it is worth specifying that, in the literature, the terms agate and carnelian are often used interchangeably; therefore, this paragraph should be consulted together with a previous one (Section 4.3. Agate provenance in archaeometry and information gained from the geological literature) and quantifications for both are presented together in Appendix 2.

5.5. Black and brown chalcedony: onyx and sardonyx

Onyx and sardonyx are banded chalcedony, or rather agates, which differ only in their black and white or brown and white colourations, respectively. In both cases, however, onyx and sardonyx may indicate chalcedonies where only the black or the brown portions are visible. This is the case for 'nicolo' (or 'niccolo') onyx, which probably corresponded to the Plinian *Aegyptilla* (*i.e.* little Egyptian stone; *N.H.* 37, 148), characterized by a black ground and an upper layer of blue.

The colours may be due to Mn-oxide or Fe-oxide inclusions (Indian and South African agates), bituminous organic substances (Brazilian agates) or carbonaceous materials incorporated in the SiO_2 matrix during formation (agates from Mali; see Götze *et al.*, 2012 with references therein).

The colouration can be induced artificially by the staining process, *i.e.* immersing an uncoloured chalcedony in an aqueous solution saturated with sugar, honey or olive oil. The organic substances will penetrate the porous silica and will undergo dehydration and oxidation reactions when heated in a sulfuric acid solution.The colour obtained will not be stable over time and it will fade. Among other 'gemstone simulants and enhancements', the onyx/sardonyx treatment is explained clearly by Whalley (2012): "the material is first boiled in a solution of sugar ($C_{12}H_{22}O_{11}$); it is then heated or soaked in concentrated sulphuric acid (H_2SO_4) which reacts with the sugar to release carbon, thus producing a black or brown body colour."

Although it was widely used for making jewellery (especially cameos, like the 'Great Cameo of France' or the "Apotheosis of Augustus"), cups, statuettes and other decorative items, Pliny briefly lingers on the description of onyx, without being particularly fascinated by it (*N.H.* 37, 90-91). He refers to the information received by his colleagues *Sudines*, *Zenothemis* and *Sotacus* about the Indian and Arabian onyx. The Arabian onyx is described as a black stone with white bands, closest to what we now refer to as onyx. Conversely, the Indian onyx was thought to have red and violet streaks, similar to those of the *carbunculus*. Apart from India and Arabia, the association of the *Aegyptilla* with the 'nicolo', should include Egypt among the ancient suppliers.

Modern supply areas are concentrated mainly in Brazil, Madagascar and Uruguay. However, the rarity of both onyx and sardonyx has prompted many manufacturers to produce it artificially.

To the best of my knowledge, there is no geochemical research on archaeological materials.

5.6. Green chalcedony: chrysoprase, Cr-chalcedony, "gem silica" or "chrysocolla chalcedony" and heliotrope

Chrysoprase (χρυσός = gold, πράσινον = leek) is characterized by an apple green hue (Fig. 15) caused by Ni-bearing inclusions rather than by substitutional Ni in the silica (Rossman, 1994 with references therein). NiO can reach up to 5.1% but it is not clear if it is always present as bunsenite (cubic NiO) or in other forms (Hyršl, 1999) such as layered Ni-rich silicates. In fact, Barsanov and Yakovleva (1981; quoted by Rossman, 1994) provided evidence of bunsenite, while a Ni-bearing phyllosilicate was found in chrysoprases from Australia, Poland and Kazakhstan by Nagase *et al.* (1997). Other authors (*e.g.* Sachanbiński *et al.*, 2001) found Ni-Mg silicate aggregate, such as garnierite (generic term, variously used as a synonym of falcondoite and/or népouite) or Ni-rich kerolite and pimelite (*i.e.* a Ni-bearing clay mineral not approved by the IMA). Furthermore, Graetsch (2011) demonstrated that the green colour of Tanzanian chrysoprase (Shigley *et al.*, 2009) was due to 0.47 wt.% of gaspeite (hexagonal nickel carbonate, $(Ni,Fe,Mg)CO_3$) while Čermáková *et al.* (2017) claimed that the presence of "pimelite" was responsible for the green colouration of the Szklary chrysoprase.

In any case, chrysoprase can be considered to be the corresponding chalcedony variety of the leek-green quartz variety called prase (quartz + chlorite); however, this latter term does not seem, at present, to be used to describe a single varietal name and it can be used for a variety of green opal as well (Shigley *et al.*, 2009). In addition to chromophores, chrysoprase can also include both tridymite and cristobalite.

Cr-chalcedony has a typical emerald green colour which is due to Cr inclusions, although minor quantities of nickel can be present as well. It is sometimes called

Figure 15. On the left, a chrysoprase from Poland (Photo by Xth-Floor – Own work, CC BY-SA 3.0, https://commons.wikimedia.org/w/index.php?curid=15062186); on the right prasiolite crystals from Torre di Rio at Rio Marina, Elba Island, Italy (Photo by Didier Descouens – Own work, CC BY-SA 4.0, https://commons.wikimedia.org/w/index.php?curid=7853444)

matauralites or mtorolite because some specimens have been found near Mtoroshanga in Zimbabwe. The presence of Cr means this variety of chalcedony approaches the quartz variety called aventurine, although with a different source of colour. Moreover, it is also possible to add that fuchsite-bearing quartzite may correspond to the Plynian *sandastros* mentioned by Pliny (*N.H.* 37, 100-102).

Based on colouring agents (Ni for chrysoprase, Cr for Cr-chalcedony), the Turkish chrysoprase reported by Ayvacıklı *et al.* (2012) and Hatipoğlu and Yardımcı (2014) should be included in this group rather than in the chrysoprase group. Similarly, the "new" bluish green variety from Africa (Renfro, 2015), which bears both Cr and V as likely chromophores, can be included in this group.

The third greenish blue variety is sold as "gem silica" or "chrysocolla chalcedony" and is coloured green by copper. Tiny inclusions of copper silicates have been indicated by Frazier and Frazier (1999) and by Laurs (2015). Also, Shen *et al.* (2006) clearly identified a Cu^{2+} band by UV-Vis-NIR spectroscopy.

Heliotrope (or bloodstone or plasma) is a dark green variety, dotted with small, hematite red, blood-like spots. The African variety is often called seftonite (Fig. 16) and frequently shows pyrite flecks. Heaney *et al.* (1994) attributed the green colour to the presence of tiny particles of chlorite or hornblende needles while Zhang and Sakurai (2009) vaguely connected the possible influence of Fe^{2+} and Cu compounds with the dark green colour, of Fe^{3+} (as hematite) with the yellow and red colours and of Fe_3O_4 with the black colour.

5.7. Pliny on green chalcedonies – description and provenance

The *chrysoprasus* mentioned by Pliny (*N.H.* 37, 77) is listed in the chapter dedicated to the varieties of *berullus*; therefore, it should be associated with a variety of emerald (chrysoberyl?) rather than chalcedony.

The gem "*iaspis*" is described as a green stone often translucent, thus being the only possible match with green chalcedony of some significance, although the term probably includes green jasper as well. According to Pliny the sources of *iaspis* were: (1) India, which provided a translucent variety resembling emerald; (2) Cyprus, with a hard and

Figure 16. Heliotrope (left, unknown photo credits) and "Seftonite" from South Africa (right, author's personal photo collection).

dull variety with greyish shades; (3) *Amisos* (modern Samsun on the south coast of the Black Sea, Turkey), where a variety similar to that from India was exploited; and (4) *Chalcedon* in Bithynia (modern Kadıköy, opposite Byzantium, south of Scutari), which supplied a "cloudy" variety (*NH* 37.115).

Regarding heliotrope, Pliny explains firstly the origin of its name (*N.H.* 37, 165): throwing the heliotrope in a container filled with water, the gemstone changed colour into blood-red when reflecting sunlight. Pliny adds that this property was particularly observed in the Ethiopian heliotrope and that the stone was often used to detect solar eclipses (this latter use was possibly correlated with the heliotropism). Lastly, Pliny ironically mocks the 'blatant effrontery' of the magicians who believed that the stone had the power to make someone invisible when used in conjunction with the homonym genus of flowering plants (*Heliotropium*, *Boraginaceae*). Regarding the distribution of heliotrope, Pliny listed Ethiopia, Africa and Cyprus as known supply areas (*NH* 37. 60.165).

The locations of European, African and Asiatic toponyms mentioned in this paragraph are shown in Fig. 17.

5.8. Modern supply areas

The main sources of chrysoprase are in Australia (Nagase *et al.*, 1997; Befi, 2009), often within weathered profiles resulting from lateritization of nickeliferous serpentinites or other ultrabasic rocks (Queensland; Krosch, 1990) but also in Kazakhstan (Sachanbinski *et al.*, 2001; Witkowski and Zabinski, 2004), Tanzania (Witkowski and Zabinski, 2004; Shigley *et al.*, 2009), Germany, Poland (Sachanbinski *et al.*, 2001; Skrzypek *et al.*, 2003, 2004; Čermáková *et al.*, 2017), Turkey (Hatipoğlu *et al.*, 2011b; Ayvacıklı *et al.*, 2012), Brazil (Komov *et al.*, 1994) and the United States (Arizona and California).

Recently, Serbia has been opening up gradually to the markets (Miladinović *et al.*, 2016). In Poland, chrysoprase was first identified in the 18th century from the Szklary mine; it was located within Ni-bearing serpentinites and has been exploited since the

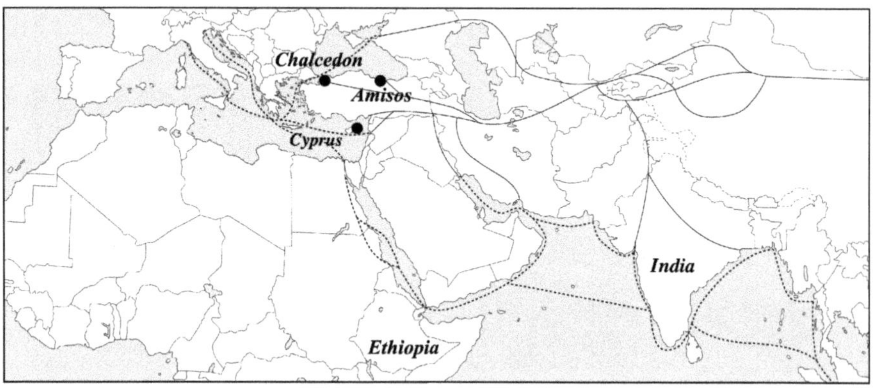

Figure 17. From the *Naturalis Historia* by Pliny the Elder: the distribution of chrysoprase, Cr-chalcedony and heliotrope. The silk road and the principal maritime routes are shown.

Middle Ages (Sachanbinski *et al.*, 2001; http://www.mindat.org/loc-232015.html). Further traces of ancient exploitation (Ottoman Empire Period) were found in the village of Dikmen, in the Biga–Canakkale region of Turkey (Hatipoğlu *et al.*, 2011b) and in the ancient Silesian territory (chiefly Frankenstein, Kosemütz Gläsendorf, Baumgarten with mines dated to the 14th century AD; Streeter, 2011).

In modern times, sources of Cr-bearing chalcedonies have been located at Saricakaya in Turkey (Eskisehir, Central Anatolia; Lule-Whipp, 2006), Tanzania (Hyršl, 2016), Zimbabwe (Smith, 1967), Bolivia (Hyršl and Petrov, 1998), the Balkans and the Ural mountains (Hyršl, 1999), as well as in the Newman area of Western Australia (formed by silicification of serpentinites; Willing and Stocklmayer, 2003), in Oregon (Riddle) and California, USA (Porterville, Exeter, Lindsay) (Frondel, 1962).

"Gem silica" or "chrysocolla chalcedony has been found in Arizona (Miami-Inspiration Mine; Globe District in Gila County, see Frondel, 1962), Peru, Namibia (Hyršl, 2016) and Spain (Laurs, 2015). Other mentioned sources are New Mexico, Mexico, Peru, Taiwan and the Philippines (https://geology.com/gemstones/gem-silica/).

Modern exploitation of heliotrope is concentrated in India (as carnelian and sard; *e.g.* the Kathiawar Peninsula mentioned by Frondel, 1962), China, Brazil, Uruguay, Madagascar, South Africa, Australia and the United States (Cavenago-Bignami Moneta, 1980). Frondel (1962) further enlisted Egypt among the areas of commercial exploitation and indicated Chatham Country in Georgia and Pony Butte in Jefferson County in Oregon as areas of occurrence.

5.9. The provenance of green chalcedonies in archaeometry and information gained from the geological literature

In the archaeometric literature, SR-XRD, Raman microspectroscopy and PIXE studies of three Cr-bearing chalcedonies and one heliotrope (from Vigna Barberini in Rome) are recorded (Gliozzo *et al.*, 2011). A single bead from Hepu county in China (one of the important ancient ports along the Maritime Silk Road) was characterized as a green chalcedony by Raman spectroscopy (Dong *et al.*, 2014) but no geochemical investigation was performed. Within the geological literature, two chrysoprases from the Biga–Çanakkale region in Turkey have been investigated by XRF and ICP-AES (Hatipoglu *et al.*, 2011b).

6. Chert, flint and jasper

"...a siliceous mineral, approaching in character flint, but less homogeneous and simple in character" (Lyell, 1830, on chert)

6.1. The distinction between chert and flint

The list of names, synonyms and definitions provided by Tarr (1938) is significant: chirt (Eng.), chert (Eng.), craie (Fre.), flint (Eng.), silex (Lat./Fre. = flint), silexite (Fre.), silex corne (Fre.) = horn flint (Eng.), silex de la craie (Fre.) = flint of the chalk (Eng.),

Hornstein (Ger.), selce (Ita. = chert = flint), Phtanite (Gre. = chert), white chert (= chert), black chert (= flint), cherty limestone = siliceous limestone; *vlint* in Dutch; *flins* in old German, *flinta* in Swedish; vlint, vlynt, flent, flynd, and flyntee in old English; *pierre a feu*' (Fre.) = firestone (Eng.) = *Feuerstein (Ger.) = flint.*

It is surprising that Tarr (1938) observed that chert was a recent name if compared to flint and that, at the beginning, both terms were used synonymously to indicate the same material, in order to distinguish the darker (flint) from the lighter (chert) coloured varieties. With the passing of time, the term chert has been increasingly used to indicate an "impure" flint since Cayeux (1929) introduced the petrological association and occurrence of chert as a distinguishing factor, hence limiting the use of this term to siliceous rocks in marine deposits (*e.g.* radiolarian cherts). Due to American usage, flint was gradually abandoned in favour of chert and reserved for artefacts.

Nowadays, the use of these terms is still controversial depending on whether archaeological or geological literature is concerned. The term flint is commonly used in archaeological papers while it is very rare in geological ones. Regarding the definitions, there is a general consensus in relation to the fact that chert is the broad term, indicating large bodies of white or lightly coloured rocks, occurring in "soft" carbonates (*e.g.* limestone, dolostones), while flint is a variety of chert, occurring as nodules of dark microcrystalline quartz and originating in chalk and marl. On the other hand, it is accepted that flint and chert cannot be distinguished based on mineralogical characteristics, although Heaney and Post (1992) found that moganite contents ranged between 13 and 17% in flint (indicated as grey varieties characterized by a greasy texture) and between 5% and 75% in chert (indicated as all other non-fibrous varieties).

Both contain microcrystalline quartz, which may be associated with amorphous silica and opal, and a subordinate content of macrocrystalline quartz (Folk and Weaver, 1952; Folk, 1980; Heaney, 1993). Both contain up to 95% silica (Hesse, 1988) and, as has already been pointed out above, variable contents of moganite.

Based on Heaney and Post (1992), the granular texture of quartz should have been a discriminating factor between chert and chalcedony; this indication is often ignored, however, and chert is thus considered to be a mixture of both granular and fibrous quartz from microcrystalline to cryptocrystalline (excluding quartzite).

6.2. The distinction between chert and jasper

Chert can be of both organic (diatomite and radiolarite on ocean floors) and inorganic origins. Its formation can be due to different types of processes: (1) accumulation of siliceous oozes on the oceanic floors (diatomites and radiolarites); (2) secondary replacement of carbonate minerals (shallow marine deposits); or (3) precipitation from silica-rich hydrothermal solutions (silica sinters). A comprehensive review of the petrogenesis of chert was provided by Knauth (1994) while a clear nomenclature based on chert formation processes was adopted by Ledevin *et al.* (2014).

As stated above, chert can have grain size from macro- to cryptocrystalline and may include several types of silica polymorphs. Chert varieties include jasper, radiolarite and porcellanite.

Jasper is frequently defined as a variety of chalcedonic silica (*e.g. Oxford dictionary of Earth Science*); however, chalcedony is constituted by micro- to cryptocrystalline quartz and moganite only, while jasper may include both a mixture of micro- to cryptocrystalline quartz polymorphs and a significant amount of other materials (*e.g.* hematite, goethite, magnetite, chlorite, *etc.*; see Kostov, 2010); therefore, it is more appropriately classified as a variety of chert. Despite the numerous attempts made by many authors to outline the classification of jasper (see Kostov, 2010), there are still many unclear features regarding its composition (very heterogeneous) and origin (metamorphic or sedimentary; see Kostov, 2010). In recent publications, the use of the term jasper seems very limited and restricted to bedded chert, associated with banded iron formations.

Among chert varieties, the jasper/radiolarites can be further distinguished into black phtanite and lydite (or lyddites). Phtanite has been used from prehistory (see *e.g.* Tsobgou and Dabard, 2010 and Mosin and Nikolsky, 2010) as well as lyddite (also named Lydian stone), the latter being commonly identified with the touchstone.

6.3. Pliny on chert, flint and jasper

Pliny uses the term *silex* to indicate both marbles (*luniensem silicem*; *N.H.* 36, 135) and several varieties of tufa (*N.H.* 36, 168-169); similarly, the term *iaspis* includes both jasper and green-chalcedony, but also other pink (rose quartz?) and blue gems (blue quartz?, sodalite?). Similarly, it is difficult to propose a reliable match for several gemstones mentioned by Pliny such as the blood-red coloured *haematiti*, which may correspond to the red jasper.

As for touchstone, the composition of this stone is controversial because Pliny (*N.H.* 33, 135) refers to it as *basanites* and noted that some people call it 'Lydian' stone ('Heraclian' was omitted based on Cohen and Drabkin, 1958). Citing Theophrastos (*De Lap.* 47, 46), Pliny further specified that touchstone was initially found along the river *Tmolus*, but the *Tmolus* is not a river but a Lydian mountain (modern Bozdağ), famous for the electrum placers of the rivers *Pactolus* (see Midas Legend) and *Hermus*. The geology of the Bozdağ area is not characterized mainly by basalts; instead granitic augen gneisses, paragneisses and schists are common, together with minor migmatites (Hetzel *et al.*, 1998; van Hinsbergen, 2010).

Hence, the correlation between touchstone and basalt becomes weaker, although several Greek authors called it βάσανος (see *e.g.* Theognis 417). In this regard, the most interesting evidence has been provided by Moore and Oddy (1985), who determined petrographically that most touchstones were made of tuff, chert and siltstone.

On a final note, it is worth adding that: (1) the term lydite or 'Lydian stone' today indicates a variety of jasper/radiolarite; and (2) assuming that *haematitis* and red jasper are the same, Ethiopia, Arabia and Egypt should be listed among the supply areas.

6.4. Chert and flint provenance in archaeometry

The archaeometric literature is strongly orientated towards functional (*e.g.* Milot *et al.*, 2017), morphological (*e.g.* Rapp, 2002) and social features (*e.g.* Schmidt *et al.*, 2016) rather than compositional ones. This situation is probably a consequence of the limited results obtained so far in flint provenancing.

Archaeological and geological samples investigated scientifically amount to >792 samples of chert (~181 archaeological and ~611 geological) and 532 of flint (>220 archaeological and >309 geological) (Appendices 3 and 4), without counting those reviewed by Luedtke (1992).

In both France and the UK, chert/flint provenance has represented an attractive research topic for a long time but, apart from the systematic work performed by Bressy (2002) on French chert, most publications have presented the results of statistical treatments only. This is the case of Ferguson (1980) who applied data coding to the analytical results obtained by the geochemical examination of 15 British flint mines. Similarly, a geochemical study performed on black chert by means of ICP-AES, ICP-MS and LA-ICP-MS was performed by Evans *et al.* (2007) but raw data were not made available. The authors described black chert as a material "with similar qualities to flint but of inland limestone origin" and sampled archaeological artefacts from Mesolithic sites and two geological sources, respectively, in the northern (Fremmington Edge and Fell End Hush) and southern (Home Farm and Middleton by Youlgrave) Pennine limestones. Likewise, an interesting comparison between geological and archaeological flints was performed on 10 Israeli samples by means of instrumental neutron activation analysis (Frachtenberg and Yellin, 1992) but the raw data were not shown.

The same problem occurs with the geochemical investigations performed on: chert from various localities analysed by Brandl (2014, 2016) and Brandl *et al.* (2011, 2014); Montana chert (Roll *et al.*, 2005); "black chert" from northern England (Evans *et al.*, 2007, 2010); chert sources from the area of New York (Rafferty *et al.*, 2007); German cherts from the area of the Swabian Jura and Franconia Alb (Bressy and Floss, 2006); flint from several sites in the UK (Pettitt *et al.*, 2012) and many others.

The most substantial and conspicuous set of data concerns cherts from France (cited above) and North America, with more than 569 specimens investigated geochemically. Among this wide group, ~124 samples were of chert stone tools while the other 445 samples were representative of several chert sources. The former group includes cherts from the US states of Indiana (Glascock, 2004), Nebraska (Hoard *et al.*, 1993), Nevada (Morgenstein, 2006), North Dakota (Huckell *et al.*, 2011), Oregon (Hess, 1996; Morrow *et al.*, 1992) and Washington (Morgenstein, 2006).

The second group includes chert sources cropping out the US states of California (Murray *et al.*, 1991), Colorado (Hoard *et al.*, 1993; republished in Huckell *et al.*, 2011), Illinois (Hess, 1996; Morrow *et al.*, 1992), Indiana (Glascock, 2004), Kansas (Hoard *et al.*, 1993), Michigan (Luedtke, 1978), Nebraska (Hoard *et al.*, 1993; republished in Huckell *et al.*, 2011), Nevada (Morgenstein, 2006), North Dakota (Huckell *et al.*, 2011), Ohio (Luedtke, 1978; Glascock, 2004), Oregon (Hess, 1996), South Dakota (Hoard *et al.*, 1993; republished in Huckell *et al.*, 2011), Wahington (Morgenstein, 2006) and Wyoming (Hoard *et al.*, 1993; republished in Huckell *et al.*, 2011).

7. Quartz varieties

All quartz varieties are characterized by a macrocrystalline grain size (>1 μm). An exhaustive discussion on quartz would be too long for this review. In the field of quartz studies, the numerous and varied research studies are readily available; however, the studies by Frondel (1962), Heaney (1994) and Götze and Möckel (2012) are the key texts from which to start.

The following sections will focus on hyaline quartz, milky quartz, smoky quartz, amethyst, citrine, prasiolite, and rose, pink and blue quartz.

7.1. Hyaline quartz (rock crystal), milky quartz, smoky quartz (morion)

These three varieties are characterized equally by the presence of macrocrystalline quartz, the latter being the sole component of rock crystal. Rock crystal has been used for a long time as a gemstone, as is well testified by ancient literature (see below) as well as by more recent appellatives – often indicative of their provenances – such as the "Herkimer diamonds" (hereafter diamonds abbreviated as d.), "Lake George d." (New York) or "Marmaros d." (from the Maramaros County of the Kingdom of Hungary now corresponding to part of Romania and Ukraine), "Cape May d." and "Arkansas d." (New Jersey and Arkansa, USA), "Bristol d." and "Cornish d." (United Kingdom), "Briançon d." and "Alençon d." (France), "Schaumberg d." (Germany), *etc.* (Frondel, 1962).

The appearance of milky quartz is due to fluid inclusions entrapped in the crystal (Götze and Möckel, 2012), while the appearance of smoky quartz is due to aluminium (Al^{3+}) impurities substituting for silicon (Si^{4+}); the charge compensators being H^+, Li^+ or Na^+. The colour centre is indicated as $[Al_{Si}O_4/M^+]^0$, where Al is a substitute for Si, while M^+ stands for H^+, Li^+ or Na^+ (Martins Nunes and Soares Lameiras, 2005 with references therein). Only Li and Na can stabilize the colour centre, given that H^0 is not stable at room temperature (Götze and Möckel, 2012).

Further details and previous studies of smoky quartz were discussed by Frondel (1962) and Rossman (1994) while a comprehensive discussion of colour enhancement and dyeing of quartz was provided by Scholz *et al.* (2012).

7.2. Pliny on hyaline, milky and smoky quartz: description and provenance

Pliny provides a vast quantity of useful information about the use, formation and provenance of rock crystal but information about the other two varieties is very sparse.

Smoky quartz (*mormorion*: Pliny's misspelling for morion) is described briefly as a very dark translucent stone from India, also known as *promnion* or *Alexandrion* or *Cyprium*, depending on the presence of red shades (*N.H.* 37, 173).

Milky quartz may be identified with the *Galactitis*, which was an entirely milk-white stone produced along the Nile (Egypt) and the Achelous (western Greece) rivers (*N.H.* 37, 162).

Rock crystal (*crystallum* in latin) formed due to the intense cold, so that the places of discovery were limited to areas where snow freezes (Pliny, *N.H.* 37, 23). The Greek etymology of the word κρύσταλλος (meaning ice, frozen) was evidence advanced by Pliny in support of that theory.

Pliny (*N.H.* 37, 26) also questions why rock crystal forms hexagonal faces, admitting that both the lack of symmetrical terminations and the presence of completely flat faces complicate matters. Specimens without defects were called '*acenteta*' (lacking in core) by the Greeks and, in general, they were not carved.

The largest crystal that Pliny had ever seen weighed 150 pounds and was dedicated by *Livia* (wife of *Augustus*) in the Capitol; other authors referred to huge vessels made of this material, however (*N.H.* 37, 27). Besides these uses and others (*e.g.* jewellery and weapons), the practice of cauterizing wounds with rock crystal exposed to sunlight is also reported (*N.H.* 37, 28).

According to Pliny, competition from glass production had to be ruthless but uneffective; in fact, the value of rock crystal (mostly determined by its weight) increased.

Regarding the supply areas, Pliny (*N.H.* 37, 23-24) mentioned India, Asia Minor (Alabanda and Orthosia), Cyprus, the island called *Necron* (*i.e.* island of the Dead) in the Red Sea facing Arabia (maybe the island of Maceira), the neighbouring island of St. John (Egypt), Portugal (*Ammaeensibus iugis*, maybe close to the ancient town of Ammaia in the municipality of Marvão, along the border with Spain) and the European Alps.

Regarding Portuguese supplies, the geoarchaeological field campaigns performed around the imperial town of Ammaia established that rock crystal could have been extracted locally from at least three mine sites and especially in the Pitaranha granite quarry (Taelman *et al.*, 2009, 2010).

The Alpine deposits provided the best quality rock crystal but outcrops were found mostly in inaccessible places, so that the miners had to be suspended by ropes (Pliny (*N.H.* 37, 27). The ancient exploitation of the San Gottardo area is confirmed by modern geological investigations as well as by archaeological remains from both the Prehistoric and Roman ages (Donati, 1986a, 102). In this respect, the finding of two blocks of rock crystal in the Roman settlement of Locarno-Muralto is of particular interest (Donati, 1986b, 213–216). Traces of a likely Neolithic age exploitation have been further recognized in the upper Romanche Valley (Rostan, 2007), as well as in the southern department of Bouches-du-Rhône (Baroni, 2003) and in the northern and wider area of the Alpes du nord (Brisotto, 1999). Actually, the majority of archaeological findings and, as a consequence, of the archaeological studies were concentrated along the Alpine Arch from Switzerland to France (Della Casa, 2005). Other sporadic findings were included in the reports of several excavations (*e.g.* Andreozzi *et al.*, 1996).

As regards the type of outcrops, the news from *Xenocrates*, according to which rock crystal was pulled up by the plow in Asia Minor and Cyprus, left Pliny puzzled, while he was more inclined to accept the argument that it could be transported by streams (*N.H.* 37, 25). In the end, though, he returns to the initial assumption and concludes that rock crystal cannot be found in well irrigated areas but only in the presence of pure snow.

The locations of the toponyms mentioned by Pliny are shown in Fig. 18, except for Necron Island.

7.3. Past and current supply areas

As for past supply areas, traces of smoky quartz were found at Romit (Egypt) in a Roman gold mine (Aston *et al.*, 2000); however, the Alpine deposits must have been exploited since ancient times for both rock crystal and smoky quartz.

As for modern supply areas, rock crystal can be found in numerous sites all over the Mediterranean basin; major fields of extraction are located in Brazil (Minas Gerais), Russia (Ural Mountains), USA (Arkansas, Virginia, North Carolina and California) and, to a lesser extent, in Guatemala, Colombia, Peru, China, Nepal, Burma, Madagascar and Australia (see also Stoiber, 1945 and Frondel, 1962).

Beyond its wide distribution, the best smoky quartz gems are extracted in Brazil, USA (Colorado, New Hampshire and Arkansas), European Alps, Scotland, Madagascar and Mozambique; however, most of those currently available commercially have been irradiated. Occurrences have been reported in Piemonte, Switzerland, Scotland, Russia, Brazil, Madagascar, Elba island, Mozambique, and from Arkansas, California, Colorado, Maine, Montana, New England, New Hampshire, North Carolina, Texas and Utah in the USA (Frondel, 1962).

Milky quartz also occurs worldwide but its exploitation for commercial purposes is not only very limited but also scarcely known from the literature.

7.4. Hyaline, milky and smoky quartz provenance in archaeometry and the information gained from the geological literature

Archaeometry has scarcely approached the provenance issue of rock crystal. The studies conducted by Taelman *et al.* (2009, 2010) are of fundamental importance

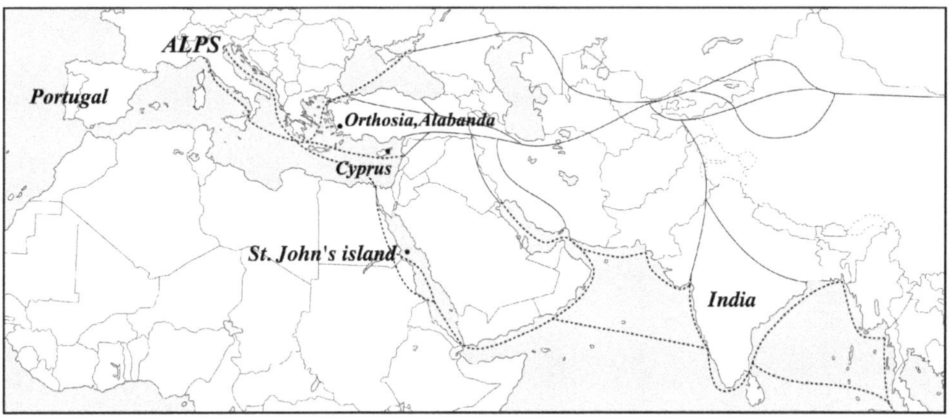

Figure 18. From the *Naturalis Historia* by Pliny the Elder: the distribution of hyaline and milky quartz. The silk road and the principal maritime routes are shown.

because they provide clear evidence of rock-crystal extraction sites, mentioned explicitly by ancient sources. Similarly, the studies conducted on lithic materials cropping out across the Alps are of great interest because they reconstruct the exploitation of this gemstone in a specific territory during a long time period such as Prehistory.

However, there has never been a specific focus on rock crystal and geochemical investigations are lacking; conversely, archaeometric studies focused mainly on fluid inclusions (see *e.g.* Cousseran *et al.*, 1998; Cousseran, 2000). Even the geological literature is of little help in this case.

7.5. Rose and pink quartz

There is a general tendency to distinguish the euhedral variety of pink quartz from massive rose quartz. The former is rare and clear while the latter is common and turbid.

The main distinguishing feature between these two varieties is represented by the sources of rose colour. As for pink quartz, the colour has been attributed to $Al\text{-}O^{-}\text{–}P$ hole centres induced by ionizing radiation (Maschmeyer and Lehmann, 1983a).

As for massive rose quartz, instead, several hypotheses have been disputed before arriving at a general consensus. The first hypothesis attributed the rose colouration to Mn^{3+} pigmentation (Holden, 1924a,b). The following hypothesis was based on the presence of rutile (Vultee and Lietz, 1956; Dennen and Puckett, 1971) and also Frondel (1962) assumed that rose quartz "typically" contained needle-like inclusions of rutile. Rutile, however, which is sometimes still mentioned in the literature, does not seem to be present in rose quartz. Between the 1960s and 1980s, research focused mainly on the presence of Ti^{3+} (Wright *et al.*, 1963; Cohen and Makar, 1984) and, especially, on the intervalence charge transfer between Ti or Fe and Ti as the main feature responsible for the rose colour (Smith *et al.*, 1978; Cohen and Makar, 1985).

Following studies by Ignatov *et al.* (1990), Smith *et al.* (1978), Alexander *et al.* (1986; on dumortierite), Hemingway *et al.* (1990) and Applin and Hicks (1987), Goreva *et al.* (2001) hypothesized that the $Fe^{2+}\text{–}Ti^{4+}$ intervalence charge transfer in dumortierite was responsible for the rose colour. Furthermore, they indicated iron reduction as being responsible for the colour changes from pink (Fe^{2+}) to colourless (Fe^{3+}) and to pink again (Fe^{2+}); the latter change being obtained by heating the colourless fibres under reducing conditions. However, the identification of dumortierite was not straightforward; the authors generally indicated this phase as the one that better matched the patterns obtained by means of various techniques (Goreva *et al.*, 2001). Only a detailed HRTEM study was able to provide an unambiguous characterization of these fibres, confirming their identification as dumortierite, characterized by a large amount of Fe substituting for Al in the M1 sites (Ma *et al.*, 2002).

Both rose and pink quartz are photochemically unstable (see also Rossman, 1994) and, thus, subject to fading. In pink quartz, fading was attributed to the $Al\text{-}O^{-}\text{–}P$ hole centres. The same explanation was given for rose quartz, although it cannot explain why soaking in water should return the rose colour to this quartz (Goreva *et al.*, 2001). The other possible explanation involved the expansion and shattering of CO_2 inclusions due to solar heat and the consequent scattering of light on a microfractured surface (Goreva *et al.*, 2001).

Further characterizing features are asterism and the tendency to darken when exposed to X-rays. Macroscopic inclusions of associated bismuthinite (Bi_2S_3) and chalcopyrite grains were reported in a rose quartz from Madagascar only (Hyršl and Danet, 2005).

Rose and pink quartz can be found in hydrothermal veins but they are mostly concentrated in granitic pegmatites, often associated with tourmaline, cassiterite, beryl and phosphate minerals (Cassedanne and Roditi, 1991; Goreva *et al.*, 2001). Major sources include Brazil, India, Madagascar and Sri Lanka, while Frondel (1962) further listed the occurrences reported in Canada (New England), California (Pala and Mesa Grande), Colorado (Scott quarry at Custer), Maine (Newry; Bumpus mine near Stoneham; Albany, Paris), South Dakota (Black Hills), South Africa (Rossing), Germany (Rabenstein) and Sweden (Långban; this containing durmortierite).

To the author's knowledge, there is no archaeometric research which has focused on these materials.

7.6. Pliny on rose and pink quartz

An exact match to the gems listed by Pliny is not traceable with certainty; therefore, it is only possible to make some assumptions.

It is reasonable to think that rose and pink quartz correspond to one of those gemstones that Pliny describes with various shades of colour. For instance, these varieties could be identified as a variety of amethyst, or a variety of sapphire, or the "mild carbuncle", *i.e.* a sub variety of the *carbunculus lychnis*, resembling the pink colour of the "Flower of Jove".

Rose and pink quartz could also be identified with one of the many Plinian gemstones that we cannot identify. This might be the case for the so-called *Aphrodisiaca* that had to be pink (Pliny, *N.H.* 37, 148: *Aphrodisiaca ex candido rufa est*). In fact, ancient and modern beliefs attribute the capacity to foster the search for a soul-mate for quartz. This property could be reflected in the name of the gem, which is referred to as "love stone" nowadays. Moreover, ancient mythology indicates rose/pink quartz as a tool in the hands of Cupid and Eros.

It is further known that rose quartz was used by the Assyrians to make beads and seals (Harper *et al.*, 1995) and by the Egyptians to make funerary vessels (Aston *et al.*, 2000) and talismans with the power to prevent aging (http://www.gia.edu/rose-quartz-history-lore). However, there are no evocative gemstone names related to these properties in the Plinian prose.

7.7. Amethyst, citrine and prasiolite

Amethyst is the purple variety of quartz and might be one of the most appreciated varieties in Antiquity. Recent X-ray absorption spectroscopy experimental data have established that Fe replaces Si in its tetrahedral site and occurs mainly in the trivalent state, although minor amounts of Fe^{4+}, responsible for the purple colour, may be present (Di Benedetto *et al.*, 2010 with references therein). Conversely, Czaja *et al.* (2017) have found Fe^{3+} only ("in interstitial positions in channels parallel or perpendicular to the *c* axis") in Polish amethyst and prasiolite. An earlier hypothesis and further details on amethyst's iron centres, synthesis, colour distribution and stability are included in Rossman (1994).

The yellow variety of quartz – citrine – is rare in nature and this is the reason why most citrine on the market is actually heat-treated (~500°C) amethyst. Numerous authors agree in stating that the incorporation of trivalent iron (Fe^{3+}) and the colour centres are basically responsible for the yellow colour and are often associated with Al impurities (Samoilovich *et al.*, 1969; Lehmann, 1971; Nassau and Prescott, 1977; Maschmeyer *et al.*, 1980; Maschmeyer and Lehmann, 1983b; Balitsky and Balitskaya, 1986; Pan *et al.*, 2008). Other authors, instead, are inclined to believed that ~100 nm-sized iron particles play the major role in producing the yellow colour of natural citrine (Lehmann and Moore, 1966). Aines and Rossmann (1986) reported the selective dominance of molecular water or hydroxide ion in the citrine and amethyst sections of a dichromatic quartz, respectively. A comprehensive review of all these aspects was carried out by Rossman (1994) and, more recently, Nunes *et al.* (2013) stated that natural citrine is not affected or strongly affected by UV or gamma irradiation, respectively.

Prasiolite is the green variety of quartz. Today's prasiolite is generally obtained from heated amethyst (from Arizona, Brazil and Zambia; see Rossman, 1994 for more details); however, the same process is observed in nature (Paradise, 1982). In this regard, Herbert and Rossman (2008) established that naturally occurring green prasiolite is a radiation-induced colour variant of quartz (*i.e.* not just a heated pre-existing amethyst), which develops in the presence of specific chemical constituents and H_2O inclusions (from the nano- to the macro-scale).

Hence, irradiation of the system is a major factor in the development of amethyst, smoky and green colours in quartz while the presence of H_2O within the quartz structure has been found to be detrimental to the development of only certain colours (Aines and Rossman, 1986).

7.8. Pliny on amethyst

Pliny (*N.H.* 37, 121) narrates that the name amethyst was related strictly to its colour, being similar to that of wine; in fact, the Greek word αμέθυστος (= α μεθύω) literally means "not drunk". In Greek mytology, Amethyst was a nymph loved by Bacchus. She did not return his love and asked Diana to help her. The goddess of the woods transformed the nymph into a crystal and when Bacchus found her, he tipped his cup of wine over her in irritation, conferring the violet colour to that beautiful crystal. It is maybe not a coincidence that modern crystal therapy uses amethyst to fight against obsessive craving for alcohol, tobacco and drugs, as well as to control the temperament and to reduce work stress and insomnia.

Pliny further adds that this gemstone was violet and transparent but not so dazzlingly brilliant as *carbunculus* and was difficult to engrave (*N.H.* 37, 121).

A variety was named *socondios* by the Indians as it had the characteristic colour of the hyacinth (*N.H.* 37, 122 and 125). This variety may correspond to the actual wisteria amethyst, considering that wisteria/glicine had not yet arrived in Europe at the time of Pliny. A fainter variety was the *sapenos* or *pharanitis*, the latter taking its name from a tribe on the border of Arabia where it was produced. This reference is likely to be linked with the town of Pharan, which was 6 miles distant from Mount Choreb, at the foot of Mount Serbal, in the Sinai Peninsula (Cosmas Indicopleustes, *Christian Topography*, Book V). A further variety is represented by a bi-coloured amethyst which turns from violet to colourless, thus resembling rock crystal (*N.H.* 37, 123).

Pliny further mentions (*N.H.* 37, 123) at least three synonyms for amethyst which were *paederos* (*i.e.* lovely youth or favourite; already used for opal at *N.H.* 37, 84), *anteros* (*i.e.* avenger of slighted love or love unrequited) and *Veneris gena* (*i.e.* Venus eyelid).

Similarly to the heavy criticism expressed during the description of agate, Pliny railed against the widespread belief of magicians whereby amethyst would prevent drunkenness, ensure access to the presence of kings, avert hail as well as the attacks of locusts (if combined with a special prayer) and protect from spells; in the latter case, however, the gemstone would have been inscribed with the names of the sun and the moon, suspended from the neck, and worn together with some hair of the cynocephalus (a kind of baboon) and swallow feathers (*N.H.* 37, 124).

As for provenance, Pliny stated that it mainly came from India but also from the region of Arabia known as Petra (maybe Jordan), Lesser Armenia, Egypt, Galatia, Thasos and Cyprus; the last two areas providing the worst quality and the least valued amethyst (*N.H.* 37, 121).

The locations of the toponyms mentioned by Pliny are shown in Fig. 19, except for the town of Pharan.

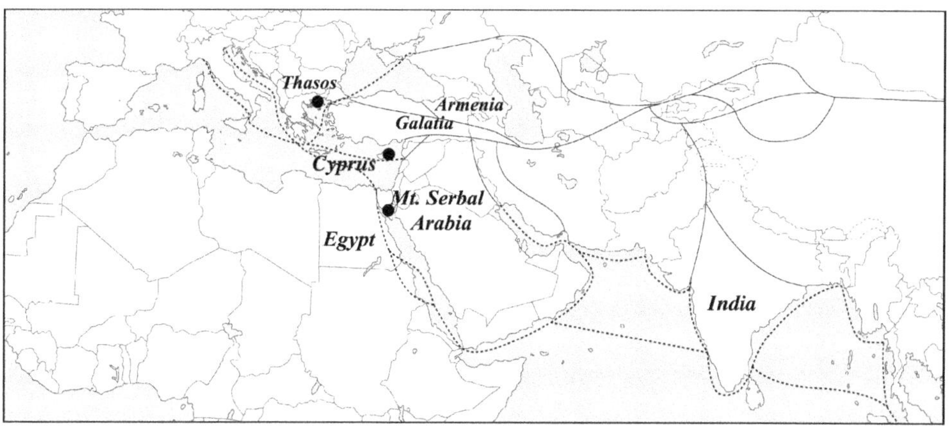

Figure 19. From the *Naturalis Historia* by Pliny the Elder: the distribution of amethyst. The silk road and the principal maritime routes are shown.

7.9. Past and current supply areas

Amethyst is a very common gemstone which occurs in deposits of varied nature throughout the world. It can be found in hydrothermal veins (*e.g.* Guanajuato in Mexico and Porkura and Macskamezö in Romania; reported by Frondel, 1962), in Alpine type veins, in cavities of pegmatites and granites, in hematite deposits and, especially, in basic igneous flow rocks such as basalts (Frondel, 1962).

Russian (Ural mountains, *e.g.* Olkhovka and the Reshev and Alapayev districts around Mursinsk mentioned by Frondel, 1962) and German (Idar-Oberstein) amethyst, now largely exhausted, has given way to Brazilian amethyst (especially from Rio Grande do Sul and Bahia; see Scholz *et al.*, 2012 for further details), at least since 1727 (date reported by Frondel, 1962). Past exploitation has also been recorded in Bohemia (Krušné hory mountains; see Řehoř, 2009), France (Puy-de-Dôme), Nova Scotia and Ontario (Thunder Bay) in Canada, and many other localities in the UK, Georgia, Arizona, Colorado, North Carolina, Pennsylvania, Virginia and Wyoming (Frondel, 1962) while current deposits are also located in Zambia, Mexico, and Uruguay. Smaller deposits are located in Australia, Sri Lanka, India, Madagascar, southwest Africa and the United States (Webster, 1983; Epstein, 1988).

Similarly, most citrine and prasiolite – both very rare – are derived mainly from Brazil (Bahia, Minas Gerais) but other occurrences have been registered in Colombia (San Cristobal, Santander), Spain (Salamanca) and France (Dauphine). As for the zones where natural citrine mainly occurs, Rossmann (1994) also listed Hyderabad in India and the Anahí mine in Bolivia while Frondel (1962) also mentioned Madagascar, the Isle of Arran in Scotland, the province of Salamanca in Spain, the Dauphiné in France, the territory of Mursinsk in Russia and North Carolina in the United States.

Prasiolite has also been found in Poland (Suszyna–Mrówieniec area and Płóczki Górne; Platonov and Szuszkiewicz, 2015 with references therein), Canada (Thunder Bay), Namibia (Farm Kos and Farm Rooisand) and Zambia (Kalomo-Mapatiqya), respectively; see Herbert and Rossman, 2008).

7.10. Amethyst, citrine and prasiolite provenance in archaeometry and the information gained from the geological literature

Geochemical studies of these gemstones are very rare to the author's knowledge; an accurate study was performed on a single sample of amethyst from Lauterbach in Germany (Götze *et al.*, 2001, reporting neutron activation analysis data from Haake and Holzhey, 1989).

7.11. Blue quartz

Blue quartz is relatively rare, generally translucent and with a waxy or greasy lustre. The origin of the blue colour remains a subject of active debate. Rayleigh scattering by submicron-sized inclusions (Jayaraman, 1939) is generally regarded as being responsible for the blue colour (see also Frondel, 1962 and Rossman, 1994 for further details). The inclusions can be of a very different nature but, among the numerous possibilities, rutile seems to be by far the most common type (Seifert *et al.*, 2011 with references therein). Among other possible inclusions, aerinite, apatite, biotite, ilmenite, muscovite, rutile, tourmaline and zircon have been mentioned (Iddings, 1904; Frondel, 1962; Parker, 1962; Barker and Burmester, 1970; Zolensky *et al.*, 1988; Romero Silva, 1996; Bartovics and Beane, 2007).

Seifert *et al.* (2009, 2010, 2011) provided extensive results on blue quartz inclusions (types, sizes, and frequencies, origin and thermal history). Those authors found abundant submicron-sized (1 μm–100 nm) inclusions of (probably syngenetic) mica, ilmenite and rutile and nanometre-sized (<100 nm) inclusions of (probably epigenetic) rutile; which of these might be responsible for the blue colour was not established, however.

The occurrence of blue quartz is limited to igneous rocks, mostly felsic, and their mildly-to-moderately metamorphosed equivalents (Seifert *et al.*, 2011). There are only a few areas where it has been recognized such as Spain (Romero Silva, 1996), Scandinavia, Norway (Goldschmidt, 1954), Virginia, USA (Wise, 1981) and India (Jayaraman, 1939). Frondel (1962) further stated that blue quartz was widespread in both igneous and metamorphic rocks from India (especially in the charnockite series, Champion gneisses of Mysore), Massachusetts, Pennsylvania (Baltimore gneiss), Rhode Island (Milford granite), Virginia (Ti-bearing rocks in the Amherst and Nelson counties; quartz diorite in Stafford county), New England (all USA), Sweden (granites), Finland (rapakivi) and Australia (albitite).

7.12. Pliny on blue quartz

Blue quartz may have been included in the description of other gems with the same colour. For instance, it may correspond to blue *sardonyx* (*N.H.* 37, 87 and 89) as well as to certain varieties of blue *iaspis* (*caeruleus iaspis*; NH 37, 115-116).

The problem is complicated further by the identification of lapis lazuli and azurite. It is well known, in fact, that lapis lazuli was called *sappirus* while azurite was called *lapis Armenius*; many modern authors share the view, however, that lapis lazuli was also called *Cyanus*, based on a well known Plinian text (*N.H.* 37, 119):

> "*We shall now give a separate account of cyanus, for a short time ago we applied the name to a iaspis owing to its blue colour. The best kind is the Scythian, then comes the Cyprian and lastly there is the Egyptian. It is very commonly counterfeited by tinting other stones, and this is a famous achievement of the kings of Egypt, whose records also mention the name of the king who first tinted stones in this way. Cyanus too is divided into male and female varieties. Sometimes inside cyanus there is a golden dust, which, however, differs from that which occurs inside the sappirus; for there the gold glistens as dots*" (Eichholz, 1962).

Indeed, this passage is controversial because it seems that Pliny – here summarizing Theophrastos (*De Lap.* 55 on ultramarine) – confuses the Egyptian production of the blue azurite pigment with both lapis lazuli and maybe the other varieties (the 'female' for instance) for which he fails to provide additional descriptions. However, it seems that Pliny does not want *cyanus* and *sappirus* to be considered as the same gemstone, although both may show golden inclusions.

The supply areas mentioned by Pliny are Scythia (*i.e.* the ancient region including part of Afghanistan, Bulgaria, Iran, Kazakhstan, Pakistan, Poland, Romania, Russia, Tajikistan, Ukraine, Uzbekistan), Cyprus and Egypt; however, the latter can be ruled out on the basis of the clear reference to azurite. Actually, Persia was the supply area of lapis lazuli, Scythia of cianos and sucina, and Cyprus of *adamas*, amethyst, *cyanus*, rock crystal, *iaspis* and *paederos*.

Also this indication of the supply areas does not find an exact match and then leaves open the suspicion that *cyanus*, like *caeruleum iaspis*, can actually correspond to blue quartz (but also to sodalite).

7.13. Current supply areas, archaeometric studies and geological literature.

Today's extraction sites are located in Brazil, Austria, South Africa and the USA (Virginia). Small occurrences have been documented close to Bologna (Val del Sillaro) in Italy but also in Spain, Sweden, Switzerland and the United Kingdom, and in numerous other countries (http://www.mindat.org/min-26723.html).

As for archaeometry, an exemplar of a blue quartz engraved gemstone (heron and palm) was found at Sevilla, during the archaeological excavation of a burial site dated to the end of the first century CE (years 80–90). The ring was investigated by Enrique Magario *et al.* (2002) using several techniques (optical microscopy, SEM-EDX, XRD, XRF and PIXE).

8. Final remarks

First of all, it seems clear that there should be a greater focus on the use of terms and in the determination of moganite. Ignoring this phase has clearly led to confused results concerning the identification, origin and occurrence of these materials. Also the definition of chromophores (elemental state/mineralogical phases) and of colour centres is not conclusive, especially as regards the quartz varieties.

As for archaeometry, the characterization and provenance issues have reached different stages of development.

Characterization is typically performed by means of a plethora of analytical techniques which are able to ensure accurate results, mostly in a non-destructive operational mode. However, the striking advances made in thermoluminescence and optically stimulated luminescence dating and use-wear of flint are counterbalanced by a far more simplistic approach in terms of the material characterization. In fact, it is common practice to limit research to phase identification – generally *via* Raman spectroscopy or portable X-ray fluorescence – neither of which is able to characterize the materials in an adequate and complete manner. Only in combination with geochemical analysis can sufficient information about the constituent elements be obtained – which may prove to be discriminant for provenance issues – and the possible treatments undergone by the gemstones established (*e.g.* use of metal tools used for surface treatments).

As for provenance, the scarce amount of geochemical data available from silica-based gemstones and chert does not allow the scientific community to achieve significant results in this field. Moreover, the reluctance to publish raw data and the large number of analytical techniques used by authors has complicated matters. As a result, a certain scepticism has spread among the community about the possibility and/or utility of performing such studies.

The handful of references quoted in the various paragraphs above should lead us to avoid reaching pessimistic conclusions and to encourage exploration in this lesser-known field of study.

A common analytical procedure should be followed in order to produce comparable results. In this instance, the main criteria should be the accuracy of the results, the non-destructive or minimally-destructive character of the analytical techniques and the instruments availability. The selection may be thus narrowed to the following techniques:

- Raman microspectroscopy and XRD for phase identification and quartz:moganite ratio quantification;
- Field emission scanning electron microscopy (FESEM), Environmental scanning electron microscopy (ESEM), or Scanning electron microscopy (SEM) for the determination of the amount of SiO_2;
- Laser ablation inductively coupled plasma mass spectrometry (LA-ICP-MS), for the determination of minor and trace elements; the *REE*s being especially important as they are mostly immobile and their patterns provide evidence of their environment of deposition.

Undoubtedly, there are some interchangeable techniques; *e.g.* SEM could be replaced by electron probe microanalysis (EPMA), which is destructive, however, or better, Particle Induced X-ray Emission (PIXE), or a non-destructive technique, able to provide accurate results, reaching a greater penetration depth than SEM-based techniques but definitely less widespread.

The determination of the quartz:moganite ratio may prove to be an important tool for the creation of a systematic classification of both chalcedony and chert but it may also provide important discriminant features to be used for provenance issues.

A further key point is represented by the need to create reference data and a database. The operation is complex and laborious and therefore opposed by many but it is undeniable that, in terms of provenance issues, archaeometry is able to provide answers only by comparison. Moreover, given the widespread distribution of these materials (as with many other objects of archaeometric interest), it is unthinkable that each individual researcher or research group should analyse each archaeological object so as to be able to compare data obtained personally only. Such an attitude is detrimental to the progress of a discipline (not only gemstones and chert) where there is a much greater need for accurate data than for review.

A further aid arrives from geological studies, which can provide useful data on natural outcrops to be used for comparison. Moreover, a better understanding of several issues involving precipitation, nucleation, diagenetic transformation and alteration of silica-based varieties, together with a more extensive and in-depth knowledge of the outcrops could provide important clues on raw material exploitation and marketing.

Meanwhile, Pliny has left us much information which should be autoptically verified, with targeted 'gem-hunting' campaigns. Pliny had no intimate knowledge of minerals and this is the reason why he followed the observations made by his colleagues and used the same name for different minerals. A revision of ancient gemstone names mentioned by both Latin and Greek authors will surely provide further insight into this topic; however, we cannot expect total clarity by ancient authors when such clarity is not even present in scientific texts today.

The Indian provenance of most precious gemstones cited by Pliny deserves a final comment. This should be considered as nothing more than a stereotype invented by merchants and, as such, should not be given much credibility. Based on the popular belief according to which the stones 'grew' better in warm climates, merchants tended to indicate India as the place of origin of the most beautiful stones. Hence, we have a well designed merchandising business which, on the one hand, exploited what today would be the name of a famous brand (*i.e.* the good name of India) and, on the other hand, helped to increase public awareness of the brand itself and, especially, line the pockets of merchants. Moreover, it is worth observing that most supply areas were: (1) repeated frequently by Pliny; and (2) situated along the principal commercial trade routes. Indeed, it is surprising that inland areas are almost never mentioned by Pliny when in fact the results of the archaeometric investigations clearly attest to their exploitation. This scenario seems to indicate that, in addition to popular belief, the origins of the gems were mostly related to the trade routes involved in their commercialization as well as to the nationality of the traders.

Acknowledgements

The author is greatly indebted to Anna Bellini who made this study possible by providing time and care in the literature search. Grateful acknowledgement for proofreading and correcting the English language goes to Helen Carruthers and Robert Preston. The author apologises to all other studies which she may have neglected to quote. This review was prepared in 2015, whereupon the bibliographic search was not systematic.

Abbreviations and references

Nat.Hist. (*N.H.*) = *Naturalis Historia* (Pliny the Elder)

Lapid. = *De Lapidibus* (Damigeron).

ORPH.L. = *Orpheos Lithika Kerigmata* (in Greek). 1881. – In: Orphei Lithica. Accedit Damigeron De Lapidibus (Rec. E. Abel). 1881. Berolini, Calvary.

Aines, R.D. and Rossman, G.R. (1986) Relationships between radiation damage and trace water in zircon, quartz, and topaz. *American Mineralogist*, **71**, 1186–1986.

Alexander, V.D., Griffen, D.T. and Martin, T.J. (1986) Crystal chemistry of some Fe- and Ti-poor dumortierites.

American Mineralogist, **71**, 786–794.

Allchin, E. (1979) The agate and carnelian industry of Western India and Pakistan. Pp. 91–107 in: *South Asian Archaeology 1975*, (J.E. Van Lohuizen-De Leeuw, editor). Brill, Leiden, The Netherlands.

Amelin, Y. and Back, M. (2006) Opal as a U-Pb geochronometer: Search for a standard. *Chemical Geology*, **232**, 67–86.

Andreozzi, G.B., Graziani, G. and Saguì, L. (1996) Gems from archaeological excavations at Rome (Cripta Balbi). *Zeitschrift der Deutschen Gemmologischen Gesellschaft*, **4**, 49–72.

Anghel, D., Har, N., Şuteu, C. and Ursu, M. (2016) Metode de investigare mineralogice şi petrografice ale unor intalii din colecţia muzeului naţional al unirii din Alba Iulia. *Terra Sebus*, **8**, 83–104.

Applin, K.R. and Hicks, B.D. (1987) Fibers of dumortierite in quartz. *American Mineralogist*, **72**, 170–172.

Armstrong, H.A., Owen, A.W. and Floyd, J.D. (1999) Rare earth geochemistry of Arenig cherts from the Ballantrae Ophiolite and Leadhills Imbricate Zone, southern Scotland: Implications for origin and significance to the Caledonian orogeny. *Journal of the Geological Society of London*, **156**, 549–560.

Aspinall, A. and Feather, S.W. (1972) Neutron activation analysis of prehistoric flint mine products. *Archaeometry*, **14**, 41–53.

Aston, B.G., Harrel, J.A. and Shaw, I. (2000) Stone. Pp. 5–77 in: *Ancient Egyptian Materials and Technology* (P.T. Nicholson and I. Shaw, editors). Cambridge University Press, UK.

Ayvacıklı, M., Garcia-Guinea, J., Jorge, A., Akalın, İ., Kotan, Z. and Can, N. (2012) Luminescence and micro-Raman investigations on inclusions of unusual habit in chrysoprase from Turkey. *Journal of Luminescence*, **132**, 1750–1758.

Baldwin, G.J., Thurston, P.C. and Kamber, B.S. (2011) High-precision rare earth element, nickel, and chromium chemistry of chert microbands pre-screened with in situ analysis. *Chemical Geology*, **285**, 133–143.

Balitsky, V.S. and Balitskaya, O.V. (1986) The amethyst-citrine dichromatism in quartz and its origin. *Physics and Chemistry of Minerals*, **13**, 415–421.

Banfield, J.F., Barker, W.W., Welch, S.A. and Taunton, A. (1999) Biological impact on mineral dissolution: application of the lichen model to inderstanding mineral weathering in the rhizosphere. *Proceedings of the National Academy of Sciences – PNAS*, **96**, 3404–3411.

Barker, D.S. and Burmester, R.F. (1970) Leaching of quartz from Precambrian hypabyssal rhyolite porphyry, Llano County, Texas. *Contributions to Mineralogy and Petrology*, **28**, 1–8.

Baroni, I. (2003) Le quartz hyalin du gisement de l'Adaouste. Contribution → l'étude des circulations de matières premières au néolithique. *Bulletin du Musée d'anthropologie préhistorique de Monaco*, **43**, 36–44.

Barsanov, G.P. and Yakovleva, M.E. (1981) Mineralogical investigations of some precious and semi-precious varieties of cryptocrystalline silica. In: *New Mineral Data*. USSR Academy of Sciences, **29** (in Russian; quoted by Rossman 1994).

Bartovics, S. and Beane, R. (2007) Analysis of blue color in quartz grains from Cushing Formation, Peaks Island, Maine. *The Maine Geologist*, **36**, 4.

Becq-Giraudon, J.-F., Bouillé, S. and Chauvel, J.-J. (1992) Genesis and significance of the silico-aluminous nodules in the Ordovician of the Montagne Noire and the Massif Armoricain (France). *Sedimentary Geology*, **77**, 77–87.

Befi, R. (2009) Australian chrysoprase with dendritic inclusions. *Gems & Gemology*, **45**, 71.

Behar, A.E., Albaz, S., Shai, I., Maeir, A.M. and Greenfield, H.J. (2016) Faience beads from Early Bronze Age contexts at Tell es-Safi/Gath, Israel. *Journal of Archaeological Science: Reports*, **7**, 609–613.

Bellina, B. (2003) Beads, social change and interaction between India and South-east Asia. *Antiquity*, **77**, 285–297.

Bennett, P.C. (1991) Quartz dissolution in organic-rich aqueous systems. *Geochimica et Cosmochimica Acta*, **55**, 1782–1797.

Bersani, D. and Lottici, P.P. (2010) Applications of Raman spectroscopy to gemology. *Analytical and Bioanalytical Chemistry*, **397**, 2631–2646.

Bettermann, P. and Liebau, F. (1975) The transformation of amorphous silica to crystalline silica under hydrothermal conditions, *Contributions to Mineralogy and Petrology*, **53**, 25–36.

Boardman, J. (1970) *Greek Gems and Finger Rings*. Thames and Hudson, London.

Brandl, M. (2014) Genesis, provenance and classification of rocks within the chert group in Central Europe. *Archaeologia Austriaca*, **97-98**, 33–58.

Brandl, M. (2016) The multi layered chert sourcing approach (mla) analytical provenance studies of silicite raw materials. *Archeometriai mühely*, **13**, 145–156.

Brandl, M., Hauzenberger, C., Postl, W., Modl, D., Kurta, C. and Trnka, G. (2011) Repolust Cave (Austria) revisited: Provenance studies of the chert finds. *Quartär*, **58**, 51–65.

Brandl, M., Hauzenberger, C., Postl, W., Martinez, M.M., Trnka, G. and Filzmoser, P. (2014) Radiolarite Studies at Krems-Wachtberg (Lower Austria): Northern Alpine versus Carpathian lithic resources. Pp. 146–162 in: *Loess and the record of Upper Palaeolithic cultures in the Danube Basin* (C. Neugebauer-Maresch, U. Hambach and M. Anghelinu, editors). *Quaternary International*, **351**.

Bressy, C. (2002) Caracterisation et gestion du silex des sites mesolithiques et neolithiques du nord-ouest de l'arc alpin. Une approche pétrographique et géochimique. *Social Anthropology and Ethnology*. Université de Provence – Aix-Marseille I, 2002.

Bressy, C. and Floss, H. (2006) Multiparametric characterisation of Southwestern German cherts: Application to the study of raw material circulation during Upper Paleolithic period. Pp. 131–136 in: *Notions de territoire et de mobilité. Exemples de l'Europe et des premières nations en Amérique du Nord avant le contact Européen* (C. Bressy, A. Burke, P. Chalard and H. Martin, editors). Université de Liège (Études et Recherches Archéologiques de l'Université de Liège, **116**). Liège, Belgium.

Brisotto, V. (1999) Quartz hyalin et obsidienne dans les sé¡ries néolithiques entre Rhône moyen et Alpes du Nord: poids et signification. *Travaux du Centre d'Archéologie Préhistorique de Valence*, **2**, 211–230.

Cackler, P.R., Glascock, M.D., Neff, H., Iceland, H., Pyburn, K.A., Hudler, D., Hester, T.R. and Chiarulli, B.M. (1999) Chipped stone artefacts, source areas, and provenance studies of the northern Belize chert-bearing zone. *Journal of Archaeological Science*, **26**, 389–397.

Cady, S.L., Wenk, H.-R. and Downing, K.H. (1996) HRTEM of microcrystalline opal in chert and porcelanite from the Monterey Formation, California. *American Mineralogist*, **81**, 1380–1395.

Cady, S.L., Wenk, H.-R. and Sintubin, M. (1998) Microfibrous quartz varieties: characterization by quantitative X-ray texture analysis and transmission electron microscopy. *Contributions to Mineralogy and Petrolology*, **130**, 320–335.

Capel Ferrón, C., León-Reina, L., Jorge-Villar, S., Compaa, J. M., Aranda, M. A. G., López Navarrete, J. T., Hernández, V., Medianero, F.J., Ramos, J., Weniger, G.-C., Domínguez-Bella, S., Linstaedter, J., Cantalejo, P., Espejo, M. and Durán Valsero, J.J. (2014) Combined Raman spectroscopic and Rietveld analyses as a useful and nondestructive approach to studying flint raw materials at prehistoric archaeological sites. *Archaeological and Anthropological Sciences*, **7**, 235–243. DOI 10.1007/s12520-014-0189-0.

Carr, R.M. and Fyfe, W.S. (1958) Some observations on the crystallization of amorphous silica. *American Mineralogist*, **43**, 908.

Carter, A.K. and Dussubieux, L. (2016) Geologic provenience analysis of agate and carnelian beads using laser ablation-inductively coupled plasma-mass spectrometry (LA-ICP-MS): A case study from Iron Age Cambodia and Thailand. *Journal of Archaeological Science: Reports*, **6**, 321–331.

Casanova, A. (2004) Osservazioni sul lessico scientifico ed i neologismi del nuovo Posidippo. *Prometheus*, **30**, 217–234.

Cassedanne, J.P. and Roditi, M. (1991) Crystallized and massive rose quartz deposits in Brazil. *Journal of Gemmology*, **22**, 273–286.

Caucia, F., Marinoni, L., Bordoni, V., Ghisoli, C. and Adamo, I. (2012) Physical and chemical properties of some Italian opals. *Periodico di Mineralogia*, **81**, 93–106.

Caucia, F., Marinoni, L., Leone, A. and Adamo, I. (2013) Investigation on the gemological, physical and compositional properties of some opals from Slovakia ("Hungarian" opals). *Periodico di Mineralogia*, **82**, 251–261.

Cavenago-Bignami Moneta, S. (1980) *Gemmologia. Pietre preziose ed ornamenti naturali, trasparenti, semitrasparenti ed opache, perle, cammei di conchiglia, opercula, ammoniti e conchiglie varie*. Hoepli, Milan, Italy.

Cayeux, L. (1929) *Les roches sedimentaires de France, roches siliceous*. Paris.

Čermáková, Z., Hradil, D., Bezdička, P. and Hradilová, J. (2017) New data on "kerolite–pimelite" series and the colouring agent of Szklary chrysoprase, Poland. *Physics and Chemistry of Minerals*, **44**, 193–202.

Charpentier, V., Brunet, O., Mery, S. and Velde, C. (2017) Carnelian, agate, and other types of chalcedony: the prehistory of Jebel al-Ma'taradh and its semi-precious stones, Emirate of Ra's al-Khaimah. *Arabian Archaeology and Epigraphy*, **28**, 175–189.

Chauviré, B., Rondeau, B. and Mangold, N. (2017a) Near infrared signature of opal and chalcedony as a proxy for their structure and formation conditions. *European Journal of Mineralogy*, **29**, 409–421.

Chauviré, B., Rondeau, B., Mazzero, F. and Ayalew, D. (2017b) The precious opal deposit at Wegel Tena, Ethiopia: Formation via successive pedogenesis events. *The Canadian Mineralogist*, **55**, 701–723.

Chang, H.J., Chu, X.L., Feng, L.J. and Huang, J. (2009) Terminal Ediacaran anoxia in deep-ocean: Trace element evidence from cherts of the Liuchapo Formation, South China. *Science in China Series D: Earth Sciences*, **52**, 807-822

Chen, C.T.A. and Marshall, W.L. (1982) Amorphous silica solubilities – IV. Behavior in pure water and acqueous sodium chloride, sodium sulphate, magnesium chloride, and magnesium sulphate up to 350ºC. *Geochimica et Cosmochimica Acta*, **46**, 279–287.

Chen, J., Blume, H.P. and Beyer, L. (2000) Weathering of rocks induced by lichen colonization – a review. *Catena*, **39**, 121–146.

Chernenko, Z.I. and Chlachula, J. (2017) Precious and decorative non-metallic minerals from east Kazakhstan: Geological deposits and present utilisation, International Multidisciplinary Scientific GeoConference Surveying Geology and Mining Ecology Management, SGEM Albena, Bulgaria, 29 June–5 July 2017), vol. **17**, 447–454.

Chevalier, J., Inizan, M.L. and Tixier, J. (1982) Une technique de perforation par percussion de pérles en cornaline. *Paléorient*, **8**, 55–65.

Clark, R. (2002) *Fairburn Agate. Gem of South Dakota.* Silverwind Agates, Publishers, Appleton, Wisconsin, USA, 104 pp.

Cohen, M.R. and Drabkin, I.E. (1958) *A Source Book in Greek Science.* Harvard University Press, Harvard, Massachusetts, USA.

Cohen, A.J. and Makar, L.N. (1984) Differing effects of ionizing radiation in massive and single crystal rose quartz. *Neues Jahrbuch für Mineralogie-Monatshefte*, **11**, 513–521.

Cohen, A.J. and Makar, L.N. (1985) Dynamic biaxal absorption spectra of Ti^{3+} and Fe^{2+} in a natural rose quartz crystal. *Mineralogical Magazine*, **49**, 709–715.

Cousseran, S. (2000) L'étude des inclusions fluides appliquée au problème de la circulation des quartz archéologiques dans les Alpes occidentales. *Revue d'Archéométrie*, **24**, 169–177.

Cousseran, S., Pecher, A. and Bintz, P. (1998) Application de l'étude des inclusions fluides aux quartz taillés de quel¡ques sites préhistoriques dans les Alpes du nord. *Revue d'Archéométrie*, **22**, 103–109.

Crassard, R. (2008) *La Préhistoire du Yémen. Diffusions et diversités locales, á travers l'étude d'industries lithiques du Hadramawt.* (BAR, British Series 1842). Archaeopress, Oxford, UK.

Crawford, H. (1998) *Dilmun and its Gulf neighbours.* Cambridge University Press, Cambridge, UK.

Crerar, D.A., Axtemann, E.V. and Axtemann, R.C. (1981) Growth and ripening of silica polymers in aqueous solutions. *Geochimica et Cosmochimica Acta*, **45**, 1259–1266.

Cummings, M.L., Trone, P.M. and Pollock, J.M. (1989) Geochemistry of colloidal silica precipitates in altered Grande Ronde Basalt, Northeastern Oregon, USA. *Chemical Geology*, **75**, 61–79.

Czaja, M., Kądziołka-Gaweł, M., Konefał, A., Sitko, R., Teper, E., Mazurak, Z. and Sachanbiński, M. (2017) The Mössbauer spectra of prasiolite and amethyst crystals from Poland. *Physics and Chemistry of Minerals*, **44**, 365–375.

Darragh, P., Gaskin, A., Terrell, B. and Sanders, J. (1966) Origin of precious opal. *Nature*, **209**, 13–16.

De Bruin, M., Korthoven, P.J.M., Bakels, C.C. and Groen, F.C.A. (1972) The use of non-destructive activation analysis and pattern recognition in the study of flint artefacts. *Archaeometry*, **14**, 55–63.

De Waele, A. and Haerinck, E. (2006) Etched (carnelian) beads from northeast and southeast Arabia. *Arabian Archaeology and Epigraphy*, **17**, 31–40.

Del Bufalo, D. (2016) *Murrina vasa.A luxury of Imperial Rome.* L'erma di Bretschneider, Rome.

Della Casa, Ph. (2005) Lithic resources in the early prehistory of the Alps. *Archaeometry*, **47**, 221–234.

Dennen, W.H. and Puckett, A.M. (1971) On the chemistry and color of rose quartz. *Mineralogical Record*, **3**,

226–227.

Di Benedetto, F., D'Acapito, F., Fornaciai, G., Innocenti, M., Montegrossi, G., Pardi, L.A., Tesi, S. and Romanelli, M. (2010) A Fe K-edge XAS study of amethyst. *Physics and Chemistry of Minerals*, **37**, 283–289.

Di Leo, P., Dinelli, E., Mongelli, G. and Schiattarella, M. (2002) Geology and geochemistry of Jurassic pelagic sediments, Scisti silicei Formation, southern Apennines, Italy. *Sedimentary Geology*, **150**, 229–246.

Donati, P. (1986a) Bellinzona, Castel Grande. 6000 anni di storia. *Archeologia Svizzera*, **9**, 94–109.

Donati, P. (1986b) Il cristallo di rocca, prodotto delle Alpi centrali per i Romani. Actes du IVème colloque sur les Alpes dans l'Antiquité, Martigny. *Bulletin d'etudes Prehistotiques Alpines*, **18**, 213–216.

Dong, J., Han, Y., Ye, J., Li, Q., Liu, S. and Gu, D. (2014) In situ identification of gemstone beads excavated from tombs of the Han Dynasties in Hepu county, Guangxi Province, China using a portable Raman spectrometer. *Journal of Raman Spectroscopy*, **45**, 596–602.

Dove, P.M. and Rimstidt, J.D. (1994) Silica-water interactions. Pp. 259–308 in: *Silica. Physical Behavior, Geochemistry, and Materials Applications* (P.J. Heaney, C.T. Prewitt and G.V. Gibbs, editors). Reviews in Mineralogy, **29**. Mineralogical Society of America, Chantilly, Virginia, USA.

Duffin, C.J. (2006) Silicate Minerals. An overview, *BMS Occasional Paper* no **21**, British Micromount Society.

Dumańska-Słowik, M., Powolny, T., Sikorska-Jaworowska, M., Gaweł, A., Kogut, L. and Poloński, K. (2017) Characteristics and origin of agates from Płóczki Górne (Lower Silesia, Poland): A combined microscopic, micro-Raman, and cathodoluminescence study. *Spectrochimica Acta A*, **192**, 6–15.

Eck, W. (2000) Emperor, Senate and Magistrates. Pp. 217–218 in: *Cambridge Ancient History: The High Empire A.D. 70–192*, Cambridge University Press, vol. **11**. Cambridge, UK.

Eckert, A.W. (1997) *The World of Opals*. John Wiley & Sons, New York..

Eichholz, D.E. (Translator) (1962) *Pliny. Natural History*, books 36-37, Loeb Classical Library.

Eker, C.S., Sipahi, F. and Kaygusuz, A. (2012) Trace and rare earth elements as indicators of provenance and depositional environments of Lias cherts in Gumushane, NE Turkey. *Chemie der Erde*, **72**, 167–177.

Ekshtain, R., Ilani, S., Segal, I. and Hovers, E. (2017) Local and nonlocal procurement of raw material in Amud Cave, Israel: the complex mobility of Late Middle Paleolithic groups. *Geoarchaeology*, **32**, 189–214.

Elzea, J.M. and Rice, S.B. (1996) TEM and X-ray diffraction evidence for cristobalite and tridymite stacking sequences in opal. *Clays and Clay Minerals*, **44**, 492–500.

Elzea J.M., Odom, I.E. and Miles, W.J. (1994) Distinguishing well-ordered opal-CT and opal-C from high temperature cristobalite by X-ray diffraction. *Analytica Chimica Acta*, **286**, 107–116.

Enrique Magario, M.E., Ramírez De Arellano, A., Tabales Rodríguez, M.A., Ruiz Conde, A. and Sánchez Soto, P.J. (2002) Aplicacion de la tecnica PIXE (Particle Induced X-Ray Emision) a la caracterización no destructiva de una pieza de ajuar funerario (siglo I d.C.) procedente de una excavacion arqueologica (antiguo Hospital de las Cinco Llagas, Sevilla). *Boletín de la Sociedad Espaola de Cerámica y Vidrio*, **41**, 531–538.

Epstein, D.S. (1988) Amethyst mining in Brazil. *Gems & Gemology*, 214–228.

Ernst, W.G. and Calvert, S.E. (1969) An experimental study of the recrystallization of porcelanite and its bearing on the origin of some bedded cherts. *American Journal of Science*, **267-A**, 114–133.

Evans, A.A., Wolframm, Y.B., Donahue, R.E. and Lovis, W.A. (2007) A pilot study of "black chert" sourcing and implications for assessing hunter-gatherer mobility strategies in Northern England. *Journal of Archaeological Science*, **34**, 2161–2169.

Evans, A.E., Langer, J.L., Donahue, R.E., Wolframm, Y.B. and Lovis, W.A. (2010) Lithic raw material sourcing and the assessment of Mesolithic landscape organization and mobility strategies in northern England. *Holocene*, **20**, 1157–1163.

Eversull, L.G. and Ferrell, R.E. (2008) Disordered silica with tridymite-like structure in the Twiggs clay. *American Mineralogist*, **93**, 565–572.

Fallick, A.E., Jocelyn, J., Donnelly, T., Guy, M. and Behan, C. (1985) Origin of agates in volcanic rocks from Scotland. *Nature*, **313**, 672–674.

Fang, X., Zhou, Y., Liu, M., Wusiman, J.-N.-T.-G.-L. and Liu, H. (2017) Geochemical characteristics and petrogenesis of siliceous rocks from Shangsi section in Guangyuan, Sichuan Province. *Kuangwu Yanshi/ Journal of Mineralogy and Petrology*, **37**, 93–102.

Ferguson, J. (1980) Application of data coding to the differentiation of British flint mine sites. *Journal of Archaeological Science*, **7**, 277–286.
Filin, S.V., Puzynin, A.I. and Samoilov, V.N. (2003) Some aspects of precious opal research. *The Australian Gemmologist*, **21**, 278–282.
Flörke, O.W., Jones, J.B. and Segnit, E.R. (1973) The genesis of hyalite. *Neues Jahrbuch für Mineralogie Monatshefte*, **2**, 82–89.
Flörke, O.W., Jones, J.B. and Segnit, E.R. (1975) Opal-CT crystals. *Neues Jahrbuch für Mineralogie Monatshefte*, 369–377.
Flörke, O.W., Jones, J.B. and Schmincke, H.U. (1976) A new microcrystalline silica from Gran Canaria. *Zeitschrift für Kristallographie*, **143**, 156–165.
Flörke, O.W., Köhler-Herbertz, B., Langer, K. and Tönges, I. (1982) Water in microcrystalline quartz of volcanic origin: Agates. *Contributions to Mineralogy and Petrology*, **80**, 324-333.
Flörke, O.W., Flörke, U. and Giese, U. (1984) Moganite: a new microcrystalline silica mineral. *Neues Jahrbuch für Mineralogie Abhandlungen*, **149**, 325–326.
Flörke, O.W., Flux, S. and Schroder, B. (1985) Hyalith vom Steinwitzhügel bei Kulmain (W-Teil des Egergrabebs). *Neues Jahrbuch für Mineralogie Abhandlungen*, **151**, 87–97.
Flörke, O.W., Graetsch, H., Martin, B., Röller, K. and Wirth, R. (1991) Nomenclature of micro- and non-crystalline silica minerals, based on structure and microstructure. *Neues Jahrbuch für Mineralogie Abhandlungen*, **163**, 19–42.
Folk, R.L. (1980) *Petrology of Sedimentary Rocks*. Hemphill, Austin, Texas, USA.
Folk, R.L. and Pittman, J.S. (1971) Length-slow chalcedony: a new testament for vanished evaporates. *Journal of Sedimentary Petrology*, **41**, 1045–1058.
Folk, R.L. and Weaver, C.E. (1952) A study of the texture and composition of chert. *American Journal of Science*, **250**, 498–510.
Fournier, R.O. (1977) Chemical geothermometers and mixing models for geothermal systems. *Geothermics*, **5**, 41–50.
Fournier, R.O. (1985) The behavior of silica in hydrothermal solutions. Pp. 45–61 in: *Geology and Geochemistry of Epithermal Solutions* (B.B. Berger and P.M. Bethke, editors). Reviews in Economic Geology, Volume **2**. The Economic Geology Publishing Company, El Paso, Texas, USA.
Fournier, R.O. and Marshall, W.L. (1983) Calculation of amorphous silica solubilities at 25° to 300°C and apparent cation hydration numbers in aqueous salt solutions using the concept of effective density of water. *Geochimica et Cosmochimica Acta*, **47**, 587–596.
Fournier, R.O. and Rowe, J.J. (1966) The solubility of amorphous silica in water at high temperature and high pressures. *American Mineralogist*, **62**, 1052–1056.
Frachtenberg, F. and Yellin, J. (1992) Preliminary study of flint sources in Israel by neutron activation analysis. Pp. 149–154 in: *Ancient Stones: Quarrying, Trade and Provenance* (M. Waelkens, N. Herz and L. Moens, editors). *Acta Archaeologica Lovaniensis*, Monograph **4**.
Frazier, S. and Frazier, A. (1999) The chrysocolla-chalcedony connection. *Lapidary Journal*, **52**, 34–40.
Frazier, S., Frazier, A. and Lehrer, G. (1998–1999) A history of gem beadmaking in Idar-Oberstein. *Beads*, **10/11**, 35–47.
French, M.W., Worden, R.H. and Lee, D.R. (2003) Electron backscatter diffraction investigation of length-fast chalcedony in agate: implications for agate genesis and growth mechanisms. *Geofluids*, **13**, 32–44.
Fritsch, E. and Rossmann, G. (1988) An update on color in gems. Part 3: colors caused by band gaps and physical phenomena. *Gems & Gemology*, **24**, 81–102.
Fritsch, F., Gaillou, E., Ostroumov, M., Rondeau, B., Devouard, B. and Barreau, A. (2004) Relationship between nanostructure and optical absorption in fibrous pink opals from Mexico and Peru. *European Journal of Mineralogy*, **16**, 743–752.
Frondel, C. (1962) *The System of Mineralogy. Volume III. Silica Minerals*. John Wiley & Sons, New York and London.
Frondel, C. (1978) Characters of quartz fibers. *American Mineralogist*, **63**, 17–27.
Fyfe, W.S. and McKay, D.S. (1962) Hydroxyl ion catalysis of the crystallization of amorphous silica at 330°C. *American Mineralogist*, **47**, 83–89.

Gaillou, E., Fritsch, E., Aguilar-Reyes, B., Rondeau, B., Post, J., Barreau, A. and Ostroumov, M. (2008a) Common gem opal: An investigation of micro- to nano-structure. *American Mineralogist*, **93**,1865–1873.

Gaillou, E., Delaunay, A., Rondeau, B., Bouhnik-Le Coz, M., Fritsch, E. and Cornen, G. (2008b) The geochemistry of gem opals as evidence of their origin. *Ore Geology Reviews*, **34**, 113–126.

Garlick, G.D. and Jones, F.T. (1990) Deciphering the origin of plume textured geodes. *Journal of Geological Education*, **38**, 298–305.

Garrels, R.M. and Christ, C.L. (1965) *Solutions. Minerals and Equilibria.* Harper and Row, New York, 450 pp.

Gíslason, S.R., Heaney, P.J., Oelkers, E.H. and Schott, J. (1997) Kinetic and thermodynamic properties of moganite, a novel silica polymorph. *Geochimica et Cosmochimica Acta*, **61**, 1193–1204.

Glascock, M.D. (2004) Neutron activation analysis of chert artifacts from a Hopewell mound. *Journal of Radioanalytical and Nuclear Chemistry*, **262**, 97–102.

Gliozzo, E., Grassi, N., Bonanni, P., Meneghini, C. and Tomei, M.A. (2011) Gemstones from Vigna Barberini at the Palatine hill (Rome, Italy). *Archaeometry*, **53**, 469–489.

Gliozzo, E., Mattingly, D.J., Cole, F. and Artioli, G. (2014) In the footsteps of Pliny: Tracing the sources of Garamantian carnelian from Fazzan, south-west Libya. *Journal of Archaeological Science*, **52**, 218–241.

Goldschmidt, V.M. (1954) *Geochemistry*. Oxford University Press, London.

Gołyźniak, P., Natkaniec-Nowak, L., Dumańska-Słowik, M. and Naglik, B. (2016) A multidisciplinary study of a group of post-classical cameos from the National Museum in Krakow, Poland. *Archaeometry*, **58**, 413–426.

Gorelick, L. and Gwinnett, A.J. (1990) Innovative lapidary craft techniques in Neolithic Jarmo. *Archaeomaterials*, **4**, 25–32.

Goreva, J.S., Ma, C. and Rossman, G.R. (2001) Fibrous nanoinclusions in massive rose quartz: The origin of rose coloration. *American Mineralogist*, **86**, 466–472.

Götze, J. and Möckel, R. (editors) (2012) *Quartz: Deposits, Mineralogy and Analytics*. Springer.

Götze, J., Nasdala, L., Kleeberg, R. and Wenzel, M. (1998) Occurrence and distribution of "moganite" in agate/chalcedony: a combined micro-Raman, Rietveld, and cathodoluminescence study. *Contributions to Mineralogy and Petrology*, **133**, 96–105.

Götze, J., Tichomirowa, M., Fuchs, H., Pilot, J. and Sharp, Z.D. (2001) Geochemistry of agates: a trace element and stable isotope study. *Chemical Geology*, **175**, 523–541.

Götze, J., Möckel, R., Kempe, U., Kapitonov, I. and Vennemann, T. (2009) Characteristics and origin of agates in sedimentary rocks from the Dryhead area, Montana, USA. *Mineralogical Magazine*, **73**, 673–690.

Götze, J., Nasdala, L., Kempe, U., Libowitzky, E., Rericha, A. and Vennemann, T. (2012) The origin of black colouration in onyx agate from Mali. *Mineralogical Magazine*, **76**, 115–127.

Götze, J., Gaft, M. and Möckel, R. (2015) Uranium and uranyl luminescence in agate/chalcedony. *Mineralogical Magazine*, **79**, 983–993.

Götze, J., Möckel, R., Vennemann' T. and Müller, A. (2016) Origin and geochemistry of agates in Permian volcanic rocks of the Sub-Erzgebirge basin, Saxony (Germany). *Chemical Geology*, **428**, 77–91.

Graetsch, H. (1994) Structural characteristics of opaline and microcrystalline silica minerals. Pp. 209–232 in: *Silica. Physical Behavior, Geochemistry, and Materials Applications* (P.J. Heaney, C.T. Prewitt and G.V. Gibbs, editors). Reviews in Mineralogy, **29**. Mineralogical Society of America, Chantilly, Virginia, USA.

Graetsch, H. (2011) Microstructure and origin of colour of chrysoprase from Haneti (Tanzania). *Neues Jahrbuch für Mineralogie Abhandlungen*, **188**, 111–117.

Graetsch, H. and Ibel, K. (1997) Small angle neutron scattering by opals. *Physics and Chemistry of Minerals*, **24**, 102–108.

Graetsch, H., Flörke, O.W. and Miehe, G. (1985) The nature of water in chalcedony and opal-C from Brazilian agate geodes. *Physics and Chemistry of Minerals*, **12**, 300–306.

Graetsch, H., Flörke, O.W. and Miehe, G. (1987) Structural defects in microcrystalline silica. *Physics and Chemistry of Minerals*, **14**, 249–257.

Graetsch, H., Topalovic, I. and Gies, H. (1994) NMR spectra of moganite and chalcedony. *European Journal of Mineralogy*, **6**, 459–464.

Gratuze, B., Blet-Lemarquand, M. and Barrandon, J.-N. (2001) Mass spectrometry with laser sampling: A new tool to characterize archaeological materials. *Journal of Radioanalytical and Nuclear Chemistry*, **247**,

645–656.

Groman-Yaroslavski, I. and Bar-Yosef Mayer, D.E. (2015) Lapidary technology revealed by functional analysis of carnelian beads from the early Neolithic site of Nahal Hemar Cave, southern Levant. *Journal of Archaeological Science*, **58**, 77–88.

Guthrie Jr., G.D., Bish, D.L. and Reynolds Jr., R.C. (1995) Modeling the X-ray diffraction pattern of opal-CT. *American Mineralogist*, **80**, 869–872.

Haake, R. and Holzhey, G. (1989) Achate in kugelformigen Rhyoliten des Rotliegenden im sachsisch-thuringischem Raum. *Chemie Erde*, **49**, 173–183.

Halamić, J., Marchig, V. and Goričan, S. (2001) Geochemistry of triassic radiolarian cherts in North-Western Croatia. *Geologica Carpathica*, **52**, 327–342.

Hampton,W.A.,White, G.P., Hoskin, P.W.O., Browne, P.R.L. and Rodgers, K.A. (2004) Cinnabar, livingstonite, stibnite and pyrite in Pliocene silica sinter from Northland, New Zealand. *Mineralogical Magazine*, **68**, 191–198.

Hand, R.J., Stevens, S.J. and Sharp, J.H. (1998) Characterization of fired silicas. *Thermochimica Acta*, **318**, 115–123.

Hardgrove, C. and Rogers, A.D. (2013) Thermal infrared and Raman microspectroscopy of moganite-bearing rocks. American Mineralogist, **98**, 78–84.

Harper, P.O., Klengel-Brandt, E., Aruz, J. and Benzel, K. (1995) *Assyrian origins: Discoveries at Ashur on the Tigris: Antiquities in the Vorderasiatisches Museum, Berlin*. The Metropolitan Museum of Art, Berlin.

Harris, C. (1989) Oxygen-isotope zonation of agates from Karoo volcanics of the Skeleton Coast, Namibia. *American Mineralogist*, **74**, 476–481.

Hatipoğlu, M. and Yardımcı, Y. (2014) Optical and cathodoluminescence investigations of the green microcrystalline (chrysoprase) quartz. *Journal of Luminescence and Applications*, **1**, 87–104.

Hatipoğlu, M., Aj, D. and Sezai Kırıkoğl, M. (2011a) Cathodoluminescence (CL) features of the Anatolian agates, hydrothermally deposited in different volcanic hosts from Turkey. *Journal of Luminescence*, **131**, 1131–1139.

Hatipoğlu, M., Oren, U. and Kibici, Y. (2011b) Micro-Raman spectroscopy of gem-quality chrysoprase from the Biga–Canakkale region of Turkey. *Journal of African Earth Sciences*, **61**, 273–285.

Heaney, O.J. (1993) A proposed mechanism for the growth of chalcedony. *Contributions to Mineralogy and Petrology*, **115**, 66–74.

Heaney, P.J. (1994) Structure and chemistry of the low-pressure silica polymorphs. Pp. 1–40 in: *Silica. Physical Behavior, Geochemistry, and Materials Applications* (P.J. Heaney, C.T. Prewitt and G.V. Gibbs, editors). Reviews in Mineralogy, **29**. Mineralogical Society of America, Chantilly, Virginia, USA.

Heaney, P.J. (1995) Moganite as an indicator for vanished evaporites: a testament reborn? *Journal of Sedimentary Research*, **A65**, 633–638.

Heaney, P.J. and Davis, A.M. (1995) Observation and origin of self-organized textures in agates. *Science*, **269**, 1562–1565.

Heaney, P.J. and Post, J.E. (1992) The widespread distribution of a novel silica polymorph in microcrystalline quartz varieties. *Science*, **255**, 441–443.

Heaney, P.J. and Post, J.E. (2001) Evidence for an *I2/a* to *Imab* phase transition in the silica polymorph moganite at ~570 K. *American Mineralogist*, **86**, 1358–1366.

Heaney, P.J., Veblen, D.R. and Post, J.E. (1994) Structural disparities between chalcedony and macrocrystalline quartz. *American Mineralogist*, **79**, 452–460.

Heaney, P.J., Mckeown, D.A. and Post, J.E. (2007) Anomalous behavior at the *I2/a* to *Imab* phase transition in SiO_2-moganite: An analysis using hard-mode Raman spectroscopy. *American Mineralogist*, **92**, 631–639.

Hemingway, B.S., Anovitz, L.M., Robie, R.A. and McGee, J.J. (1990) The thermodynamic properties of dumortierite $Si_3B[Al_{6.75}\text{-}_{0.25}O_{17.25}(OH)_{0.75}]$. *American Mineralogist*, **75**, 1370–1375.

Hebert, L.B. and Rossman, G.R. (2008) Greenish quartz from the Thunder Bay Amethyst Mine Panorama, Thunder Bay, Ontario, Canada. *The Canadian Mineralogist*, **46**, 111–124.

Herdianita, N.R., Rodgers, K.A. and Browne, P.R.L. (2000a) Routine procedures to characterise the mineralogy of modern and ancient silica sinter deposits. *Geothermics*, **29**, 367–375.

Herdianita, N.R., Browne, P.R.L., Rodgers, K.A. and Campbell, K.A. (2000b) Mineralogical and morphological changes accompanying aging of siliceous sinter and silica residue. *Mineralium Deposita*, **35**, 48–62.

Hess, S.C. (1996) Chert provenance analysis at the Mack Canyon Site, Sherman County, Oregon: an evaluative study. *Geoarchaeology*, **11**, 51–81.

Hesse, R. (1988) Origin of chert: diagenesis of biogenic siliceous sediments. *Geoscience Canada*, **15**, 171–192.

Hetzel, R., Romer, R.L., Candan, O. and Passchier, C.W. (1998) Geology of the Bozdag area, central Menderes massif, SW Turkey: Pan-African basement and Alpine deformation. *Geologische Rundschau*, **87**, 394–406.

Hoard, R.J., Bozell, J.R., Holen, S.R., Glascock, M.D., Neff, H. and Elam, J.M. (1993) Source determination of White River Group Silicates from two archaeological sites in the Great Plains. *American Antiquity*, **58**, 698–710.

Holden, E.F. (1924a) The cause of color in rose quartz. *American Mineralogist*, **9**, 75–89.

Holden, E.F. (1924b) The cause of color in rose quartz (concluded). *American Mineralogist*, **9**, 101–108.

Huckell, B.B. Kilby, J.D., Boulanger, M.T. and Glascock, M.D. (2011) Sentinel Butte: neutron activation analysis of White River Group chert from a primary source and artifacts from a Clovis cache in North Dakota, USA. *Journal of Archaeological Science*, **38**, 965–976.

Hughes, R.E., Högberg, A. and Olausson, D. (2012) The chemical composition of some archaeologically significant flint from Denmark and Sweden. *Archaeometry*, **54**, 779–795.

Hyršl, J. (1999) Chrome chalcedony – A review. *Journal of Gemmology*, **26**, 364–370.

Hyršl, J. (2016) Chrome chalcedony from Tanzania. *Journal of Gemmology*, **35**, 189–190.

Hyršl, J. and Petrov, A. (1998) Gemstones of Bolivia, a review. *Journal of Gemmology*, **26**, 41–47.

Hyršl, J. and Danet, F. (2005) Bismuthinite inclusions in rose quartz from Madagascar. *Gems & Gemology*, **41**, 181–182

Iddings, J.P. (1904) Quartz-feldspar-porphyry (graniphyro liparose-alaskose) from Llano, Texas. *Journal of Geology*, **12**, 225–231.

Ignatov, S.I., Platonov, A.N., Sedenko, V.S. and Taran, M.N. (1990) Rose color of quartz caused by dumortierite microinclusions. *Dopovidi Akademii Nauk Ukrainskoi RSR* Seriya **B7**, 23-27 (in Ukranian).

Inizan, M.-L. (1995) Cornaline et agates: Production et circulation de la préhistoire → nos jours. Pp. 21–24 in: *Les pierres précieuses de l'Orient ancien: des Sumériens aux Sassanides* (F. Tallon, editor). Paris: Éditions de la Réunion des musées nationaux.

Inizan, M.-L. (1999) La cornaline de l'Indus et la voie du Golfe au IIIème millénaire. Pp. 125–138 in: *Cornaline et pierres précieuses. La Méditerranée de l'Antiquité á l'Islam* (A. Caubet, editor). La Documentation française, Musée du Louvre, Paris.

Inizan, M.-L. (2007) Peuplements → l'Holocène. Pp. 23–38 in: *Art rupestre et peuplements préhistoriques au Yémen* (M.-L. Inizan and M. Rachad, editors). Sanaa: CEFAS.

Inizan, M.L. and Tixier, J. (2000) Emergence des arts du feu: le traitement thermique des roches siliceuses. *Paléorient*, **26**, 23–36.

Insoll, T., Polya, D.A., Bhan, K., Irving, D. and Jarvis, K. (2004) Towards an understanding of the carnelian bead trade from Western India to sub-Saharan Africa: The application of UV-LA-ICP-MS to carnelian from Gujarat, India, and West Africa. *Journal of Archaeological Science*, **31**, 1161–1173.

Jansen, J.H.F. and Van der Gaast, S.J. (1988) Accumulation and dissolution of opal in Quatemary sediments of the Zaire deep-sea fan (N.E. Angola Basin). *Marine Geology*, **83**, 1–7.

Jayaraman, N. (1939) The cause of colour of the blue quartzes of the charnockite of south India and of the champion gneiss and other related rocks of Mysore. *Proceedings of the Indian Academy of Sciences*, **A9**, 265–285.

Johnson, M.L., Kammerling, R.C., DeGhionno, D.G. and Koivula, J.I. (1996) Opal from Shewa province, Ethiopia. *Gems & Gemology*, **32**, 112–120.

Jones, J.B. and Segnit, E.R. (1971) The nature of opal. Part 1: Nomenclature and constituent phases. *Journal of the Geological Society of Australia*, **18**, 57–68.

Julig, P.J., Pavlish, L.A. and Hancock, R.G.V. (1991) Chemical characterization and sourcing of Upper Great Lakes cherts by INAA. *Ontario Archaeology*, **54**, 37–50.

Kahraman, S., Önal, M., Sarıkaya, Y. and Bozdoğan, İ., 2005. Characterization of silica polymorphs in kaolins by X-ray diffraction before and after phosphoric acid digestion and thermal treatment. *Analytica Chimica Acta*, **552**, 201–206.

Kano, K. and Taguchi, K. (1982) Experimental study on the ordering of opal-CT. *Geochemical Journal*, **16**, 33–41.

Karacik, Z., Genc, S.C., Esenli, F. and Goller, G. (2011) The Gumuldur Fire Opal: mode of occurrence and mineralogical aspects. *Turkish Journal of Earth Sciences*, **20**, 99–114.

Kastner, M. (1980) Length-slow chalcedony: The end of the New Testament. *Transactions of the American Geophysical Union EOS*, vol. **61**, p. 399.

Kastner, M., Keene, J.B. and Gieskes, J.M. (1977) Diagenesis of siliceous oozes – 1. Chemical controls on rate of Opal-A to Opal-CT transformation – an experimental-study. *Geochimica et Cosmochimica Acta*, **41**, 1041–1059.

Kasztovszky, Z.S., Biró, K.T., Markó, A. and Dobosi, V. (2008) Cold neutron prompt gamma activation analysis – A non-destructive method for characterization of high silica content chipped stone tools and raw materials. *Archaeometry*, **50**, 12–29.

Kenoyer, J.M., Vidale, M. and Bhan, K.K. (1994) Carnelian bead production in Khambhat, India: an ethnoarchaeological study. Pp. 281–306 in: *Living Traditions* (B. Allchin, editor). Studies in the Ethnoarcaheology of South Asia, Oxford publishing.

Kenoyer, J.M. (2005) Bead technologies at Harappa, 3300–1900 BC: A comparative summary. Pp. 157–170 in: *South Asian Archaeology* 2001 (C. Jarrige and V. Lefèvre, editors). Editions Recherche sur les Civilisations – ADPF, Paris.

Kenoyer, J.M. (2008) Indus and Mesopotamian trade networks: new insights from shell and carnelian artifacts. Intercultural relations between south and southwest Asia. Pp. 19–28 in: *Studies in Commemoration of E.C.L. During-Caspers (1934–1996)* (E. Olijdam and R.H. Spoor, editors). BAR International Series 1826, Oxford, UK.

Kenoyer, J.M., Price, J.D. and Burton, J.H. (2013) A new approach to tracking connections between the Indus Valley and Mesopotamia: initial results of strontium isotope analyses from Harappa and Ur. *Journal of Archaeological Science*, **40**, 2286–2297.

Kiefert, L. and Karampelas, S. (2011) Use of the Raman spectrometer in gemmological laboratories: Review. *Spectrochimica Acta Part A*, **8**, 119–124.

Kiefert, L., Epelboym, M., Kan-Nyunt, H.-P. and Paralusz, S. (2012) Applications to the study of gems and jewellery. Pp. 455–468 in: *Infrared and Raman Spectroscopy in Forensic Science* (J.M. Chalmers, H.G.M. Edwards, M.-D. Hargreaves, editors). Wiley, Chichester, UK.

Kingma, K.J. and Hemley, R.J. (1994) Raman spectroscopic study of microcrystalline silica. *American Mineralogist*, **79**, 269–273.

Klamár, R., Matlovič, R., Mitura, T., Buczek-Kowalik, M. and Kopor, I. (2017) Development of geoturism and mining heritage on the examples of Slovak opal mines and oil mine Bóbrka. Pp. 241–248 in: *International Multidisciplinary Scientific GeoConference Surveying Geology and Mining Ecology Management, SGEM*, **17(52)**.

Knauth, L.P. (1994) Petrogenesis of chert. Pp. 233–258 in: *Silica. Physical Behavior, Geochemistry, and Materials Applications* (P.J. Heaney, C.T. Prewitt and G.V. Gibbs, editors). Reviews in Mineralogy, **29**. Mineralogical Society of America, Chantilly, Virginia, USA.

Komov, I.L., Lukashev, A.N. and Koplus, A.V. (1994) *Geochemical Methods of Prospecting for Non-metallic Minerals*. VSP BV. The Netherlands.

Koralay,T. and Kadıoğlu, Y.K. (2015) The origin and determination of silica types in the silica occurrences from Altintaş region (Uşak–Western Anatolia) using multianalytical techniques. *Spectrochimica Acta Part A*, **137**, 516–526.

Kostov, R.I. (2008) Orphic lithica as a source of late antiquity mineralogical knowledge. *Annual of the University of Mining and Geology "St. Ivan Rilski"*, vol. **51**, Part I, Geology and Geophysics, pp. 109–115.

Kostov, R.I. (2010) Review of the mineralogical systematics of jasper and related rocks. *Archeometriai Mühely*, **3**, 209–214.

Krauskopf, K.B. (1956) Dissolution and precipitation of silica at low temperature. *Geochimica et Cosmochimica Acta*, **10**, 1–27.

Krosch, N.J. 1990, Queensland chrysoprase. *The Australian Gemmologist*, **17**, 303–306.

Landmesser, M. (1995) "Mobility by metastability": silica transport and accumulation at low temperatures.

Chemie der Erde, **55**, 149–176.

Landmesser, M. (1998) "Mobility by metastability" in sedimentary and agate petrology: applications. *Chemie der Erde*, **58**, 1–22.

Langer, K. and Flörke, O.W. (1974) Near infrared absorption spectra (4000–9000 cm^{-1}) of opals and the role of "water" in these $SiO_2 \cdot nH_2O$ minerals. *Fortschritte der Mineralogie*, **52**, 18.

Laurs, B.M. (2015) Chrysocolla chalcedony from Spain. *Journal of Gemmology*, **34**, 472.

Law, R., Carter, A., Bhan, K., Malik, A. and Glascock, M.D. (2012) INAA of agate sources and artifacts from the Indus, Helmand, and Thailand regions. Pp. 177–184 in: *South Asian Archaeology 2007 - Proceedings of the 19th International Conference of the European Association of South Asian Archaeology (Ravenna, Italy, 2–6 July 2007)* (D. Frenez and M. Tosi, editors). *British Archaeological Reports*, International Series, 2454.

Ledevin, M., Arndt, N., Simionovici, A., Jaillard, E. and Ulrich, M. (2014) Silica precipitation triggered by clastic sedimentation in the Archean: New petrographic evidence from cherts of the Kromberg type section, South Africa. *Precambrian Research*, **255**, 316–334.

Lee, D.R. (2007) Characterisation of silica minerals in a banded agate: Implications for agate genesis and growth mechanisms. Unpublished Master's thesis, University of Liverpool. (http://www.geos.ed.ac.uk/homes/s0789516/agate.pdf, 18 pp.)

Lehmann, G. (1971) Yellow color centers in natural and synthetic quartz. *Physik der Kondensierten Materie*, **13**, 297–306.

Lehmann, G. and Moore, W.J. (1966) Optical and paramagnetic properties of iron centers in quartz. *Journal of Chemical Physics*, **44**, 1741–1745.

Liesegang, M. and Milke, R. (2018) Silica colloid ordering in a dynamic sedimentary environment. *Minerals*, **8**, 12; doi:10.3390/min8010012.

Lovering, T.G. (1962) The origin of jasperoid in limestone. *Economic Geology*, **57**, 861–889.

Luedtke, B.E. (1978) Chert sources and trace-element analysis. *American Antiquity*, **43**, 413–423.

Luedtke, B.E. (1992) An archaeologist's guide to chert and flint. *Archaeological Research Tools*, 7. Cotsen Institute of Archaeology Press, UCLA, California, USA.

Lule-Whipp, C. (2006) Chromium chalcedony from Turkey and its possible archaeological connections. *Gems & Gemmology*, **42**, 115.

Lyell, C. (1830) *Principles of Geology*, vol. 3, appendix, p. 65.

Lynne, B.Y. and Campbell, K.A. (2004) Morphologic and mineralogic transitions from opal-A to opal-CT in low-temperature siliceous sinter diagenesis, Taupo Volcanic Zone, New Zealand. *Journal of Sedimentary Research*, **74**, 561–579.

Ma, C., Goreva, J.S. and Rossman, G.R. (2002) Fibrous nanoinclusions in massive rose quartz: HRTEM and AEM investigations. *American Mineralogist*, **87**, 269–276.

Makrides, A.C., Turner, M. and Slaughter, J. (1980) Condensation of silica from supersaturated silicic acid solutions. *Journal of Colloid and Interface Science*, **73**, 345–367.

Mallard, E. (1890) Sur la tridymite et la cristobalite. *Bulletin de la Société française de Minéralogie et de Cristallographie*, **13**, 161–180.

Manetti P., Peccerillo, A. and Poli, G. (1979) Rare earth element distribution in Jurassic siliceous rocks from northern Apennines (Italy). *Mineralogica et Petrographica Acta*, **23**, 87–98.

Marshall, W.L. (1980) Amorphous silica solubilities – III. Activity coefficient relations and predictions of solubility behavior in salt solutions, 0–350°C. *Geochimica et Cosmochimica Acta*, **44**, 925–931.

Marshall, W.L. and Chen, C.T.A. (1982a) Amorphous silica solubilities V. Predictions of solubility behavior in aqueous mixed electrolyte solutions to 300°C. *Geochimica et Cosmochimica Acta*, **46**, 289–291.

Marshall, W.L. and Chen, C.T.A. (1982b) Amorphous silica solubilities – VI. Postulated sulfate-silicic acid solution complex. *Geochimica et Cosmochimica Acta*, **46**, 367–370.

Marshall,W.L. and Warakomski, J.M. (1980) Amorphous silica solubilities. II Effect of aqueous salt solutions at 25°C. *Geochimica et Cosmochimica Acta*, **44**, 915–924.

Martins Nunes, E.H. and Soares Lameiras, F. (2005) The optical absorption of gamma irradiated and heat-treated natural quartz. *Materials Research*, **8**, 305–308.

Maschmeyer, D. and Lehmann, G. (1983a) A trapped-hole center causing rose coloration of natural quartz.

Zeitschrift für Kristallographie, **163**, 181–196.
Maschmeyer, D. and Lehmann, G. (1983b) New hole centers in natural quartz. *Physics and Chemistry of Minerals*, **10**, 84–88.
Maschmeyer D., Niemann, K., Hake, H., Lehmann, G. and Riiuber, A. (1980) Two modified smoky quartz centers in natural citrine. *Physics and Chemistry of Minerals*, **6**, 145–156.
McOrist, G.D. and Smallwood, A. (1995) Trace elements in colored opals using neutron activation analysis. *Journal of Radioanalytical and Nuclear Chemistry*, **198**, 499–510.
McOrist, G.D. and Smallwood, A. (1997) Trace elements in precious and common opals using neutron activation analysis. *Journal of Radioanalytical and Nuclear Chemistry*, **223**, 9-15.
McOrist, G.D., Smallwood, A. and Fardy, J.J. (1994) Trace elements in Australian opals using neutron activation analysis. *Journal of Radioanalytical and Nuclear Chemistry*, **185**, 293–303.
Merino, E., Wang, Y. and Deloule, E. (1995) Genesis of agates in flood basalts: twisting of chalcedony fibers and trace-element geo-chemistry. *American Journal of Science*, **295**, 1156–1176.
Michel-Levy, A. and Munier-Chalmas, C.P.E. (1892) Mémoire sur les diverses formes affectées par le réseau élémentaire du quartz. *Bulletin de la Société française de Minéralogie*, **15(7)**, 159–195.
Miehe, G. and Graetsch, H. (1992) Crystal structure of moganite: a new structure type of silica. *European Journal of Mineralogy*, **4**, 693–706.
Miehe, G., Graetsch, H. and Flörke, O.W. (1984) Crystal structure and growth fabric of length-fast chalcedony. *Physics and Chemistry of Minerals*, **10**, 197–199.
Miehe, G., Graetsch, H., Flörke, O.W. and Fuess, H. (1988) Die monokline Kristallstruktur des SiO_2-Minerals Moganit. *Zeitschrift für Kristallographie*, **182**, 183–184.
Migaszewski, Z.M., Gałuszka, A., Durakiewicz, T. and Starnawska, E. (2006) Middle Oxfordian-Lower Kimmeridgian chert nodules in the Holy Cross Mountains, south-central Poland. *Sedimentary Geology*, **187**, 11–28.
Miladinović, Z., Simić, V., Jelenković, R. and Ilić, M. (2016) Gemstone deposits of Serbia. *Geologica Carpathica*, **67**, 211–222.
Miles, W.J. (1994) Crystalline silica analysis of Wyoming bentonite by X-ray diffraction after phosphoric acid digestion. *Analytica Chimica Acta*, **286**, 97–105.
Milot, J., Siebenaller, L., Béziat, D., Léa, V., Schmidt, P. and Binder, D. (2017) Formation of fluid inclusions during heat treatment of Barremo-Bedoulian Flint: Archaeometric implications. *Archaeometry*, **59**, 417–434.
Misik, M. (1996) Silica spherulites and fossil silcretes in carbonate rocks of the Western Carpathians. *Geologica Carpathica*, **47**, 91–105.
Mitchell, R.S. (1967) Tridymite pseudomorphs after wood in Virginian Lower Cretaceous sediments. *Science*, **158**, 905–906.
Mitchell, R.S. and Tufts, S. (1973) Wood opal – A tridymite-like mineral. *American Mineralogist*, **58**, 717–720.
Mizutani, S. (1966) Transformation of silica under hydrothermal conditions. *The Journal of Earth and Planetary Sciences, Nagoya University*, **14**, 56–88.
Mizutani, S. (1970) Silica mineral in the early stage of diagenesis. *Sedimentology*, **15**, 419–436.
Mizutani, S. (1977) Progressive ordering of cristobalite in the early stage of diagenesis. *Contributions to Mineralogy and Petrology*, **61**, 129–140.
Möckel, R., Götze, J., Sergeev, S.A., Kapitonov, I.N., Adamskaya, E.V., Goltsin, N.A. and Vennemann, T. (2009) Trace-element analysis by laser ablation inductively coupled plasma mass spectrometry (LA-ICP-MS): a case study for agates from Nowy Kościoł, Poland. *Journal of Siberian Federal University. Engineering & Technologies*, **2**, 123–138.
Moore, D.T. and Oddy, W.A. (1985) Touchstones: some aspects of their nomenclature, petrography and provenance. *Journal of Archaeological Science*, **12**, 59–80.
Moreau, L., Brandl, M., Filzmoser, P., Hauzenberger, C., Goemaere, E., Jadin, I., Collet, H., Hauzeur, A. and Schmitz, R.W. (2016) Geochemical sourcing of flint artifacts from western Belgium and the German Rhineland: Testing hypotheses on Gravettian Period mobility and raw material economy. *Geoarchaeology*, **31**, 229–243.

Morey, G.W., Fournier, R.O., Hemley, J.J. and Rowe, J.J. (1961) Field measurements of silica in water from hot springs and geysers in Yellowstone National Park. Pp. C333–336 in: Short papers in the geologic and hydrologic sciences: U.S. Geological Survey Prof. Paper 424-C.

Morgenstein, M. (2006) Geochemical and petrographic approaches to chert tool provenance studies: Evidence from two Western USA Holocene archaeological sites. Pp. 307–321 in: *Geomaterials in Cultural Heritage* (M. Maggetti and, B. Messiga, editors). Geological Society of London, Special Publication, **257**. London.

Moroni, B. and Petrelli, M. (2005) Geochemical characterisation of flint artifacts by inductively coupled plasma-mass spectrometry with laser sampling (LA-ICP-MS): Results and prospects. *Mediterranean Archaeology and Archaeometry*, **5**, 49–62.

Morrow, C.A., Elam, J.M. and Glascock, M.D. (1992)The use of blue-grey chert in Midwestern Prehistory. *Midcontinental Journal of Archaeology*, **17**, 166–197.

Mosin, V.S. and Nikolsky, V.Y. (2010) Siliceous rock of the Southern Urals: Distribution and usage in the Stone Age. *Archaeology Ethnology and Anthropology of Eurasia*, **38**, 2–9.

Moxon, T. (1991) On the origin of agate with particular reference to fortification agate found in the Midland Valley, Scotland. *Chemie der Erde*, **51**, 251–260.

Moxon, J.T. (1996) *Agate: Microstructure and Possible Origin*. Terra, Doncaster, UK.

Moxon, T. (2017) A re-examination of water in agate and its bearing on the agate genesis enigma. *Mineralogical Magazine*, **81**, 1223–1244.

Moxon, T. and Reed, S.J.B. (2006) Agate and chalcedony from igneous and sedimentary hosts aged 13 to 3480 Ma: a cathodoluminescence study. *Mineralogical Magazine*, **70**, 485–498.

Moxon, T. and Ríos, S. (2004) Moganite and water content as a function of age in agate: an XRD and thermogravimetric study. *European Journal of Mineralogy*, **16**, 269–278.

Moxon, T., Nelson, D.R. and Zhang, M. (2006) Agate recrystallization: evidence from samples found in Archaean and Proterozoic host rocks, Western Australia. *Australian Journal of Earth Sciences*, **53**, 235–248.

Moxon, T., Reed, S.J.B. and Zhang, M. (2007) Metamorphic effects on agate found near the Shap granite, Cumbria, England: as demonstrated by petrography, X-ray diffraction and spectroscopic methods. *Mineralogical Magazine*, **71**, 461–476.

Moxon, T., Petrone, C.M. and Reed, S.J.B. (2013) Characterization and genesis of horizontal banding in Brazilian agate: An X-ray diffraction, thermogravimetric and electron microprobe study. *Mineralogical Magazine*, **77**, 227–248.

Muldarisnur, M. and Marlow, F. (2017) Spectroscopic Investigation of Opal Formation from Suspensions, *Journal of Physical Chemistry C*, **121**, 8274–18279.

Mukhopahyay, A. and Chanda, S.K. (1972) Silica diagenesis in the banded hematite jasper and bedded chert associated with the iron ore group of Jamda-Koira Valley, Orissa, India. *Sedimentary Geology*, **8**, 113–135.

Murata, K.J. and Larsen, R.R. (1975) Silica mineralogy and structure of Monterey shale, Temblor range, California. *Journal of Research of the U. S. Geological Survey*, **3**, 553–566.

Murata, K.J. and Nakata, J.K. (1974) Cristobalitic stage in the diagenesis of diatomaceous shale. *Science*, **184**, 567–568.

Murata, K.J. and Randall, R.G. (1975) Silica mineralogy and structure of the Monterey Shale, Temblor Range, California. *U. S. Geological Survey Journal of Research*, **3**, 567–572.

Murata, K.J., Friedman, I. and Gleason, J.D. (1977) Oxygen isotope relations between diagenetic silica minerals in Monterey Shale, Temblor Range, California. *American Journal of Science*, **277**, 259–272.

Murray, R.W., Buchholtz Ten Brink, M.R., Gerlach, D.C., Price Russ II, G. and Jones, D.L. (1991) Rare earth, and trace elements in chert from the Franciscan Complex and Monterey Group, California: assessing REE sources to fine-grained marine sediments. *Geochimica et Cosmochimica Acta*, **55**, 1875–1895.

Murray, R.W., Buchholtz Ten Brink, M.R., Gerlach, D.C., Price Russ II, G. and Jones, D.L. (1992) Rare earth, major and trace element composition of Monterey and DSDP chert and associated host sediment: assessing the influence of chemical fractionation during diagenesis. *Geochimica et Cosmochimica Acta*, **56**, 2657–2671.

Muyal, J. (2017) Ethiopian precious OPAL with a dyed brown bodycolor. *Gems & Gemology*, **53**, 362–363.

Nacken, R. (1948) Über die Nachbildung von Chalcedon-Mandeln. *Natur und Volk, B*, **78**, 2–8.

Nagase, T. and Akizuki, M. (1997) Texture and structure of opal-CT and opal-C in volcanic rocks. *The Canadian Mineralogist*, **35**, 947–958.

Nagase, T., Akizuki, M., Onoda, M. and Sato, M. (1997) Chrysoprase from Warrawanda, Western Australia. *Neues Jahrbuch für Mineralogie Monatshefte*, **7**, 289–300.

Nassau, K. and Prescott, B.E. (1977) Smoky, blue, greenish yellow, and other irradiation-related colors in quartz. *Mineralogical Magazine*, **41**, 301–312.

Nathan, Y., Segal, I. and Delage, C. (1999) Geochemical characterization of cherts from northern Israel (western Galilee). *Israel Journal of Earth Sciences*, **48**, 235–245.

Natkaniec-Nowak, L., Dumanska-Słowik, M., Pršek, J., Lankosz, M., Wróbel, P., Gaweł, A., Kowalczyk, J. and Kocemba, J. (2016) Agates from Kerrouchen (The Atlas Mountains, Morocco): Textural types and their gemmological characteristics. *Minerals*, **6**, 77; doi:10.3390/min6030077.

Nunes E.H.M., Lameiras, F.S., Houmard, M. and Vasconcelos, W.L. (2013) Spectroscopic study of natural quartz samples. *Radiation Physics and Chemistry*, **90**, 79–86.

Oehler, J.H. 1976. Hydrothermal crystallization of silica gel. *Geological Society of America Bulletin*, **87**, 1143–1152.

Olofsson, A., Rodushkin, I. (2011) Provenancing flint artefacts with ICP–MS using REE signatures and Pb isotopes as discriminants: preliminary results of a case study from northern Sweden. *Archaeometry*, **53**, 1142–1170.

Önal, M., Kahraman, S. and Sarikaya, Y. (2007) Differentiation of α-cristobalite from opals in bentonites from Turkey. *Applied Clay Science*, **35**, 25–30.

Pan, Y., Nilges, M.G. and Mashkovtsev, R.I. (2008) Radiation-induced defects in quartz. II. Single-crystal W-band EPR study of a natural citrine quartz. *Physics and Chemistry of Minerals*, **35**, 387–397.

Paradise, T.R. (1982) The natural formation and occurrence of green quartz. *Gems & Gemmology*, **18**, 39–42.

Parali, L., Garcia Guinea, J., Kibar, R., Cetin, A. and Can, N. (2011) Luminescence behaviour and Raman characterization of dendritic agate in the Dereyalak village (Eskişehir), Turkey. *Journal of Luminescence*, **131**, 2317–2324.

Parker, R.B. (1962) Blue quartz from the Wind River Range, Wyoming. *American Mineralogist*, **47**, 1201–1202.

Parthasarathy, G., Kunwar, A.C. and Srinivasan, R. (2001) Occurrence of moganite-rich chalcedony in Deccan flood basalts, Killari, Maharashtra, India. *European Journal of Mineralogy*, **13**, 127–134.

Patrizi, G., Vagnini, M., Vivani, R., Fiorini, L. and Miliani, C. (2016) Archaeometric study of Etruscan scarab gemstones by non-destructive chemical and topographical analysis. *Journal of Archaeological Science: Reports*, **8**, 381–391.

Pecover, S.R. (2007) Australian opal resources: outback spectral fire. *Rocks & Minerals*, **82**, 102–115.

Pelto, C.R. (1956) A study of chalcedony. *American Journal of Science*, **254**, 32–50.

Peng, Jun, Yi, H. and Xia, W. (2000) Geochemical criteria of the Upper Sinian cherts of hydrothermal origin on the southeast continental margin of the Yangtze Plate. *Chinese Journal of Geochemistry*, **19**, 217–226.

Pettitt, P., Rockman, M. and Chenery, S. (2012) The British final Magdalenian: Society, settlement and raw material movements revealed through LA-ICP-MS trace element analysis of diagnostic artefacts. *Quaternary International*, **272–273**, 275–287.

Pewkliang, B., Pring, A. and Brugger, J. (2008) The formation of precious opal: Clues from the opalisation of bone. *The Canadian Mineralogist*, **46**, 139–149.

Platonov, A.N. and Szuszkiewicz, A. (2015) Green to blue-green quartz from Rakowice Wielkie (Sudetes, south-western Poland) – a re-examination of prasiolite-related colour varieties of quartz. *Mineralogia*, **46**, 19–28.

Pop, D., Constantina, C., Dan, T. and Wolfgang, K. (2004) Raman spectroscopy on gem-quality microcrystalline and amorphous silica varieties from Romania. *Studia UBB Geologia*, **49**, 41–52.

Prassas, M., Phalippou, J., Hench, L.L. and Zarzycki, J. (1982) Preparation of xNa$_2$O-(1-x)SiO$_2$ gels for the gel-glass process. I. Atmospheric effect on the structural evolution of the gels. *Journal of Non-Crystalline Solids*, **48**, 79–95.

Pretola, J.P. (2001) A feasibility study using silica polymorph ratios for sourcing chert and chalcedony lithic materials. *Journal of Archaeological Sciences*, **28**, 721–739.

Prinsloo, L.C., van der Merwe, E. and Wadley, L. (2017) The thermal behaviour of silica varieties used for tool making in the Stone Age. *Journal of Thermal Analysis and Calorimetry*, **131**, 1135–1145. DOI 10.1007/s10973-017-6602-z.

Prudêncio, M.I., Roldán, C., Dias, M.I., Marques, R., Eixea, A. and Villaverde, V. (2016) A micro-invasive approach using INAA for new insights into Palaeolithic flint archaeological artefacts. *Journal of Radioanalytical and Nuclear Chemistry*, **308**, 195–203.

Rafferty, S.M., Wood, C. and Rieth, C.B. (2007) A trace element analysis of New York chert sources. *North American Archaeologist*, **28/2**, 167–186.

Rapp, G.R. (2002) *Archaeomineralogy*, Springer-Verlag, Berlin.

Řehoř, M. (2009) Historically important discoveries of precious stones in the Krušné hory and the České středohoří mountains, deposit present situation and protection. Pp. 509–516 in: *9th International Multidicsciplinary Scientific Geoconference and EXPO - Modern Management of Mine Producing, Geology and Environmental Protection*, SGEM 2009, Vol. **1**.

Renfro, N. (2015) A new natural-color bluish green chalcedony. *Gems & Gemology*, **51**.

Rey, P.F. (2013) Opalisation of the Great Artesian Basin (central Australia): An Australian story with a Martian Twist. *Australian Journal of Earth Sciences*, **60**, 291–314.

Rice, S.B. and Elzea, J.M. (1993) Stacking disorder in the crystalline opals. The Clay Minerals Society Annual Meeting, Abstract no. 137.

Rice, S.B., Freund, H., Huang, H.L., Clouse, J.A. and Isaacs C.M. (1995) Application of Fourier Transform Infrared Spectroscopy to silica diagenesis: the opal-A to opal-CT transformation. *Journal of Sedimentary Research*, **A65**, 639–647.

Rodgers, K.A. and Cressey, G. (2001) The occurrence, detection and significance of moganite (SiO_2) among some silica sinters. *Mineralogical Magazine*, **65**, 157–167.

Rodgers, K.A. and Hampton, W.A. (2003) Laser Raman identification of silica phases comprising microtextural components of sinters. *Mineralogical Magazine*, **67**, 1–13.

Roll, T., Neeley, M., Speakman, R.J. and Glascock, M.D. (2005) Characterization of Montana cherts by LA-ICP-MS. Pp. 58–76 in: *Laser Ablation-ICP-MS in Archaeological Research* (R.J. Speakman and H. Neff, editors). University of New Mexico Press, Albuquerque, New Mexico, USA.

Romero Silva, J.C. (1996) Blue quartz: from the Antequera-Olvera Ophite, Malaga, Spain. *Mineralogical Record*, **27**, 99–103.

Rondeau, B., Fritsch, E., Guiraud, M. and Renac, C. (2004) Opals from Slovakia ("Hungarian" opals): a re-assesment of the conditions of formation. *European Journal of Mineralogy*, **16**, 789–799.

Rondeau, B., Fritsch, E., Mazzero, F., Gauthier, J.P., Cenki-Tok, B., Eyassu Bekele, E. and Gaillou, E. (2010) Play-of-color opal from Wegeltena, Wollo Province, Ethiopia. *Gems & Gemology*, **46**, 90–105.

Rondeau, B., Cenki-Tok, B., Fritsch, E., Mazzero, F., Gauthier, J.-P., Bodeur, Y., Bekele, E. and Gaillou, E. (2012) Geochemical and petrological characterization of gem opals from Wegel Tena, Wollo, Ethiopia: opal formation in an Oligocene soil. *Geochemistry: Exploration, Environment, Analysis*, **12**, 93–104.

Rosenfeld, A. and Dvorachek, M. (2003) Roman wheel-cut engraving, dyeing and painting microquartz gemstones. *Journal of Archaeological Science*, **30**, 227–238.

Rossmann, G.R. (1994) Colored varieties of silica minerals. Pp. 433–467 in: *Silica. Physical Behavior, Geochemistry, and Materials Applications* (P.J. Heaney, C.T. Prewitt and G.V. Gibbs, editors). Reviews in Mineralogy, **29**. Mineralogical Society of America, Chantilly, Virginia, USA.

Rostan, P. (2007) First data on the exploitation of hyaline quartz crystals in the upper Romanche (Isère and Hautes Alpes, France). *Preistoria Alpina*, **42**, 75–82.

Rothbaum, H., Anderton, B., Harrison, R., Rohde, A. and Slatter, A. (1979) Effect of silica polymerisation and pH on geothermal scaling. *Geothermics*, **8**, 1–20.

Roux, V. (2000) *Cornaline de l'Inde. Des pratiques techniques de Cambay aux techno-systemes de l'Indus*. Editions de la MSH, Paris.

Sachanbinski, M., Janeczek, J., Platonov, A. and Rietmeijer, F.J.M. (2001) The origin of colour of chrysoprase from Szklary (Poland) and Sarykul Boldy (Kazakhstan). *Neues Jahrbuch fur Mineralogie Abhandlungen*, **177**, 61–76.

Samoilovich, M.I., Tsinober, L.I. and Kreiskop, V.N. (1969) The nature of radiation-produced citrine coloration in quartz. *Soviet Physics, Crystallography*, **13**, 626–628.

Sanders, J.V. (1964) Color of precious opal. *Nature*, **204**, 1151–1153.

Sanders, J.V. (1968) Diffraction of light by opals, *Acta Crystallographica*, **A24**, 427–434.

Sanders, J.V. and Darragh, P.J. (1971) The microstructure of precious opal. *The Mineralogical Record*, **2**, 261–268.

Sanders, J.V. and Dietz, R.W. (1967) Precious opal from California. *Gems and Minerals*, 16–19.

Saunders, J.A. (1990) Oxygen-isotope zonation of agates from Karoo volcanics of the Skeleton Coast, Namibia: discussion. *American Mineralogist*, **75**, 1205 1206.

Schaefer, D.W. (1989) Polymers, fractals, and ceramic materials. *Science*, **243**, 1023–1027.

Schaefer, D.W. and Keefer, K.D. (1984) Fractal geometry of silica condensation polymers. *Physical Review Letters*, **53**, 1383–1386.

Schindler, M., Fayek, M., Courchesne, B., Kyser, K. and Hawthorne, F.C. (2017) Uranium-bearing opals: Products of U-mobilization, diffusion, and transformation processes. *American Mineralogist*, **102**, 1154–1164.

Schmidt, P., Badou, A. and Fröhlich, F. (2011) Detailed FT near-infrared study of the behaviour of water and hydroxyl in sedimentary length-fast chalcedony, SiO_2, upon heat treatment. *Spectrochimica Acta A*, **81**, 552–559.

Schmidt, C., Pettke, T., Preusser, F., Rufer, D., Uwe Kasper, H. and Hilgers, A. (2012a) Quantification and spatial distribution of dose rate relevant elements in silex used for luminescence dating. *Quaternary Geochronology*, **12**, 65–73.

Schmidt, P., Bellot-Gurlet, L. Slodczyk, A. and Fröhlich, F. (2012b) A hitherto unrecognised band in the Raman spectra of silica rocks: influence of hydroxylated Si–O bonds (silanole) on the Raman moganite band in chalcedony and flint (SiO_2). *Physics and Chemistry of Minerals*, **39**, 455–464.

Schmidt, P., Slodczyk, A., Léa, V., Davidson, A., Puaud, S. and Sciau, P. (2013) A comparative study of the thermal behaviour of length-fast chalcedony, length-slow chalcedony (quartzine) and moganite. *Physics and Chemistry of Minerals*, **40**, 331–340.

Schmidt, P., Bellot-Gurlet, L., Leá, V. and Sciau, P. (2014) Moganite detection in silica rocks using Raman and infrared spectroscopy. *European Journal of Mineralogy*, **25**, 797–805.

Schmidt, P., Paris, C. and Bellot-Gurlet, L. (2016) The investment in time needed for heat treatment of flint and chert. *Archaeological and Anthropological Sciences*, **8**, 839–848.

Scholz, R., Chaves, M.L.S.C., Krambrock, K., Pinheiro, M.V.B., Barreto, S.B. and de Menezes, M.G. (2012) Brazilian quartz deposits with special emphasis on gemstone quartz and its color treatment. Pp. 138–159 in: *Quartz: Deposits, Mineralogy and Analytics* (J. Götze and R. Möckel, editors). Springer-Verlag, Berlin, Heidelberg.

Seifert, W., Rhede, D., Förster, H.-J. and Thomas, R. (2009) Accessory minerals as fingerprints for the thermal history and geochronology of the Caledonian Rumburk granite. *Neues Jahrbuch für Mineralogie, Abhandlungen*, **186**, 215–233.

Seifert, W., Th.omas, R., Rhede, D. and Förster, H.-J. (2010) Origin of coexisting wüstite, Mg-Fe and REE phosphate minerals in graphite-bearing fluorapatite from the Rumburk granite. *European Journal of Mineralogy*, **22**, 495–507.

Seifert, W., Rhede, D., Thomas, R., Förster, H.J., Lucassen, F., Dulski, P. and Wirth, R. (2011) Distinctive properties of rock-forming blue quartz: inferences from a multi-analytical study of submicron mineral inclusions. *Mineralogical Magazine*, **75**, 2519–2534.

Shen, A., Fritz, E., De Ghionno, D. and McClure, S. (2006) Identification of dyed chrysocolla chalcedony. *Gems & Gemology*, **42**, 140.

Shigley, J.E., Laurs, B.M. and Renfro, N.D. (2009) Chrysoprase and prase opal from Haneti, central Tanzania. *Gems & Gemology*, **45**, 271–279.

Simoni, M., Caucia, F., Adamo, I. and Galinetto, P. (2010) New occurrence of fire opal from Bemia, Madagascar. *Gems & Gemology*, **46**, 114–121.

Sindelar, F.W., Silva, L.F.O., Machado, V.R., dos Santos, L.C.M. and Stülp, S. (2016) Treatment of effluent from the agate dyeing industry using photodegradation and electrodialysis processes. *Separation Science and Technology*, **50**, 142–147.

Skrzypek, G., Sachanbinski, M. and Jedrysek, M.O. (2003) Oxygen isotope evidence for low temperature formation of chrysoprase. *Polskie Towarzystwo Mineralogiczne–Prace Specjalne* (Mineralogical Society of Poland–Special Papers), Zeszyt 22, 2003, pp. 204–206.

Skrzypek, G., Jedrysek, M.O. and Sachanbinski, M. (2004) Oxygen stable isotope geochemistry of chrysoprase from wiry and Szklary Mines (SE Poland). Pp. 470–472 in: *Conference papers of International Symposium on Isotope Hydrology and Integrated Water Resources Management*, 19–23 May 2003, Vienna, Austria.

Smith, C.C. (1967) A preliminary account of Rhodesia's new gemstone – chrome chalcedony. *Chamber of Mines Journal*, December, pp. 31–34.

Smith, D.K. (1998) Opal, cristobalite and tridymite: non-crystallinity versus crystallinity, nomenclature of the silica minerals and bibliography. *Powder Diffraction*, **13**, 2–19.

Smith, G., Vance, E.R., Hasan, Z., Edgar, A. and Runciman, W.A. (1978) A charge transfer mechanism for the colour of rose quartz. *Physica Status Solidi (A)*, **46**, K135–K140.

Sodo, A., Casanova Municchia, A., Barucca, S.,Bellatreccia, F., Della Ventura, G., Butini, F. and Ricci, M.A. (2016) Raman, FT-IR and XRD investigation of natural opals. *Journal of Raman Spectroscopy*, **47**, 1444–1451.

Spencer, R.J., Levinson, A.A. and Koivula, J.I. (1992) Opal from Querétaro, Mexico: Fluid inclusions study. *Gems & Gemology*, **28**, 28–34.

Speer, C.A. (2014) LA-ICP-MS analysis of Clovis period projectile points from the Gault Site. *Journal of Archaeological Science*, **52**, 1–11.

Spier, J. (1992) *Ancient Gems and Finger Rings. Catalogue of the Collections of the J. Paul Getty Museum*. The Paul Getty Trust, 200 pp.

Stein, C.L. and Kirkpatrick, R.J. (1976) Experimental porcelanite recrystallization kinetics: a nucleation and growth model. *Journal of Sedimentary Petrology*, **46**, 430–435.

Stoiber, R.E. (1945) Geology of quartz crystal deposits. *American Mineralogist*, **30**, 219–229.

Streeter, E.W. (2011) *Semi-precious Stones – A Historical Article on Agate, Amber, Amethyst and Many Other Varieties of Gemstones*. Read Books Ltd., Redditch, UK.

Taelman, D., Deprez, S., Vermeulen, F. and De Dapper M. (2009) Granite and rock crystal quarrying in the *Civitas Ammaiensis* (north-eastern Alentejo, Portugal): a geoarchaeological case study. *BABesch*, **84**, 177–188.

Taelman, D., Corsi, C., De Dapper, M., Deprez, S., Verdonck, L. and Vermeulen, F. (2010) Geoarchaeological research in the Roman town of *Ammaia* (Alentejo, Portugal). In: XVII International Congress of Classical Archaeology (Roma 22–26 Sept. 2008), *Bollettino di Archeologia online.*

Tarr, W.A. (1938) Terminology of the chemical siliceous sediments. *Report of the Committee on Sedimentation, 1937.1938 - National Research Council, Washington, 1938* [http://citeseerx.ist.psu.edu/viewdoc/download?doi=10.1.1.819.4291105

Thiry, M., Fernandes, P., Milnes, A. and Raynal, J.-P. (2014) Driving forces for the weathering and alteration of silica in the regolith: Implications for studies of prehistoric flint tools. *Earth-Science Reviews*, **136**, 141–154.

Tsobgou, A.R. and Dabard, M.-P. (2010) Petrographical, structural, and mechanical analysis of Armorican phtanites: A key raw material for the Mesolithic in Western France. *Geoarchaeology*, **25**, 327–351.

van Hinsbergen, D.J.J. (2010) A key extensional metamorphic complex reviewed and restored: The Menderes Massif of western Turkey. *Earth-Science Reviews*, **102**, 60–76.

Vultee, J. von and Lietz, J. (1956) Über die Rolle des Titans als Farbungsursache von Blau- und Rosenquarzen. *Neues Jahrbuch für Mineralogie Monatshefte*, 49–58.

Wadley, L., de la Pea, P. and Prinsloo, L.C. (2017) Responses of South African agate and chalcedony when heated experimentally, and the broader implications for heated archaeological minerals. *Journal of Field Archaeology*, **42**, 364–377.

Walters, H.B. (1897) On some antiquities of the Mycenaean Age recently acquired by the British Museum. *Journal of Hellenic Studies*, **17**, 63–77.

Walther, J.V. and Helgeson, H.C. (1977) Calculation of the thermodynamic properties of aqueous silica and the solubility of quartz and its polymorphs at high pressures and temperatures. *American Journal of Science*, **277**, 1315–1351.

Wang, Y. and Merino, E. (1990) Self-organizational origin of agates: banding, fiber twisting, composition, and dynamic crystallization model. *Geochimica et Cosmochimica Acta*, **54**, 1627–1638.

Wang, Y. and Merino, E. (1995) Origin of fibrosity and banding in agates from flood basalts. *American Journal of Science*, **295**, 49–77.

Wang, J., Chen, D., Wang, D., Yan, D., Zhou, X. and Wang, Q. (2012) Petrology and geochemistry of chert on the marginal zone of Yangtze Platform, western Hunan, South China, during the Ediacaran–Cambrian transition. *Sedimentology*, **59**, 809–829.

Webster, R. (1983) *Gems: Their Sources, Descriptions and Identification*, 4th ed. Butterworths, London.

Weres, O., Yee, A. and Tsao, L. (1982) Equations and type curves for predicting the polymerization of amorphous silica in geothermal brines. *Society of Petroleum Engineers Journal*, February, 9–16.

Whalley, J. (2012) Faded glory: Gemstone simulants and enhancements. *Studies in Conservation*, **57 supl**, S313–S321.

White, J.F. and Corwin, J.F. (1961) Synthesis and origin of chalcedony. *American Mineralogist*, **46**, 112–119.

White, D.E., Hem, J.D. and Waring, G.A. (1963) Chemical composition of subsurface waters. In: *Data of Geochemistry*, 6th ed. (M. Fleischer, editor). U.S. Geological Survey Professional Paper 440-F.

Whitehouse, D. (1975) Carnelian in the Persian Gulf. *Antiquity*, **49**, 129–130.

Williams, L.A. and Crerar, D.A. (1985) Silica diagenesis. II. General mechanisms. *Journal of Sedimentary Petrology*, **55**, 312–321.

Williams, R., Parks, G.A. and Crerar, D.A. (1985) Silica diagenesis. I. Solubility controls. *Journal of Sedimentary Petrology*, **55**, 301–311.

Willing, M.J. and Stocklmayer, S.M. (2003) A new chrome chalcedony occurrence from Western Australia. *Gems & Gemology*, **23**, 265–279.

Wilson, W. (1976) Famous mineral localities: Saint John's Island, Egypt. *Mineralogical Record*, **7**, 310–314.

Wilson, M.J. (2014) The structure of opal-CT revisited. *Journal of Non-Crystalline Solids*, **405**, 68–75.

Wilson, M.J., Russell, J.D. and Tait, J.M. (1974) A new interpretation of the structure of disordered α-cristobalite. *Contributions to Mineralogy and Petrology*, **47**, 1–6.

Wise, M.A. (1981) Blue quartz in Virginia. *Virginia Minerals*, **27**, 9–13.

Witkowski, S. and Zabinski, W. (2004) Application of the TPR method for the studies of Ni(II) in chrysoprase. *Journal of Thermal Analysis and Calorimetry*, **77**, 143–149.

Wright, PM., Weil, J.A., Buch, T. and Anderson, J.H. (1963) Titanium colour centers in rose quartz. *Nature*, **197**, 246–248.

Zhang, Y. and Sakurai, K. (2009) Intercorrelation of impurity trace elements in bloodstone rocks: X-ray fluorescence mapping study. *Journal of Analytical Atomic Spectromety*, **24**, 1579–1583.

Zolensky, M.E., Sylvester, P.J. and Paces, J.B. (1988) Origin and significance of blue coloration in quartz from Llano rhyolite (llanite), north-central Llano County, Texas. *American Mineralogist*, **73**, 313–332.

Appendix 1. Geochemical analyses available for 287 samples of opals (25 archaeological; 262 geological). In the first column: A indicates archaeological objects; G indicates geological literature. The analytical technique used for opal characterization is specified in the last column.

	No. of samples investigated	Country	Reference	Analytical technique
A	25	USA	Hess (1996); Cummings *et al.* (1989)	NAA
G	150	Australia	McOrist *et al.* (1994); McOrist and Snallwood (1997); Gaillou *et al.* (2008a,b)	NAA
G	6	Brazil	Gaillou *et al.* (2008a,b)	ICP-MS
G	1	Canada	Amelin and Back (2006)	ICP-MS
G	8	Ethiopia	Gaillou *et al.* (2008a)	ICP-MS
G	7	Honduras	Gaillou *et al.* (2008a,b)	ICP-MS
G	5	Italy	Caucia *et al.* (2012)	ICP-MS
G	3	Kazakhstan	Gaillou *et al.* (2008a,b)	ICP-MS
G	4	Madagascar	Simoni *et al.* (2010)	LA-ICP-MS
G	11	Mexico	Gaillou *et al.* (2008a,b)	ICP-MS
G	2	Peru	Gaillou *et al.* (2008a,b)	ICP-MS
G	11	Slovakia	Gaillou *et al.* (2008a,b); Caucia *et al.* (2013)	ICP-MS; LA-ICP-MS
G	2	Tanzania	Gaillou *et al.* (2008a,b)	ICP-MS
G	2	Turkey	Karacik *et al.* (2011)	ICP-MS
G	50	USA	Cummings *et al.* (1989); Murray *et al.* (1992); Amelin and Back (2006); Gaillou *et al.* (2008a,b)	INAA; XRF; ICP-MS

Appendix 2. Geochemical analyses available on ~354 samples (183 archaeological and 171 geological) of agates and chalcedony varieties (mostly carnelians). In the first column: A indicates archaeological objects; G indicates geological literature. The analytical technique used for opal characterization is specified in the last column.

	No. of samples investigated	Country	Reference	Analytical technique
A	15	Afghanistan	Law *et al.* (2012)	NAA
A	52	Cambodia	Carter and Dussubieux (2016)	LA-ICP-MS
A	30	India	Law *et al.* (2012)	NAA
A	15	Iran	Law *et al.* (2012)	NAA
A	11	Italy	Gliozzo *et al.* (2011)	PIXE
A	16	Lybia	Gliozzo *et al.* (2014)	LA-ICP-MS
A	3	South Africa	Wadley *et al.* (2017); Prinsloo *et al.* (2017)	XRF
A	41	Thailand	Law *et al.* (2012); Carter and Dussubieux (2016)	NAA; LA-ICP-MS
G	1	Brazil	Götze *et al.* (2001)	NAA
G	4	Canada	Götze *et al.* (2001)	NAA
G	8	Czech Republic	Götze *et al.* (2001)	NAA
G	34	Germany	Götze *et al.* (2001, 2009, 2015, 2016)	NAA; ICP-MS; LA-ICP-MS
G	30	India	Carter and Dussubieux (2016)	LA-ICP-MS
G	17	Iran	Carter and Dussubieux (2016)	LA-ICP-MS
G	2	Lesotho	Prinsloo *et al.* (2017)	XRF
G	2	Mexico	Götze *et al.* (2001, 2015)	NAA; ICP-MS; LA-ICP-MS
G	1	Namibia	Götze *et al.* (2001)	NAA; LA-ICP-MS
G	6	Poland	Möckel *et al.* (2009); Götze *et al.* (2015)	LA-ICP-MS
G	17	Thailand	Carter and Dussubieux (2016)	LA-ICP-MS
G	27 or more	Turkey	Parali *et al.* (2011); Hatipoglu *et al.* (2011a); Koralay and Kadıoğlu (2015)	XRF; ICP-AES; XRF/ICP-MS?
G	2	UK	Götze *et al.* (2001)	NAA
G	20	USA	Götze *et al.* (2009, 2015)	LA-ICP-MS

Appendix 3. Geochemical analyses available for ~814 samples of chert (181 archaeological; 633 geological). In the first column: A indicates archaeological objects; G indicates geological literature. The analytical technique used for opal characterization is specified in the last column. Unclear quantification in Hoard *et al.* (1993).

	No. of samples investigated	Country	Reference	Analytical technique
A	11	Belize	Cackler *et al.* (1999)	INAA
A	33	France	Speer (2014)	LA-ICP-MS
A	13	Lybia	Gliozzo *et al.* (2014)	LA-ICP-MS
A	~124	USA	Julig *et al.* (1991; Morrow *et al.* (1992; Hoard *et al.* (1993); Hess (1996); Glascock (2004); Morgenstein (2006); Huckell *et al.* (2011)	INAA; ICP, ICP-MS
G	3	Canada	Baldwin *et al.* (2011)	LA-ICP-MS
G	17	China	Chang *et al.* (2009)	XRF; ICP-MS
G	17	China	Peng *et al.* (2000)	NAA
G	25	China	Wang *et al.* (2012)	XRF; ICP-MS
G	8	China	Fang *et al.* (2017)	ICP
G	42	Croatia	Halamić *et al.* (2001)	XRF, ICP-MS
G	5	France	Becq-Giraudon *et al.* (1992)	XRF; ICP
G	14	Italy	Manetti *et al.* (1979); Di Leo *et al.* (2002)	INAA; XRF
G	9	Poland	Migaszewski *et al.* (2006)	ICP-AES; FAAS
G	2	Slovakia	Misik (1996)	Spectral analysis
G	1	South Africa	Prinsloo *et al.* (2017)	XRF
G	23	Turkey	Eker *et al.* (2012)	ICP-MS
G	22	UK	Armstrong *et al.* (1999)	ICP-MS
G	~445	USA	Luedtke (1978); Julig *et al.* (1991); Morrow *et al.* (1992); Hoard *et al.* (1993); Hess (1996); Glascock (2004); Morgenstein (2006); Huckell *et al.* (2011)	INAA; ICP, ICP-MS

The numerous cherts analysed by Bressy (2002) are not listed.

Appendix 4. Geochemical analyses available for >532 samples of flint (>220 archaeological; >309 geological). In the first column: A indicates archaeological objects; G indicates geological literature. The analytical technique used for opal characterization is specified in the last column. Data review in Luedtke (1992) not included.

	No. of samples investigated	Country	Reference	Analytical technique
A	35	Belgium	Moreau *et al.* (2016)	LA-ICP-MS
A	~32	France	Gratuze *et al.* (2001); Bressy (2002)	LA-ICP-MS; ICP-MS
A	4	Germany	Moreau *et al.* (2016)	LA-ICP-MS
A	45	Israel	Nathan *et al.*, 1999; Ekshtain *et al.* (2017)	ICP-MS, ICP-OES; ACP-AES, ICP-MS
A	8	Italy	Moroni and Petrelli (2005)	LA-ICP-MS
A	18	Russia	Olofsson and Rodushkin (2011)	ICP–SFMS, MC–ICP–MS
A	17	Spain	Prudêncio *et al.* (2016)	INAA
A	35	The Netherlands	De Bruin *et al.* (1972)	INAA
A	>26	UK	Aspinall and Feather (1972)	INAA
A	3	Ukraine/Moldavia	Kasztovszky *et al.* (2008)	PGAA
G	7 (sources)	Denmark and Sweden	Hughes *et al.* (2012)	EDXRF
G	~104	France	Bressy (2002)	ICP-MS
G	183	Israel	Nathan *et al.*, 1999; Ekshtain *et al.* (2017)	ICP-MS, ICP-OES; ACP-AES, ICP-MS
G	2	Italy	Schmidt *et al.* (2012a)	LA-ICP-MS
G	6	Poland	Kasztovszky *et al.* (2008)	PGAA
G	7	Spain	Prinsloo *et al.* (2017); Prudêncio *et al.* (2016)	XRF; INAA

EMU Notes in Mineralogy, Vol. 20 (2019), Chapter 3, 87–150

Glass and other vitreous materials through history

Ivana ANGELINI[1], Bernard GRATUZE[2] and Gilberto ARTIOLI[3]

[1]Department of Cultural Heritage, Università di Padova, Piazza Capitaniato 7, 35139 Padova, Italy
ivana.angelini@unipd.it
[2]CNRS, Université d'Orléans, IRAMAT-CEB, 3D rue de la Férollerie, F-45071, Orléans Cedex 2, France
gratuze@cnrs-orleans.fr
[3]Department of Geosciences and CIRCe Centre, Università di Padova, Via Gradenigo 6, 35131 Padova, Italy
gilberto.artioli@unipd.it

Early vitreous materials include homogeneous glass, glassy faience, faience and glazed stones. These materials evolved slowly into more specialized substances such as enamels, engobes, lustres, or even modern metallic glass. The nature and properties of vitreous materials are summarized briefly, with an eye to the historical evolution of glass production in the Mediterranean world. Focus is on the evolution of European, Egyptian, and Near East materials. Notes on Chinese and Indian glass are reported for comparison. The most common techniques of mineralogical and chemical characterization of vitreous materials are described, highlighting the information derived for the purposes of archaeometric analysis and conservation.

1. Introduction: chemistry, mineralogy and texture of vitreous materials

Glass is a solid material that does not have long-range order in the atomic arrangement, as opposed to crystalline solids having ordered atomic configurations on a lattice (Doremus, 1994; Shelby, 2005). It has been shown experimentally (Huang *et al.*, 2012) that amorphous solids can be described adequately by the model proposed by Zachariasen, the so-called random network theory (Zachariasen, 1932). Because of the contribution of configurational entropy, glass has a higher Gibbs free energy than a solid with the same composition. To decrease its free energy, glass tends inevitably to transform into the related crystalline form with time: the process is called devitrification. However, the kinetics of transformation at ambient temperature is very slow, so that the glass appears thermodynamically stable over the human life time. Glass can also be considered to be a metastable undercooled liquid, that is a liquid that has solidified below its melting temperature but did not have enough time to order (*i.e.* crystallize) because of the slow diffusion kinetics of the atoms. There is no critical temperature for the melt–solid transformation such as is present for crystalline solids, rather there is a temperature region (defined as *T*g or the glass transition temperature)

DOI: 10.1180/EMU-notes.20.3

where the thermodynamic variables (volume, enthalpy and entropy) change continuously without showing sharp discontinuities. Ancient glasses were invariably produced by high-temperature melting technology, whereas nowadays there are several ways of producing glass materials at low temperature, such as vapour deposition, sol–gel processing of solutions, or neutron irradiation of crystalline solids.

From the point of view of composition, there are few oxides of 3- or 4-coordinated cations that can form a continuous network of oxygen-sharing polyhedra and produce glassy materials: they are B_2O_3, SiO_2, GeO_2, P_2O_5. The cation–oxygen bonds in good glass networks should have nearly 50% of covalent bond character. B, Si, Ge and P are the cations that form oxides satisfying these conditions, and therefore they are all natural 'network formers' although throughout history only silica-based glasses are found. Virtually, glass science is the mineralogy, chemistry and physics of silica. Natural glasses are formed at high temperature by rapid quenching of a silica-rich melt. There are three main geological processes which form glass: rapid cooling of a lava flow (obsidian, pumice), jet cooling of ejecta from a meteoritic impact (tektites) and quenching of the molten layer at the site of the shock impact (impactite, desert glass). All three types of natural glass have been used by humans to make implements or decorations (Fig. 1).

Because it is virtually impossible with old pyrotechnologies to reach the melting point of quartz (1610°C) or the high equilibrium temperature of the polymorphs of silica (trydimite 1703°C, cristobalite 1723°C), in order to produce silica glass with ancient technologies, the melting point had to be lowered substantially. This was achieved by adding elements with a very low electronegative character that form highly ionic bonds with oxygen, commonly alkali elements (Na, K) or Pb. These elements are called network modifiers (or fluxants). However, silica glass that contains only alkali

Figure 1. (a) Head of the Central Mexican god Xochipilli-Macuilxochitl made of black obsidian. Late Aztec sculptural tradition dated to the 15th–early 16th century AD; (b) a modern gem in moldavite, that is an olive green tectite found in southern Germany and the Czech Republic, formed about 15 million years ago; (c) Tutankhamen's pectoral ornament decorated with precious stones (such as carnelian, lapis lazuli, turquoise). The yellow central scarab is carved from desert glass, Egypt 18th Dynasty. [Figure credits: (a) Metropolitan Museum of Art (MET) website, public domain; (b) retrieved from wikipedia public domain; (c) retrieved from www.egyptarchive.co.uk, Copyrighted free use].

cations is easily degraded, and it must be stabilized by cations with a greater ionic strength, most commonly alkaline earths (Mg, Ca), or stabilizers. Some other elements with intermediate electronegative character between the network formers and the network modifiers are called intermediates (Al, Ti, Zr). They can never act as network formers, though they can substitute Si in the tetrahedral sites of existing networks.

Finally, the only other intentional components present in the formulation of ancient glass are the colourants: commonly oxides of the transition elements (V, Cr, Mn, Fe, Co, Ni, Cu); and the decolourants: Sb, Mn (Bamford, 1977; Gliozzo, 2017). If the colourants are added in small amounts they are dispersed and diluted in the glass network, so that the material is coloured, transparent and homogeneous. However, if they are added in substantial amounts they may form specific crystalline compounds and modify the optical properties of the glass. For example, Cu-coloured glass is blue, but it may turn red if cuprite crystals or copper metal are formed (Fig. 2). Such glass is commonly called aventurine glass.

Bulk glass is thus composed of amorphous matter, although it may contain variable amounts of crystal phases. Crystals in glass may originate from three main processes (Artioli *et al.*, 2008): (1) the incomplete melting of the starting raw materials employed as network formers, commonly quartz sand or quartzite rocks; (2) deliberate addition as opacifying agents; and (3) nucleation and crystal growth from the molten glass during cooling. The latter process depends heavily on the chemical composition of the system, the solubility properties of the components employed, and the kinetics of cooling.

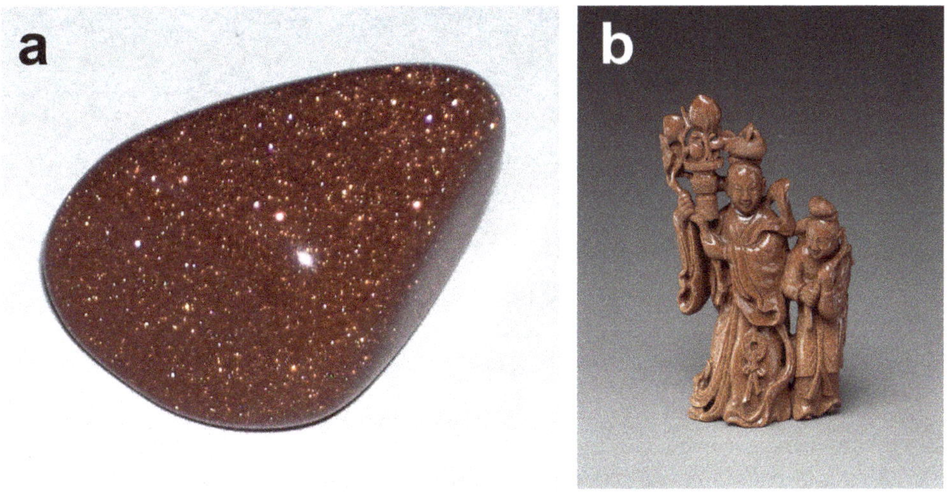

Figure 2. (a) A mass of goldstone, or aventurine glass from the Italian *avventurina*. The name derives from the term *avventura* that means 'adventure' or 'chance' (due to the difficulty in obtaining the right red colour); (b) small (17.8 cm in height) statue in aventurine glass from Qing dynasty (1644–1911 AD), China, 18th century. [(a) Retrieved from commons.wikimedia.org/w/index.php?curid=2419956, under CC BY-SA 3.0; (b) retrieved from the MET website www.metmuseum.org/art/collection/search/43374, public domain].

Based on the final proportions between the glass matrix and the crystalline phases, and the composition, size and distribution of the crystal phases, the glass may be transparent, translucent, or opaque.

The general term 'vitreous materials' refers to a wide range of materials that besides true glass includes glazed lithics, specific synthetic pigments such as Egyptian blue and Egyptian green, enamels, frits, faience and glassy faience. 'Frit' is a processed material that is commonly manufactured as an intermediate during the production of glass and faience: it is obtained by the partial melting and/or sintering of the raw materials, and it must undergo further grinding and furnace (or crucible) heating to yield the final product (Tite and Shortland, 2008).

Egyptian blue and Egyptian green are synthetic pigments that will be described in detail below. They are considered to be among the earliest man-made pigments and they are actually mixtures of specific mineral phases (cuprorivaite, wollastonite, cristobalite, *etc.*) and a minor amount of amorphous glass. The Cu-containing cuprorivaite causes the blue colour in Egyptian blue, whereas the Cu,Fe-containing glass causes the green tinge in Egyptian green (Jaksch *et al.*, 1983; Pagès-Camagna and Colinart, 2003; Tite *et al.*, 2008g). When coloured glass is produced specifically for application onto a metallic substrate it is called 'enamel'. This is generally achieved by grinding the glass to a fine powder that is then fused to the metallic substrate by high-temperature firing.

Faience is defined as a glazed non-clay ceramic material (Nicholson, 1998; Nicholson and Peltenburg, 2000) having a body (or core) composed of quartz sand that may contain small quantities of interparticle glass. The body is covered by a coloured glaze and a layer comprising a mix of glass and crystalline phases, the 'interaction layer' (IAL) is present between the glaze and the core. The thickness of the IAL and the small amount and distribution of glass in the core are related to the glazing method used in the faience production (Tite and Bimson, 1986; Tite, 1987; Tite *et al.*, 2008a). The structure and the texture of faience are completely different from those of glassy faience, the latter being a vitreous material often confused with faience or with opaque glass. Glassy faience, named by Lucas also as 'faience Variant E' (Lucas and Harris, 1962), is a homogeneous material that may be considered as a hybrid of faience and glass (Tite *et al.*, 2008a), with roughly equal proportions of amorphous and crystalline phases. Lucas described glassy-faience objects from Egypt lacking a surface glaze layer; nevertheless some objects show a glassy faience texture but also a distinct blue glaze on the surface. Several ornaments produced in Italy during the early phases of the middle Bronze Age present such features (Angelini *et al.*, 2002; Tite *et al.*, 2008b). These Italian materials do not fit properly with the 'faience Variant E' class as described by Lucas. Considering the measured proportion between the glass matrix and the crystalline phases, however, and taking into account that the assumed production technique is necessarily different from those used for the production of faience present in the same area, the material is also described as glassy faience (Angelini *et al.*, 2002; Bellintani *et al.*, 2006a). The characteristic inner textures of objects made of faience, glassy faience and glass are reported in Table 1.

Table 1. Typical examples of faience, glassy faience and glass textures from Bronze Age Italian vitreous materials.

Object type	Image	Micro-texture (SEM-BSE images)	Proportion of amorphous glass *
Faience: biconical bead (Lavagnone – BS)		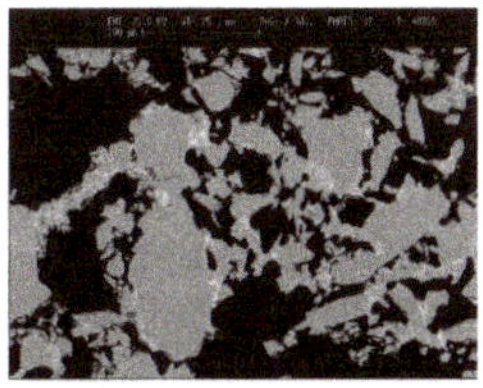	0.03
Glassy faience: bead spacer (Prato di Frabulino – VT)		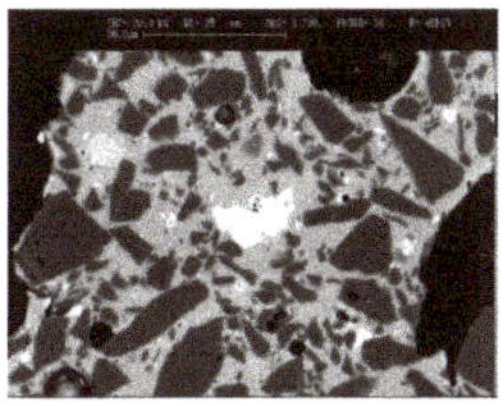	0.44
Opaque glass: slice of raw glass scrap (Frattesina – RO)		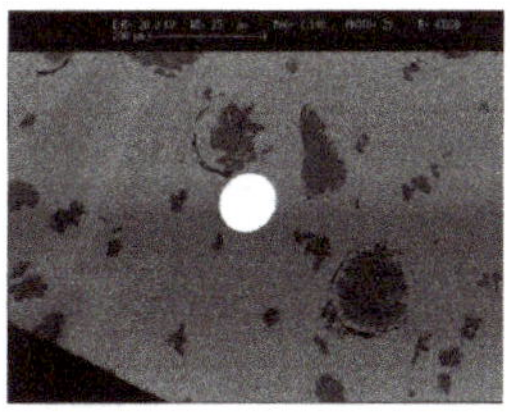	0.88
Transparent glass: flattened globular bead (Alba – CN)		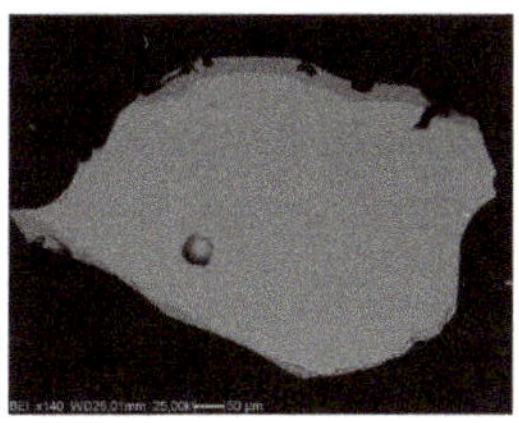	1.00

* The proportion is reported as [(volume of glass)/(volume of glass + volume of crystalline phases)].

The terms 'paste', 'glass paste' or 'glassy paste' are often used in a colloquial and non-rigorous way, and do not relate to the texture, the nature or the composition of the vitreous materials. In order to avoid ambiguity and misunderstanding, these terms should not be used even when the identification of the proper type of vitreous materials is difficult or in the presence of weathering and leaching of the glass phase.

Based on the archaeological and historical evidence, the time of appearance, diffusion and the extent and duration of use of the different types of vitreous materials (faience, glass, pigments, *etc.*) is characteristic of different geographical and cultural contexts. Vitreous objects and evidence of glass-production technology, therefore, may be regarded as excellent indicators of cultural and technological diffusion, exchanges and trades.

Iconic illustrations of such processes are, for example, the introduction of 'trade beads' produced in Europe, the Islamic world and India into Africa in exchange for goods and slaves (Fage, 1962; Alpern, 1995; Lankton and Dussubieux, 2006; Robertshaw *et al.*, 2010; Wood, 2011), or the introduction of glass by traders and explorers to indigenous people in North America (Hancock *et al.*, 1996; Sempowski *et al.*, 2001). Of course investigation of the cultural, social and personal use of the glass objects may also transcend the simple materiality of the objects themselves, providing insights into the intangible values of heritage materials (Yentsch, 1995).

A brief overview and some of the main features of physicochemical and textural variations of glass in time will be attempted below.

2. Composition of vitreous materials through time in the Mediterranean and Indian worlds

Ancient glass is generally classified into compositional groups, based mostly on the relative amount of fluxant (Na, K, Pb) and stabilizing elements (Mg, Ca). The observed chemical-compositional groups correspond roughly to chronologically and/or geographically distinct glass productions, and reflect the use of different components in the glass formulation and production technology. Sayre and Smith (1961) distinguished several compositional groups, based on the chemical and archaeological evidence available at the time. Their observations are still largely valid (Henderson, 2001, 2013a,b), although subsequent analytical data allowed refinement of the distribution and evolution of early vitreous materials (Rehren and Freestone, 2015). Table 2 reports the main compositional groups of ancient glass. Modern ordinary soda-lime silica glass is very similar to standard Roman glass.

2.1. Bronze Age

Glazed stones are the earliest vitreous materials invented by ancient artisans. They may be classified in two categories: 'glazed steatite' and 'glazed quartz', the latter including quartzite, chert and crystalline quartz.

Small objects made of glazed quartz were produced in Egypt (Fig. 3a) and the Near East from about 4000 BC, as suggested by Beck (1935). However, due to the large uncertainty in dating the materials it is not known if glazed quartz predated glazed steatite production (Tite *et al.*, 2008c). Glazed quartz has been reported from Nubia (nowadays North Sudan) since the A-group period, about 3100 BC, and production is suggested to occur in Kerma in the flowering period of the Kerma Culture (corresponding approximately to the late Second Intermediate Period of Egypt)

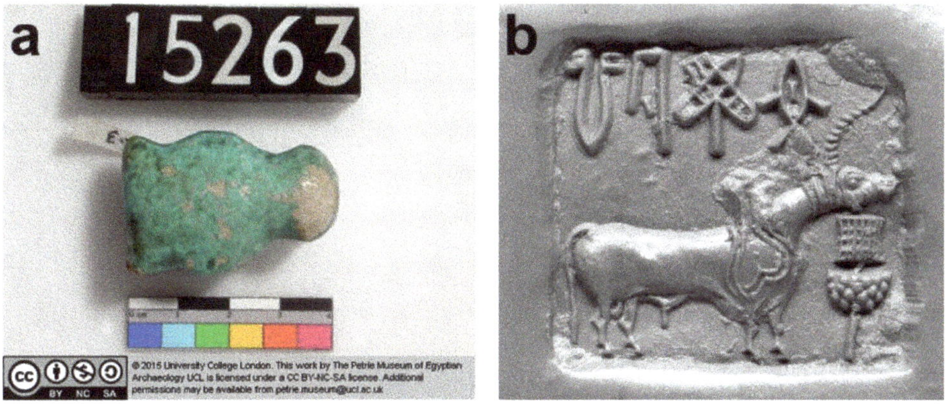

Figure 3. (a) An Egyptian blue bucranium amulet in glazed quartz, dated to the Early Dynastic Period; (b) stamp seal in burnt steatite, Harappan culture, Indus Valley (~2600–1900 BC) [(a) The Petrie Museum of Egyptian Archaeology UCL website; (b) from the MET website].

based on the archaeological objects, even if no direct evidence of furnace or workshop has been found (Lacovara, 1998). To date, no chemical analyses of glazed quartz are known, therefore the raw materials used and the production technology are still hypothetical.

Early glazed steatite production, mainly of small beads, is dated to about the end of the 5th millennium BC, and was in use in Egypt, the Near East and the Indus Valley (Tite *et al.*, 2008c). Steatite is formally a soft compact rock composed mainly of talc (hardness of talc is 1 on the Mohs scale) and minor amounts of other minerals, such as chlorite, mica, tremolite, quartz and magnetite. In the archaeological literature, steatite is frequently termed 'soapstone', whereas formally soapstone should be characterized by larger contents of chlorite, serpentine and sometimes dark mica. However, both rocks are characterized by a low hardness (<3–4) and may be cut and engraved easily. Because of their soft state, they are still used widely today to carve ornaments, decorations and tourist souvenirs. In the early stages of glass/glaze development, steatite and soapstone were, therefore, ideal to carve and shape the core of small objects that was subsequently stabilized by glazing.

In fact after the shaping of the objects the hardness could be increased up to 5–7 by heating at >900°C (Rapp, 2009). The talc undergoes high-temperature decomposition reactions and it is transformed into enstatite and cristobalite (Nakahira and Kato, 1963; Wesolowski, 1984; Zhang *et al.*, 2006):

$$Mg_3Si_4O_{10}(OH)_2 \text{ talc} \leftrightarrow 3\ MgSiO_3 \text{ enstatite} + SiO_2 \text{ cristobalite} + H_2O$$

Small ornaments made of heated steatite were used widely since prehistory in numerous archaeological cultures in North America, Europe and Africa (Rapp, 2009), although the early use of glazed steatite was limited to three regions: Egypt, the Near East and the Indus Valley. On the surface of a steatite body a glazing slurry was applied

Table 2. Major compositional groups of archaeological and historic silica-based glass. Modified from Artioli (2010, table 3.12). The significant chemical signatures distinctive of the group are in bold.

Glass type	Chemical signature (oxide wt.%) Na_2O	K_2O	MgO	CaO	Other	Note	Alkali source
HMG	8–20	**0–3**	**2–10**	3–10		**High-Mg glass** – Near East (1500–800 BC)	plant ash
LMHK	0–8	**4–18**	**0–1**	0–4		**Low-Mg High-K glass** – Europe (Final Bronze Age), China-Thailand (400 BC–200 AD)	plant ash
LMG (SLS)	13–20	**0–1**	**0–1**	5–10	0–1 Fe_2O_3 0–2 MnO	**Low-Mg glass** – Near East (after 800 BC) and standard Roman-type **soda-lime silica** glass (50–300 AD)	natron salts
HIMT	16–20	**0–1**	1–2	5-9	**1–3** Fe_2O_3 **1–3** MnO **0–1** TiO_2	**High-Fe-Mn-Ti glass** – Mediterranean area and Northwestern provinces (300–500 AD)	natron salts
LEG	10–15	**0–1**	0–1	8–12	**1-3** Al_2O_3	**Levantine-type** – Near East, Mediterranean area, Northwestern provinces (500–800 AD)	natron salts
HSB	15–21	**0–1**	0–1	4–6	**0–2** Sb_2O_3	**High-Sb glass** - Near East, Mediterranean area (100–300 AD)	natron salts
HKEG	0–14	**2–14**	0–5	6–20		**High-K Medieval glass** – Europe (stained window glass)	plant ash
HLLA	2–4	3–6	2–5	**16–25**		**High-lime Low-alkali glass** - Britain (17th–18th cent AD)	plant ash

HLEG	0–1	**3–10**	1–3	4–16	**20–65** PbO	**High-Pb Medieval glass** – Europe	plant ash
HMEIG	10–18	1–3	**3–7**	6–12	**0–1** MnO	**High-Mg Early Islamic** glass (840–1000 AD)	natron/ plant-ash
HBHA	17–21	4–5	1–5	2–6	**0.1–0.2** B **8–12** Al_2O_3	High-B High-Al Byzantine Western Anatolia	mineral flux
HLIG	8–10	0–2	0–1	4–5	**30–65** PbO	**High-Pb Islamic glass** (1000–1400 AD)	natron/ plant ash
HLHB	2–9	0–4	0–1	0–3	**18-44** PbO **6-22** BaO	**High-Pb High-Ba glass** – China (Han Dynasty 206 BC-221 AD)	witherite
HSHA	15–20	0–5	0–3	0–6	**10–15** Al_2O_3	**High-Na High-Al glass** Asia (China, Korea, Southeast Asia)	mineral flux
HAG	2–12	**4–16**	1–2	2–6	**2–4** Al_2O_3	**High-Al glass** – India (1st millennium AD) – maybe several groups	plant ash
HLHA	0–7	3–8	0–1	**11–31**	**11–18** Al_2O_3	**High-Ca High-Al glass** – West Africa (9th–13th cent AD)	plant ash
FDV	12–15	2–4	1–3	4–10		**Venetian** "cristallo" and **Dutch** "façon de Venise" (16th–17th cent AD)	soda-rich ash

before the heating process in order to obtain a glaze with the preferred colour covering the object. The slurry always contains some kind of fluxing agent, with different overall composition specifically depending on the geographical area. The thermally stimulated reaction between the glaze and the talc core may form specific mineral phases, mostly forsteritic olivine (Mg_2SiO_4) or enstatitic pyroxene ($MgSiO_3$), though the local concentration of alkali may stimulate the crystallization of unusual phases of the osumilite–milarite group of minerals (Artioli *et al.*, 2013). The morphology and distribution of the crystals in the internal part of the glaze is indicative of the glazing process that could be either direct application or cementation (Tite and Bimson, 1989).

In Mesopotamia the early known glazed steatite is dated to 5500–4000 BC, and was found in Arpachiya, Niniveh and Tepe Gawra. Production continued to be in use in the Early Dynastic period (3000–2350 BC) and later, as testified by the finds of some seals in Ur and Kish (Beck, 1934; Moorey, 1994; Tite *et al.*, 2008c). The steatite generally is white, displaying at best a poorly glazed surface layer, therefore these materials were named by Beck (1934) and Morey (1994) as 'burnt steatite'. However, the strongly weathered state of the finds does not allow exclusion of the presence of an original fully glazed layer that is now totally lost (Tite *et al.*, 2008c). To date, no analyses of Mesopotamian burnt steatite are known, adding to the uncertain interpretation of these materials and their production technique. In the Levant, ~200 white beads dated to the 5th millennium BC were found in the Chalcolithic burial cave of Peqi'in (northern Israel). X-ray diffraction (XRD) and scanning electron microscopy-energy dispersive spectrometry (SEM-EDS) analyses performed on some of the beads (Bar-Yosef Mayer *et al.*, 2004) confirmed that they are made of glazed steatite. Traces of Cu are present suggesting an original blue colour of the glaze, although no information about the fluxing agents or other chemical species are reported. Based on the dimensions and the orientation of the enstatite crystals, the authors suggested that the beads were produced by a sort of 'steatite faience' process, *i.e.* a faience production made with crushed steatite instead of sand or quartz. To date, steatite faience is known from pre-Indus and Indus civilizations (Tite *et al.*, 2008c), but it has not been found in the Egyptian and Mesopotamian world. Moreover, Bar-Yosef Mayer *et al.* (2004) do not report SEM or optical microscopy (OM) images of sections of the beads, so the question of the fabrication technique of such materials requires further investigation.

Steatite faience is rare, and is known mainly from South Asia: in Mehrgarh (pre-Indus period) and in Nausharo (Indus period). The steatite faience from Mehrgarh seems to be produced by the efflorescence or the application method and the amorphous phase has a composition similar to that observed in glazed steatite (discussed below). From Nausharo analytical data are available only for one sample, and its peculiar composition suggests a different origin (Barthélemy de Saizieu and Bouquillon, 1997, 2001; Tite *et al.*, 2008c).

Glazed steatite is found largely in the Indus Valley during the Chalcolithic (about 4500–3500 BC) and the Indus period (2500–2000 BC), and its use disappears from 1900 BC onwards (Barthélemy de Saizieu and Bouquillon, 2001). The glaze has a greater Al content than local sands (a similar Al content is known also for later Indian glass – see below). The ratio of the monovalent alkali contents suggests that *ghar* (a local type of plant ash) is used as flux or, more rarely, *reh* a mix of efflorescent salts, is suggested as flux (Tite *et al.*, 2008c). An important production centre of glazed steatite, other stone beads and faience, is Harappa, where excavation provided a large number of materials and finds related to production activities (Kenoyer, 1994, 1997; Miller, 1999) (Fig. 3b).

Egyptian glazed steatite was widely diffuse both in terms of geographical distribution and time range (Fig. 4). Referring to the standard Egyptian chronology as reported in the Cambridge Ancient History volume, the earliest glazed Egyptian steatite is dated to the late 5th millennium BC (Predynastic Period) and it was found in graves at Qau-Badari (Badarian Culture). It comprises mainly small coloured beads that continued to be produced in the Naqada I and II phases of the Predynastic Period (~4000–3000 BC). From the Early Dynastic (3000–2570 BC) to the Middle Kingdom (1991–1750 BC) not only beads, but also amulets and scarabs were produced. Since the New Kingdom (1540–1070 BC) and until the early Late Period (747–332 BC) the production of glazed steatite continued: amulet scarab, seals, small and large decorative objects were produced commonly, whereas the manufacture of beads was rare (Vandiver, 1983). Tite and Bimson (1989) analysed a few Badarian beads and decorative object of the New Kingdom finding that different recipes and glazing methods were used. The glaze of the Badarian beads shows large Mg, Na and Cu contents, suggesting that a mix of natron and copper minerals is used to prepare the slurry and create a blue glaze. Forsterite crystals are present in the glaze with a preferred dimensional distribution: the smaller are close to the surface and the larger are formed in the area of interaction with the talc core: this suggests to the author the probable use of the cementation process.

The glazed steatite of the New Kingdom is characterized by a glaze with large Na and K and small Mg and Al contents; therefore the likely use of powdered quartz or pure sand mixed with soda plant ash (or sometimes mixed alkali plant ash) can be assumed. The glaze is coloured by Cu, although found in smaller concentrations than in the Badarian glazed steatite (Tite and Bimson, 1989; Tite *et al.*, 2008c).

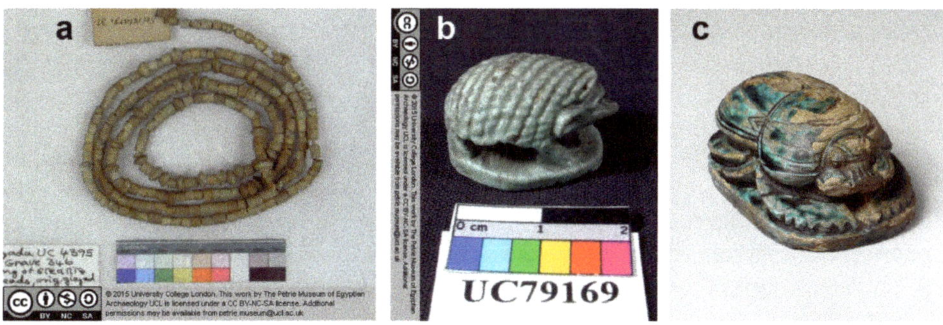

Figure 4. (a) Glazed steatite beads from Naqada Tomb-346, Egypt (Naqada II, ~3500 –3200 BC). The whitish colour is due to the loss of glaze; (b) blue-green glazed steatite scaraboid in form of hedgehog (18th Dynasty, 1540–1295 BC); (c) commemorative scarab of Amenhotep III recording a Lion Hunt. Glazed steatite, 18th Dynasty, from Thebes. [(a) and (c) from The Petrie Museum of Egyptian Archaeology UCL website; (b) from the MET website].

An important and widely found vitreous material is faience. Faience first appeared at the end of the 5th millennium BC in the Near East and then in Egypt (Nicholson, 1998; Nicholson and Peltenburg, 2000; Tite *et al.*, 2008d). In the Caucasian regions, faience has been known since the middle–end of the 3rd millennium BC (Shortland *et al.*, 2007a) and seems to predate the presence of the earliest faience in Europe in the BzA1, about 2200–1900 BC (Bellintani *et al.*, 2006b). In the Aegean world, early faience is known from the Middle Chalcolitic cemetery in Cyprus (3200–2900 BC), at approximately the same time at which they appeared in Crete. Some time later it is present in Rhodes and mainland Greece (Tite *et al.*, 2008e). In the Indus Valley there is an independent development of a faience industry that takes place and prospers during the entire Indus Civilizations (from the beginning of the 3rd millennium BC to the early 2nd millennium BC). Some of these faience productions and the early known or possible production centres will be discussed in detail.

In the Near East, the oldest known faience materials are dated to the "final phase of the Ubadi period (5500–4000 BC)" (Tite *et al.*, 2008f). Throughout this time, beads and vessels were produced in blue-green Cu-coloured faience. During the 3rd millennium BC (Early Dynastic and Akkadian periods), a number of different object types were produced, such as amulets, seals, moulded vessels and others. The same types of objects were still manufactured in Mesopotamia during the first half of the 2nd millennium BC, although new colours and decorations were introduced. Black-brown decorations obtained from Fe-Mn ores are specifically to be noted. In the same period, faience was introduced in the Levant, probably from Egypt, although the possibility of local production is not to be excluded (Tite *et al.*, 2008f). The second half of the 2nd millennium BC was a time of significant changes in the production of vitreous materials thanks to technical developments in the glass industry. Nevertheless, faience continued to be produced, largely in the Near East, and new colours such as yellow (due to Pb antimonate) and different hues of green obtained by mixing Cu and Pb antimonate were used. Glazed pottery products were created now and their production increased progressively during the 1st millennium BC. Small objects and faience vessels continued to be produced during the 1st millennium BC up to the Sassanian Period (637 AD) (Tite *et al.*, 2008f). Despite the importance of the Near East faience production, archaeometric investigations have been extremely scarce. Information about the colouring agents and opacifiers is available based on the analyses of ~300 objects (Kaczmarczyk, 2007) using non-invasive analysis. Chemical data for the amorphous phase are available only for a few objects from Tell Taya, dated to 2200 BC (Tite *et al.*, 2008f) (Fig. 5a). The data suggest that sand was used for the production of faience, but due to the weathering of the glass phase it is difficult to evaluate the fluxant used. Large K contents suggest the use of mixed alkali plant ash, but K may also be derived from feldspar present in the sand, therefore the use of soda plant ash is also possible (Tite *et al.*, 2008f).

More data are available for Egyptian production thanks to the large number of excavations and archaeometric studies. For the manufacture of the body a paste composed of sand or ground quartz mixed with water and flux was prepared (flux is

Figure 5. (a) Early Bronze Age faience ornament from Tell Taya, Mesopotamia, *ca.* late 3rd millennium BC. The white colour is typical of the strong weathering of these faience; (b) faience bowl from Mesopotamia, 16th–12th centuries BC [Images from the MET website].

added only if the glaze is obtained by efflorescence). The body of the faience could be shaped in different ways: manual modelling, moulding, core forming, rod forming, using different techniques of inlay and by less common methods (Vandiver, 1982; Nicholson, 1998; Nicholson and Peltenburg, 2000). Three main glazing techniques are known:

(1) direct application: a vitrifying slurry is prepared with water, flux, lime, colouring agent (generally Cu) and eventually ground quartz or sand and it is applied to the surface before firing;
(2) efflorescence: the vitrifying slurry is mixed in the body paste before the shaping phase, then the faience is dried in air (when the alkali salts effloresce on the surface) and fired;
(3) cementation: the shaped nucleus is immersed in a powder of lime, flux and colouring agent, and then fired.

The glazing techniques have been discussed in detail in the literature and experimental replicas have been performed to investigate the correlation between the resulting microstructure of the faience and the production techniques (Vandiver, 1982; Tite and Bimson, 1989; Nicholson, 1998; Nicholson and Peltenburg, 2000; Shortland and Tite, 2005; Tite *et al.*, 2008d). The modelling and glazing techniques changed over time, as did the variety of objects produced and the colours used. The main changes (Vandiver, 1982; Nicholson, 1998; Patch, 1998; Nicholson and Peltenburg, 2000; Tite *et al.*, 2008d) are summarized here.

Early Egyptian faience, mainly beads and amulets, is known from Badarian burials in the 4th millennium BC, but it is not clear whether they were an independent local invention or the knowledge of faience production derived from the Near East. Predynastic faience is generally coloured by copper and it seems that all three glazing techniques were actually in use (Fig. 6a). During the Early Dynastic period the same type of faience was produced, and vessels, tiles and statuettes (dog, bird, baboon, *etc.*) were manufactured. The variety of faience items produced during the Old Kingdom and

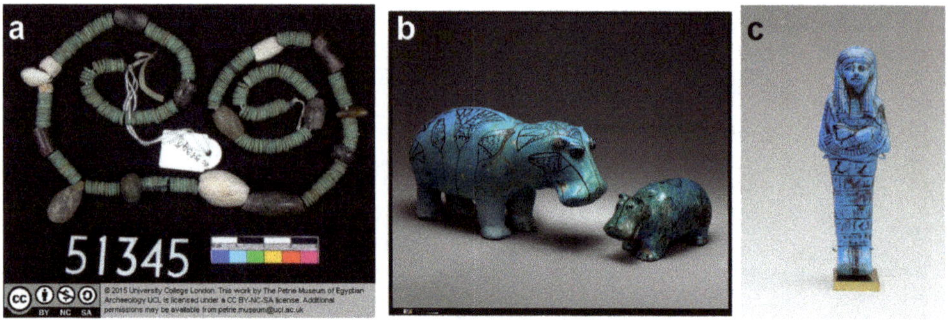

Figure 6. (a) Faience beads dated to the Predynastic Period, from Diospolis Parva (Egypt); (b) faience hippo from Meir (Egypt), Tomb B3, Khashaba excavations, 1910. Dated to the Middle Kingdom, 12th Dynasty, reign of Senwosret I–Senwosret II (*ca.* 1961–1878 BC); (c) Shabti of Paser, the Vizier of Seti I and Ramesses II. Dated to *ca.* 1294–1213 BC, 19th Dynasty, New Kingdom; [(a) from The Petrie Museum of Egyptian Archaeology UCL website; (b,c) from the MET website].

the First Intermediate Period is even wider, and includes plaques, game pieces and boat models, though the main products are beads and especially tiles. A massive amount of high-quality blue faience tiles was found in Djoser's step Pyramid in Saqqara (3rd Dynasty) and in the Reneferef's Pyramid in Abusir (5th Dynasty). In this phase new colours were used: black-brown decorations obtained by the use of Mn-Fe pigments and white objects resulting from faience prepared without colouring agents. By the 12th Dynasty the variety of faience objects increased again and small statues of all scales were produced. In the Middle Kingdom and the Second Intermediate Period a wider range of faience objects was created, including jewellery, vessels and animal figurines such as hippos, apes, cats and hedgehogs (Fig. 6b). Composite objects were prepared by joining different faience pieces using quartz-paste slurry. The apogee of faience production may be considered the New Kingdom and in particular the 18th Dynasty. At this time a multitude of object types was created: shabtis, sistra, furniture models, game boards, statuettes, plaques, bowls, canopic jars, boxes and others. The most colourful faience was produced thanks to the discovery of new colourants: Co for deep dark blue; Pb antimonate for yellow and in combination with Cu for green; iron oxide for red. As noted for coeval Near East faience, the development in this chronological phase of an evolved glass industry had a positive influence on faience production. The link between faience and glass production is supported by the simultaneous presence of both types of product in the same site, such as in Amarna and Quantir (Table 3 and references therein). During the Third Intermediate Period the same types of faience objects continued to be produced, especially shabtis, beads, amulets and chalices. The colour palette was limited: cobalt blue disappeared and Pb antimonate was scarcely used.

The earliest production centre of faience excavated in Egypt is Abydos. The site dates from the middle of the Old Kingdom to the early Middle Kingdom. Remains of circular features and bricks are associated with the probable presence of a kiln (Nicholson, 1998; Nicholson and Peltenburg, 2000). Evidence for the existence of a faience workshop in Lisht comes from the reading of an epigraphic note on a sarcophagus reporting the existence of an "overseer of faience workers" (dated to the 13th Dynasty, 1750–1640 BC). Moreover, excavation campaigns of the Metropolitan Museum of Art at the beginning of the 19th century unearthed the remains of a kiln (Nicholson, 1998; Nicholson and Peltenburg, 2000). However, at the Lisht site, glass finds and waste connected to glass production are present: glass slags, cullets, crucibles and glass fragments of objects were found, together with faience beads and pendants, half-finished glass and faience objects. The site of Lisht has been established since the 12th Dynasty (1985–1956 BC), however the finds connected to the vitreous material production do not have precise archaeological context and, based on the typologies of finished objects, the glass-related activity at the site may span from 1295 BC to 1070 BC (Keller, 1983). Possible faience production has also been suggested in Kerma (north Sudan) due to the presence of quartzite pebbles and other remains related to faience production. No kiln or furnace has been identified (Reisner, 1923; Nicholson, 1998; Lacovara, 1998; Nicholson and Peltenburg, 2000). Confirmed evidence of faience production is known from the excavations of the New Kingdom sites in Malqata, Amarna and Qantir. The excavations at Malqata in the palace of Amenhotep III (18th Dynasty, 1390–1352 BC) yield scraps of faience, finished objects and ceramic moulds, some of them with the faience paste still pressed in the mould (Nicholson, 1998; Nicholson and Peltenburg, 2000). At the same site, the presence of a crucible with dark blue glass and glass rods may be indicative of the production of glass. Unfortunately, some of the rods are now lost and the production of glass at the site is still debated (a summary in Pusch and Rehren, 2007). In the site of Amarna the excavation started in the end of the 18th century with the works of Petrie (1894), and modern campaigns are progressing thanks to the numerous activities of Nicholson and others (Amarna Project website: www.amarnaproject.com/pages/publications/excavation.shtml). Remains of kilns and materials related to the production of faience, glass and Egyptian blue have been found (Nicholson, 1998, 2007; Nicholson and Peltenburg, 2000). During the excavation at Qantir, an incredible number of faience moulds and also Egyptian blue cakes, tiles and faience objects were discovered. For this reason the production of faience is suggested, although the kiln has not been found (Hamza, 1930; Nicholson and Peltenburg, 2000). The production of glass at Qantir is attested by the presence of firing crucibles, raw glass remains and glass objects (Pusch and Rehren, 2007). Considering the presence of finished objects in vitreous materials from Kom Medinet Ghurab, the possible production of faience and/or glass in the site has been suggested, but to date no evidence has been found of a firing structure or remains related to the production process (Nicholson, 1998; Nicholson and Peltenburg, 2000; Pusch and Rehren, 2007). In Memphis, excavation by Petrie discovered one of the most important sites for the production of faience in Greek-Roman time (Petrie, 1909, 1911). Six kilns were identified, probably related to the production of faience and Egyptian blue: lumps of the pigment were actually found in the site. The presence of glazed pottery also suggests that this material could have been produced at Memphis. A furnace dated to the Roman period was discovered in excavation at the area of Kóm Helul, in Memphis (Nicholson, 2002, 2013). In Naukratis, a Greek settlement dated to

Table 3. Main archaeological sites showing evidence of working or production activity of faience, Egyptian blue or glass in Bronze Age Egypt and Near East.

Area	Site	Age	Materials /activity	Presence of kiln	Selected references
Egypt	Abydos	Middle Old Kingdom – early Middle Kingdom	Faience production	Trace and possible remains of a kiln	Nicholson (1998); Nicholson and Peltenburg (2000)
Egypt	Lisht	Faience: Middle Kingdom (13th Dyn. 1750–1640 BC), Glass: 20th Dyn (1985–1956 BC)	Faience production Glass working and production	Remains of a kiln	Keller (1983); Nicholson (1998); Nicholson and Peltenburg (2000); Mass *et al.* (2002); Shortland and Eremin (2006); Shortland *et al.* (2007b); Smirniou *et al.* (2018)
North Sudan	Kerma	Middle Kingdom (1991–1640 BC)	Faience production	No conclusive evidence	Reisner (1923); Nicholson (1998); Lacovara (1998); Nicholson and Peltenburg (2000)
Egypt	Malqata (Thebes)	New Kingdom, Amenhotep III (18th Dyn., 1390–1352 BC)	Faience production. Glass working- Glass production?	No remains	Nicholson (1998); Nicholson and Peltenburg (2000); Mass *et al.* (2002); Shortland and Eremin (2006); Shortland *et al.* (2007b); Degryse *et al.* (2010)
Egypt	Amarna	New Kingdom, Akhenaten (18th Dyn., 1352–1336 BC)	Faience, Egyptian blue and glass production	Several structures are present	Petrie, (1894); Brill (1999); Nicholson (1998); Nicholson and Peltenburg (2000); Shortland (2000); Nicholson (2007); Henderson *et al.* (2010); Shortland and Eremin (2006); Shortland *et al.* (2007b); Degryse *et al.* (2010); Smirniou and Rehren (2011)

Egypt	Qantir	New Kingdom (1540–1070 BC)	Faience, Egyptian blue and glass production	No remains	Hamza (1930); Rehren (1997); Rehren and Pusch (2005); Pusch and Rehren (2007)
Egypt	Kom Medinet Ghurab	New Kingdom, Amenhotep III (1390–1352 BC) – Ramesess II (1279–1213 BC)	Faience production? Glass production?	No remains	Nicholson (1998); Nicholson and Peltenburg (2000); Pusch and Rehren (2007)
Egypt	Memphis	Greek – Roman	Faience and Egyptian blue production	Several kilns	Petrie (1909, 1911); Nicholson (2002, 2013)
Egypt	Naukratis	Greek – Roman	Faience production?	No remains	Petrie and Gardner (1886); Nicholson (1998)
Egypt	Buto	Greek – Roman	Faience production?	Base of the furnace	Nicholson (1998); Shortland and Tite (2005)
Syria	Tell Brak	15th – 14th century BC	Glass working Glass production?	No evidence	Oates *et al.* (1997); Brill (1999); Shortland and Eremin (2006); Shortland *et al.* (2007b); Degryse *et al.* (2010); Henderson (2010, 2013b)
Turkey	Alalakh, Tell Atchana	Late Bronze Age I (*ca.* 1500–1450 BC) – Late Bronze Age II (1400–1300/1200 BC)	Glass production and working	A kiln	Barag (1970, 1985); Bimson and Freestone (1985); Henderson (2013b); Dardeniz (2018)

the Late Period, Petrie reported the presence of a scarab factory possibly related to faience production (Petrie and Gardner, 1886). However, recent information about the materials and the work of Petrie is lacking, so that the production of faience at the site is unconfirmed (Nicholson, 1998).

Chemical analyses of a very large number of Egyptian faience pieces were performed by Kaczmarczyk and Hedges (1983) in non-invasive mode by X-ray fluorescence (XRF). The data are useful mainly for identifying the type of colouring agent and opacifier (discussed above). Chemical data measured on sections of the beads by SEM are reported by Shortland (2000); Tite and Shortland (2003); Shortland and Tite (2005); and Tite *et al.* (2007). The results were well summarized and discussed by Tite *et al.* (2008d). The analyses show that the flux used in faience production from the Middle Kingdom to the New Kingdom (18th Dynasty) is soda-rich plant ash, that was replaced by natron from the beginning of the 3rd Intermediate Period onwards. Concerning the source of silica, the level of Al and other elements, suggests that in the majority of cases quartz sand was used. In a small number of objects from Amarna, ground quartz pebbles were preferred. The Co ores used to colour the faience come from the Western Desert of Egypt, specifically from the Kharga and Dakhla Oases, where cobaltiferous alum ores are present. The origin of the Co ores may be identified in the glass and glaze by the high Al and Mn levels and by the presence of traces of Ni and Zn. All these elements are characteristic of the Co-ores from the Western Egyptian Desert. In contrast, an Eastern source of Co from Iran seems to be associated to a trace of As (Tite *et al.*, 2008d; Henderson, 2013b).

The technology of faience spread from the Near East and Egypt to the Mediterranean world around the end of the 3rd millennium BC or the beginning of the 2nd millennium BC. In the northern Caucasus, faience beads appear probably earlier than in Europe: cylindrical and tubular faience beads, white or coloured, are present at the end of the Catacomb Culture (~2600–2200 BC) and in the Lola Culture (~2000–1900 BC). A small set of Caucasian pearls has been analysed by wavelength dispersive spectroscopy (WDS) (Shortland *et al.*, 2007a). The results show a variable composition: some faience is soda-rich, some potash-rich and some has a peculiar composition. Cu is present as a colouring metal, whereas Co and Sb were not identified. Based on their chemistry, some of these beads are similar to Egyptian analogues, whereas others are probably made locally because of the peculiar composition (Shortland *et al.*, 2007a). Because of such early occurrences, it is suggested that the spread of the faience technique to Europe may have occurred through the Caucasus.

In Central Europe, small cylindrical, biconical, annular and segmented beads have been present since the very start of the Early Bronze Age (Bz A1), ~2200–1900 BC (Bellintani *et al.*, 2006b). During the beginning of 2nd millennium BC faience was present in the Carpathian basin (Angelini *et al.*, 2006), Switzerland (Henderson, 1993; Angelini and Olmeda, 2018), Poland (Robinson *et al.*, 2004), Holland, France (Gratuze *et al.*, 1998, 2013; Gratuze and Janssens, 2004; Gratuze and Billaud, 2014), northern Italy (Angelini *et al.*, 2006; Bellintani *et al.*, 2006b) and Great Britain (Aspinal *et al.*, 1972; Sheridan *et al.*, 2005). The general shape of the beads is similar, but local typological variations are present suggesting the existence of different production centres. The early analyses of European faience were done by neutron activation analysis, a non-invasive technique producing bulk chemical data. In this case the comparison between the composition of faience from different sites is problematic because the amorphous and crystalline part are analysed together. Nevertheless, Harding and Warren (1973) were able to distinguish some compositional ranges of trace metals that allowed the discrimination of British faience (with high Sn) from the Carpathian (with higher Co and Sb). The chemical SEM/EDS and WDS data available for faience found in central and western Europe have been collected and compared in Tite *et al.* (2008b). A mixed alkali flux was used in the production of all the analysed European faience, suggesting a rather different formulation from those used in Egypt and the Near East. The large Al content in the amorphous phase and mineralogical studies (when available) suggest the use of sand as the source of silica. Generally, the beads are copper-blue coloured. The analyses of minor and trace elements suggest the existence of many different production centres in Europe, but to date no hard archaeological evidence has been discovered.

The diffusion of faience throughout Europe decreased later, and virtually stopped at the end of the Early Bronze Age. A small group of ornaments in glassy faience that are considered a local production dated to the early phases of the Middle Bronze Age 1–2 (~1700/1650–1450 BC) have been found in northern and central Italy. They are characterized by the 'typical' European low magnesia/high potash (LMHK) composition (Table 2) produced using mixed alkali as flux. These glassy faience items are mainly conical buttons (Fig. 7c), a typology known only in the area (Angelini *et al.*, 2005; Bellintani *et al.*, 2006a,b). Both chemical and typological variations within this class of materials, and other glassy faience spread in the same area suggest that possibly two different production centres were active (Angelini, 2014).

Egyptian blue is another type of vitreous material sometimes confused with faience or glassy faience. Egyptian blue is a synthetic pigment composed mainly of cuprorivaite ($CaCuSi_4O_{10}$) that is responsible for the typical blue colour. The cuprorivaite crystals are invariably associated with unreacted or partially transformed quartz, cristobalite, wollastonite and an alkali-rich glass phase depending on the

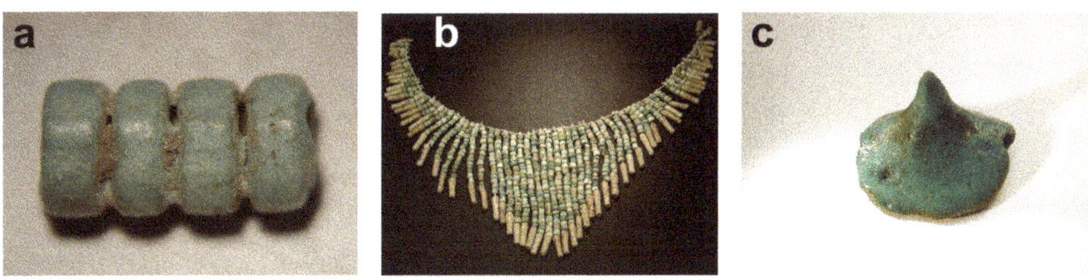

Figure. 7. (a) Segmented faience beads from the pile dwelling of Lavagnone (northern Italy), dated to the Early Bronze Age IA–IC (2100–1800 BC); (b) a faience collar from the cemetery of Nižná Myšl'a, Slovakia (about 1800–1600 BC); (c) glassy faience conical button from Mercurago, Middle Bronze Age 1–2 (~1700/1650–1450 BC).

production process. To obtain Egyptian blue, ground quartz, lime, copper compounds and small quantities of alkali were mixed and fired at a temperature in the range 850–1000°C (Tite *et al.*, 2008g). The dimensions of the crystals and the amount of the amorphous phase seem to be related to the different blue hues of the pigments (Tite *et al.*, 1984, 2008g; Henderson, 2013b).

Egyptian green (or green frit) is a pigment similar to Egyptian blue. It is synthetized using the same raw materials, although mixed in different proportions and fired under different condition with respect to the production of Egyptian blue. In particular, lime must exceed the Cu content in the production of Egyptian green, which is composed of large amounts of wollastonite and Cu-coloured glass, with residues of unreacted quartz (Tite *et al.*, 2008g).

Both Egyptian green and Egyptian blue were used mainly as pigments; however, Egyptian blue was also used in massive form in the production of small objects (Fig. 8). Egyptian blue was first produced in the early 3rd millennium BC in the Near East and in Egypt. The most ancient evidence of Egyptian blue in Egypt is from the painting of a tomb of the 1st Dynasty in Saqqara (~2900 BC), but its use became widespread only with the 4th Dynasty (2570–2450 BC). Egyptian green was used mainly in Egypt and had a very limited diffusion outside. The beginning of the use of Egyptian green is debated in the literature and spans from the First Intermediate Period to 18th Dynasty depending on the author (Tite *et al.*, 2008g). In the Near East, small beads from the Ur cemetery dated to the Early Dynastic III (~2600–2305 BC) are the oldest witnesses of

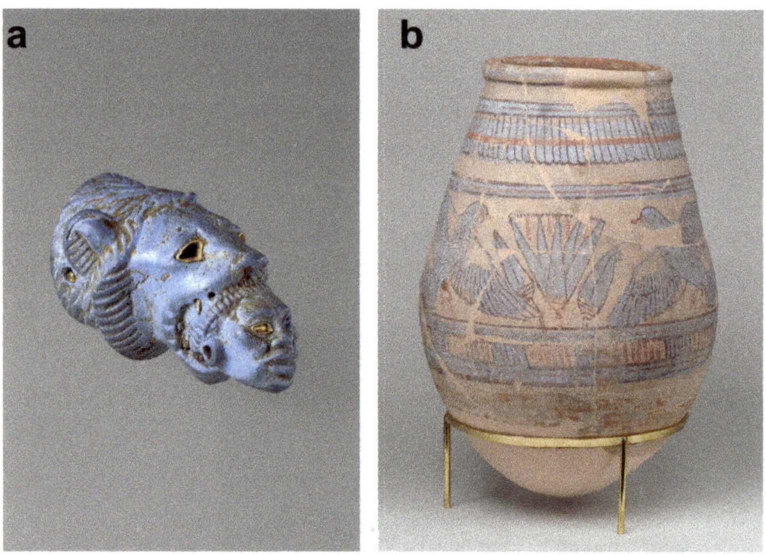

Figure 8. (a) Handle depicting a lion subduing a Nubian head, dated to the 19th Dynasty, reign of Ramesses II (*ca.* 1279–1213 BC). Egyptian blue, said to be from Eastern Delta, Qantir (Piramesse), Egypt; (b) jar painted with Egyptian blue from Malqata, New Kingdom, 18th Dynasty, reign of Amenhotep III (*ca.* 1390–1353 BC). [Images from the MET website].

the use of Egyptian blue. In the Near East therefore Egyptian blue was at the beginning employed for the production of objects and only later on as a pigment. Small ornaments in Egyptian blue are present in numerous sites around the middle of the 2nd millennium BC, such as Ugarit (Syria) and Nuzi and Tell Brak (Iran). The use of Egyptian blue for the production of small items continued in Niniveh and Nimrud at the beginning of the 1st millennium BC, and throughout the Achaemenid period (540–330 BC) in Iraq and Iran (Moorey, 1994; Tite *et al.*, 2008g). By the 2nd millennium BC the use of Egyptian blue spread in the Mediterranean area: small ornaments were found in Knossos (19th century BC) and in Middle Minoan–Late Minoan tombs in Crete (~1700–1200 BC). In Mycenaean Greece, Egyptian blue appeared only later, around 1400–1200 BC, but examples of Egyptian blue in wall paintings are known in Crete and in mainland Greece (Tite *et al.*, 2008g). Egyptian blue is of course well known in the Roman world, where it was traded as small balls and largely used as pigment. It was used extensively in the Pompeii frescos, where lumps of this pigment were found in a villa that was under restoration at the time of the Vesuvian eruption (Varone and Beart, 1996). Production centres of Egyptian blue are known in Egypt and were discussed before (see also Table 3).

The most ancient known glasses from the Near East and Syria are dated to the middle of the 3rd millennium BC, or later (Moorey, 1994). Based on linguistic analyses of cuneiform texts reporting instruction for glass production and the colouring process, Oppenheim (1970) suggested that glass was invented in northern Syria. In the Near East, some of the very early finds are: a pale-yellow bead dated to the early 3rd millennium BC from Tell Judeideh (Syria); a glass head pin from Nuzi (Iraq), ~2350–2150 BC; a chipped glass rod from Tell Asmar, pale blue-green in colour and dated to the 23rd century BC; a block of blue glass found in Eridu (Iraq) and dated approximately to 2300 BC. The analyses of the sample from Eridu show that the glass is coloured by Co and produced with plant ash flux and sand (Garner, 1956; Henderson, 2013b). After this early start glass production remained more or less inactive for ~1000 years, and the really important development in the production of glass started, both in the Near East and in Egypt, around 1500 BC. In the Mesopotamian area, new techniques of glass working were implemented, *e.g.* at the end of the 16th–15th century BC glass vessels were first made. Glass vessels have been found in the Near East in: Tell Brak (Syria), Chagar Bazar (Syria), where a polychrome vase of the 15th–14th century BC and a mosaic vessel of the 1350–1000 BC were found; Ashur and Niniveh (Assiria); Alalakh (Turkey); Nuzi, Assur, Tell el-Rimah (Mesopotamia, Iraq). Mosaic vessels were probably invented by Mesopotamian glass workers because they have a peculiar design typical of this area and age (~15th–13th century BC). Mosaic vessels are known from: Rimah, Aqar Qüf (Mesopotamia) and Marlik (Iran; Henderson, 2013b), although vessels are not the only glass objects found in these sites. Considering the presence of raw glass and other evidence that may be related to glass production, several sites need to be highlighted:

(1) Tell Brak (north Mesopotamia, Syria): glass vessels, mosaic glass vessels, polychrome beads, core-moulded glass bottles, ingots and fragments of

ingots mainly of pale blue glass. The finds are dated between the 15th century (vessels and beads) and the 14th century BC (ingots) (Oates *et al.*, 1997; Henderson, 2013b).

(2) Alalakh, Tell Atchana (Turkey): decorated plaques and vessels dated from the late 16th to the 13th century BC. The vessels include bottles, bowls and goblets, generally blue with white, opaque yellow and brown translucent decoration. Three pieces of raw glass, dated to the 15th century BC are also present (Barag, 1970; Henderson, 2013b). Two ingots are red glass coloured by cuprite crystals and covered with a green layer of weathered glass (Bimson and Freestone, 1985; Barag, 1985). In recent excavations numerous glass, faience objects, possible remains of crucibles and a kiln were discovered, proving the production activity at the site (Dardeniz, 2018). The workshop area is dated to Late Bronze Age I (~1500–1450 BC) - Late Bronze Age II (~1400–1300/1200 BC). Chemical and isotopic investigations show that the glass objects were made using plant ash glass, with a peculiar chemical fingerprint. Two objects were used for isotopic analysis; the results are different: the blue glass data plot in the isotopic field together with those from Tell Brak (Degryse *et al.*, 2010; Henderson *et al.*, 2010) and may be a signal characteristic of the Alalakh glass; the amber glass is close to a small group of glasses from Nuzi (Degryse *et al.*, 2010). As suggested also by Dardeniz (2018), more data are necessary for a better understanding of glass production and trade in the area.

(3) Assur (Iraq): many vessels were found at the site during old excavations, but the majority are still unpublished (Henderson, 2013b). Considering the typology of the vessels the context may be dated to the 15th–14th centuries BC (Barag, 1970) or to the 13th century BC (Moorey, 1994), depending on the author. A cullet of probable raw glass is also present. Particularly significant is the presence of numerous mosaic glasses. Due to the type of items discovered and to their large number, a local glass workshop or even a larger glass production has been suggested (Moorey, 1994).

(4) Ugarit, Ras Shamra (Syria): from the excavations of the site ~20,000 objects were discovered, mainly faience and glass, but also Egyptian blue and glazed pottery. A large variety of vessels, beads, pendants and other items was unearthed. Moreover, the presence of raw glass and ingots suggests that a production or working centre is present at the site (Matoïan, 2000; Henderson, 2013b). The site is dated to the 14th–12th centuries BC.

(5) Tell el-Rimah (Iraq): the excavations discovered a large quantity of vessels (also made of mosaic glass) that may be stratigraphically dated to 1350–1250 BC. Frit and faience manufactures were also found (Oates, 1968; Henderson, 2013b).

(6) Tell Ashar-Terqa (Iraq): at the site a large ingot of blue colour was found, ~5 kg in weight (Matoïan and Bouquillon, 2000; Henderson, 2013b).

(7) Nuzi (Iraq): an incredible quantity of glass was found in the site: beads, amulets, moulded objects, vessels, plaques and also chunks and chips of raw glass. The materials may be dated to ~1450–1350 BC (Barag, 1970; Henderson, 2013b).

Other sites (*e.g*. Ur and Nippur, Iraq and Tchoga Zanbil, Iran) have been suggested to have glass working centres, but with very little archaeological evidence. Other than at Alalakh, no remains of kilns have been found at these sites, so even if in some cases the glass working activity is confirmed, no archaeological evidence of glass production has been found. Glass has been found commonly at numerous Syrian-Palestinian sites and among them there are also Mesopotamian glass objects (Henderson, 2013b).

Another invention of Mesopotamian glass production is the core-forming vessel, that is the making of vessels by winding molten glass around a sand core supported by a rod. After forming, the object is removed from the rod and annealed. After annealing, the core is removed by scraping. The most ancient products of this kind were found in the late 16th–15th centuries BC and are soda-lime-silica glass fluxed with plant ash (Henderson *et al.*, 2010).

Chemical investigations of Near East glasses are available for several sites, the main set of data is related to: Nuzi, Tell Brak, Alalak, Nippur (Brill, 1999; Shortland and Eremin, 2006; Shortland *et al.*, 2007b; Henderson *et al.*, 2010; Degryse *et al.*, 2010; Walton *et al.*, 2012; Dardeniz, 2018). Brill (1999) also performed chemical analyses of the major and minor element contents of glass dated to Bronze Age and Iron Age from: Tell Al-Rimah, Tchoga Zanbil, Nimrud, Hasanlu, Marlik and Persepolis. The general composition of the glass from all these sites is similar, all the glasses are prepared using plant ash. The variation in minor elements and trace elements helps in the identification of similarities and differences. It is important to note that coeval Egyptian (Fig. 9) and Greek glass were also produced using plant ash, and the general composition is similar.

Chemical analyses of Egyptian glass are available for numerous samples from Amarna, Malkata, Lisht and Qantir (Rehren, 1997; Shortland, 2000; Mass *et al.*, 2002;

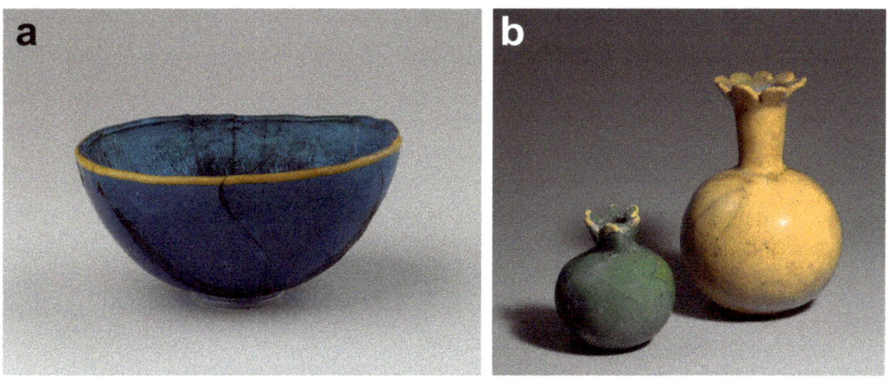

Figure 9. (a) Bowl in transparent blue glass from Egypt, New Kingdom, 19th–20th Dynasty (*ca*. 1295–1070 BC); (b) opaque glass bottles in the form of pomegranates, Egypt, New Kingdom, 19th–20th Dynasty (*ca*. 1295–1070 BC). [Images from the MET website].

Rehren and Pusch, 2005; Pusch and Rehren, 2007; Shortland and Eremin, 2006; Shortland *et al.*, 2007b; Jackson and Nicholson, 2007; Henderson *et al.*, 2010; Degryse *et al.*, 2010; Smirniou and Rehren, 2011, 2013). The production of glass in Quantir (Rehren and Pusch, 2005; Pusch and Rehren, 2007), Amarna (Smirniou and Rehren, 2011; Shortland and Eremin, 2006; Shortland *et al.* 2007b) and Lisht (Smirniou *et al.*, 2018) has been discussed and supported by in depth analysis of major-, minor- and trace-element contents. It is generally accepted that these sites are not simply workshops, but also important glass production centres. On a similar basis, it has been demonstrated recently that trace elements are able to distinguish Egyptian glass from Mesopotamian, mainly by the variations of Ti, Cr, Zr and La (Shortland *et al.*, 2007b; Walton *et al.*, 2012; Smirniou and Rehren, 2013; Dardeniz, 2018; Smirniou *et al.*, 2018). Also the colouring agents have been studied in detail, as for example the Co source in the blue coloured glass from Nippur is clearly different from the one used in the Egyptian Co-coloured glass (Walton *et al.*, 2012).

The early glass in Egypt appeared during the reign of Thutmose III (1479–1425 BC) (Lilyquist and Brill, 1993). The origin of the invention of glass is not certain, even if the presence of early glass in Mesopotamia, as discussed above, suggests that it originated in that area, or in northern Syria as proposed by Oppenheim (1970). Thutmose probably imported glass into Egypt for the first time at the beginning of his reign (Shortland, 2009). The identification of early glass production or working sites in Egypt is difficult. Confirmed glass working sites have been identified in Malqata, Amarna, Lisht and Quantir (Table 3), but recently primary glass production has been proposed for Amarna, Qatir and Lish (see above). In order to confirm whether glass production in Egypt and Mesopotamia was different, isotopic analyses were performed on samples from Tel Brak and Nuzi (Mesopotamia) and from Amarna and Malkata (Egypt) (Degryse *et al.*, 2010; Henderson *et al.*, 2010). The results show that the Egyptian glasses have a signal which is clearly different form those of the Near Eastern materials. The use of diverse raw material is responsible for this difference. Moreover the Mesopotamian data seem to group in three classes, suggesting the existence of several production centres. Based on trace-element analysis, a recent study demonstrated that two coloured rods from Amarna are actually made with Mesopotamian glass (Varberg *et al.*, 2016). Notably, these data seem to prove for the first time that even in an important Egyptian production centre such as Amarna, glass derived from trade with distant countries may be present and worked together with local products.

Glass was introduced in the Aegean during the 15th century BC and may be found both in Minoan Crete and in Mycenaean territories. Around 1450 BC the Mycenaean products dominate over those from Minoan Crete and their vitreous material products spread in the Aegean Sea and in mainland Greece. Both faience and glass are used, often associated with gold. Even if glass vessels are known, the vitreous materials were mainly used for jewellery. Blue glass is used to create beads and pendants with peculiar typologies. Decorated relief beads (Fig. 10) are crafted by moulding the glass and sometimes covering the bead with gold foil. During the three centuries of life of the Mycenaean culture, beads and plaques continued to be manufactured, not only in glass

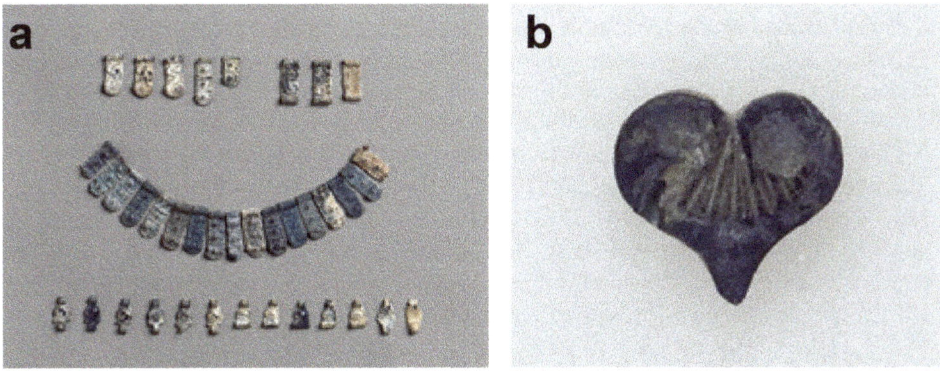

Figure 10. (a) Ornamental moulded plaques in blue glass, Late Helladic III (*ca.* 1370–1200 BC); (b) ivy-leaf-shaped ornament made with moulded glass. Late Helladic III (*ca.* 1370–1200 BC). [Images from the MET website].

and faience, but also in Egyptian blue and cobalt blue glassy faience (Panagiotaki *et al.*, 2004; Tite *et al.*, 2005). Due to the specific typologies of the items, the existence of local glass working centres is clear. But the possibility of the presence of local glass production has been debated in the literature for a long time, especially after the discovery of the Uluburun shipwreck. The shipwreck is located off the coast of Turkey (near Kas) and in its cargo incredible quantities of metal and glass were found, both as ingots and objects. Among the glass there were 175 ingots of different colours (mainly blue) and a Canaanite jar full of beads (Pulak, 1998). The early analyses performed on the glass ingots from Uluburun show a similarity with some Egyptian and generally also with other Mycenaean and Mesopotamian glasses (Brill, 1999). The final answer to the origin of the Uluburun glasses comes from recent works that analyse major and trace elements of the ingots (Jackson and Nicholson, 2010; Smirniou and Rehren, 2013). Comparison of Egyptian and Mesopotamian data has proved clearly the Egyptian origin of both Cu-coloured and Co-coloured glass. In a similar way Walton *et al.* (2009) analysed 11 Mycenaean beads (seven from unknown origin and four from Tiryns). The results show that six Co and Cu coloured beads are from Egypt, whereas the other five Cu-coloured ones are from Mesopotamia. In order to investigate the origin of the glass, isotopic analyses were also used. Henderson *et al.* (2010) proved the Egyptian origin of nine beads dated to the 13th and the 12th centuries BC from Tebe, Elateia and Atlanti-Spartia (Greece). Only one of the ornaments analysed originated from Mesopotamia: a bead from a cemetery in Athens with a "Nuzi" typology and dated to the 15th century BC (Henderson *et al.*, 2010). The Egyptian origin has been identified also for 11 Co-coloured beads from the Mycenaean tholos tomb of Kazanaki in eastern Thessaly (Smirniou *et al.*, 2012). Even though only a few Mycenaean glasses have been analysed to date , it is possible to draw a general picture: Cu-coloured glass seems to be imported both from Egypt and Mesopotamia, whereas the Co-coloured glass comes only from Egypt. The local production of glass in the Aegean and in Greece in the Bronze Age has still to be proven.

Early 'true glass' appears in south and central Italy in the Middle Bronze Age 1–2 (~1650–1450 BC) and are small beads obtained using plant ash as flux (HMG – high magnesia glass). They are mainly beads with simple typology (such as annular and globular beads) and arrive more or less at the same time as the arrival of glass in the Aegean and Greek world. It is not possible to say if the beads arrive by direct contact with the Egypt or the Levantine world or thanks to middle-men, such as Mycenaean or Aegean traders (Angelini *et al.*, 2005; Bellintani *et al.*, 2006b; Angelini, 2011; Artioli and Angelini, 2013). At the same age in northern Italy, particular LMHK glassy faience that is probably of local production is present (see above). In any case, it was only with the end of the Middle Bronze Age and the Recent Bronze Age (1450–1200 BC) that HMG glasses are became widespread throughout the whole Italian peninsula (Fig. 11a,b). They consist mainly of beads of blue glass, sometimes with white decoration and the general composition is comparable with all the coeval plant-ash glass from Egypt and Mesopotamia (Angelini *et al.*, 2005; Bellintani *et al.*, 2006b; Angelini, 2011; Artioli and Angelini, 2013). At the moment no trace-element analyses are available for these materials. The range of shape of the glass beads is now wider and many typologies are similar to those of Mycenaean and Levantine. Beside glass, glassy faience and some faience with typical Mycenaean, Egyptian and Levantine typologies are also present (Bellintani *et al.*, 2006b; Bellintani, 2011). The glassy faience of this age shows a variable composition: it has an amorphous phase with HMG glass, or a peculiar composition with low K, Ca and Mg, and some show a chemistry very similar to Mycenaean coeval glassy faience (Angelini *et al.*, 2005; Tite *et al.*, 2005). So, considering both the shape and composition, the glassy faience seems to prove contact with the Mycenaean world.

During the Bronze Age in central and northern Europe a large number of glass beads made with plant ash as flux (HMG glass) have been identified. As far as we know from archaeometric analyses, they are spread throughout France (Guilaine *et al.*, 1991; Gratuze *et al.*, 1998, 2013; Plouin *et a*l., 2012), in Denmark (Varberg *et al.*, 2015, 2016); Poland (Purowski *et al.*, 2014, 2018), Germany (Hartmann *et al.*, 1997; Varberg *et al.*, 2016; Mildner *et al.*, 2010, 2014, 2018) and Romania (Varberg *et al.*, 2015). The majority of the HMG glasses are dated between the 14th and the 13th century BC, but their appearance and disappearance seem to vary within the various regions. In Italy they are present only until the end of the 13th century BC, then are substituted completely by the LMHK glass (mixed alkali glass), a typical recipe of Final Bronze Age Europe (Bellintani *et al.*, 2006b; Angelini, 2011; Artioli and Angelini, 2013). Considering the analysed beads, Bronze Age HMG glass is reported from Denmark in1400–1100 BC (Varberg *et al.*, 2015, 2016), in Romania in the 14th–12th century BC (Varberg *et al.*, 2016), in Germany in 1500–1050 BC (Mildner *et al.*, 2014). In Poland early HMG beads are reported in the BzB (roughly 1600–1500 BC), but the majority are present in the BzC-HaA1 (~1500–1025 BC) (Purowski *et al.*, 2018). In France HMG glass is described from Ancient Bronze Age/Middle Bronze Age to the Bronze Final IIb-IIIa (~1600–1000 BC) (Guilaine *et al.*, 1991; Gratuze *et al.*, 1998; Plouin *et al.*, 2012), This seems to be the wider chronological range identified for this glass in a region.

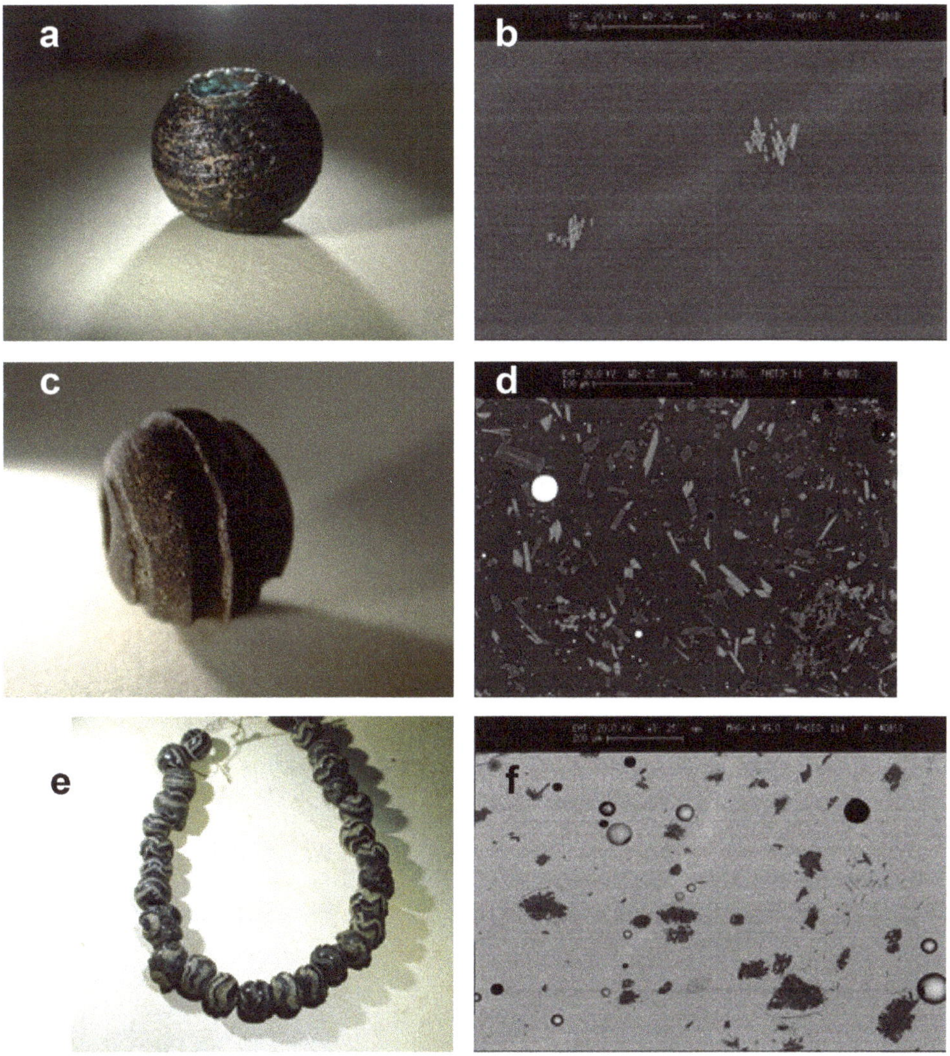

Figure 11. Examples of glass beads from Italy consisting of glass with the typical compositions present in the Middle and Final Bronze Age: (a) HMG blue bead dated to the 13th century BC, Franzine (VR); (b) SEM-BSE image of an HMG glass with the typical homogenous texture. Here in white a few devitrification crystals of Na-Ca silicate; (c) HMBG brown glass bead: a composition present only in northern Italy (age and origin as in a); (d) SEM-BSE image of an HMBG glass with devitrification crystals of diopside (pale grey), zoned augite (dark grey) and Cu sulfides (white); (e) LMHK beads with Co-blue body and white decorations from Bismantova (RE) – Final Bronze Age; (f) typical texture of LMHK glass rich in SiO_2 inclusions that may be quartz, trydimite or cristobalite (see Section 5).

In order to define a clear picture of the early use of HMG in Europe, their diffusion *via* possible trade routes, and especially to understand when HMG was replaced by LMHK glass, it will be important in the future to link the analytical data with a more detailed archaeological characterization of the finds in all regions. From what we know at the present time, it seems that in central and northern Europe HMG glass was in use for a longer time than in Italy and that there is a coexistence of HMG and LMHK glass for a greater range of time than in southern Europe. In Italy HMG glass was rapidly replaced by LMHK glass at the beginning of the 12th century BC. This was probably related to the existence in northeast Italy of working/production centres of LMHK, as discussed below.

Recently, trace element analysis has provided interesting new information about the possible origin of the European HMG glass. According to Varberg *et al.* (2015, 2016), the majority of the Danish, Romanian and German beads that they analysed are from Mesopotamia. Co-blue glasses have been identified both from Mesopotamia and in minor amounts from Egypt. Those authors also identified specific compositions that may be interpreted as a mix of Mesopotamian and Egyptian glasses. The HMG from Poland has been considered to originate in Mesopotamia too, except for one Co-coloured glass that is considered to be from Egypt (Purowski *et al.*, 2018). From these data, Mesopotamia seem to be the principal supplier of HMG glass to Europe, apart from the Co-coloured glass that may be somehow derived from Egypt. A different interpretation has been proposed by Mildner *et al.* (2010, 2014, 2018) both for their data and for the previously published data from Denmark. It has been noted that some samples from Germany have different Ti, Zr, La, Cr and Nd contents with respect to Egyptian and Mesopotamian glass. This group of glasses has a different and unknown origin. Many of the other glasses have contents of these elements that show a trend comparable with the Mesopotamian one, but with generally higher concentration of the elements. Also the quantities of P and B seem to differentiate the European HMG from the Near Eastern and Egyptian ones. Mildner *et al.* (2018) observed the presence of several HMG compositional groups, possibly associated with different production centres. Based on the analysis of later glass from Turkey, they suggested a probable location of some of these productions in Anatolia.

If raw imported glass was worked in Europe or if we take into account the possible recycling of the glass, it is clear that some elements may be derived from the interaction of the melt with the crucible and this may change the trace-element pattern of the original glass. Besides the Mycenaean contexts, the only known evidence of remains connected to the process of working HMG glass in Europe is from the Anzola *terramara* (northern Italy) and is dated to the 13th century BC. It consists of a pottery tray with an encrustation of a layer of HMG glass that has been analysed recently (Cupit *et al.*, 2018). Indirect evidence of possible glass production comes from a small group of brown beads spread throughout central and northern Italy (Fig. 11c,d). These beads are HMG glasses but with higher contents of Al, Fe and Ca than the blue ones. Moreover they are characterized by the presence of numerous spherical segregations of copper sulfides and devitrification crystals of diopside and augite. Similar types of beads with comparable chemical and textural features are unknown outside this area; therefore, up to now, they have been considered to have been produced somewhere in the area

(Angelini *et al.*, 2005; Angelini, 2011; Artioli and Angelini, 2013). Interestingly, such beads are also found in the beginning of the 12th century BC, a time when in Italy the production of another type of glass starts: the mixed alkali glass (LMHK, Fig. 11e,f).

In the Final Bronze Age, LMHK glass was widespread in Europe, used mainly for the productions of beads. Archaeometric investigation shows its presence in Italy (among other locations: Brill, 1992; Santopadre and Verità, 2000; Towle *et al.*, 2001; Angelini *et al.*, 2002, 2004, 2010; Conte *et al.*, 2019), France (Guilaine *et al.*, 1991; Gratuze *et al.* 1998, 2013; Gratuze and Billaud, 2014; Séguier *et al.*, 2010; Plouin *et al.*, 2012), Bohemia (Venclová *et al.*, 2011), Switzerland (Henderson, 1993; Angelini and Olmeda, 2018), Germany (Hartmann *et al.*, 1997; Lorenz, 2006; Mildner *et al.*, 2014, 2018), Poland (Purowski *et al.*, 2018), Greece (Nikita and Henderson, 2006), Ireland (Henderson, 1988) and also England (Paynter, 2014). The typologies of the beads are similar throughout the whole of Europe: annular, globular or barrel-shaped blue body, sometimes decorated with white points, a spiral or eyes (Bellintani *et al.*, 2006b; Nikita and Henderson, 2006; Bellintani and Stefan, 2009; Venclová *et al.*, 2011; Gratuze and Billaud, 2014; Purowski *et al.*, 2018). The blue colour is obtained from Cu and/or Co and the Co is found with associated traces of Ni and As. All the researchers that have analysed LMHK glass agree that the source of Co is different from those used in the Egyptian or Mesopotamian glasses, but possible provenance is still unknown. Despite the large diffusion of LMHK glass there are only a few sites with evidence of working and possible production activities and all are located in northeastern Italy, in the Veneto region. Among these, Frattesina is the most investigated and best known site. Analyses of glass from Frattesina have been done by several researchers (Brill, 1992; Santopadre and Verità, 2000; Towle *et al.*, 2001; Angelini *et al.*, 2004, 2010; Henderson *et al.*, 2015). Less known are other sites where scraps of coloured glass and glass wastes have been found: Mariconda di Melara (Italy), Montagnana (Italy) and Fondo Paviani (Italy). Moreover, in Frattesina and Mariconda, fragments of crucibles or ceramic trays with a glass layer on the surface are present. Considering also the large amount of glass finds, the LMHK glass in Frattesina and Mariconda was probably produced and not just worked (Towle *et al.*, 2001; Bellintani *et al.*, 2006b; Bellintani 2011; Angelini *et al.*, 2004, 2010; Artioli and Angelini, 2013; Angelini, 2019). Actually, in these two settlements the bead typologies (Bellintani and Stefan, 2009) and the variety of glass colour are wider than in the rest of Europe. In particular beads with red surface or with red decorations that are not spread in the rest of Europe were found (only a few exceptions are known; Angelini and Olmeda, 2018; Angelini, 2019). In the glass from Frattesina different techniques seem to have been used in order to obtain the same colour, *e.g.* based on the chemical and mineralogical features two types of white and two types of red glass have been identified (Angelini *et al.*, 2004, 2010; Angelini, 2019).

Beside the LMHK glass, potash glass is present in small quantities in Frattesina and in other sites in Italy, but also in numerous sites in France, Germany, Bohemia and England (Towle *et al.*, 2001; Angelini *et al.*, 2004, 2010; Séguier *et al.*, 2010; Venclová *et al.*, 2011; Plouin *et al.*, 2012; Gratuze and Billaud, 2014; Mildner *et al.*, 2014, 2018; Conte *et al.*, 2019). The type of flux used in this glass is not known nor is their origin clear.

Isotopic analyses have been done on a few LMHK and one potash glass from Frattesina (Henderson *et al.*, 2015). The two type of glass present diverse isotopic ratios suggesting the use of different sands in their production. Henderson *et al.* suggest that one source is compatible with local sand and the second one is possibly located in central Italy. Actually, there is no evidence of glass production in central Italy and the finds of LMHK glass in the area are scarce. As also noted by those authors, there are too few data for a definitive answer. The database of mineralogical, chemical and isotopic data of fluvial sand also needs to be improved for proper comparison with the archaeometric data.

2.2. Iron Age

Since the beginning of the Iron Age glass production was characterized by the change in the flux used: plant ash was replaced by a soda mineral source: natron. Natron is an evaporitic rock with variable composition, consisting mainly of: (1) sodium carbonate and bicarbonate essentially in the form of trona ($Na_2CO_3.NaHCO_3.2H_2O$) or natron ($Na_2CO_3.10H_2O$), and with lesser amounts of nahcolite ($NaHCO_3$) or thermonatrite ($Na_2CO_3{\cdot}H_2O$); (2) mixed Na carbonate and sulfate, *e.g.* burkeite ($Na_6CO_3.2SO_4$); (3) Na sulfate as thenardite (Na_2SO_4) and mirabilite ($Na_2SO_4{\cdot}10H_2O$); (4) and Na chloride, halite (NaCl). Potassium may also be present as a substitute for Na. The composition of natron may vary depending on the geographical origin, the season of collection, and the weather and changes in hydrography through time (Shortland 2004; Shortland *et al.*, 2006, 2011). The best natron for the production of glass has to be rich in carbonate and bicarbonate and poor in chloride and sulfate. It is important therefore to know not only the origin of the natron (related to the composition), but also how the deposits were exploited. It has been suggested that some treatment of the "raw" natron was probably applied also in order to increase the carbonate content and reduce the sulfate content (Shortland *et al.*, 2011). Experimental and analytical work confirms the exploitation in antiquity of Wadi Natrum (northwest of Cairo) and al-Barnuj (Western Nile Delta), both in Egypt. These, and especially Wadi Natrum are considered the major sources of natron in pre-Roman times onwards (Shortland *et al.*, 2006). The exploitation of Wadi Natrun in Roman times is testified by the remains of large glass-making factories located close to some lakes, established in the area before the 1st century BC (Nenna, 2003, 2007). Nevertheless other areas are reported in the historical literature as sources of natron (in particular in the *Natural History*, *book XXXI*, by Pliny the Elder): a salt lake near al-Jabbul (northern Syria); Lake Van in Armenia; and Lake Pikrolimni in Macedonia (Shortland, 2004; Shortland *et al.*, 2006). These lakes are known to have been exploited in different historic times, but no reliable information is available for ancient times. Recent analyses of evaporite samples from Lake Pikrolimni (Macedonia) show its compatibility with its use in glass production (Dotsika *et al.*, 2009). This has to be considered in the study of Hellenistic glass production.

Natron has been used in the production of vitreous material since the 4th millennium BC. In the Badarian Culture, blue glazed steatite was produced by a mixture of natron and Cu applied to the steatite body (Tite *et al.*, 2008c). Then the use of natron in the vitreous production seems to have been ignored up to the beginning of the 1st millennium BC when natron glass appeared.

Natron glass is also known as LMG (Low Magnesium Glass) for its small Mg content compared to plant ash glass (Table 2). Compositionally, LMG shows high levels of Na, low K and trace of Cl and S, chemical characteristics derived from the use of natron. LMG glass is present in Egypt, Near East, Europe and generally in the entire Mediterranean area. In the Near East, especially in Mesopotamia, the traditional plant ash glass continued to be produced in the Iron Age (Shortland *et al.*, 2006; Henderson, 2013b).

The early known natron glass dates from the 10th–9th centuries BC, but it is only from 800 BC that it is systematically produced in large quantities. The most ancient confirmed date for LMG glass is a core-formed vessel from the tomb of Nesikhons, in Egypt dated to the 10th century BC (Schlick-Nolte and Werthmann, 2003). LMG Co-coloured glasses come from 8th–9th century BC Nimrud and prove the early use of natron in the Near East, even if plant ash glass is also present at the site (Reade *et al.*, 2005). Early natron glasses from Nimrud and Egypt show such a small amount of lime that their stability and durability are seriously compromised. Ca and Mg stabilize the glass structure (Section 1), but in the LMG glass the low Mg introduced with the flux needs to be balanced by higher Ca in order to prevent the weathering and 'dissolution' of glass. If the sand used does not have the proper quantity of shells and/or lime is not voluntarily added, the resulting glass may be easily degraded with time. As suggested by some authors (Sayre and Smith, 1961; Reade *et al.*, 2005; Shortland *et al.*, 2006) this may be the reason for the lack in the archaeological record of a large amount of glass in the early two centuries of the first millennium BC, both in Egypt and the Near East. The spread of LMG is rapid and LMG glass is known in Europe as early as the 9th century BC (see Section 5). From the 6th–5th centuries BC natron is "the flux used west of the Euphrates in the great majority of glass" (Shortland *et al.*, 2006). In the east natron did not displace totally plant ash and during the Iron Age HMG glass are still present in Mesopotamia, Iran and Central Asia. Whether the plant ash glass had such a long life in the east due to difficulty in the supply of natron in the area or because the glassmakers preferred to comply with the old and traditional glass-making recipes is a matter of debate (Shortland *et al.*, 2006).

From the 7th–6th centuries BC to Early Roman time the variety of beads, pendants, arming, moulded vessels, core-formed vases and bottles that circulate in the Mediterranean area, Egypt, Near East and Europe is extremely wide and also the variety of colour and shapes increased significantly with respect to the Bronze Age.

Among this large variety of glass objects the rod-formed head pendants (Fig. 12) merit mention because of how widespread they are and the significant aesthetic and technological values. The head pendants have a Phoenician style but are present largely in the Mediterranean world also in non-Phoenician sites. There is no archaeological evidence of the pendants' origin and the existence of Phoenician production is not confirmed. Beside Phoenicia, Rhodes and Carthage have also been suggested as production provenance. The Phoenicians play an important role in the spreading of

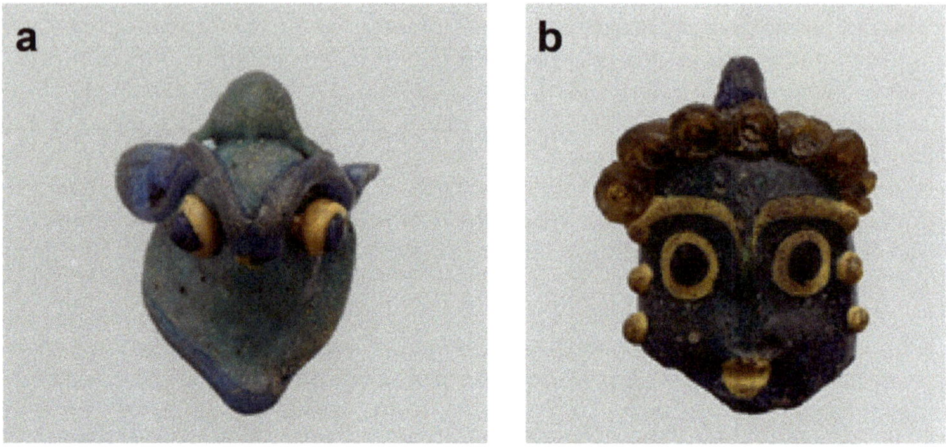

Figure 12. (a) Rod-formed glass pendant in the form of a demonic mask (type 1 in the text). Phoenician style dated to the 6th–5th century BC; (b) rod-formed glass pendant with an elaborate male face (type 4 in the text). Phoenician or Carthaginian, 5th century BC. [Images from the MET website].

glass throughout the Mediterranean (Spaer, 2001). The head pendants are known between the 7th/6th centuries BC to the 4th/3rd centuries BC but the different types have specific distribution zones and ages. The rod-formed head pendants are classified in four main group: (1) demon mask or grotesque; (2) double-faced cylindrical male heads; (3) single-faced cylindrical male heads; and (4) elaborate heads, male and female (Spaer, 2001). Only a few head pendants have been analysed to date: two are known from Mozia (Sicily) and are LMG glass with blue body coloured by traces of Co and large amounts of Fe (Arletti *et al.*, 2012). The yellow decoration is composed of LMG-Pb antimonate glass. A systematic archaeometric study of these wonderful ornaments is still lacking.

The production of core-formed vases, known since the Bronze Age (Section 2) was revitalized between the 6th and the 1st century BC when small polychrome vases were one of the most widespread glass objects throughout the entire Mediterranean area. According to their shape and decorative style they are divided in three groups, labelled Mediterranean groups I, II and III. The studies of these classes of objects are numerous and according to some of the more quoted works (Barag, 1970; Harden 1981; McClelland, 1984; Grose, 1989; Stern and Schlick-Nolte, 1994; Triantafyllidis, 2003), the characteristics of the Mediterranean Groups may be summarized as follows:

(1) Group I (late 6th century–mid-4th century BC; Fig. 13a). This comprises four main types of vase: *aryballoi*, *oinochoai*, *alabastra* and *amphoriskoi*. It is the largest of the three groups and is characterized by homogeneous production. The body is generally dark blue, blue or opaque white and they are decorated by threads in yellow, turquoise, purple or white opaque glass.

(2) Group II (mid-4th–late 3rd centuries BC; Fig. 13b). The same type of vases present in Group I were also produced in Group II, with stylistic

Figure 13. (a) Core-formed vessel of the Mediterranean Group I; Greek, Eastern Mediterranean, late 6th–5th century BC. From left to right: an *aryballos*, an *alabastron*, an *amphoriskos* and one *oinochoe*. (b) Hellenistic core-formed *hydriske* belonging to the Mediterranean Group II, late 4th–3rd century BC. (c) Hellenistic core-formed *alabastron* of the Mediterranean Group III, 2nd –mid-1st century BC. [Images from the MET website].

differentiation. In Group II, new shapes were created, such as *hydriai*. These vases usually have, in general, a dark blue body with white and yellow decoration. Some authors note the existence of a chronological gap of about two generations between Group I and Group II (McClelland, 1984; Grose, 1989; Henderson, 2013b). The beginning of production of Group II is more or less coeval with the conquest of the Persian Empire by Alexander the Great.

(3) Group III (mid-2nd century BC–early 1st century AD; Fig. 13c). The types of vessel produced was reduced, *alabastra* and *amphoriskoi* being used mainly. The body is generally made with translucent blue or green glass, decorated with opaque yellow, white, blue and green threads.

Over the past ten years numerous chemical analyses of major, minor and trace elements have been done so that a good set of data is now available in the literature. The finds analysed are: Mediterranean Group I and II vases from: Italy (Arletti *et al.*, 2011a, 2012; Panighello *et al.*, 2012; Oikonomou *et al.*, 2018); Rhodes, Greece (Triantafyllidis *et al.*, 2012); mainland Greece (Blomme *et al.*, 2017; Oikonomou, 2018) and Georgia (Shortland and Schroeder, 2009). The glass used for the body and decoration of the vases is always LMG glass produced using sand and natron. The range of colourants and opacifiers used is usually the same as used in the previous glass production tradition: Pb antimonate, Ca antimonate, Cu and Co minerals and Mn. Considering the results of the archaeometric studies, both major and trace elements seem to form different chemical groups that provide links between finds from different regions. The authors still debate the origin of the production, but generally it has been suggested that the Mediterranean Group I, II and III vessels may be produced in Rhodes, southern Italy and the Syro-Palestinian (or Syro-Cypriot) regions, respectively (Harden, 1981; McClelan, 1984; Grose, 1989; Oikonomou, 2018). Nevertheless, isotopic study performed on some Group I vessels from Pydna and Methoni (northwestern Greece) suggest an origin from the Syro-Palestinian coast, at least of the raw glass (Blomme *et al.*, 2017). To date no confirmed evidence of primary or secondary production sites are known for these classes of objects.

Hellenistic glass production flourished between the 3rd and 1st centuries BC. The core-formed vessels of the Mediterranean Groups II and III are also considered to be Hellenistic productions. Besides the core-formed vessels, an incredible variety of moulded vessels was produced during the Hellenistic time using the slumping and casting techniques (Fig. 14). A variety of colours and decoration motifs was created, such as the mosaic vessels, gold sandwich glass, *millefiori* glass, lacework or network glass. Based on historical texts and archaeological evidence, production centres of Hellenistic vases are said to be present in Rhodes, Sidon (or generally on the Syro-Palestinian coast), Alexandria (Egypt), Tyre (Lebanon) and southern Italy (Henderson, 2013b). Once again, no confirmed archaeological data have been found to prove the origin of these Hellenistic vessels.

Chemical analyses of Hellenistic glasses prove that they are composed of natron glass with Sb used as decolourant (Henderson, 2013b, and literature quoted therein). Sb is employed to obtain colourless glass from the 7th century BC to the 1st century BC. Mn started to replace Sb as a decolourant from the 2nd century BC. A large set of Hellenistic mosaic vessel fragments from nine sites in France, Italy, Greece and from Tebtynis in Egypt have been analysed recently using LA-ICP-MS in order to investigate the composition of the glass and compare the materials from different sites. In their summary, the authors outlined some interesting preliminary results (Nenna and Gratuze, 2009). The majority of the glasses are LMG, but plant ash glasses are also present. In some cases, the composition seems to present a homogeneity within materials from the same site, such as the mosaic glass fragments from Delos. Generally, a correlation between the composition and the colour is observed: red and orange glass are mainly plant ash glass, and ~1/4 or1/3 of green and amber colour glass are also produced with plant ash. With a few exceptions, the other glasses are usually natron glass.

During the Hellenistic phase new capabilities in glass-working and colouring techniques were developed. This is evident in the production of mosaic glass and in the

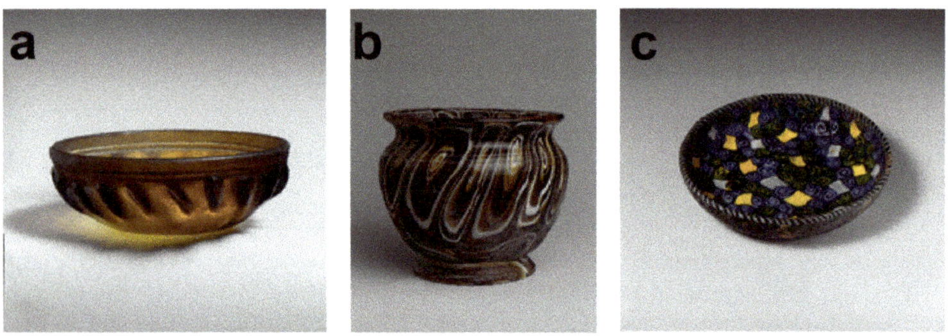

Figure 14. (a) Glass-ribbed bowl cast, tooled and cut. Late Hellenistic, 1st century BC; Greek, Eastern Mediterranean; (b) glass mosaic jar cast and cut ('agate' glass). Hellenistic, 2nd–early 1st century BC; Greek, probably Eastern Mediterranean; (c) cast glass mosaic dish. Hellenistic, 2nd–mid-1st century BC; Greek, probably Eastern Mediterranean. [Images from the MET website].

creation of a large variety of inlays applied to wooden furniture and architectural elements (Fig. 15). In addition, the glass inlays are often made of mosaic glass. The glass inlays are also referred to as enamels by numerous authors. In a general sense 'enamel' is used for any type of glass applied on a body made with different materials, such as metals (as in Byzantine enamel), or wooden (as in the Egyptian furniture) or even plasters or stone. The inlay is glass, often opacified and coloured. Glass inlay appears in Egypt as early as the New Kingdom, but it is in the Ptolemaic and Early Roman time that their production is developed strongly. Secondary workshops related to the activities of temples are present along the whole Nile. In a recent study, new data from the inlay workshop of Tebtynis (Fayum, Egypt) have been obtained by a review of the literature, a study of the unpublished excavation reports and thanks to archaeometric analysis of ~130 glasses (Bettineschi, 2018; Bettineschi *et al.*, 2019a,b). In Egypt, secondary workshops are present: between the 5th and the 4th centuries BC in Ayn Manawir; in Ptolemaic and probably Early Roman time in Tebtynis, Gumaiyima, Tanis, Soknopaiou Nesos, Karnak and Dyonisias; after the 2nd century AD in Apollinopolis Magna; and in Late Antiquity (and some sites also in later time) in Alexandria, Akoris, Antinopolis and Hermothis (Bettineschi, 2018; Bettineschi *et al.*, 2019a). The site of Tebtynis is particularly interesting because during the excavation in 1931, besides the glass inlay and finished objects, raw glass, glass cans, ceramic moulds, small ceramic trays, weights, and bronze tools were also found in the workshop. Moreover, the presence of a furnace has been properly located thanks to the old excavation diary and satellite images of the sites (Bettineschi *et al.*, 2019b). Similar to what was observed by Nenna and Gratuze (2009) the glasses are both plant-ash and natron based. The red and the orange colours are mainly plant ash-lead

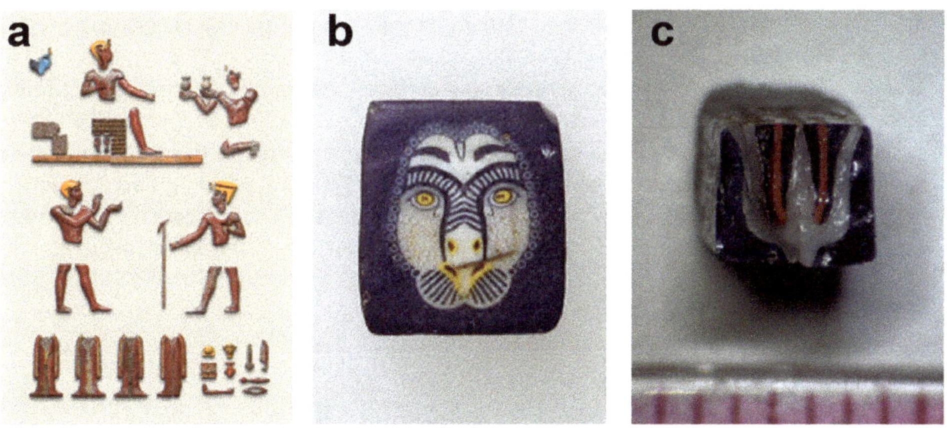

Figure 15. (a) Inlays and shrine elements from Egypt, Late Period–Ptolemaic Period, (~380–30 BC); (b) mosaic glass inlay probably Egyptian, Late Hellenistic or Early Imperial (~1st century BC–1st century AD); (c) 3D micro-photography of a small mosaic cane slice from Tebtynis with millimetric scale, Ptolemaic [(a) and (b) images from the Metropolitan Museum of Art website; (c): image courtesy of C. Bettineschi, object Museo Egizio, Torino, Tebtynis collection].

glass coloured by cuprite and/or Cu metallic segregation. HMG glasses are observed in some green glasses and in a few blue ones; however, they are generally natron glass. The colouring techniques used are a mix of the recipes related to the old glass-making traditions and the new knowledge and techniques that were to be used in Roman times. For example, yellow glasses are coloured both by the use of Pb antimonate or by the use of very small cuprite crystals (Bettineschi, 2018).

Glass production was revolutionized completely at the end of the 1st century BC when the Hellenistic phase faded and glass blowing was discovered.

2.3. Antiquity and Late Antiquity

At the end of the Iron Age, natron glass continued to prevail. As the volume of glass recovered, taken as indicative of the volume of glass produced, increased tremendously, large groups of glass with more homogeneous compositions were observed. The discovery of glass blowing in the 1st century BC is possibly linked to the rapid increase in glass production capacity (Israeli, 1991; Stern, 1999; Cummings, 2002). However, it is still possible to distinguish fairly distinct groups among the glasses named as Syro-Palestinian or Roman and a diachronic evolution of the composition is observed during Antiquity (Grose, 1979; Jackson-Tal, 2004; Henderson, 2013b).

A large number of secondary glass workshops have been unearthed in western Europe . Thanks to the excavation of different shipwrecks in the western Mediterranean Sea (Les Embiez, Cap Corse, Julia Felix; see for example: Fontaine and Foy, 2007; Silvestri *et al.*, 2008) and of several primary glass workshops in the Near East (Freestone *et al.*, 2000; Freestone, 2006; Rosenow and Rehren, 2014), the supply of these European secondary glass workshops is well documented. It has thus been shown that finished objects as well as raw glass and recycled glass wastes were exported from the Near-East towards western Europe (Baxter *et al.*, 2005; Freestone, 2015; Foy, 2018). From the Augustan Age to Late Antiquity, different groups of glass were identified and named according to the raw materials used or by the decolourizing recipes (Table 2). In addition to these well defined groups the presence of mixed glasses resulting from the recycling of the prevailing groups is also observed. The recycling and the mixing of the different glass groups are highlighted particularly by secondary glass workshop studies where analysis reveals the presence of pure glass groups but also of mixed glass (Foster and Jackson, 2009; Jackson and Paynter, 2016; Schibille *et al.*, 2017).

From a recipes point of view, starting from the Augustan period, blue-green un-decoloured glass, manganese decoloured glass and antimony decoloured glass are encountered (Foy *et al.*, 2004; Jackson, 2005; Gliozzo, 2017). In the same way, during the 4th century, antimony was replaced progressively by tin as the main opacifier for white and yellow opaque glass. Consider the raw materials used, and considering that sand is characterized mainly by the contents of alumina, lime, titanium and iron, the main recognized groups of glass are mainly: Syro-Palestinian, Levantine I and II, or Egyptian HIMT I and II (for high iron, magnesium and titanium) (Freestone *et al.*, 2000; 2005; Rosenow and Rehren, 2014). According to recent isotopic studies all these glass objects were produced in the Near East, from the Syrian coast to the Egyptian coast (Degryse and Shortland, 2009; Brems and Degryse, 2014). For that period other specific geographic groups of natron glass of less importance have also been recognized such as Egypt I and II, as well as the continuous presence of plant-ash glass in some Roman glass production. Analytical work carried out on Sasanian glass attests to the continuous production of plant-ash glass to the east of the Euphrates during that period (Mirti *et al.*, 2008; Rehren and Freestone, 2015).

At the end of Late Antiquity, an intensification of recycling is observed, resulting in increasing concentrations of copper, lead, tin and antimony in the glass composition. Some new groups of glass such as the high-boron high-alumina glass identified in Turkey at Pergamon also appear (Schibille, 2011; Swan *et al.*, 2018).

2.4. Early Middle Ages and later

Depending on the different regions, the production of natron glass in the Near East and its importation into Europe seems to break off progressively from the 9th to the 11th or 13th centuries AD. In the Near East, glasses produced from soda coastal plant ash replace natron glass (Phelps *et al.*, 2016). In western Europe, the end of the first millennium was also a period of transition resulting in the emergence of a new glass production period: Near Eastern imported glass is partially replaced by new locally produced glass types, containing potash, lime or lead. Thus, in the Mediterranean world during the 9th century a radical change is observed in glass production recipes resulting in the systematic use of plant ash instead of natron (Gratuze and Barrandon, 1990).

In the course of the 9th century, in the eastern Mediterranean and Mesopotamia, soda lime glass, produced from the ashes of halophytic plants, became the prevailing type of glass. This recipe was adopted all around the Mediterranean and became predominant before the end of the 12th century (Henderson, 2013b).

At the same period in western continental Europe, the development of the production of potash-lime glass using forest plant ashes was observed (Hartmann, 1994; Hartmann and Grünewald, 2010). The main fluxing agent is no longer soda but a mixture in variable proportions of potash, lime containing large amounts of magnesia, phosphorus and also manganese (Jackson and Smedley, 2004, 2008). However, although these types of glass became the prevailing types in continental Europe in the 10th century, the supply to some glass workshops of natron glass seems to have persisted until the end of the 12th century as is highlighted by some typical glass productions (Pactat *et al.*, 2017), *e.g.* cobalt blue stained glass or cobalt blue glass vessels such as the bowl of Saint-Savin (Simon-Hiernard and Gratuze, 2011).

The existence of another type of glass in which the calc-alkaline flux is partly or totally replaced by lead oxide must also be pointed out (Table 2). These types of glass start to develop from the 7th century AD, in the eastern part of Europe (Wedepohl *et al.*, 1995), and form two main families:

(1) The lead glasses, strictly speaking, made only from lead and sand, characterized by a lead oxide content generally $\geqslant 60\%$, and the absence of alkaline elements.

(2) The lead-alkali glasses, made from a ternary mixture of lead oxide, sand and ash. Among them can be distinguished lead-sodium glass, lead-potassium glass and glass mixed lead-soda-potash.

Until the 12th century, lead glass seems to have been distributed mainly in eastern Europe (Poland, Russia) and in the Caucasus region. Within Fatimid glass, around the late 10th and early 11th centuries, an emerald green glass category is also made up of >60% lead (Henderson *et al.*, 2004; Henderson, 2013b). But it is not known whether there are links between the productions of the Christian and Muslim worlds.

2.5. Modern period

Between the 12th and 15th centuries, plant ash soda glass characterizes the Mediterranean coastal regions while forest plant ash glass is produced in continental Europe. Since the beginning of the Modern period, sodium plant ash glass, which characterizes the Venetian productions (*e.g.* Common Glass, Vitrum Blanchum and Cristallo) were then adopted gradually throughout Europe. Its use is well illustrated by the production and spread of façon-de-Venise glass vessels (Šmit *et al.*, 2004). However, several other different glass types were still in use or under development in Europe during the same period. If, until the Modern period, glass compositions were linked mostly to the geographical location of the glass workshops rather than to the types of production, a progressive specialization of glass compositions according to their uses is observed. Typical glass types of the period are the HLLA glass (for High Lime Low Alkali) used mainly for window panes (Schalm *et al.*, 2007; Dungworth, 2012; Kunicki-Goldfinger *et al.*, 2014); a high-alumina HLLA glass was used for bottle production from the 17th century (Dungworth, 2003; Dungworth *et al.*, 2006).

From the end of the 16th century many innovations were developed in the glass industry. The main ones were probably related to the development of lead glass known as crystal or flint glass and to the progressive use of coal fuel. But a lot of less well known innovations were developed such as the use of kelp ash or of basalt as ingredients in glass manufacturing and at the end of the eighteenth century the synthesis of artificial soda which replaced plant ashes.

2.6. Notes on Indian and Chinese glasses

This section provides a rapid overview of Indian Ocean glass compositions compared to those circulating in the Mediterranean world. Note that in contrast with the variety of glass finds recovered from Mediterranean archaeological sites, glass material recovered from early South Asian archaeological sites consist mainly of beads, bangles and other ornaments. Glass vessels are rare and often identified as imported. Based on pioneering work by Dussubieux (2001) and others, Lankton and Dussubieux (2006) proposed a basic scheme for the compositions, based on major and minor elements, of early glasses found in the maritime trade from India to southern China. Although high-alumina soda glasses seemed to be characteristic of Indian productions, many other different glass types were identified in south Asia and southeast Asia. More recent data can be found in Lankton and Dussubieux (2013) and Dussubieux and Gratuze (2013). Fuxi *et al.* (2009, 2016) provide novel data on Chinese materials. Among the most famous and peculiar are the typical lead-barium Chinese glasses (Table 2). Potash glasses are also found commonly at early southeastern Asian sites and it is possible to suggest that at least three types of potash glass, based on levels of lime and alumina were produced. Different types of mixed alkali glass have also been identified. More recently, the availability of high-quality trace element analyses using LA-ICPMS has led to the identification of at least five groups of Asian alumina soda glasses based on levels of uranium and cesium (Dussubieux *et al.*, 2010; Dussubieux and Gratuze, 2013). Using these trace elements, it is possible to distinguish between north Indian production (high uranium, high cesium) and south Indian production (moderate uranium, low cesium). For potash glass, the situation is less clear, although the ratio of rubidium to strontium does appear to help separate groups in the low-lime range. From a chronological point of view, a diachronic evolution model of the composition, beginning with potash glass and ending with alumina soda glass can be sketched. The recent discovery of glass ingots in the Godavaya shipwreck, dated between the 1st century BC and the 1st century AD, has provided physical evidence for the early exchange of raw glass in south Asia (Muthucumarana *et al.*, 2014). The analytical results show that the glass in the ingots belongs to the m-Na-Al1 type defined by Dussubieux (2001). Its trace element contents appear to share many chemical features with glasses found or produced on the Tamil coast of south India. These samples help open new doors to the study of glass exchange and production in these regions.

3. Glass-making transition at the beginning of the 1st millenium BC

The revolution in the glass-making recipes occurred at the beginning of the Iron Age: the introduction of natron as flux, as described previously. This change started in the 10th century BC and involved the whole Egyptian, European, Aegean, Levantine and Near Eastern world. The questions of how and when the introduction of natron occurred, and what type of glass compositions were present at the same time vary region by region, especially in the early Iron Age. General trends may be established, as discussed above (Section 2.2), but in some geographical areas the chemical composition and the shape typologies of the glass are so variable that many questions about the origins and the production of some glass recipes are still far from being understood. Europe is one area where variations in the glass composition in the early Iron Age are particularly complex. During the Final Bronze Age up to its end (beginning at or near the end of the 9th century BC depending on the zone) the glass had essentially LMHK composition. In Italy, the mixed alkali glasses disappear in the early Iron Age: only a few beads are known from the north (Golasecca Culture, Angelini *et al.*, 2011; and Villanovan Culture, Arletti *et al.*, 2011b) and only one in the south (Conte *et al.*, 2019). In Poland, LMHK glass are present from HaA2 to HaB2 (about end of the 12th–middle of the 9th century BC), but only a few are known from the HaB3 and none is dated to the HaC, so from the 8th century BC they ceased to exist (Purowski *et al.*, 2018). The early Iron Age LMHK beads are few in number and spread over a large geographical area, therefore they are more likely to be related to reuse of older materials than to local production. The only exception seems to be the 9th–7th centuries BC LMHK glass from Rathgall (Ireland) where a significant number of beads is present (Henderson, 1988).

In early Iron Age Europe, LMG is not the only class of glass that appeared with respect to the Final Bronze Age production; HMG glass also began to be present. Plant-ash glasses re-appear, because they are typical in Europe during the 14th–13th centuries BC (Section 2.1), but with a slightly different chemistry. In Poland HMG are present in the early and full Iron Age, in the HaC–beginning HaD (~8th–end of the 6th century BC), and the authors do not report significant differences from the oldest materials (Purowski *et al.*, 2014, 2018). The oldest HMG in Poland have been interpreted mainly as having a Mesopotamian origin and a few with an Egyptian origin. The few differences in the chemistry between the Polish HMG glass and the Egyptian and Mesopotamian ones are interpreted as variations introduced during the recycling or working process (Purowski *et al.*, 2018). On the contrary, the HMG from Germany dated to 800–750 BC show a composition different from the oldest German HMG, in particular the trace elements pattern suggests a different origin and also the possibility of local production (Mildner *et al.*, 2014). In Italy, HMG glasses were present throughout the 9th to 7th centuries BC in northern (Villanovian Culture, Polla *et al.*, 2011; Arletti *et al.*, 2011b) and in southern Italy (in the Campania and Calabria regions, Conte *et al.*, 2018, 2019). Some of these HMG glass are characterized by a lower K content with respect to the classical Near Eastern or Egyptian glass, so that a different production is possible.

Natron glass dated as early as the 9th–7th centuries BC is known from France (Gratuze, 2009) and Italy, both in the north and in the south of the peninsula (Polla *et al.*, 2011; Arletti *et al.*, 2011b; Conte *et al.*, 2018, 2019). During the early Iron Age in France and Italy, glasses with particular compositions are also present: (1) natron glass with Co, high Al and Mg; (2) natron black glass with very high Fe contents (FeO > 7 wt.%, generally around 10%) (Gratuze, 2009; Conte *et al.*, 2018, 2019). For the natron glass from Italy differences in the trace elements suggest the existence of diverse production centres, but there are no confirmed indications of the origin. Isotopic analysis of some natron black beads from Bologna (northern Italy) are compared by the authors with sand and sandstone from Egypt and a possible Egyptian origin of the beads is suggested (Conte *et al.*, 2018).

Other peculiar vitreous materials are known from northern Italy and central Europe. For example there are beads dated to the 7th–6th centuries BC and made in glassy faience rich in Pb present in the Golasecca Culture in northen Italy (Angelini *et al.*, 2011) and in Poland (Purowski *et al.*, 2014). These beads have a black or dark blue body and are decorated with circles of yellow glass, coloured with Pb antimonate. This bead typology is widespread in central Europe, northern Italy and the Balkans, so the production had to be located somewhere in this area (Angelini *et al.*, 2011; Purowski *et al.*, 2014).

During HaC and HaD (about the 8th century to the middle of the 5th century BC) in Poland natron glass, natron glass with Pb and glass with low Mg and K (named LMMK, Purowski *et al.*, 2012, 2014) were also present. The LMMK glassy faience shows a small Ca content and the type of flux used is unknown, possibly plant ash. Its unique composition is not reported from Egypt or the Near East so that a local production is probable (Purowski *et al.*, 2012, 2014).

From the 6th century BC up to early Roman time in the whole of Europe mainly natron glass is found, even if the composition may vary slightly depending on the region and age. 'Classical' natron glass is typical *e.g.* of second Iron Age Italy (Olmeda *et al.*, 2015; Arletti *et al.*, 2010), France (see a summary in Gratuze, 2009), and in northern and central Europe during the La Tène Culture. The Celtic glass production of beads and bangles is all natron-based and presents a high variety of colour and shape.

Considering the previous discussion it is clear that the study of the glass composition in Europe during the early part of the 1st millennium BC is an extremely complex topic. The open questions are still many and varied and the dataset available is limited, often restricted to some geographical areas, and needs to be improved.

4. Glass-making transition at the end of the 1st millenium AD

As discussed above, important changes were observed in western Europe concerning glass manufacture during the Carolingian period. While an increase in soda glass recycling was observed throughout the Merovingian period, a slow but constant increase of the potash content in the glass was observed from the end of the 8th century in continental Europe. Within the same period of time, a shift from natron glass towards soda plant-ash glass is also observed in Mediterranean Europe, while the sporadic presence of lead glass is encountered according to the different European and Near Eastern regions. In recent years, the increasing number of analyses carried out on glass materials dated to this transition period has highlighted the presence of a great variety of different compositions. This important variability seems to reflect the use by glassmakers of local raw materials which enabled them to face the decreasing availability of natron glass (both raw and recycled).

Three main examples will illustrate the different solutions developed by European glassmakers to face the lack of natron glass at the end of the 8th century in western Europe.

The first two originate from the studies carried out on two glass workshops, dated from the 9th–10th centuries and located in the northeast and in the west of France. The third one shows how glassmakers have recycled metallurgical glassy slag to make glass objects.

In 2008, an archaeological evaluation conducted in the municipality of Meru (Oise, between Paris and Beauvais) revealed the presence of a mediaeval site of ~3000 m^2 in area characterized by a glass workshop, which was probably abandoned after a fire and was not subsequently reoccupied. Despite the relative isolation of that site, far from large towns or important religious sites, it is located along a major thoroughfare and has all the necessary resources (wood, clay and water). Although the site has not been fully excavated, the evaluation enabled recovery of different elements which prove undoubtedly the presence of a glass workshop: vitreous slag and vitrified furnace walls, crucibles (with three complete ones), several glass production wastes and glass objects. According to the ceramics and to the glass vessels, the site can be reliably dated to the 9th century. The presence among the finds of a lead glass smoother (see below, the production of glass around Melle) corroborates this dating (Pactat *et al.*, 2015).

In order to understand which types of glass were worked and produced at Meru, 58 samples were analysed using LA-ICP-MS. They represent the whole variability of the finds associated with glass working (raw glass, cullet, glass working waste, glass drops, crucibles, glass vessels and glass panels). The presence on the site of Roman tesserae, glass beads, Roman and Merovingian glass vessels (such as bottles and mosaic glass bowl), indicates recycling activities. Only few tiny raw glass fragments have been unearthed on the site, therefore, the main raw material might have been glass cullet.

The analysis reveals the presence of two main glass compositional groups which correspond to two different glass recipes. They are mainly distinguished by their contents of soda (Na_2O), potash (K_2O), lime (CaO) and phosphorus oxide (P_2O_5). The first recipe is in continuity with Antique glassmaking traditions and is based on reused natron glass. It is, however, probable that during the 9th century, this way of working was not able to meet the demand for glass, therefore glass-workers had to adapt by developing new recipes using local raw materials to increase their production. The second recipe, which corresponds to a potash-lime glass containing fairly large amounts of soda, is based partly on the reuse of natron glass, but the glass has increasing contents of potash, magnesia and phosphorus. This particular feature could be interpreted either by the mixing of two different type of glass (natron and potash-lime), or by the addition of increasing amount of forest plant ash into the reused natron glass cullet. The ratio between silica and the fusing elements (Na+Mg+K+Ca+P+Cl) tends to decrease with the potash content for the potash-lime-soda glass (2.7 to 1.6, while the average value is 2.7 for natron glass). This second compositional group can be further divided into two sub-groups. In the first there is an inverse relationship between soda on the one hand and, potash magnesia and phophorus on the other: soda decreases from 14% to 5.5% while potash increases from 1.7% to 9%. This glass corresponds to a gradual transition between natron glass and forest plant-ash glass. The second sub-group has more stable contents of soda and potash (Na_2O 4.1% and K_2O 10.5%) and seems to characterize the forest plant-ash glass produced and worked in Meru. While all the raw glass and the glass cullet belong to the natron group, the other categories of objects (crucible, glass wastes and glass vessels) are represented in the two main groups and sub-groups.

Among the different glass vessels unearthed at Meru, one, a funnel beaker made with three different glasses (colourless, cobalt-blue and light-blue), is particularly interesting. The analyses reveal that the different parts of the beaker have different soda, potash, lime and magnesia contents (Fig. 16). The presence of that particular object and the repartition of the different types of objects (crucible, glass wastes and glass vessels) among the different compositional groups of glass show undoubtedly that the transition in glass making (from natron glass towards forest plant-ash glass) occurred at Meru. The important recycling of natron glass is shown by the presence of glass cullet and a block of raw glass with that composition. The discovery of a blown sleeve glass window shows that the production of stained glass was probably also one of the vocations of that workshop.

The archaeological site of Bois Beslan at La Milesse (Sarthe, 10 km northwest of Le Mans) was discovered and excavated by the National Institute for Preventive Archaeology in 2012–2013. It consists mainly of an iron metallurgical site (extraction and production) dated from Late Iron Age to Antiquity (Raux *et al.*, 2015). A glass workshop implemented in the northeast part of the site, in the dumps of abandoned mining, has also been identified. In one of the pits excavated in that part of the site, archaeologists have recovered abundant elements characteristics of glass-working activity such as glass fragments of crucibles and processed glass waste. Seven crucibles, with the presence of glass or vitrified matter on their internal or external walls have been identified, and >20 kg of glass waste has been recovered. Half of the weight consists of waste originating from glass crafts (wires and rods of drawn glass, with round or flat sections, glass drops, small raw glass fragments and shapeless drips of various sizes), the other half being composed of scoriaceous glassy material.

According to radiocarbon analyses and to the typologies of the glassware and ceramic containers originating from the pit, the workshop can be dated from the 10th century AD. The most documented forms are goblets, with straight or returning edge, concave base and ovoid body. They are decorated with horizontal or spiral, threaded inserts. They are similar to several exemplaries dated from the 9th–10th centuries. All the glass samples have a deep green colour.

Fourteen samples, representative of the variability of the finds (crucibles, raw glass, glass waste and vessels) were analysed using LA-ICP-MS. The various samples form a homogeneous chemical group, characteristic of lime-potash glass, obtained from forest plant-ash glass. Compared with coeval lime-potash glass objects, the glass objects originating from La Milesse are characterized by larger amounts of alumina, iron, titanium and zirconium and smaller sodium concentrations. The chemical homogeneity

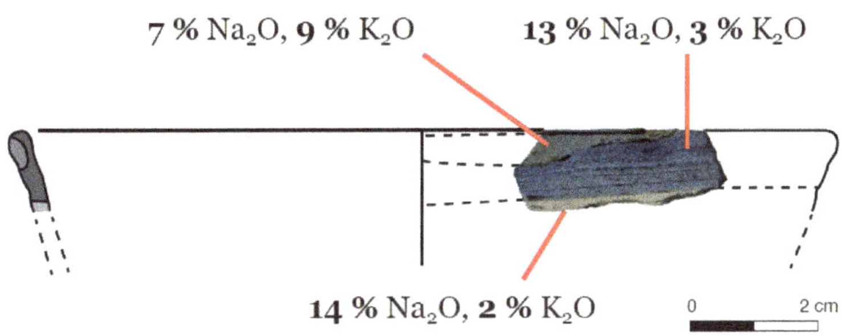

Figure 16. Soda/potash ratios measured in one funnel beaker edge from Meru. © MSHE, I. Pactat.

of this group of objects favours the presence of primary glass production on the site. The large iron content of these glasses shows that the glass-workers used the ferrous sand present at the site. From the small soda content of La Milesse's glass (0.54±0.10%), it can be concluded that the recycling of natron glass was not carried out by La Milesse's glass-workers, due either to the total lack of this sort of glass in that area or to its later production period than Meru's glass workshop.

In 1992, Julia Scapova published an uncommon composition for a 10th century linen glass smoother discovered in Novgorod (Russia) (Scapova, 1992). This object was made from a silica-lime-alumina-lead glass characterized by the presence of large amounts of iron, barium and antimony. Over the following decades, several linen glass smoothers of similar composition were described from Germany, France, England, Ireland and Norway and more recently Denmark and Belgium. All these objects, which are distributed along Viking trade routes, share the same type of composition and are dated between the end of the 8th century and the end of the 10th century. Until 2013, this composition seemed to be specific to the smoothers as no other analysed lead-glass objects of this period were found with a similar composition. At the end of the 1990s, based on chemical and lead isotope analyses, a study carried out on the Carolingian lead-silver mines of Melle (Deux-Sevres, west of France, 30 km south of Niort and 50 km southwest of Poitiers) allowed us to establish a relationship between the glassy slag produced on that mining site and the silica-lime-alumina-lead of all the linen glass smoothers originating from Europe. From 2012, thanks to the systematic analysis of the glass objects originating from different Carolingian archaeological sites located to the north and west of Melle (Bressuires, Fayes-sur-Ardin, Pussigny, Poitiers) it became possible to identify several glass vessels sharing this particular composition. Most of these objects are beakers made of strongly coloured green glass with an externally thickened rim and decorated by several rows of opaque white glass on the neck. The site of Faye-sur-Ardin also yielded many different glass fragments, with that composition. However, it was not possible to relate them to well established typologies (Gratuze *et al.*, 2014).

The analyses carried out on these different objects show that all of them were made by recycling the glassy slag produced during the metallurgical process leading to the production of silver from galena. The comparison between the composition of the glassy slag and the glass of the beakers and the linen smoothers reveals the following features: (1) soda glass cullet was added systematically to the melting batch; (2) soda glass cullet was added in a lesser extent for the production of linen smoothers (Na_2O 1.3±0.3%) and to a greater extent for the production of glass vessels (2.4±1.3%); and (3) the white glass was produced by adding tin oxide to the same base glass used to make the beaker, therefore the tin white opacified glass was produced directly in the glass workshop.

This workshop, which has not yet been discovered, is up to now known only by its products (glass smoothers and vessels), but we can hypothesize that it was located around the mining site of Melle.

This direct use of lead slag in glass production remains to this day without equivalent for that period. The valorization of metallurgical byproducts presents two main advantages for Melle's glass-workers. On one side, it allowed them to obtain a cheap raw material for dealing with the natron glass shortage encountered at that time in the whole of western Europe. Secondly, it allowed them to make a significant saving of combustible material. Glass making requires a significant amount of energy. The manufacture of potash glass, developed at that time in western Europe, needed a large amount of wood or other plants. These are necessary on the one hand as a raw material for the preparation of ash used as fusing agent, and secondly as a fuel to melt the mixture of sand and ash. The valorization of the slag significantly reduced the wood needs of the glass-workers, which are then limited to the re-melting of the slag and the natron glass cullet. The use of slag could thus be considered as the response of Melle's glass-workers to the two main challenges they faced: the lack of natron glass at the European scale and the fierce local competition for the exploitation of combustible materials due to mining and metallurgical activities on the site. These activities are also major consumers of wood (mining using fire-setting, roasting and reduction of ore; coin minting).

Thanks to the discovery at Meru of a glass linen smoother produced by the Melle glass-workshop, it can be hypothesized that these two glass workshops are coeval.

The adaptability of glass workers and their search for new recipes in order to continue to produce glass is also highlighted by the studies carried out on several typical Carolingian productions (bowls, beakers and vessels decorated with reticulated glass, and/or rows of opaque glass; Pactat *et al.*, 2017) and by the composition of two coeval glass objects, unearthed at Vernou-sur-Brenne (Indre-et-Loir, 15 km east of Tours, France), and dated from the 9th–10th centuries. These two pieces which were produced by an unknown glass workshop might, as in the case of Meru, represent a transition between natron soda lime glass and forest plant-ash glass. While the first is a soda lime glass containing a small amount of plant ash (3% K_2O and 1.4% MgO), the second is a lime potash glass containing a large amount of soda (5%). Based on the analytical data from Meru and other consumption sites, it seems there was a two-step process: at first, an increase in potash concentration, up to 3–4%, a change that is not correlated with an increase in magnesia, phosphorus and lime, which may be the consequence of pollution by wood's fly ash, and, secondly, a correlated increase in potash, magnesia, phosphorus and lime, which really indicates the introduction of forest plant ash.

This transition period seems to end between the end of the 12th century and the beginning of the 13th century. From that period, several glass production areas, characterized by well defined chemical compositions, can be identified in western Europe: plant-ash soda-lime glass on the Mediterranean coastal areas, potash-lime glass in northwestern Europe and high-lime glass in continental Europe.

The variations in chemical composition of glass through time are summarized in Fig. 17.

5. Techniques for the investigation of glass

Several analytical methods are used commonly for chemical analysis of ancient glass. A rapid survey of recent publications dealing with this subject (*e.g.* Barca *et al.*, 2009; Giussani *et al.*, 2009; van der Werf *et al.*, 2009; Walton *et al.*, 2009; Carmona *et al.*,

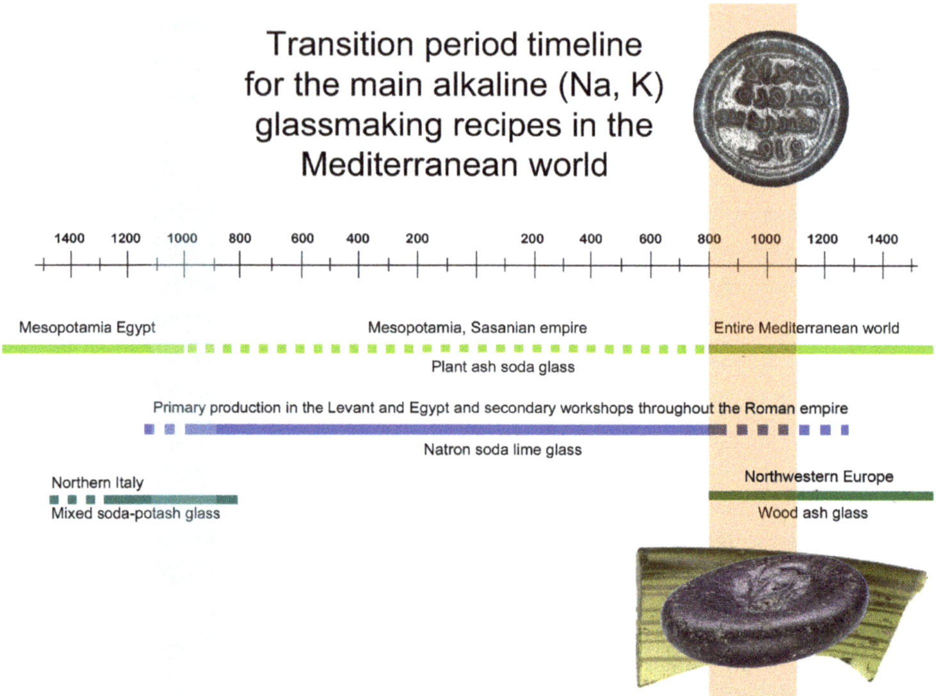

Figure 17. Timeline of the main glass-making recipes based on alkaline fluxes used in the Mediterranean and European world between the second half of the 2nd millennium BC and the first half of the 2nd millennium AD. Adapted from Wedepohl (1997). The recipes based on lead oxide fluxes are not illustrated.

2010a; Wedepohl *et al.*, 2011a,b) shows that the most common analytical methods are electron probe micro-analysis (EPMA) and scanning electron microscopy with energy or wavelength dispersive X-ray Spectroscopy (SEM-EDX or WDX). Then there are all the other methods, X-ray fluorescence spectroscopy (XRF), particle-induced X-ray emission (PIXE) which is often associated with particle-induced γ-ray emission (PIGE), ICP-MS (either in its liquid mode or in its laser mode, LA-ICP-MS) and to a lesser extent instrumental thermal neutron activation analysis (INAA) and laser induced breakdown spectroscopy (LIBS). Of course isotopic characterization towards glass provenancing, diffusion, and interpretation of glass technology needs measurements performed by mass spectrometry (Degryse *et al.*, 2009). Spectroscopic characterization of structural or chromophore elements in glass is performed commonly by X-ray absorption spectroscopy (XAS), either in extended fine-structure mode (EXAFS) or in near-edge mode (XANES) (see *e.g.* Quartieri *et al.*, 2002; Veiga and Figueiredo, 2008; Quartieri and Arletti, 2013). Such investigations are fundamental to understanding the colouring/decolouring mechanisms of the chromophore ions and the structural role (*i.e.* network forming or network modifier) of the chemical species in the glass.

The investigation of glass materials typically poses peculiar problems, related to the facility of weathering, and to the substantial heterogeneity of the glassy materials, which often contain crystalline components of various natures (Artioli *et al.*, 2008), as discussed above (Table 1). Figure 18 shows a representative example of the micro-structural complexity of ancient glass materials.

Mineralogical, crystallographic and textural analyses are therefore important to observe and decode the single crystalline components, and thus obtain information on production processes, recycling, phase transitions, physicochemical properties, and degradation. Bulk X-ray diffraction (XRD) or neutron diffraction (ND) allow detection and quantification of the crystalline components, though spatial information is lost, unless measurements are carried out with microbeams to map the distribution of the crystal phases, *e.g.* with synchrotron micro- or nano-X-ray beams or with electron diffraction (ED). XRD is of course mandatory when investigating *in situ* the mechanisms of crystallization of opacifiers in the glass matrix, simulating ancient production techniques (Lahlil *et al.*, 2008, 2010). Characterization performed by Raman microbeams, including point analyses or maps (Figs 19 and 20) may also be very important for identifying the nature of the crystalline phases present in the glass matrix and to extract mineralogical, chemical and crystallographic information such as crystal

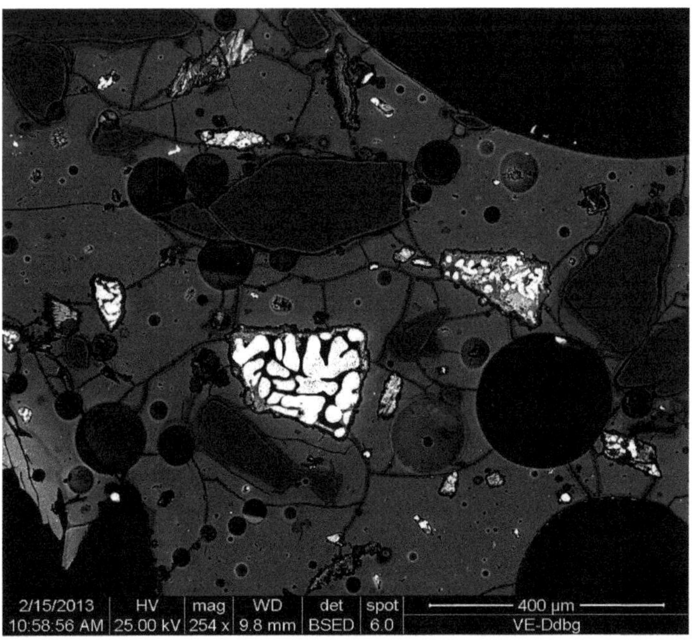

Figure 18. SEM backscattered electron image of a fragment of an Etruscan blue-green glass bead from Verucchio, Italy (Phase V, 7th century BC). The image clearly shows the complex micro-structure composed of small fragments of wüstite-rich slags from iron metallurgy (white), quartz fragments (dark grey) and gas bubbles embedded in the glass matrix. (Image I. Angelini).

morphology, chemical zoning, phase transformations, *etc.* (see *e.g.* Colomban, 2003; Ricciardi *et al.*, 2009). Of course multi-analytical characterization yields complementary information and therefore a more complete interpretation of complex materials (see *e.g.* Welter *et al.*, 2007).

Whenever complex multiphasic and heterogeneous materials are to be investigated, imaging techniques are mandatory to retrieve spatially resolved information. Optical microscopy (OM) operated in transmitted light, reflected light, or confocal modes is of course a primary tool for the visualization of the micro-structure (Fig. 21), though SEM imaging is also a very versatile and widespread tool (Fig. 18). Optical coherence tomography (OCT) has been used to analyse the interaction layer in faience materials (Kunicki-Goldfinger *et al.*, 2009; Liang *et al.* 2012), though the silica grains, the voids, and the permeation of the glass matrix in the interstices of the faience can be visualized efficiently by high-resolution computed X-ray tomography (CT) (Fig. 22).

Imaging methods reaching atomic or molecular resolution, such as high-resolution transmission electron microscopy (HRTEM) or atomic force microscopy (AFM) are seldom applied, though they may be important for decoding the fine features of the glassmaking processes. TEM techniques have been applied to investigate lustre, glazes and glass coloured by nanoparticles (Angelini *et al.*, 2004; Darque-Ceretti *et al.*, 2005;

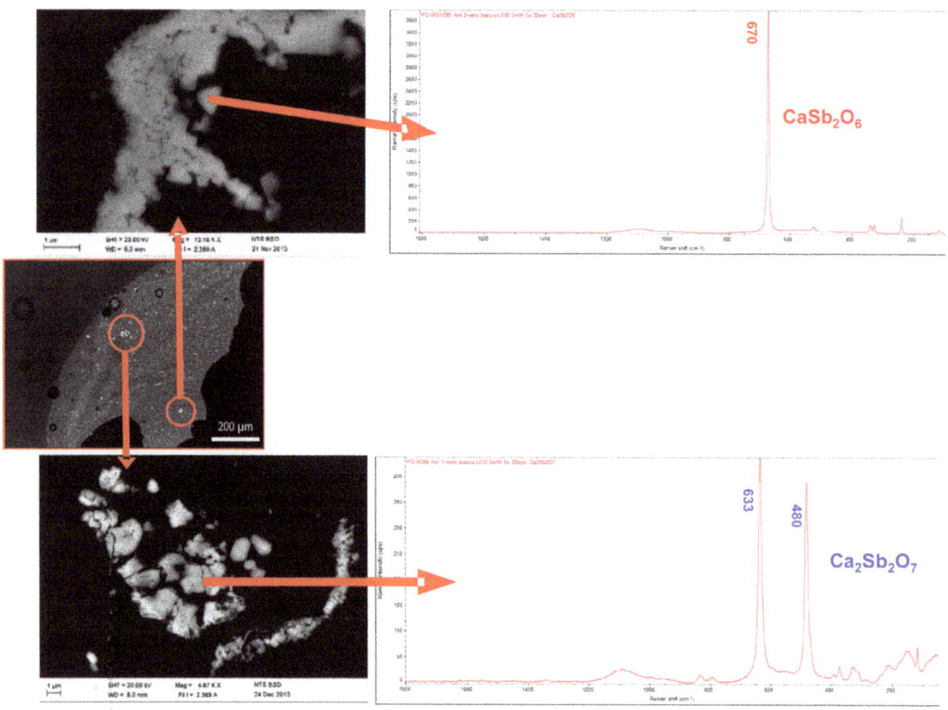

Figure 19. SEM backscattered electron images and Raman characterization of the Ca antimonate crystals present in the matrix of a white glass bead from Villa di Villa, Treviso, Italy (Iron Age). (Image and data: I. Angelini and G. Olmeda).

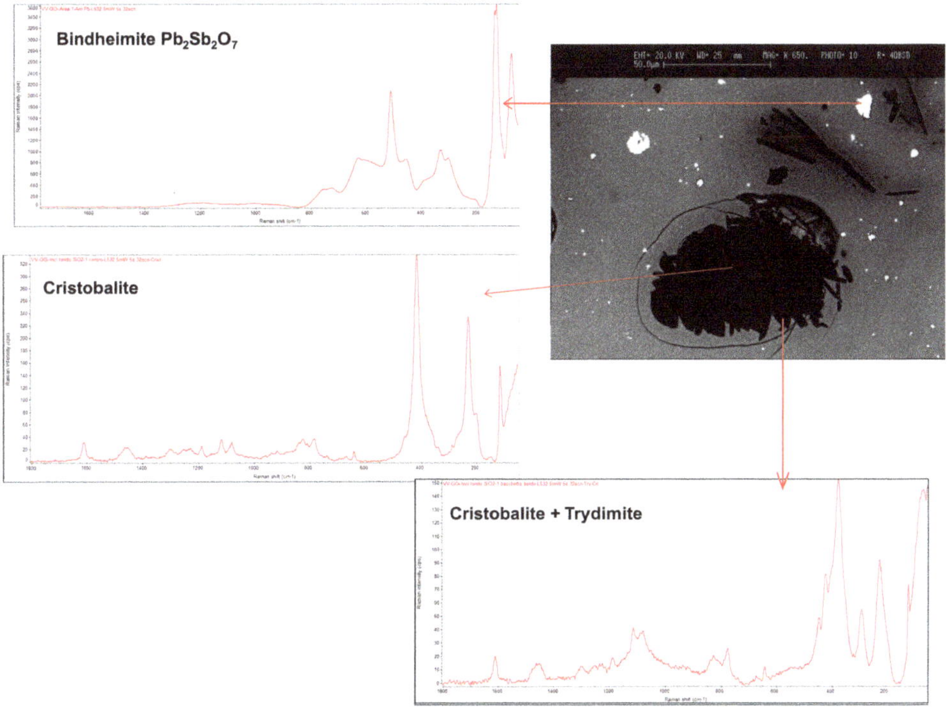

Figure 20. SEM backscattered electron image and Raman characterization of the crystals present in the matrix of a yellow glass eye-bead from Villa di Villa, Treviso, Italy (Iron Age). The white inclusions are Pb antimonate crystals accounting for the yellow colour of the bead, whereas the dark crystal clusters are composed of cristobalite and trydimite. The volume reduction during the phase transition in the silica cluster causes the circular fracture in the glass matrix. (Image and data: I. Angelini and G. Olmeda).

Roqué *et al.* 2006; Colomban, 2009). Recent developments of advanced crystallographic TEM analysis (by 3D electron diffraction tomography) open the possibility of identifying, characterizing, and even determining the crystal structure of nm-sized particles in the glass (Zacharias *et al.*, 2018). AFM microscopy has been used to understand surface alteration processes at the molecular level (Carmona *et al.*, 2010b).

5.1. Surface vs. bulk chemical analysis

Ancient glass artefacts are often weathered, thus the corroded surface of these objects has a different composition from the intact glass underneath. Depending on the type of glass and on the intensity of the weathering, this perturbed layer may be several hundreds of μm thick. Most X-ray methods only measure material to a depth of a few microns, and thus typically do not reach intact glass. Moreover, these methods are highly sensitive to surface effects and need to be applied on very flat surfaces. Embedding glass samples in resin and polishing is often necessary, which can be considered as destructive, time consuming and expensive. It should also be noted that

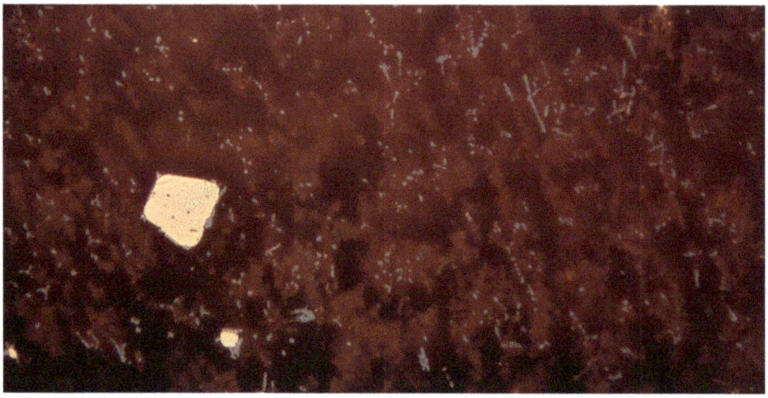

Figure 21. Deep-focused image of a red Pb glass from Tebtunis, Fayum, Egypt (Hellenistic Age). The combination of stacked OM images focused at different depths allows the perception of the cuprite dendrites embedded in the transparent glass matrix. A couple of copper metal droplets are surfacing. (Image courtesy of C. Bettineschi, UNIPD).

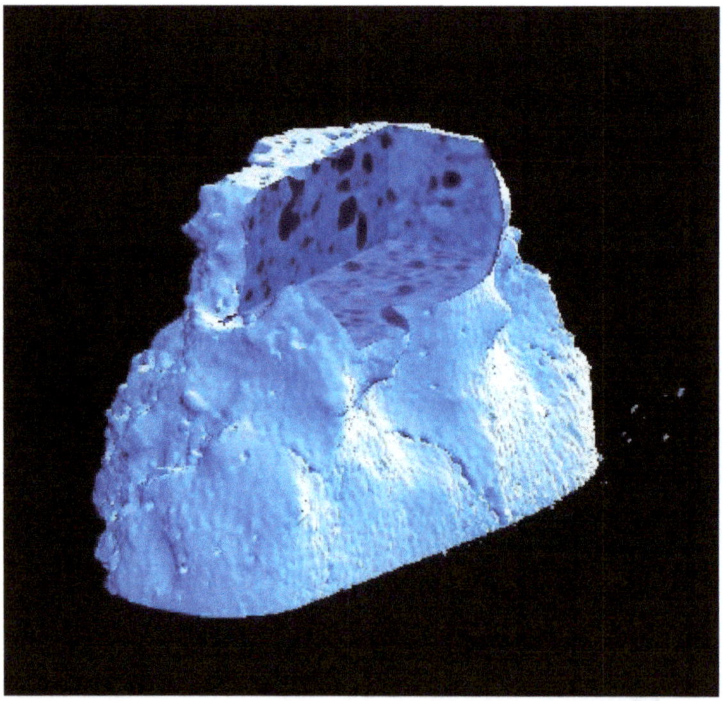

Figure 22. Computed X-ray micro-tomographic image of a faience bead from Lavagnone, Brescia, Italy (Early Bronze Age) (Angelini *et al.* 2006). The virtual sectioning of the reconstructed 3D image allows visualization and quantification of the voids, the quartz grains, and the glass interaction layer. (Image: G. Artioli).

energy-dispersive X-ray methods suffer matrix effects and interferences. For example, detection limits of bismuth or gold in glass are affected when lead is also present in high concentrations. Compared to X-ray methods, LA-ICP-MS analysis can be done on unprepared objects and allows the measurement of a large number of chemical elements (present either as major, minor or trace components) with only a few chemical or spectrometric interferences (Gratuze, 2016). LA-ICP-MS also offers lower detection limits than any energy- or wavelength-dispersive X-ray method. However, the spatial resolution of LA-ICP-MS analysis is not as good as that of EPMA, and LA-ICP-MS is also less relevant and selective for characterizing mineral inclusions.

As mentioned above, glass is a complex material made by mixing at least two main compounds: sand or crushed quartz pebbles (for silica) and plant ashes, mineral deposit or other oxides (for the fluxing agent: soda, potash or lead) to which different materials can also be added in order to modify its physical properties (colourants and opacifiers such as metallic oxides).

Until the 19th to 20th centuries, all these components had frequently relatively high levels of impurities, which reflected their origin or their manufacturing processes. The major elements typically characterize the recipes followed by the glass workers, and the minor- and trace-element associations reflect the origin of the materials used, thus representing the chemical fingerprints of the glass.

Chemical analysis of ancient glass items should, therefore, on the one hand, allow identification of the different recipes used for their production (soda, potash, lime, lead or other type of glass) and, if possible, the nature and origin of the raw materials (sand, quartz pebbles, fluxing agent from mineral deposits or from plant ash, colouring and opacifing agents), and on the other, allow us to follow the objects from the workshop to the consummation site. However glass objects, like metals, can be recycled and mixed easily, which makes interpretation of the analytical data more delicate, and often limits the conclusions of glass provenance studies.

More recently, the geochemical signature (isotopic composition) of some glass components (such as lead, strontium, neodymium, boron, oxygen) has been used to identify the origin of the raw materials (Degryse *et al.*, 2009). Isotopic analysis can also help to follow the diffusion of the manufactured objects or to locate geographically the primary glass workshops.

According to the analytical methods used, some or all of the different types of information are accessible. Using scanning electron spectroscopy coupled with X-ray energy-dispersive analysis (SEM-EDX), only major elements are determined, therefore this method allows only classification of the main glass type according to the fluxing agents used. However, due to the different imaging techniques (backscattered electron, secondary electron and X-ray mapping) it provides good identification of all the different mineralogical phases present in the glass (glass, devitrification structures, crystals, bubbles ...), giving the possibility to distinguish between unmelted relict phases (quartz, zircon, feldspars, metallic particles, opacifier ...) and *in situ* newly formed crystals (opacifier, wollastonite, metallic crystals ...). All this information provides a valuable tool to understand glass-making processes and opacifying recipes (*in situ* crystallization, addition of previously formed crystals, or mixing of clear and opaque glass). Compared to SEM-EDX, electron probe micro analysis offers the same imaging facilities with better detection limits (WDS improves detection limits by an order of magnitude or so, *vs.* EDS), giving access to the main minor elements (*i.e.* colouring element such as Co, and minor compound such as Ti, Sr, Ba ...). One of the main interests of SEM-EDX and EPMA is the option to carry out both bulk and point analyses.

Other X-ray and associated γ-ray methods (XRF, PIXE, PIGE) often offer better detection limits (ppm level instead of hundreds of ppm), but they are not associated with imaging facilities. These methods are particularly relevant for provenance study based on trace-element bulk analysis, but suffer the needs of expensive and complex analytical facilities such as Van de Graaff accelerators or Synchrotron lines. For similar reasons, access to facilities and cost, instrumental neutron activation analysis (INAA), which has been largely used to determine rare-earth elements and other trace elements for glass provenance studies, is used less nowadays.

More recently developed, inductively coupled plasma mass spectrometry techniques (either in liquid mode or laser ablation mode) offer a good compromise for trace-element analyses. Different options are offered according to the type of mass analyser (quadrupole, magnetic sector or time of flight) and the introduction mode (liquid, laser wavelength: Nd-YAG or Excimer laser). If ICP-MS in liquid mode offers only the possibility of bulk analysis on dissolved glass samplings, LA-ICP-MS offers more possibilities. Compared to EPMA or SEM-EDX analysis, analysis can be done directly on the sample without any preparation. Removal of the corrosion layer is easily carried out with a pre-ablation of the area to be analysed, regardless of the thickness of the corrosion layer. It is possible to obtain reliable results down to 1 mm below the surface. As the laser sampling is invisible to the naked eye, it is thus possible to analyse a greater number of objects and to get a better representation of compositional variation within a population of such glass objects. But laser ablation techniques can be applied with a great flexibility on a large range of sample types, it is hence also possible to analyse objects or sections of samples embedded in resin previously characterized by SEM-EDX or EPMA and to carry out either point analysis of inclusions or bulk analysis of the whole glass. It is thus possible to characterize fully (for major, minor and trace elements) the glass but also the different large inclusions present inside its structure. To characterize inclusions with great selectivity and good sensitivity, their size should not be <40–50 μm. Characterization of smaller inclusions (down to 10 μm) is possible, but it is done to the detriment of detection limits and selectivity.

EPMA and LA-ICP-MS appear then as perfect complementary techniques. The first allows us to carry out a full analysis of the different phases present in the glass (morphologies and compositions) as well as to determine the bulk composition of the major and main minor elements of the glass, while the second offers the options of achieving trace-elements analysis of the various structures encountered for trade and provenance studies.

References

Alpern, S.B. (1995) What Africans got for their slaves: A master list of European trade goods. *History in Africa*, **22**, 5–43.

Angelini, I. (2011) Archaeometry of Bronze Age and Early Iron Age Italian vitreous materials: a review. Pp. 17–23 in: *Proceedings of the 37th International Symposium on Archaeometry*, 13th–16th May 2008. Springer, Siena, Italy.

Angelini, I. (2014) Analisi dei materiali vetrosi. Pp. 74–89 in: *La necropoli con tombe a camera del Bronzo Medio* (N. Negroni Catacchio, M. Aspesi, C. Metta and G. Pasquini, editors). Ricerche e scavi del Centro Studi di Preistoria e Archeologia. Roccoia, Milan, Italy.

Angelini, I. (2019) Il vetro di Frattesina: composizione e tecniche di colorazione come deducibili dai dati chimici, mineralogici e tessiturali In: *Frattesina: un centro internazionale di produzione e di scambio dell'Età del Bronzo* (A.M. Bietti Sestieri, P. Bellintani and C. Giardino, editors). Atti dell'Accademia Nazionale dei Lincei, Anno CDXV-2018, Bardi Edizioni, Rome.

Angelini, I. and Olmeda, G. (2018) Archaeometric study of the vitreous materials beads. Pp. 323–339 in: *Tolochenaz (VD) - La Caroline. Du Mésolithique à l'époque romaine en passant par la nécropole du Boiron* (A. Gallay, E. Burri-Wyser, F. Menna and M. David-Elbiali, editors). Cahiers D'archéologie Romande 168, Lausanne, Switzerland.

Angelini, I., Artioli, G., Bellintani, P., Diella, V., Polla, A. and Residori, G. (2002) Project "Glass materials in the protohistory of North Italy": a first summary. Pp. 581–585 in: *Atti del Secondo Congresso Nazionale AIAr, Bologna 29 Gennaio-1 Febbraio 2002* (C. D'Amico, editor). Pàtron Editor, Bologna, Italy.

Angelini, I., Artioli, G., Bellintani, P., Diella, V., Gemmi, M., Polla, A. and Rossi, A. (2004) Chemical analyses of Bronze Age glasses from Frattesina di Rovigo, northern Italy. *Journal of Archaeological Science*, **31**, 1175–1184.

Angelini, I., Artioli, G., Bellintani, P. and Polla, A. (2005) Protohistoric vitreous materials of Italy: from early faience to final bronze age glasses. Pp. 32–36 in: *Proceedings of the AIHV 2003, 16e Congrès de l'Association Internationale pour l'Histoire du Verre*, 7–13 September 2003, London.

Angelini, I., Artioli, G., Polla, A. and De Marinis, R.C. (2006) Early Bronze Age faience from North Italy and Slovakia: A comparative archaeometric study. Pp. 371–378 in: *Proc. 34th Int. Symposium on Archaeometry*, Zaragoza, Spain, 3–7 May 2004.

Angelini, I., Polla, A. and Molin, G. (2010) Studio analitico dei vaghi in vetro provenienti dalla necropoli di Narde. Pp. 105–134 in: *La fragilità dell'Urna. I recenti scavi a Narde, Necropoli di Frattesina (XII-IX sec. a.C.)*. Soprintendenza per i Beni Archeologici del Veneto, Comune di Fratta Polesine, Italy.

Angelini, I., Nicola, C., Artioli, G., De Marinis, R.C., Rapi, M. and Uboldi, M. (2011) Chemical, mineralogical and textural characterization of the Early Iron Age vitreous materials from the Golasecca Culture (northern Italy). Pp. 25–32 in *Proceedings of the 37th International Symposium on Archaeometry* 13th–16th May 2008. Springer, Siena, Italy.

Arletti, R., Maiorano, C., Ferrari, D., Vezzalini, G. and Quartieri, S. (2010) The first archaeometric data on polychrome Iron Age glass from sites located in Northern Italy. *Journal of Archaeological Science*, **37**, 703–712.

Arletti, R., Rivi, L., Ferrari, D. and Vezzalini, G. (2011a) The Mediterranean group II: analyses of vessels from Etruscan contexts in northern Italy. *Journal of Archaeological Science*, **38**, 2094–2100.

Arletti, R., Bertoni, E., Vezzalini, G. and Mengoli, D. (2011b) Glass beads from Villanovan excavation in Bologna (Italy): an archaeometrical investigation. *European Journal of Mineralogy*, **23**, 959–968.

Arletti, R., Ferrari, D. and Vezzalini, G. (2012) Pre-Roman glass from Mozia (Sicily-Italy); the first archaeological data. *Journal of Archaeological Science*, **39**, 3396–3401.

Artioli, G. (2010) *Scientific Methods and Cultural Heritage: An Introduction to the Application of Materials Science to Archaeometry and Conservation Science*. Oxford University Press, Oxford, UK.

Artioli, G. and Angelini, I. (2013) Evolution of vitreous materials in Bronze Age Italy. Pp 355–368 in: *Modern Methods for Analysing Archaeological and Historical Glass*, Vol. II (K. Janssens, editor). John Wiley & Sons, Chichester, UK.

Artioli, G., Angelini, I. and Polla, A. (2008) Crystals and phase transitions in protohistoric glass materials. *Phase*

Transitions, **81**, 233–252.

Artioli, G., Angelini, I. and Nestola, F. (2013) New milarite/osumilite-type phase formed during ancient glazing of an Egyptian scarab. *Applied Physics A*, **110**, 371–377.

Aspinall, A., Warren, S.E., Crummett, J.G. and Newton, R.G. (1972) Neutron activation analysid of the faience beads. *Archaeometry*, **14**, 27–40.

Bamford, C.R. (1977) *Colour generation and control in glass*. Glass Science and Technology, 2. Elsevier, Amsterdam.

Barca, D., Abate, M., Crisci, G.M. and De Presbiteris, D. (2009) Post-medieval glass from the Castle of Cosenza, Italy: chemical characterization by LA-ICP-MS and SEM-EDS. *Periodico Di Mineralogia*, **78**, 49–64.

Barthélémy de Saizieu, B. and Bouquillon, A. (1997) Evolution of glazed materials from the chalcolithic to the Indus period based on the data of Mehrgarh and Nausharo. Pp. 63–76 in: *South Asian Archaeology 1995* (R. Allchin and B. Allchin, editors). The Ancient Indian and Iran Trust, Cambridge, UK.

Barthélémy De Saizieu, B. and Bouquillon, A. (2001) Émergence et évolution des matériaux vitrifiés dans la région de l'Indus du 5ᵉ au 3ᵉ millénaire (Merhgarh-Nausharo). *Paléorient*, **26/2**, 93–111.

Bar-Yosef Mayer, D.E., Porat, N., Gal, Z,. Shalem, D. and Smithline, H. (2004) Steatite beads at Peqi'in: long distance trade and pyro-technology during the Chalcolithic of the Levant. *Journal of Archaeological Science*, **31**, 493–502.

Barag, D. (1970) Mesopotamian Core-Formed Vessels 1500–500 BC. Pp. 131–199 in: *Glass and Glassmaking in Ancient Mesopotamia* (A.L. Oppenheim, R.H. Brill, D.P. Barag and A. Von Saldern, editors). The Corning Museum of Glass [reprint 1988], Corning, New York.

Barag, D. (1985) *Catalogue of the western Asiatic glass in the British Museum*, Volume 1. British Museum Publications, London.

Baxter, M.J., Cool, H.E.M. and Jackson, C.M. (2005) Further studies in the compositional variability of colourless Romano-British vessel glass. *Archaeometry*, **47**, 47–68.

Beck, H.C. (1934) Notes on glazed stones: Part I. Glazed steatite. Pp. 69–88 in: *Ancient Egypt and the East*. British School of Archaeology in Egypt, London.

Beck, H.C. (1935) Notes on glazed stones: Part II. Glazed quartz Pp. 19–37 in: *Ancient Egypt and the East*. British School of Archaeology in Egypt, London.

Bellintani, P. (2011) Progetto "Materiali vetrosi della protostoria italiana". Aggiornamento e stato della ricerca. *Rivista di Scienze Preistoriche*, **LXI**, 257–284.

Bellintani, P. and Stefan, L. (2009) Nuovi dati sul primo vetro europeo: il caso di Frattesina. Pp. 71–86 in: *Atti del primo convegno interdisciplinare sul vetro nei Beni Culturali e nell'Arte ieri e oggi*, Parma, 27–29 Novembre 2008, Tipocrom s.r.l., Parma, Italy.

Bellintani, P., Angelini, I., Artioli, G. and Polla, A. (2006a) Bottoni conici e perle in glassy faience delle fasi iniziale e piena della media età del Bronzo dell'Italia centrale tirrenica: archeologia e archeometria. *Padusa*, **XII**, 2005, 223–230.

Bellintani, P., Angelini, I., Artioli, G. and Polla, A. (2006b) Origini dei materiali vetrosi italiani: esotismi e localismi. *Atti XXXIX Riunone Scientifica* "Materie prime e scambi nella preistoria Italiana". Pp. 1495–1531 in: Istituto Italiano di Preistoria e Protostoria, Firenze 25–27 Novembre 2004. IIPP, Florence, Italy.

Bettineschi, C. (2018) *Archaeometric study from Egyptian vitreous materials from Tebtynis: Integration of analytical and archaeological data*. PhD thesis, Università degli Studi di Padova, Italy.

Bettineschi, C., Angelini, I. and Molin, G. (2019a) Contestualizzazione degli intarsi in vetro da Tebtynis nell'ambito dell'Egitto Greco-Romano. Pp. 457–482 in: *Proceedings of the Conference* "Anti. Archeologia. Archivi", Venice, June 2017, IVSLA Edition.

Bettineschi, C., Deotto, G., Begg, I., Molin, G., Zanovello, P., Fassone, A., Greco, C., Borla, M. and Angelini, I. (2019b) A dig through archives and depots: rediscovering the inlay workshop of Tebtynis and its materials **Nuncius**, 2019/3. In press.

Bimson, M. and Freestone, I.C. (1985) Appendix 3, 'Scientific examination of opaque red glass of the second and first millennia BC. Pp. 119–122 in: *Catalogue of the western Asiatic glass in the British Museum*, Volume 1 (D. Barag, editor). British Museum Publications, London.

Blomme, A., Degryse, P., Dotsika, E., Ignatiadou, D., Longinelli, A. and Silvestri, A. (2017) Provenance of polychrome and colourless 8th–4th century BC glass from Pieria, Greece: a chemical and isotopic approach. *Journal of Archaeological Science*, **78**, 134–146.

Brems, D. and Degryse, P. (2014) Trace element analysis in provenancing Roman glass-making. *Archaeometry*, **56**, 116–136.

Brill, R.H. (1992) Chemical analyses of some glasses from Frattesina. *Journal of Glass Studies*, **34**, 11–22.

Brill, R.H. (1999) *Chemical Analyses of Early Glasses*, The Corning Museum of glass, Corning, New York, USA.

Carmona, N., Ortega-Feliu, I., Gomez-Tubio, B. and Villegas, M.A. (2010a) Advantages and disadvantages of PIXE/PIGE, XRF and EDX spectrometries applied to archaeometric characterisation of glasses. *Materials Characterization*, **61**, 257–267.

Carmona, N., Kowal, A., Rincon, J.M. and Villegas, M.A. (2010b) AFM assessment of the surface nano/microstructure on chemically damaged historical and model glasses. *Materials Chemistry and Physics*, **119**, 254–260.

Colomban, P. (2003) Polymerization degree and Raman identification of ancient glasses used for jewelry, ceramic enamels and mosaics. *Journal of Non-Crystalline Solids*, **323**, 180–187.

Colomban, P. (2009) The use of metal nanoparticles to produce yellow, red and iridescent colour, from Bronze Age to present times in lustre pottery and glass: solid state chemistry, spectroscopy and nanostructure. *Journal of Nano Research*, **8**, 109–132.

Conte, S., Arletti, R., Henderson, J., Degryse, P. and Blomme, A. (2018) Different glassmaking technologies in the production of Iron age black glass from Italy and Slovakia. *Archaeological and Anthropological Science*, **10**, 503–521.

Conte, S., Matarese, I., Vezzalini, G., Pacciarelli, M., Scarano, T., Vanzetti, A., Gratuze, B. and Arletti, R. (2019) How much is known about glassy materials in Bronze and Iron Age Italy? New data and general overview. *Archaeological and Anthropological Sciences*, **11**, 1813–1841.

Cummings, K. (2002) *A History of Glassforming*. A and C Black, London.

Cupitò, M., Angelini, I., Vidale, M. and Dalla Longa, E. (2018) The plate for glass working from the terramara site of Anzola dell'Emilia (Bologna). Pp. 26–28 in: *5th Annual Meetings of Prehistory and Protohistory, Precious, semi-precious and unusual materials in Copper and Bronze Age Italy. Archaeology, archaeometry and paleotechnology*, Abstract Book, Istituto Italiano di Preistoria e Protostori, Florence, Italy. http://iapp2018padova.beniculturali.unipd.it/

Dardeniz, C. (2018) The preliminary archaeological and scientific evidence for glass making at Tell Atchana/Alalakh, Hatay (Turkey). Pp. 95–110 in: *Aspects of Late Bronze Age glass in the Mediterranean, Proceedings of JIAA Late Bronze Age Glass Workshop, 27th–28th September 2014, Kaman, Turkey* (J. Henderson and K. Matsumura, editors). Anatolian Archaeological Studies vol. **XXI**.

Darque-Ceretti, E., Helary, D., Bouquillon, A. and Aucouturier, M. (2005) Gold-like lustre: nanometric surface treatment for decoration of glazed ceramics in ancient Islam, Moresque Spain and Renaissance Italy. *Surface Engineering*, **21**, 352–358.

Degryse, P. and Shortland, A.J. (2009) Trace elements in provenancing raw materials for Roman glass production (an inaugural lecture to the society). *Geologica Belgica*, **12**, 135–143.

Degryse, P., Henderson, J. and Hodgins, G. (editors) (2009) *Isotopes in Vitreous Materials – A State-of-the-art and Perspectives*. Leuven University Press, Leuven, Belgium.

Degryse, P., Boyce, A., Erb-Satullo, N., Eremin, K., Kirk, S., Scott, R.B., Shortland, A.J., Schneider, J. and Walton, M., (2010) Isotopic discriminants between Late Bronze Age glasses from Egypt and the Near East. *Archaeometry*, **52**, 380–388.

Doremus, R.H. (1994) *Glass Science*. John Wiley & Sons, New York.

Dotsika, E., Poutoukis, D., Tzavidopoulos, I., Maniatis, Y., Ignatiadou, D. and Raco, B. (2009) A natron source at Pikrolimni Lake in Greece? Geochemical evidence. *Journal of Geochemical Exploration*, **103**, 133–143.

Dungworth, D. (2003) *Scientific examination of glass and glassworking materials from Silkstone, Yorkshire*. English Heritage, UK.

Dungworth, D. (2012) Historic window glass: The use of chemical analysis to date manufacture. *Journal of Architectural Conservation*, **18**, 7–25.

Dungworth, D., Cromwell, T., Ashurst, D., Cumberpatch, C., Higgins, D. and Willmott, H. (2006) Glass and pottery manufacture at Silkstone, Yorkshire. *Post-Medieval Archaeology*, **40**, 160–190.

Dussubieux, L. (2001) *L'Apport de l'ablation laser couplée a l'ICP-MS à la caractérisation des verres: Application a l'étude du verre de l'océan Indien*, Ph.D. dissertation, Université d'Orléans, France.

Dussubieux, L. and Gratuze, B. (2013). Glass in South Asia. *Modern Methods for Analysing Archaeological and Historical Glass*, **1**, 399–413.

Dussubieux, L., Gratuze, B. and Blet-Lemarquand, M. (2010) Mineral soda alumina glass: occurence and meaning. *Journal of Archaeological Science*, **37**, 1646–1655.

Fage, J.D. (1962) Some remarks on beads and trade in lower Guinea in the sixteenth and seventeenth centuries. *The Journal of African History*, **3**, 343–347.

Fontaine, S. and Foy, D. (2007) L'épave Ouest-Embiez 1, Var: le commerce maritime du verre brut et manufacturé en Méditerranée occidentale dans l'Antiquité. *Revue archéologique de Narbonnaise*, **40**, 235–265.

Foster, H.E. and Jackson, C.M. (2009) The composition of 'naturally coloured'late Roman vessel glass from Britain and the implications for models of glass production and supply. *Journal of Archaeological Science*, **36**, 189–204.

Foy, D. (2018) An overview of the circulation of glass in antiquity. Pp. 265–299 in: *Trade, Commerce, and the State in the Roman World* (A. Wilson and A.K. Bowman, editors). Oxford University Press, Oxford, UK.

Foy, D., Thirion Merle, V. and Vichy, M. (2004) Contribution à l'étude des verres antiques décolorés à l'antimoine. *ArchéoSciences, revue d'Archéométrie*, **28**, 169–177.

Freestone, I.C. (2006) Glass production in Late Antiquity and the Early Islamic period: a geochemical perspective. Pp. 201–216 in: *Geomaterials in Cultural Heritage* (M. Maggetti and B. Messiga, editors). Geological Society Special Publications, **257**. Geological Society, London.

Freestone, I.C. (2015) The recycling and reuse of Roman glass: analytical approaches. *Journal of Glass Studies*, **57**, 29–40.

Freestone, I.C., Gorin-Rosen, Y. and Hughes, M.J. (2000) Primary glass from Israel and the production of glass in late antiquity and the early Islamic period. *Publications de la Maison de l'Orient et de la Méditerranée*, **33**, 65–83.

Freestone, I., Wolf, S. and Thirlwall, M. (2005) The production of HIMT glass: elemental and isotopic evidence. Pp. 153–157 in: *Annales du 16e Congrès de l'Association Internationale pour l'Histoire du Verre: London 2003.*

Fuxi, G., Brill, R.H. and Shouyun, T. (editors) (2009) *Ancient Glass Research along the Silk Road.* World Scientific, Singapore.

Fuxi, G., Li, Q. and Henderson, J. (editors) (2016) *Recent Advances in the Scientific Research on Ancient Glass and Glaze.* World Scientific, Singapore.

Garner, H. (1956) An early piece of glass from Eridu. *Iraq*, **18**, 147–149.

Giussani, B., Monticelli, D. and Rampazzi, L. (2009) Role of laser ablation-inductively coupled plasma-mass spectrometry in cultural heritage research: a review. *Analytica Chimica Acta*, **635**, 6–21.

Gliozzo, E. (2017) The composition of colourless glass: a review. *Archaeological and Anthropological Sciences*, **9**, 455–483.

Gratuze, B. (2009) Les premiers verres au natron retrouvés en Europe occidentale: composition chimique et chronotypologie. Pp. 8–14 in: *Annales of the 17th Congress of the International Association for the History of Glass* (K. Janssens, P. Degryse, P.J. Cosyns, J. Caen and L. Van't dack, editors). University Press Antwerp, Antwerp.

Gratuze, B. (2016) Glass characterization using laser ablation-inductively coupled plasma-mass spectrometry methods. Pp. 179–196 in: *Recent Advances in Laser Ablation ICP-MS for Archaeology*. Springer, Berlin, Heidelberg.

Gratuze, B. and Barrandon, J.N. (1990) Islamic glass weights and stamps: analysis using nuclear techniques. *Archaeometry*, **32**, 155–162.

Gratuze, B. and Billaud, Y. (2014) Inventaire des perles en verre et en faïence de l'Age du Bronze originaires des ateliers de la région de Frattesina retrouvées en France. Pp. 25–38 in: *Atti delle XVI Giornate Nazionali di Studio sul Vetro, Adria (RO), Museo Archeologico Nazionale, 12–13 maggio 2012, Association Internationale pour l'Histoire du Verre - Comitato Nazionale Italiano* (S. Ciappi, A. Larese and M. Uboldi, editors).

Gratuze, B. and Janssens, K. (2004) Provenance analysis of glass artefacts. Pp. 663–712 in: *Non-destructive analysis of Cultural Heritage Materials* (K. Janssen and R.van Griecken, editors). Elsevier, Amsterdam.

Gratuze, B., Louboutin, C. and Billaud, Y. (1998) Les perles protohistoriques en verre du Musée des Antiquités Nationales. *Antiquités Nationales*, **30**, 13–24.

Gratuze, B., Koenig, M.P., Plouin, S. and Treffort, J.M. (2013) Les perles en faïence et en verre de l'âge du Bronze: contextes archéologiques et analyses pour l'Alsace et la Lorraine. *Cahiers alsaciens d'archéologie d'art et d'histoire, Société pour la conservation des monuments historiques d'Alsace*, **LVI**, 21–52.

Gratuze, B., Guérit, M., Simon, L., Villaverde, L. and Barbier, E. (2014) Des verres carolingiens de compositions atypiques à Bressuire et Faye-sur-Ardin (nord de la région melloise, Deux-Sèvres). *Bulletin de l'Association Française pour l'Archéologie du Verre*, 114–120.

Grose, D.F. (1979) The Syro-Palestinian glass industry in the later Hellenistic period. *Muse. Annual of the Museum of Art and Archaeology, University of Missouri Columbia, Columbia, Missouri*, **13**, 54–67.

Grose, D.F. (1989) *Early Ancient Glass: Core-Formed, Rod-Formed, and Cast Vessels and Objects from the Late Bronze Age to the Early Roman Empire, 1600 BC to 50 AD*. The Toledo Museum of Art. Hudson Hills Press in Association with the Toledo Museum of Art, New York.

Guilaine, J., Gratuze, B. and Barrandon, J.N. (1991) Les perles de verre du Calcolithique et de l'Age du Bronze. Pp. 255–266 in: *Le Bronze Atlantique*, Proceedings of the 1er Colloque de Beynac, 10–14 Septembre 1990.

Hamza, M. (1930) Excavations of the Department of Antiquites at Qantir. *Annales du Service des Antiquites de l'Égypte*, **30**, 31–68.

Hancock, R.G., Aufreiter, S. and Kenyon, I. (1996) European white glass trade beads as chronological and trade markers. *MRS Online Proceedings Library Archive*, **462**, 181.

Harden, D. (1981) *Catalogue of Greek and Roman Glass in the British Museum*, vol. I. British Museum, London.

Harding, A. and Warren, S.E. (1973) Early Bronze Age faience beads from Central Europe. *Antiquity*, **47**, 64–66.

Hartmann, G. (1994) Late-medieval glass manufacture in the Eichsfeld region (Thuringia, Germany). *Chemie der Erde*, **54**, 103–128.

Hartmann, S. and Grünewald, M. (2010) The late antique glass from Mayen,(Germany): First results of chemical and archaeological studies. Pp. 15–28 in: *Glass Along the Silk Road from 200 BC to AD 1000: International Conference Within the Scope of the "Sino-German Project on Cultural Heritage Preservation" of the RGZM and the Shaanxi Provincial Institute of Archaeology, December 11th-12th, 2008* (B. Zorn and A. Hilgner, editors). Verlag des Römisch-Germanischen Zentralmuseum, Germany.

Hartmann, G., Kappel, I., Grote, K. and Arndt, B. (1997) Chemistry and technology of prehistoric glass from Lower Saxony and Hesse. *Journal of Archaeological Science*, **24**, 547–559.

Henderson, J. (1988) Electron probe microanalysis of mixed-alkali glasses. *Archaeometry*, **30**, 77–91.

Henderson, J. (1993) Chemical analysis of the glass and faience from Hauterive-Champréveyeres, Switzerland. Pp. 111–117 in: *Hauterive-Champréveyeres, 9: Mètal et Parure au Bronze Final, Neuchatel, Musèe Cantonal d'Archèologie* (A.M. Rychner-Faraggi, editor).

Henderson, J. (2001) Glass and glazes. Pp. 471–482 in: *Handbook of Archaeological Sciences* (D.R. Brothwell and A.M. Pollard, editors). John Wiley & Sons, Chichester, UK.

Henderson, J. (2013a) *The Science and Archaeology of Materials: An Investigation of Inorganic Materials*. Routledge, UK.

Henderson, J. (2013b) *Ancient Glass: An Interdisciplinary Exploration*. Cambridge University Press, Cambridge, UK.

Henderson, J., McLoughlin, S.D. and McPhail, D.S. (2004) Radical changes in Islamic glass technology: evidence for conservatism and experimentation with new glass recipes from early and middle Islamic Raqqa, Syria. *Archaeometry*, **46**, 439–468.

Henderson, J., Evans, J.A. and Nikita, K. (2010) Isotopic evidence for the primary production, provenance and trade of Late Bronze Age glass in the Mediterranean. *Mediterranean Archaeology and Archaeometry*, **10**, 1–24.

Henderson, J., Evans, J., Bellintani, P. and Bietti-Sestieri, A.M. (2015) Production, mixing and provenance of Late Bronze Age mixed alkali glasses from northern Italy: an isotopic approach. *Journal of Archaeological Science*, **55**, 1-8.

Huang, P.Y., Kurasch, S., Srivastava, A., Skakalova, V., Kotakoski, J., Krasheninnikov, A.V., Hovden, R., Mao, Q., Meyer, J.C., Smet, J., Muller, D.A and Kaiser, U. (2012) Direct imaging of a two-dimensional silica glass on graphene. *Nano Letters*, **12**, 1081–1086.

Israeli, Y. (1991) The invention of blowing. Pp. 46–55 in: *Roman Glass: Two Centuries of Art and Innovation* (M. Newby and K. Painter, editors). *Society of Antiquarians London Occasional Papers*, **13**.

Jackson, C.M. (2005) Making colourless glass in the Roman period. *Archaeometry*, *47*, 763–780.

Jackson, C.M. and Nicholson, P.T. (2007) Compositional analysis of the vitreous materials found at Amarna. Pp. 101–116 in: *Brilliant things for Akhenaten: the production of glass, vitreous materials and pottery at Amarna site 045.1*. Egypt Exploration Society, London.

Jackson, C.M. and Nicholson, P.T. (2010) The provenance of some glass ingots from the Uluburun shipwreck. *Journal of Archaeological Science*, **37**, 295–301.

Jackson, C.M. and Paynter, S. (2016) A great big melting pot: exploring patterns of glass supply, consumption and recycling in Roman Coppergate, York. *Archaeometry*, **58**, 68–95.

Jackson, C.M. and Smedley, J.W. (2004) Medieval and post-medieval glass technology: melting characteristics of some glasses melted from vegetable ash and sand mixtures. *Glass Technology*, **45**, 36–42.

Jackson, C.M. and Smedley, J.W. (2008) Theophilus and the use of beech ash as a glassmaking alkali. Pp. 117–130 in: *Archaeology, History and Science* (M. Martinon-Torres and T. Rehren, editors). Left Coast Press, Walnut Creek, California, USA.

Jackson-Tal, R.E. (2004) The late Hellenistic glass industry in Syro-Palestine: a reappraisal. *Journal of Glass Studies*, **46**, 11–32.

Jaksch, H., Seipel, W., Weiner, K.L. and El Goresy, A. (1983) Egyptian blue – cuprorivaite a window to ancient Egyptian technology. *Naturwissenschaften*, **70**, 525–535.

Kaczmarczyk, A. (2007) Historical and regional variation in composition. Pp. 29–37 in: *Faiences et matière vitreuses de 'Orient Ancien* (A. Caubert, editor).

Kaczmarczyk, A. and Hedges, R.E.M. (1983) *Ancient Egyptian faience: an analytical survey of Egyptian faience from Predynastic to Roman times*. Aris and Phillips, Warminster, UK.

Keller, C.A. (1983) Problem in dating glass industries of the Egyptian New Kingdom: examples from Malkata and Lisht. *Journal of Glass Studies*, **25**, 19–28.

Kenoyer, M.J. (1994) Faience ornaments of Harappa and the Indus Valley civilization. *Ornament*, **17**, 35–39.

Kenoyer, M.J. (1997) Trade and technology of the Indus Valley: new insights from Harappa, Pakistan. *World Archaeology*, **29**, 262–280.

Kunicki-Goldfinger, J., Targowski, P., Góra, M., Karaszkiewicz, P. and Dzierżanowski, P. (2009) Characterization of glass surface morphology by optical coherence tomography. *Studies in Conservation*, **54**, 117–128.

Kunicki-Goldfinger, J.J., Freestone, I.C., McDonald, I., Hobot, J.A., Gilderdale-Scott, H. and Ayers, T. (2014) Technology, production and chronology of red window glass in the medieval period–rediscovery of a lost technology. *Journal of Archaeological Science*, **41**, 89–105.

Lacovara, P. (1998) Nubian faience. Pp. 46–49 in: *Gift of the Nile. Ancient Egyptian Faience* (F.D. Friedman, editor). Thames and Hudson, New York.

Lahlil, S., Biron, I., Galoisy, L. and Morin, G. (2008) Rediscovering ancient glass technologies through the examination of opacifier crystals. *Applied Physics A*, **92**, 109–116.

Lahlil, S., Biron, I., Cotte, M. and Susini, J. (2010) New insight on the in situ crystallization of calcium antimonate opacified glass during the Roman period. *Applied Physics A*, **100**, 683–692.

Lankton, J.W. and Dussubieux, L. (2006) Early glass in the Asian maritime trade: a review and an interpretation of compositional analyses. *Journal of Glass Studies*, **48**, 121–144.

Lankton, J.W. and Dussubieux, L. (2013) Early glass in southeast Asia. *Modern Methods for Analysing Archaeological and Historical Glass*, **1**, 415–443.

Liang, H., Sax, M., Saunders, D. and Tite, M. (2012) Optical coherence tomography for the non-invasive investigation of the microstructure of ancient Egyptian faience. *Journal of Archaeological Science*, **39**, 3683–3690.

Lilyquist, C. and Brill, R.H., editors (1993) *Studies in early Egyptian glass*. Metropolitan Museum of Art, New York.

Lorenz, A. (2006) Der spätbronzezeitliche Hortfund von Stadtallendorf unter besonderer Berücksichtigung seiner Gläser. *Archäologische Berichte* 20. Habelt, Bonn, Germany.

Lucas, A. and Harris, J.R. (1962) *Ancient Egyptian Materials and Industries*. IV Edition. Edward Arnold and Co, London.

Mass, J.L., Wypyski, M.T. and Stone, R.E. (2002) Malkata and Lisht glassmaking technologies: towards a specific link between second millennium BC metallurgists and glassmakers, *Archaeometry*, **44**, 67–82.

Matoïan, V. (2000) Données nouvelles sur le verre en Syrie au Iie millénaire av. J.-C.: le cas de Ras Shamra-Ougarit. Pp. 23–48 in: *La Route du verre: Ateliers primaires et secondaires du second millénaire av. J.-C. au Moyen Age* (M.D. Nenna, editor). Maison de l'Orient Méditerraneéan-Jean Pouilloux, Lyon, France.

Matoïan, V. and Bouquillon, A. (2000) Le "bleu 'egyptien" 'a Ras Shamra-Ougarit (Syrie)'. Pp. 983–1000 in: *Proceedings of the First International Congress on the Archaeology of the Ancient Near East*, Rome, 18–23 May 1998 (P. Matthiae, A. Enea, L. Peyronel and F. Pinnock, editors), Rome.

McClelland, M. (1984) *Core-formed glass from dated contexts*. Unpublished Ph.D. dissertation, University of Philadelphia.

Mildner, S., Falkenstein, F., Schmidt, J.P. and Schüssler, U. (2010) Materialanalytische Untersuchungen an ausgewählten Glasperlen des bronzezeitlichen Hortfundes von Neustrelitz, Lkr. Mecklenburg-Strelitz. *Bodendenkmalpflege in Mecklenburg-Vorpommern Jahrbuch*, **57**, 43–63.

Mildner, S., Schüssler, U., Falkenstein, F. and Brätz. H., (2014) Bronzezeitliches Glas im westlichen Mitteleuropa-Funde, Zusammensetzung und die Frage nach seiner Herkunft. Pp. 100–108 in: *Ressourcen und Rohstoff ein der Bronzezeit* (B. Nessel, I. Heske and D. Brandherm, editors). Nutzung-Distribution-Kontrolle. Brandenburgisches Landesamt für Denkmalpflege und Archäologisches Landesmuseum, Wünsdorf, Germany.

Mildner, S., Schüssler, U., Falkenstein, F. and Brätz, H. (2018) Bronzezeitliches "High-Magnesium-Glass" in Mitteleuropa – Lithium und Bor als Indizien für eine mögliche Herkunft aus Westanatolien. Pp. 20–24 in: *Archäometrie und Denkmalβflege* 2018 (L. Glaser, editor). Jahrestagung am Deutschen Elektronen-Synchrotron Hamburg, Germany.

Miller H.M.L. (1999) *Pyrotechnology and society in the cities of the Indus Valley*. Ph.D. thesis, University of Wisconsin-Madison, USA.

Mirti, P., Pace, M., Negro Ponzi, M. M. and Aceto, M. (2008) ICP–MS analysis of glass fragments of Parthian and Sasanian Epoch from Seleucia and Veh Ardašir (Central Iraq). *Archaeometry*, **50**, 429–450.

Moorey, P.R.S. (1994) *Ancient Mesopotamian Materials and Industries*. Oxford University Press, Oxford, UK.

Muthucumarana, R., Gaur, A.S., Chandraratne, W.M., Manders, M., Rao, B.R., Bhushan, R., Khedekar, V.D. and Dayananda, A.M.A. (2014) An early historic assemblage offshore of Godawaya, Sri Lanka: Evidence for early regional seafaring in south Asia. *Journal of Maritime Archaeology*, **9**, 41–58.

Nakahira M. and Kato T. (1963) Thermal transformations of pyrophyllite and talc as revealed by X-ray and electron diffraction studies. *Clays and Clay Minerals*, **12**, 21–27.

Nenna, M.D. (2003) Les ateliers égyptiens à l'époque gréco-romaine. Pp. 32–33 in: *Coeur de verre. Production et diffusion du verre antique* (D. Foy, editor). Infolio, Gollion, Switzerand.

Nenna, M.D. (2007) Production et commerce du verre à l'époque impériale: nouvelles découvertes et problématiques. *Facta*, **1**, 125–47.

Nenna, M.D. and Gratuze, B. (2009) Étude diachronique des compositions de verres employés dans les vases mosaïqués antiques. Pp. 59–63 in: *Annales du 17e Congrès de l'Association Internationale pour l'Histoire du Verre. Bruxelles.*

Nicholson, P.T. (1998) Materials and technology. Pp. 50–66 in: *Gift of the Nile. Ancient Egyptian Faience* (F.D. Friedman, editor). Thames and Hudson, New York.

Nicholson, P.T. (2002) Hellenistic/Roman faience production at Memphis, Egypt. Pp. 141–145 in: *Hyalos Vitrum Glass* (G. Kordas, editor). Glasnet, Athens.

Nicholson, P.T. (2007) *Brilliant things for Akhenaten: the production of glass, vitreous materials and pottery at Amarna site O45.1*. Egypt Exploration Society, London.

Nicholson, P.T. (2013) *Working in Memphis: The production of faience at the Romano Priod Kom Helul, Excavation*. Memoir 105, Egypt Exploration Society, London.

Nicholson, P.T. and Peltenburg, E.J. (2000) Egyptian faience. Pp. 177–195 in: *Ancient Egyptian Materials and Technology* (P.T. Nicholson and I. Shaw, editors). Cambridge University Press, Cambridge, UK.

Nikita, K. and Henderson, J. (2006) Glass analyses from Mycenaean Thebes and Elateia: compositional evidence for a Mycenaean glass industry. *Journal of Glass Studies*, **48**, 71–120.

Oates, D. (1968) The excavations at Tell el Rimah. *Iraq*, **30**, 115–38.

Oates, J., Oates, D. and MacDonald, H. (1997) *Excavations at Tell Brak, vol.* 1. The Macdonald Institute of Archaeological Research, Cambridge, UK.

Oikonomou, A. (2018) Hellenistic core formed glass from Epirus, Greece. A technological and provenance study. *Journal of Archaeological Science: Reports*, **22**, 513–523.

Oikonomou, A., Henderson, J., Gnade, M., Chenery, S. and Zacharias, N. (2018) An archaeometric study of Hellenistic glass vessels: evidence for multiple sources. *Journal of Archaeological and Anthropological Sciences*, **10**, 97–110.

Olmeda, G., Prosdocimi, B., Angelini, I., Cupit, M., Molin, G. and Leonardi, G. (2015) Archeologia e archeometria delle perle in vetro (VI-IV secolo a.C.). Osservazioni sulla tecnologia della necropoli patavina del CUS-Piovego del vetro in Veneto nella piena Età del ferro. Pp. 549–557 in: *Preistoria e Protostoria del Veneto* (G. Leonardi and V. Tinè, editors). Studi di Preistoria e Protostoria 2, Grafiche Antiga, Crocetta del Montello, Treviso, Italy.

Oppenheim, A.L. (1970) Chapters 1–3. Pp. 9–10 in: *Glass and Glassmaking in Ancient Mesopotamia* (A.L. Oppenheim, R.H. Brill, D.P. Barag and A. Von Saldern, editors). The Corning Museum of Glass [reprint 1988], Corning, New York.

Pactat, I., Gratuze, B. and Derbois, M. (2015) "Un atelier de verrier carolingien à Méru,"ZAC Nouvelle France"(Oise)." *Bulletin de l'Association Française pour l'Archéologie du Verre*, 73–78.

Pactat, I., Guérit, M., Simon, L., Gratuze, B., Raux, S. and Aunay, C. (2017) Evolution of glass recipes during the Early Middle Ages in France: analytical evidence of multiple solutions adapted to local contexts. Pp. 334–340 in: *Annales du 20e Congrès de l'Association Internationale pour l'Histoire du Verre 2015* (S. Wolf and A. de Pury-Gysel, editors). AIHV, Fribourg/Romont, Switzerland.

Pagès-Camagna, S. and Colinart, S. (2003) The Egyptian green pigment: its manufacturing process and links to Egyptian blue. *Archaeometry*, **45**, 637–658.

Panagiotaki M., Maniatis Y., Kavoussanaki D., Hatton G. and Tite M.S. (2004) The production technology of Aegean Bronze Age vitreous materials. Pp. 155–180 in: *Social Context of Technological Change II* (J. Bourriau and J Phillips, editors). Oxbow Books, Oxford, UK.

Panighello, S., Orsega, E.F., van Elteren, J.T. and Šelih, V.S. (2012) Analysis of polychrome Iron Age glass vessels from Mediterranean I, II and III groups by LA-ICP-MS. *Journal of Archaeological Science*, **39**, 2945–2955.

Patch, D.C. (1998) By necessity or design: faience used in ancient Egypt. Pp. 32–45 in: *Gift of the Nile. Ancient Egyptian Faience* (F.D. Friedman, editor). Thames and Hudson, New York.

Paynter, S. (2014) Late Bronze Age glass beads. *Glass News*, nº 35, January, 8–9.

Petrie, W.M.F. (1894) Tell el Amarna. Methuen and Co, Warminster, London.

Petrie, W.M.F. (1909) Memphis I, British School of Archaeology in Egypt. London.

Petrie, W.M.F. (1911) The pottery kilns at Memphis, in Historical Studies II. Quaritch, London, 7–34.

Petrie, W.M.F. and Gardner, E.A. (1886) Naukratis I, Trubener and Co, London.

Phelps, M., Freestone, I. C., Gorin-Rosen, Y. and Gratuze, B. (2016) Natron glass production and supply in the late antique and early medieval Near East: The effect of the Byzantine-Islamic transition. *Journal of Archaeological Science*, **75**, 57–71.

Plouin, S., Koenig, M.P. and Gratuze, B. (2012) Les perles en verre de l'âge du Bronze d'Alsace et de Lorraine. Pp. 11–36 in: *Le Verre en Lorraine et dans les régions voisines* (V. Arveiller and H. Cabart, editors). Actes du colloque de l'AFAV, Metz, 18–19 Novembre 2011, Éditions Monique Mergoil, Monographies Instrumentum, 42, Montagnac, France.

Polla, A., Angelini, I., Artioli, G., Bellintani, P. and Dore, A. (2011) Archaeometric investigation of Early Iron Age glass from Bologna. Pp. 139–144 in: *Proceedings of the 37th International Symposium on Archaeometry*, 13th–16th May 2008. Springer, Siena, Italy.

Pulak, C. (1998) The Uluburun shipwreck: an overview. *The International Journal of Nautical Archaeology*, **27**, 188–224.

Purowski, T., Dzierżanowski, P., Bulska, E., Wagner, B. and Nowak, A. (2012) A study of glass beads from the Hallstatt C-D from southwestern Poland: implications for glass technology and provenance. *Archaeometry*,

54, 144–166.

Purowski, T., Wagner, B., Bulska, E., Syta, O. and Dzierżanowski, P. (2014) Glassy faience from the Hallstatt C period in Poland: a chemico-physical study. *Journal of Archaeological Science*, **50**, 288–304.

Purowski, T., Kepa, L. and Wagner, B. (2018) Glass on the Amber Road: the chemical composition of glass beads from the Bronze Age in Poland. *Archaeological and Anthropological Sciences*, **10**, 1283–1302.

Pusch, E.B. and Rehren, T. (2007) *Hochtemperatur-Technologie in der Ramses-Stadt, Rubinglas für den Pharao*. Gerstenberg, Hildesheim, Germany.

Quartieri, S. and Arletti, R. (2013). The use of X-ray absorption spectroscopy in historical glass research. Pp. 301–309 in: *Modern Methods for Analysing Archaeological and Historical Glass*, vol. 1 (K. Janssens, editor). John Wiley & Sons. New Delhi.

Quartieri, S., Triscari, M., Sabatino, G., Boscherini, F. and Sani, A. (2002) Fe and Mn K-edge XANES study of ancient Roman glasses. *European Journal of Mineralogy*, **14**, 749–756.

Rapp, G. (2009) *Archaeomineralogy*. Springer-Verlag, Berlin-Heidelberg.

Raux, S., Gratuze, B., Langlois, J.Y. and Coffineau, E. (2015) Indices d'une production verrière du Xe siècle à La Milesse (Sarthe). *Bulletin de l'Association Française pour l'Archéologie du Verre*, **2015**, 66–70.

Reade, W., Freestone, I.C. and St Simpson, J. (2005) Innovation or continuity? Early first millennium BCE glass in the Near East: the cobalt blue glasses from Assyrian Nimrud. Pp. 23–27 in: *Annales du 16e Congres de l'Association Interntionale pour l'Histoire du Verre-London*, 2003 (H. Cool, editor).

Rehren, Th. (1997) Ramesside glass-colouring crucibles. *Archaeometry*, **39**, 355–68.

Rehren, T. and Freestone, I.C. (2015) Ancient glass: from kaleidoscope to crystal ball. *Journal of Archaeological Science*, **56**, 233–241.

Rehren, Th. and Pusch, E.B. (2005) Late Bronze Age glass production at Qantir–Piramesses, Egypt. *Science*, **308**, 1756–1758.

Reisner, G.A. (1923) Excavations at Kerma IV/V. In: *Harvard African Studies* **6**. Harvard University Press, Cambridge, Massachusetts, USA.

Ricciardi, P., Colomban, P., Tournie, A., Macchiarola, M. and Ayed, N. (2009) A non-invasive study of Roman Age mosaic glass tesserae by means of Raman spectroscopy. *Journal of Archaeological Science*, **36**, 2551–2559.

Robertshaw, P., Wood, M., Melchiorre, E., Popelka-Filcoff, R.S. and Glascock, M.D. (2010) Southern African glass beads: chemistry, glass sources and patterns of trade. *Journal of Archaeological Science*, **37**, 1898–1912.

Robinson, C., Baczyńska, B. and Polańska, M. (2004) The origin of faience in Poland. *Sprawozdania Archeologiczne*, **56**, 79–154.

Roqué, J., Molera, J., Sciau, P., Pantos, E. and Vendrell-Saz, M. (2006) Copper and silver nanocrystals in lustre lead glazes: Development and optical properties. *Journal of the European Ceramic Society*, **26**, 3813–3824.

Rosenow, D. and Rehren, T. (2014) Herding cats–Roman to Late Antique glass groups from Bubastis, northern Egypt. *Journal of Archaeological Science*, **49**, 170–184.

Santopadre, P. and Verità, M. (2000) Analyses of production technologies of Italian vitreous materials of the Bronze Age. *Journal of Glass Studies*, **42**, 25–40.

Sayre, E.V. and Smith, R.W. (1961) Compositional categories of ancient glass. *Science*, **133**, 1824–1826.

Scapova, J.L. (1992) 'Un lissoir de Novgorod, réflexions sur la verrerie médiévale'. *Acta Archeologica*, **62**, 231–243.

Schalm, O., Janssens, K., Wouters, H. and Caluwé, D. (2007) Composition of 12–18th century window glass in Belgium: Non-figurative windows in secular buildings and stained-glass windows in religious buildings. *Spectrochimica Acta Part B: Atomic Spectroscopy*, **62**, 663–668.

Schibille, N. (2011) Late Byzantine mineral soda high alumina glasses from Asia Minor: A new primary glass production group. *PloS ONE*, **6** (4): e18970.

Schibille, N., Sterrett-Krause, A. and Freestone, I.C. (2017) Glass groups, glass supply and recycling in late Roman Carthage. *Archaeological and Anthropological Sciences*, **9**, 1223–1241.

Schlick-Nolte, L.A. and Werthmann, R. (2003) Glass vessels from the Burial of Nesikhons. *Journal of Glass Studies*, **45**, 11–34.

Séguier, J.M., Delattre, V., Gratuze, B., Peake, R., Viand, A., Didelot, C., Masse, F. and Relier, C. (2010) Les nécropoles protohistoriques de "La Haute Grève" à Gouaix (Seine-et-Marne). Contribution à l'étude des pratiques funéraires au cours de l'étape moyenne du Bronze final (XIIe-XIe siècle av. J.-C.) et au début du second âge du Fer (Ve-IIIe siècle av. J.-C.) dans le sud du Bassin parisien. *Supplément à la RACF n°37, FERACF*.

Sempowski, M.L., Nohe, A.W., Hancock, R.G., Moreau, J.F., Kwok, F., Aufreiter, S., Karklins, K., Baart, J., Garrad, C. and Kenyon, I. (2001) Chemical analysis of 17th-century red glass trade beads from northeastern North America and Amsterdam. *Archaeometry*, **43**, 503–515.

Shelby, J.E. (2005) *Introduction to Glass Science and Technology*. 2nd Edition. The Royal Society of Chemistry, Cambridge, UK, 291 pp.

Sheridan, A., Eremin, K. and Shortland, A. (2005) Understanding Bronze Age faience in Britain and in Ireland. Pp. 217–229 in: *Materials Issues in Art and Archaeology VII, Materials Research Society Symposium Proceeding, Vol* **852** (P.M. Vandiver, J.M. Mass and A. Murray, editors). Warrendale, Pennsylvania, USA.

Shortland, A.J. (2000) *Vitreous materials at Amarna*. Archaeopress, BAR international Series 827, Oxford, UK.

Shortland, A.J. (2004) Evaporites of the Wadi Natrun: seasonal and annual variation and its implication for ancient exploitation. *Archaeometry*, **46**, 497–516.

Shortland, A.J. (2009) *Glass Production*, UCLA Encyclopedia of Egyptology.

Shortland, A.J. and Eremin, K. (2006) The analysis of second millennium glass from Egypt and Mesopotamia, part 1: new WDS analyses. *Archaeometry*, **48**, 581–603.

Shortland, A.J. and Schroeder, H. (2009) Analysis of first millennium BC glass vessels and beads from the Pichvnari Necropolis, Georgia. *Archaeometry*, **51**, 947–965.

Shortland, A.J. and Tite, M.S. (2005) Technological study of ptolemaic – Early Roman faience from Memphis, Egypt. *Archaeometry*, **47**, 31–46.

Shortland, A., Schachner, L., Freestone, I. and Tite, M. (2006) Natron as a flux in the early vitreous materials industry: sources, beginnings and reasons for decline. *Journal of Archaeological Science*, **33**, 521–530.

Shortland, A.J., Shishlina, N. and Egorkov, A. (2007a) Origins and production of faience beads in the North Caucasus and the Northwest Caspian Sea region in the Bronze Age. Pp. 269–283 in: *Les cultures du Caucase: leur relations avec le Proche-Orient* (B. Lyonnet, editor). Éditions Recherche sur les Civilisations, CNRS Éditions, Paris.

Shortland, A.J., Rogers, N. and Eremin, K., (2007b) Trace element discriminants between Egyptian and Mesopotamian Late Bronze Age glasses. *Journal of Archaeological Science*, **34**, 781–789.

Shortland, A.J., Degryse P., Walton, M., Geer, M., Lauwers, V. and Salou, L. (2011) The evaporitic deposits of lake Fazda (Wadi Natrun, Egypt) and their use in Roman glass production. *Archaeometry*, **53**, 916–929.

Silvestri, A., Molin, G. and Salviulo, G. (2008) The colourless glass of Iulia Felix. *Journal of Archaeological Science*, **35**, 331–341.

Simon-Hiernard, D. and Gratuze, B. (2011) Le vase de Saint-Savin en Poitou et les verres médiévaux bleu-cobalt à décors blancs. *Bulletin de l'Association Française pour l'Archéologie du Verre*, 69–73.

Smirniou, M. and Rehren, Th. (2011) Direct evidence of primary glass production in late Bronze Age Amarna, Egypt. *Archaeometry*, **53**, 58–80.

Smirniou, M. and Rehren, Th. (2013) Shades of blue – cobalt-copper coloured blue glass from New Kingdom Egypt and the Mycenaean world: a matter of production or colourant source? *Journal of Archaeological Science*, **40**, 4731–4743

Smirniou, M., Rehren, Th., Adrymi-Sismani, V., Asderaki, E. and Gratuze, B. (2012) Mycenaean beads from Kazanaki, Volos: a further node in the LBA glass network. Pp. 11–18 in: *Annales du 18e Congrès de l'Association Internationale pour l'Histoire du Verre* (D. Ignatiadou and A. Antonaras, editors). Thessaloniki, Greece.

Smirniou, M., Rehren, Th. and Gratuze, B. (2018) Lisht as a New Kingdom glass-making site with its own chemical signature. *Archaeometry*, **60**, 502–516.

Šmit, Ž., Janssens, K., Schalm, O. and Kos, M. (2004) Spread of façon-de-Venise glassmaking through central and western Europe. *Nuclear Instruments and Methods in Physics Research Section B: Beam Interactions with Materials and Atoms*, **213**, 717–722.

Spaer, M. (2001) *Ancient glass in the Israel Museum. Beads and other small objects*. The Israel Museum, Jerusalem.

Stern, E.M. (1999) Roman glassblowing in a cultural context. *American Journal of Archaeology*, **103**, 441–484.

Stern, E.M. and Schlick-Nolte, B. (1994) *Early Glass of the Ancient World. 1600 B.C.–A.D. 50. Ernesto Wolf Collection*. Verlag Gerd Hatje, Ostfilderm, Germany.

Swan, C.M., Rehren, T., Dussubieux, L. and Eger, A.A. (2018) High-boron and high-alumina Middle Byzantine

(10th–12th century CE) glass bracelets: A Western Anatolian Glass industry. *Archaeometry*, **60**, 207–232.
Tite, M.S. (1987) Characterization of early vitreous materials. *Archaeometry*, **29**, 21–34.
Tite, M.S. and Bimson, M. (1986) Faience: An investigation of the microstructures associated with the different methods of glazing. *Archaeometry*, **28**, 69–78.
Tite, M.S. and Bimson, M. (1989) Glazed steatite: an investigation of the methods of glazing used in Ancient Egypt. *World Archaeology*, **21**, 87–100.
Tite, M.S. and Shortland, A. (2003) Production technology for copper- and cobalt-blue vitreous materials from the New Kingdom site of Amarna: A reappraisal. *Archaeometry*, **45**, 285–309.
Tite, M.S. and Shortland, A.J. (2008) *Production technology of faience and related early vitreous materials.* Oxford University School of Archaeology, Monograph 72, Oxford, UK.
Tite, M.S., Bimson, M. and Cowell, M.R. (1984) Technological examination of Egyptian blue. Pp. 215–242 in: *Archaeological Chemistry III* (J.B. Lambert, editor). American Chemical Society, Advances in Chemistry Series N° 205, Washington D.C.
Tite, M.S., Hatton, G., Shortland, A.J., Manatis, Y., Kavoussanaki, D. and Panagiotaki, M. (2005) Raw materials used to produce Aegean Bronze Age glass and related vitreous materials. Pp. 10–13 in: *Annales of 16th AIHV Congress* (H. Cool, editor). London.
Tite, M.S., Manti, P. and Shortland, A. (2007) A technological study of ancient faience from Egypt. *Journal of Archaeological Science*, **34**, 1568–1583.
Tite, M.S., Shortland, A.J. and Vandiver, P.B. (2008a) Raw materials and fabrication methods used in the production of faience. Pp. 37–56 in: *Production Technology of Faience and Related Early Vitreous Materials* (M.S. Tite and A.J. Shortland, editors). Oxford University School of Archaeology, Monograph 72. Oxford, UK.
Tite, M.S., Shortland, A.J. and Angelini, I. (2008b) Faience production in the Northern and Western Europe. Pp. 129–146 in: *Production Technology of Faience and Related Early Vitreous Materials* (M.S. Tite and A.J. Shortland, editors). Oxford University School of Archaeology, Monograph 72. Oxford, UK.
Tite, M.S., Shortland, A.J. and Bouquillon, A. (2008c) Glazed steatite. Pp. 23–36 in: *Production Technology of Faience and Related Early Vitreous Materials* (M.S. Tite and A.J. Shortland, editors). Oxford University School of Archaeology, Monograph 72. Oxford, UK.
Tite, M.S., Shortland, A.J., Kaczmarczyk, A. and Vandiver, P.B. (2008d) Faience production in Egypt. Pp. 57–90 in: *Production Technology of Faience and Related Early Vitreous Materials* (M.S. Tite and A.J. Shortland, editors). Oxford University School of Archaeology, Monograph 72. Oxford, UK.
Tite, M.S., Shortland, A.J., Maniatis, Y., Panagiotaki, M. and Kaczmarczyk, A. (2008e) Faience production in the Eastern Mediterranean. Pp. 111–127 in: *Production Technology of Faience and Related Early Vitreous Materials* (M.S. Tite and A.J. Shortland, editors). Oxford University School of Archaeology, Monograph 72. Oxford, UK.
Tite, M.S., Shortland, A.J., Bouquillon, A., Kaczmarczyk, A. and Vandiver, P.B. (2008f) Faience production in the Near East and the Indus Valley. Pp. 93–107 in: *Production Technology of Faience and Related Early Vitreous Materials* (M.S. Tite and A.J. Shortland, editors). Oxford University School of Archaeology, Monograph 72. Oxford, UK.
Tite, M.S., Shortland, A.J. and Hatton, G.D. (2008g) Production of Egyptian blue and green frits. Pp. 147–184 in: *Production Technology of Faience and Related Early Vitreous Materials* (M.S. Tite and A.J. Shortland, editors). Oxford University School of Archaeology, Monograph 72. Oxford, UK.
Towle, A., Henderson, J., Bellintani, P. and Gambacurta, G. (2001) Frattesina and Adria: report of scientific analyses of early glass from the Veneto. *Padusa*, **37**, 7–68.
Triantafyllidis, P. (2003) Classical and Hellenistic workshop from Rhodes. Pp. 131–138 in: *Exchanges et commerce du verre dans le monde antique* (D. Foy and M.D. Nenna, editors). Actes du colloque AFV, Aix-en-Provence et Marseille 2001. Mergoil Editor, Montagnac, France.
Triantafyllidis, P., Karatasios, I. and Andreopoulou-Magkou, E. (2012) Study of Core- Formed Glass Vessels from Rhodes. Pp. 529–544 in: *Proceedings of the 5th Symposium of HSA* (N. Zacharias, M. Georgakopoulou, K. Polikreti, G. Fakorellis and Th. Vakoulis, editors). University of Peloponnese Publications, Greece.
van der Werf, I., Mangone, A., Giannossa, L.C., Traini, A., Laviano, R., Coralini, A. and Sabbatini, L. (2009)

Archaeometric investigation of Roman tesserae from Herculaneum (Italy) by the combined use of complementary micro-destructive analytical techniques. *Journal of Archaeological Science*, **36**, 2625–2634.

Vandiver, P.B. (1982) Technological change in Egyptian faience. Pp. 167–179 in: *Archaeological Ceramics* (J.S. Olia and A.D. Franklin, editors). Smithsonian Institution Press, Washington D.C.

Vandiver, P.B. (1983) Egyptian faience technology. Appendix A. Pp. A1–A144 in: *Ancient Egyptian Faience: An Analytical Survey of Egyptian Faience from Predynastic to Roman Times* (A. Kaczmarczyk and R.E.M. Hedges, editors). Aris and Phillips, Warminster, UK.

Varberg, J., Gratuze, B. and Kaul, F. (2015) Between Egypt, Mesopotamia and Scandinavia: Late Bronze Age glass beads found in Denmark. *Journal of Archaeological Science*, **54**, 168–181.

Varberg, J., Gratuze, B., Kaul, F., Hansen, A.H., Rotea, M. and Wittenberger, M. (2016) Mesopotamian glass from Late Bronze Age Egypt, Romania, Germany, and Denmark. *Journal of Archaeological Science*, **74**, 184–194.

Varone, A. and Beart, H. (1996) Pittori Romani al lavoro, materiali, strumenti, tecniche: evidenze archeologiche e dati analiticidi un recente scavo Pompeiano lungo via dell'abbondanza (reg, IX INS. 12). Pp. 196–214 in *Proceedings of the International Workshop on Roman Wall Painting*, Fribourg, Switzerland.

Veiga, J.P. and Figueiredo, M.O. (2008) Calcium in ancient glazes and glasses: a XAFS study. *Applied Physics A*, **92**, 229–233.

Venclová, N., Hulínsk, V., Henderson, J., Chenery, S., Šulová, L. and Hložek, J. (2011) Late Bronze Age mixed-alkali glasses from Bhoemia, *Archeologické rozhledy*, **LXIII**, 559–585.

Walton, M.S., Shortland, A., Kirk, S. and Degryse, P. (2009) Evidence for the trade of Mesopotamian and Egyptian glass to Mycenaean Greece. *Journal of Archaeological Science*, **36**, 1496–1503.

Walton, M., Eremin, K., Shortland, A., Degryse, P. and Kirk, S. (2012) Analysis of Late Bronze Age glass axes from Nippur – a new cobalt colourant. *Archaeometry*, **54**, 835–852.

Wedepohl, K.H. (1997) Chemical composition of medieval glass from excavations in West Germany. *Glastechnische Berichte–Glass Science and Technology*, **70**, 246–255.

Wedepohl, K.H., Krueger, I. and Hartmann, G. (1995) Medieval lead glass from northwestern Europe. *Journal of Glass Studies*, **37**, 65–82.

Wedepohl, K.H., Simon, K. and Kronz, A. (2011a) Data on 61 chemical elements for the characterization of three major glass compositions in Late Antiquity and the Middle Ages. *Archaeometry*, **53**, 81–102.

Wedepohl, K.H., Simon, K. and Kronz, A. (2011b) The chemical composition including the Rare Earth Elements of the three major glass types of Europe and the Orient used in late antiquity and the Middle Ages. *Chemie der Erde – Geochemistry*, **71**, 289–296.

Welter, N., Schüssler, U. and Kiefer, W. (2007) Characterisation of inorganic pigments in ancient glass beads by means of Raman microspectroscopy, microprobe analysis and X-ray diffractometry. *Journal of Raman Spectroscopy*, **38**, 113–121.

Wesolowski, M. (1984) Thermal decomposition of talc: a review. *Thermochimica Acta*, **78**, 395–421.

Wood, M. (2011) A glass bead sequence for southern Africa from the 8th to the 16th century AD. *Journal of African Archaeology*, **9**, 67–84.

Yentsch, A. (1995) Beads as silent witnesses of an African-American past: Social identity and the artifacts of slavery in Annapolis, Maryland. *Kroeber Anthropological Society Papers*, **79**, 44–60.

Zacharias, N., Karavassili, F., Das, P., Nicolopoulos, S., Oikonomou, A., Galanis, A., Rauch, E., Arenal, R., Portillo, J., Roque, J., Casablanca, J. and Margiolaki, I. (2018) A novelty for cultural heritage material analysis: transmission electron microscope (TEM) 3D electron diffraction tomography applied to Roman glass tesserae. *Microchemical Journal*, **138**, 19–25.

Zachariasen, W.H. (1932) The atomic arrangement in glass. *Journal of the American Chemical Society*, **54**, 3841–3851.

Zhang, M., Hui, Q., Lou, X.J., Redfern, S.A.T., Salje, E.K.H. and Tarantino, S.C. (2006) Dehydroxylation, proton migration, and structural changes in heated talc: An infrared spectroscopic study. *American Mineralogist*, **91**, 816–825.

EMU Notes in Mineralogy, Vol. 20 (2019), Chapter 4, 151–202

The Vitruvian legacy: Mortars and binders before and after the Roman world

Gilberto ARTIOLI, Michele SECCO and Anna ADDIS

Department of Geosciences and CIRCe Centre, Università di Padova, Via Gradenigo 6, 35131 Padova, Italy
e-mail: gilberto.artioli@unipd.it, michele.secco@unipd.it, anna.addis@unipd.it

A brief history of the nature, use and technology of binders in ancient constructions and buildings is outlined, including the apparent chronological discontinuities related to technological developments. The skilled and clever use of mineral resources is at the base of the technical achievements related to architectural activities, from simple adobe to high-performance modern concrete. It is argued that among pre-industrial binders the Roman pozzolanic mortars were highly optimized materials, skillfully prepared and very durable. Their innovative use in architecture is one of the keys of the successful expansion of the Roman Empire. The role of mineralogy and mineral reactions is emphasized in terms of: (1) the preparation and manufacturing of the binding materials; (2) the hardening process and the development of the physical properties of the binder; and (3) the archaeometric reconstruction of the ancient materials.

1. Historical survey

Living in a sheltered place, permanently or temporarily, is a fundamental need of humans. Nevertheless, the use of natural materials to build shelter is not exclusive to humans: termite mounds, bird nests, beaver dams, and beehives are perfect examples of efficient architectural skills on the part of animals. According to Mike Ashby "the difference lies in the *competence* demonstrated by man in his extraordinary ability to expand and adapt that competence and development" (Ashby, 2013, p. 11). After dwelling in natural caves and an extensive period of preparation of temporary structures made of organic materials (skin, wood, leaf, *etc.*), the development of long-lasting architecture in human prehistory was based necessarily on the clever use of natural rocks and/or man-made binders (Wright, 2005).

1.1. In the beginning it was clay ...

The use of clay-rich mud to plaster huts and floors is the most direct use of natural minerals to solidify surfaces and make them impermeable (Wright, 1983; Staubach, 2013). This follows from the physical properties of clay minerals (Brown and Brindley, 1980; Bailey, 1988) which exhibit a small particle size and a marked flat morphology derived from the layered crystal structure. Clays in excess water form colloidal

DOI: 10.1180/EMU-notes.20.4

suspensions, and the settled fractions possess ideal plastic properties, so that readily formable materials can be obtained easily by soaking dry clay minerals.

Clays (and loess in China) have been used since Neolithic times to plaster walls (inside and outside), to consolidate and smooth rough floors and eventually reshape them repeatedly after use, to waterproof roofs made of organic material, and eventually as a mortar between stones and, finally, mudbricks. There is ample evidence of this architectural use of clays in the Middle East starting from the 10th millennium BC (see for example Schmandt-Besserat, 1977). In its simplest form, with no use of pyrotechnology, a masonry wall can be composed of sun-dried bricks (adobe) bound by a moist layer of mud that, on drying, makes the wall a solid mass of dry clay (Artioli and Secco, 2016). In other cases earth-based structures are formed by direct application and shaping of the constructions: these techniques are variously referred to as cob, if they have a load-bearing role, or wattle and daub if the earth has an infilling role (Friesem *et al.*, 2017). In some cases the clay slabs or blocks can be cut directly out of clay or muddy soil if they are naturally consolidated by organic material (turf, sod) or other pedogenetic processes (Huisman and Milek, 2017). The man-made clay bricks and the clay mortars regularly contain a substantial organic (straw, dung) or inorganic component (gravel, pebbles) in order to avoid cracking upon drying; a technology developed before or in parallel with pottery making. The past use of unfired or sun-baked clay products is often difficult to assess in the archaeological record, because they are subject to rapid degradation mainly by wind erosion, water and salts (Friesem *et al.*, 2017). Only in rather arid climates are ancient occurrences still recognizable (Figs. 1, 2), in most cases as structures embedded in mounds (tells) formed by layers of sequential human activity.

The earliest evidence of the use of sun-dried bricks is reported from the Pre-Pottery Neolithic layers in Jericho, Israel (9th millennium BC: Kenyon, 1981), in Nemrik, Iraq (Kozlowski and Kempisty, 1990) and in Ganj Darreh, Iran (Vatandoust *et al.*, 2011) at least from the 8th millennium BC. Evidence of the use of mudbricks in the Indus Valley (Harappa, Mohenjo-Daro) started at about the same time (end of the 8th millennium BC: Khan and Lemmen, 2013). Jericho stands as a special site combining the early systematic evidence of the use of mud-brick buildings alongside the earliest example of a Cyclopean structure, a massive round tower ~10 m in diameter and standing to almost the same height; though the significance of such massive construction is still debated (Kenyon, 1981; Bar-Yosef, 1986). There is ample use of mud-brick structures in the site of Çatalhöyük (7500–6000 BC: Hodder, 2006, 2012) and a number of other Anatolian sites, together with painted mud plaster covering internal and external structures, for both practical and symbolic purposes (Figs 3, 4). Interestingly, although there is evidence that the Çatalhöyük people could have used pyrotechnologically produced lime plaster, there is no firm evidence of it being employed. The wall surfaces as well as the interior floors of the structures at Çatalhöyük were carefully plastered and frequently re-plastered in time, but were coated in earthen plasters, which were supplemented by a thinner coat of local white marly clay (Kopelson, 1996; Arkun, 2003). The first Mesopotamian building in brick on a monumental scale occurred in the

Figure 1. The massive body of the Deffufa ('mud-brick building' in Nubian) of Kerma, Nubia. It was built entirely of sun-dried bricks and was the reference structure of the classic Kush civilization (2000–1500 BC).

Figure 2. Mud-brick qubba in the cemetery from the Islamic period near Dongola, Sudan. Qubbas are domed mausoleums which contain the grave of a saint or some important personage.

Figure 3. Sun-dried mud bricks with clay mortar in a Neolithic wall at Çatalhöyük, Turkey.

Figure 4. An internal wall of a living quarter in Çatalhöyük showing wall plastering and decorations.

Tell Halaf period (5th millenium BC: Oates, 1990; Robson, 1996). Rectangular mould-based bricks were introduced in Mesopotamia at least from the 4th or 3rd millennium BC (Tell Braq, Syria: Oates, 1990; Halaf, Syria: Davidson and McKerrell, 1976), rapidly becoming decorated or inscribed.

The tradition of using earthen architecture based on a mixture of clay, water and straw is still largely present in many rural areas of the world (Figs 5, 6), the splendid multi-story Al Mihdar Mosque of Tarim, Yemen and the Mosque of Djenne, Mali being two amazing examples of earthen architectural achievements. The technological basis of modern earth constructions is very similar to that inferred from the prehistoric record (Oates, 1990). It is worth mentioning that several trends of contemporary architecture (earth architecture, sustainable architecture, bioarchitecture, *etc.*) actively propose the modern use of clays or consolidated clays as natural materials in buildings (Minke, 2012). Furthermore, in the field of modern and ancient earthen architecture there are many unresolved issues concerning earthen conservation, capacity building and dissemination of information for appropriate conservation interventions on historic buildings, settlements and archaeological sites composed of earthen materials (Avrami *et al.*, 2008; Fratini *et al.*, 2011). The Getty Conservation Institute (GCI) through its Earthen Architecture Initiative (EAI: www.getty.edu/conservation/our_projects/field_projects/earthen/overview.html), a long-term legacy of the past Terra project, is at the forefront of active research and education on the theme.

Figure 5. Modern production of sun-dried bricks in Sudan.

Figure 6. Mud-brick walls in Sardinian houses. Traditional earthen architecture is preserved alongside to modern concrete building.

1.2. And then fire came ...

The Sumerian, Babylonian, Assyrian and Hittite civilizations used sun-dried clay tablets widely for cuneiform inscriptions, which was the written system for most of the languages of the Mesopotamian region (Walker, 1987). Large archives of tablets are available starting from about the 4th millennium BC and they present a number of conservation problems because they are fragile and salt-loaded (Organ, 1961). Only a very small number of them were originally fired, often accidentally during destructive conflagrations, and these are the most stable ones. Paradoxically, high temperature helps their preservation, so that modern electrical heating up to 740°C seems to be one of the best conservation treatments to stabilize these materials (Thickett *et al.*, 2002). By firing clay tablets at high temperature, conservators are now applying the long-known pyrotechnological process which has been the basis of production of clay figurines since Upper Palaeolithic times (see the Venus of Dolní Veštonice, Vandiver *et al.*, 1989), the making of early fired ceramics in Eastern Asia since at least ~15000 y BC (Kuzmin, 2006; Wu *et al.*, 2012; Gibbs, 2015), and the shift from sun-dried to fired bricks in Mesopotamia around the 3rd millennium BC. It is the very same firing process transforming plastic clays into stable structural products (Artioli, 2010; Heimann *et al.*, 2010; Staubach, 2013): during firing the clay minerals are dehydroxylated progressively and reactive oxides are formed, yielding glass and a variety of high-temperature crystal phases, depending on the starting clay composition and the time-temperature path followed. High temperatures and long firing times imply more complete reactions, better crystallization and highly sintered micro-structures (Cultrone *et al.*, 2004). Technically, fired bricks are ~15% denser than corresponding mudbricks of the same size, and about five times more resistant to compression. They are also lighter than natural limestone but possess almost doubled values of the compressive strength of many common carbonate stones. Bricks represent an almost ideal unit material for masonry.

Fired bricks bearing inscriptions (Fig. 7) have been found in most Middle Eastern excavations, including Babylonia (Oppenheim, 1965), Nimrud (Oates, 1961) and Nippur (McCown, 1952). The buildings in many of the Mesopotamian tells are made of mixed construction materials: mainly mudbricks, but also fired bricks and stones. In fact "...once bricks had been developed, it became general practice to build the mass of a building in sun-dried bricks, whilst facing the lower courses and paving the floors with kiln-fired bricks. In a country short of wood for fuel, baked bricks were a luxury, commonly used only where necessary to protect the unfired from erosion by wind or water." (Moorey, 1999). The large prevalence of mudbricks may prove that "burnt bricks were not as fundamental to Mesopotamian civilization as was fine stone dressing to Pharaonic civilization" (Wright, 2005, p. 110). The culmination of burnt brick construction in Mesopotamia was during Neo-Babylonian times (see the famous Ishtar Gate and Processional Way at Babylon, which also represent a great example of glaze

Figure. 7. Inscribed bricks in the wall of Chogha Zanbil, the great Elamite zigurrath in Iran, some 30 km south-east of Susa.

decoration of bricks). A Mesopotamian-type evolution of building techniques is present also in the Indus Valley, where the transition from mudbricks to fired bricks appears in the Kot-Dijan period (2800–2600 BC) and burnt bricks become common in the mature Harappan stage (2600–2450 BC; Kenoyer, 1991; Khan and Lemmen, 2013), used mostly with simple clay mortar (Mackay, 1938). There are no other known examples of structural use of fired bricks elsewhere in the world until the first half of the 1st millennium BC, whereas already in the 2nd millennium BC there is ample evidence of the use of terra-cottas for roofing tiles and temple decorations in Greece and later in Etruria. Roofing tiles were also introduced in China during the Shang Dinasty (1700–1027 BC, Sui Pheng, 2001).

It is puzzling to note that everywhere in the world the process of mudbrick firing for architectural purposes occurred several thousand years later than the use of pyrotechnology to fire lime and ceramics in the same regions. The production of fired bricks of course required large fuel resources and substantial manpower (Potts, 2014). In any case the bricks were used to produce more flexible architecture and more stable masonry, even if the brick units were not strengthened with binder. In Mesopotamia a mixture of gypsum and clay was mostly used as mortar, apparently following an interesting regional pattern (see fig. 14 of Kingery *et al.*, 1988), although the lack of scientific analyses and the use of ambiguous terminology for materials makes many of the early archaeological reports rather unreliable (Moorey, 1999, p. 330). Note that some of the Mesopotamian buildings encompass the early known water-treatment structures (Sanizadeh, 2008): in several of them some degree of waterproofing was obtained by use of fired bricks and tar plastering. In Mesopotamia the first use of bitumen (or a mixture of gypsum and bitumen) as mortar in masonry is recorded (Moorey, 1999; Sauvage, 2011), as also reported by historical sources (Herodotus 1, 179, 1–4): "Further, I must relate where the earth was used as it was dug from the moat and how the wall was constructed. As they dug the moat, they made bricks of the earth which was carried out of the place they dug, and when they had moulded bricks enough, they baked them in ovens; then using hot bitumen for cement and interposing layers of wattled reeds at every thirtieth course of bricks, they built first the border of the moat and then the wall itself in the same fashion. ... There is another city, called Is, eight days' journey from Babylon, where there is a little river, also named Is, a tributary of the Euphrates river; from the source of this river Is, many lumps of bitumen rise with the water; and from there the bitumen was brought for the wall of Babylon." The latter phrase of Herodotus indicates how common the tar material was in the area.

2. Classification of inorganic binders: their chemistry and mineralogy

Before delving into the details of binder developments through prehistory and history, it is necessary to summarize the nature and properties of the materials. Because of the ample literature available (Barnes and Bensted, 2002; Hewlett, 2003; Artioli, 2010), only a brief introduction will be given.

To avoid confusion, we will define the terms used for the binding materials following the physical condition and context of use (Hobbs and Siddall, 2011). 'Cement' is a powder material providing internal cohesion derived from some sort of chemical reaction, mostly with water; 'concrete' is a composite made by a binder and large-sized (inert) aggregate material; 'mortar' is a composite made by a binder and small-sized aggregate material, mostly used as structural binder between masonry units (bricks, stones); 'plaster' is any mortar or binder material used for wall and floor covering, mainly for smoothing, waterproofing, or preparation for paintings and decorations.

Cements based on Portland-type clinkers, mortars (pastes and plasters prepared with fine aggregates) and other inorganic binders form an important class of construction material: they are all supplied as powders and when mixed with water they form a fluid mass (paste) that can be shaped, moulded, added to other components or attached to the surface of other materials. The paste then hardens spontaneously under normal environmental conditions. Binding materials are used in buildings with the aim of (1) making structural elements for constructions; (2) increasing the resistance of the construction by linking the structural and architectural elements; (3) increasing waterproofing and protecting masonry surfaces from environmental degradation; and (4) preparing substrates for artwork and decorative purposes.

Excluding the tar products mentioned above and binders and adhesives based on polymeric compounds, virtually all binders used in antiquity were based on carbonates (calcite, dolomite), sulfates (gypsum) or alumino-silicates (cements). Table 1 provides an overall classification of inorganic binders based on their chemical nature and the main reaction process when mixed with water. The important concept is that in all cases the pyrotechnological production process yields a reactive material that transforms into a more stable product during setting and hardening. The major differences between the different binder types are: (1) the nature of the starting material that determines the chemical reaction pathway; and (2) the temperature of the firing process that controls the quality and reactivity of the starting binder.

Furthermore, a fundamental difference concerning the nature of the reaction processes of lime-saturated binders is whether they involve simple absorption of CO_2 from the gas phase to produce carbonates (aerial carbonation), or whether they also involve more complex processes of dissolution of alumino-silicate phases and precipitation of hydrated calcium-aluminium-silicate phases ('pozzolanic' reactions). The former are known as aerial binders, because they set in contact with the atmosphere. The latter are called hydraulic binders, because they may harden even under water.

In practical terms, if the binder is used as produced from the kiln with adequate grinding, then the binder/water mixture is called 'paste', *i.e.* the whole volume of the mixture comprises reactive phases and it will convert finally into a material composed entirely of the recrystallized reaction products. Therefore, if a lime paste undergoes complete aerial carbonation it will end up as a material composed totally of fine calcite crystals. A magnesian lime paste will yield a material composed of calcium carbonate and magnesium carbonate. The re-hydration of a bassanite paste will produce a plaster

Table 1. Main types of inorganic binders, their nature and reaction processes.

Type of binder		Starting material	Appx. *T* of firing (°C)	Reactive material product	Reaction process
Binders based on carbonate	Lime plaster	limestone	800–1000	Quicklime (CaO) Slaked lime ($Ca(OH)_2$)	aerial carbonation
	Hydraulic Lime plaster	limestone		Slaked lime + pozzolan	pozzolanic reaction
	Natural Hydraulic Lime plaster	limestone + clays/volcanic glass		Natural hydraulic lime	hydration reaction
	Magnesian plaster	dolomite		Slaked magnesia-lime ($Ca(OH)_2$) ($Mg(OH)_2$)	aerial carbonation
Binders based on gypsum	Gypsum plaster (Plaster of Paris)	gypsum	250–300	Bassanite $CaSO_4 \cdot 0.5H_2O$	hydration
Binders based on Portland clinker	Portland clinker (cement)	limestone + clays/marls	1400–1450	Clinker phases (alite C3S, belite C2S, Ca-aluminate C3A)	cement hydration

composed totally of gypsum. Finally the complete hydration of a clinker should yield a material composed largely of Ca-Si-hydrates (C-S-H) and calcite (from the carbonation of excess portlandite). As may be suspected, it is the recrystallization of the reaction products in the matrix and the entanglement of the crystals of the newly formed phases that confers mechanical resistance to the mature binder. The microstructure (*i.e.* the size, shape and orientation of the crystal phases) of the binder thus fundamentally controls the physical and engineering properties of the material.

In practice, the reaction normally goes to completion for gypsum and lime plasters; the reactions of these processes are kinetically quite fast at ambient conditions. However, it is often found that the kinetics of magnesian plasters are much slower, so that reaction products (MgO and $Mg(OH)_2$) are commonly present quite some time after the application. In the case of Portland cement, the hydration reaction barely goes to completion, so that a substantial part of the starting phases is invariably present with the reaction products even a long time after the mixture has been prepared.

Pure binder pastes are used rarely. It is much more common to mix part of the binder with a nominally unreactive phase (the so-called inert phase, or aggregate) such as quartz, in order to reduce volume changes during hardening and thus limit shrinkage effects. The role of the aggregate is to reduce macro-cracking during drying of the binder/water mixture, to increase the bulk modulus of the composite, and to increase the overall volume of the binder. If the aggregate is added with particle size in the sand range (generally referred to a standard with grain size in the range 0.6–0.8 mm), then the binder/aggregate/water mixture is called 'mortar'. Depending on the mineral nature of the binder, we may have lime mortars, natural hydraulic lime mortars, magnesian lime mortars, gypsum mortars, or clinker mortars. Lime mortars and gypsum mortars are the materials mostly found in the archaeological and historical record. They are still much used for small-volume applications such as decorations, panels, wall plastering and repointing.

In the case of Portland clinker, which is used as the main material in large-scale applications, there is need of large-volume aggregates in the size range of coarse sand or pebbles. The cement/large aggregate/water mixture is called 'concrete'. In most cases the various size fractions of the aggregates from fine to coarse are planned carefully through gradation models (following Bolomey, Fuller, Graf or Rissel curves; Day, 2003; Collepardi *et al.*, 2007) in order to produce an optimal volume packing of the particles in the mix. The planning of the optimal quantities and sizes of cement, aggregate and water is defined 'concrete mix design'.

3. The composition and properties of ancient mortars

3.1. Lime-based binders, technology and development

The first artificial binders used by mankind are the limestone- and gypsum-based plasters used widely in the Near and Middle East in the 7th and 8th millennia BC (Frierman, 1971; Gourdin and Kingery, 1975; Kingery *et al.*, 1988). The technological basis of plaster material is very simple: the reactive compound (quicklime in the case of lime plasters) is obtained by burning limestone at the appropriate temperature (Table 1). Then the heated block is ground to a fine powder and slaked with water to form a slurry (at high water/solid ratio) or a paste (at low water/solid ratio) composed of portlandite (calcium hydroxide, $Ca(OH)_2$). In Roman times long aging of the slaked slurry in excess water was a priority with respect to the modern attitude of using the ground powder and mixing it with water into a paste at the time of application. The effect of aging on the properties of the lime putty has been investigated thoroughly (Rodriguez-Navarro *et al.*, 1998; Cazalla *et al.*, 2000), as the issue is relevant for the preparation of restoration lime mortars (Ruiz-Agudo and Rodriguez-Navarro, 2009).

The reactions involved in the production and use of lime are:

$$CaCO_3\text{(calcite)} + \text{heat} \rightarrow CaO\text{(lime)} + CO_2\text{(carbon dioxide) [production of quicklime]}$$

$$CaO\text{(lime)} + H_2O\text{(water)} \rightarrow Ca(OH)_2\text{(Portlandite)} + \text{heat [quick hydration]}$$

$$Ca(OH)_2 + CO_2 \rightarrow CaCO_3 + H_2O \text{ [long term carbonation]}$$

The temperature needed to produce CaO should be ~850°C (Rodriguez-Navarro *et al.*, 2009), though the decomposition reaction of the carbonate can also proceed at slightly lower temperatures (780–800°C) in reducing conditions. Operational temperatures of lime-kilns are in the range 920–1000°C in order to speed up the decarbonation reaction. Excessive temperatures are avoided because they produce unreactive'dead-burned' lime. The production of lime therefore is a very energy-intensive process and it requires a substantial amount of biomass fuel. It has been estimated that the ratio of fuel biomass/quicklime is in the range 2–5 (Kingery *et al.*, 1992; Hauptmann and Yalcin, 2001), so that about 4–8 tons of wood would be required to produce the quicklime necessary for one house. Considering the diffusion of lime plaster resulting from the excavations in the Levant, the production of lime could have had a serious impact on the environment (Rollefson and Kohler-Rollefson, 1992; Redman, 1999). Similar claims were raised for the Mayan Lowlands (Wernecke, 2008).

Traditionally, the burning of carbonates (limestones, dolomites, travertine, marbles, but also shells and corals) is performed in lime-kilns, which are massive furnaces sometimes several metres high, charged from above with decimetre- to metre-sized blocks of limestone, and then fired for days by adding wood or charcoal to the combustion chamber at the base (Oates, 1998; Williams, 2004). Several ancient lime-kilns have been excavated from Roman (Dix, 1982; Coulson and Wilkie, 1986), to Late Classic Maya (Abrams and Freter, 1996), to more recent times (Williams, 2004). A detailed description of lime burning operations in Roman times was supplied by Marcus Porcius Cato (Cato the Elder: *On Agriculture*, XXXVIII).

The fired blocks are then ground to obtain the fine powdered quicklime that is, however, rather unstable in normal humidity conditions and tends to hydrate quickly to portlandite. If the CaO powder is mixed with an exact (*i.e.* stoichiometric) amount of water (lime/water = 75.7/24.3 = 3.12 by weight) the product is a fine dry powder and the process is called 'dry hydration' because there is exactly the right amount of water required to produce portlandite. If the CaO powder is mixed with excess water then a smooth paste is obtained in a slurry form, and the process is referred to as 'lime slaking'. The portlandite paste (slaked lime or lime putty) can then be used as a binder and an architectural component (filler, adhesive, cracks sealer, floor consolidant, surface smoother, *etc.*) or as a raw material for modelling objects, vessels and even artwork. After the application the paste dehydrates slowly and reacts with atmospheric CO_2 producing a hard material composed of microcrystalline calcite. The kinetics of carbonation are slow (Van Balen, 2005) so that in recent samples the reaction is not complete and residual crystals of portlandite may be observed.

The quality of the binder depends on a variety of parameters, including the composition, porosity and impurity content of the fired limestone, the maximum temperature and the time–temperature path of the firing, and the conditions of slaking. The starting limestone should have a non-carbonate mineral content (usually silicates and clays) of <5–10 wt.%, and the carbonate should be pure calcium. If the carbonate contains flints or a substantial amount of clays, then some (alumino)-silicate reactive phases may form at high temperature, the material acquires hydraulic properties, and it is called natural hydraulic lime (NHL). This material will be discussed later. If the carbonate contains magnesium, derived from the presence of magnesian calcite ($Ca_{1-x}Mg_xCO_3$ with $x < 0.1$) or dolomite ($CaMg(CO_3)_2$), then the material is a magnesian- or dolomitic-lime (Table 1). The periclase (MgO) produced together with lime during firing has much slower rehydration kinetics with respect to CaO, so that in the magnesian putty both periclase and brucite ($Mg(OH)_2$) are present together with portlandite (Blauer-Bohm and Jagers, 1997).

During carbonation and hardening the microstructure of the slaked lime paste changes significantly (Leslie and Hughes, 2002; Arandigoyen *et al.*, 2006). The identification of lime-derived calcite from an unconsolidated archaeological layer of ground calcite therefore is essentially based on the carbonate particles' dimensions (in the range 0.1–2.0 mm), its texture, and its mechanical properties (Kingery *et al.*, 1988; Affonso and Freiberg, 2001; Karkanas, 2007). Optical microscopy (Fig. 8) is therefore a very useful tool to investigate the nature and texture of ancient lime mortars (Elsen, 2006; Pecchioni *et al.*, 2014), although it may prove difficult to discriminate between fine-grained calcite of geological origin (chemically precipitated) and that derived from portlandite.

The presence and role of calcite nodules in the mortar (the so-called 'lime lumps') have been the subject of ample discussion, because of their importance in identifying ancient mortars (Hughes *et al.*, 2001; Karkanas, 2007), in the quantitative determination of the original binder/aggregate proportions (Lindqvist and

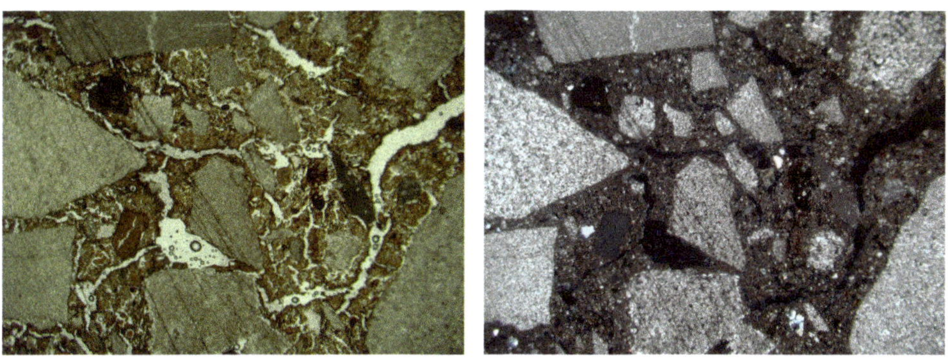

Figure 8. Optical micrograph of the thin section of a Medieval lime mortar from the Sachuidic castle, Friuli, Italy. The images in plane polarized light (a) and cross polarized light (b) show fragments of geological carbonate, lime lumps and the lime binder matrix.

Sandström, 2000; Elsen, 2004), and in the protocols sometimes employed for radiocarbon dating (Pesce *et al.*, 2012). In general, we may summarize that calcite nodules in the binder may have originated in one of three ways: as residues of unburned geological carbonate, as inhomogeneously slaked lime putty forming portlandite lumps that underwent late carbonation, or as reprecipitated late calcite. Failure to identify the proper nature of the lumps may lead to serious errors in the interpretation of the material and eventually in radiocarbon dating of the inappropriate fraction of the binder.

Very careful Fourier transform infrared spectroscopy (FTIR) work, based on the ratios of specific absorption peaks of calcite (specifically the ratio between in-plane and out-of-plane bending modes of the carbonate group), has been proposed as a technique to discriminate lime derived-calcite from natural calcite (Chu *et al.*, 2008; Regev *et al.*, 2010b). This discrimination may also be done using the luminescence properties of calcium carbonate (Machel *et al.*, 1991; El Ali *et al.*, 1993), as shown in the example of Fig. 9. Both methods rely on the different density and distribution of atomic defects in the calcite crystal structure. In principle the fractionation of carbon and oxygen stable isotopes during portlandite carbonation could also be used to test the nature of the carbonate phase (Ambers, 1987; van Strydonck *et al.*, 1989; Kosednar-Legenstein *et al.*, 2008). Once more, the issue is especially important when characterizing the carbonate fraction for radiocarbon dating of the mortar. In many cases the presence of 'anthropogenic' (*i.e.* man-made) carbonate is identified through micromorphological features (Stoops *et al.*, 2017), or by textural elements such as clear smooth surfaces representing the floor used, sometimes with a finishing coating (Affonso and Freiberg, 2001; Shahack-Gross *et al.*, 2005; Karkanas, 2007; Karkanas and Goldberg, 2007).

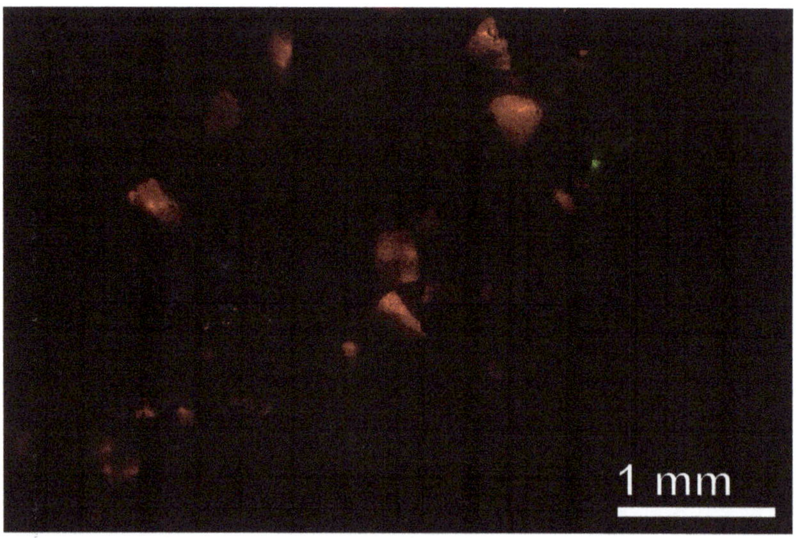

Figure 9. Cathodoluminescence image of a mortar showing sub-mm grains of geological carbonate embedded in the mortar binder.

The earliest well characterized example of quicklime production is the Hayonim Cave in Israel (Kingery *et al.*, 1988; Goldberg and Bar-Yosef, 1998), dated to the Natufian period at ~10.4–10.0 ky BC. Numerous other reports of lime-plaster in the Pre-Pottery Neolithic sites of the Levant (8.7–7.0 ky BC) were reviewed and characterized carefully by Kingery and co-workers (1988). They demonstrated beyond any doubt that there was a widespread use of pyrotechnologically produced lime-based plaster in the Near East coastal area from Palestine to Anatolia (Thuesen and Gwozdz, 1982; Garfinkel, 1987; Malinowski and Garfinkel, 1991; Poduska *et al.*, 2012). Especially striking are the plastered faces from 'Ain Ghazal in Jordan, originally modelled on human skulls (Griffin *et al.*, 1998), and the exceptional sculptured head found in Jericho (fig. 10 of Kingery *et al.*, 1988; Goren *et al.*, 2001) and Yiftahel (Milevski *et al.*, 2008).

There are numerous interesting issues concerning the prehistoric development of lime-based binders in the Pre-Pottery Neolithic of the Levant. In the first place, as already mentioned, it is curious that limestone firing and production of reactive quicklime pre-dates ceramic pyrotechnology (Frierman, 1971), although this gap is now being partially revisited (Biton *et al.*, 2014). Further, the plasters in this early phase "do not comprise the rendering of the floor, they are the floor itself" (Wright, 2005, p. 157), *i.e.* they are made as durable surfaces supporting loads and meant to be cleaned. They also rapidly extended to walls for protection and decoration (Fig. 4). It is evident, however, that at least several thousand years separate this early burst in the use of lime binders from any structural use connected with fired bricks in the region. Actually, there seems to be a puzzling discontinuity between this widespread use of lime plaster in the 8th and 7th millennia BC in the Levant (especially in skull-plastering, figurine modelling and floor plastering), and the subsequent evidence of use of lime in architecture in the 2nd millennium BC.

The case of Lepenski Vir, a small Neolithic village on the banks of the Danube, stands as an isolated example apparently unconnected with the previous Levantine experience and the subsequent reprise of lime plaster in the Minoan world. In Lepenski Vir a large number of peculiar trapezoidal huts have been excavated, each with a thick floor of red-coloured limestone plaster (Srejović, 1972, 1981; Radovanović, 2000; one XRD analysis of a Lepenski Vir sample is reported by Thuesen and Gwozdz, 1982). The settlement has been radiocarbon dated to the Early Neolithic (6200–5400 BC), though there are claims of early Mesolithic occupation (Radovanović, 1996; Borić, 2002; Bonsall *et al.*, 2008). Despite the uncertainty in the early life-span of the village and nearby settlements (Vlasac, Padina and Hajdutka Vodenica), the question of interest to us is whether the lime pyrotechnology on the Danube has anything to do to the previous Levant experience, or rather is a stand-alone development of lime production. Carefully contextualized investigations are needed to clarify the problem. The report of the use of lime at Makri, Thrace and at the Drakaina Cave, Kefallonia, Ionian Islands in the 6th millennium BC (Karkanas, 2007; Karkanas and Stratouli, 2008; Karkanas and Efstratiou, 2009) may prove to be an interesting link between the Levant and the Serbian lime experiences.

After the Lepenski Vir case, there were some four thousand years of technological discontinuity, or maybe lack of archaeological information. Chronologically, the successive occurrences of lime mortars and lime plasters are in the Minoan world (2nd millennium BC), where lime-based materials were used extensively in Crete, Cyprus and elsewhere for wall decorations and cistern waterproofing (Shaw, 1973; Wright, 2005). Without dwelling on the details too much, extensive lime plastering in the walls has been found, mainly to cover rough wall surfaces and floors, leaving the smooth ashlar unplastered. Plasters covering wall surfaces have a fine lime finish or wash as they serve as support for the famous Palatial paintings (*i.e.* Knossos, Fig. 10) and decorations. They are always lime-based, even when the wall-building stones are rock gypsum (Cameron *et al.*, 1977; Zoppi *et al.*, 2012). We stress that no hard evidence is observed in this period of the use of lime mortar as a structural component (*i.e.* to strengthen the masonry or ashlar blocks): all the Minoan, Mycenean and the later Classical Greek ashlar is dry.

A key development occurring in the Greek world is the consistent and systematic use of pozzolanic material (volcanic ash, *i.e.* 'Santorini earth', or crushed ceramics, *i.e.* 'cocciopesto') replacing inert aggregate in the mortar used for waterproofing cisterns and waterways in Bronze Age Crete and Cyprus (Maravelaki-Kalaitzaki *et al.*, 2003; Moropoulou *et al.*, 2005; Theodoridou *et al.*, 2013). The tradition continued in pre-Roman Rhodes (Koui and Ftikos, 1998). The later use in Laurion, Greece (4th century

Figure 10. The partially reconstructed Throne Room of King Minos in Knossos, Crete showing the ample use of lime plaster for wall decorations.

BC) of a very peculiar waterproofing layer obtained by mixing Pb-rich oxides (litharge) and Fe,Mn-rich ores with lime is also reported (Papadimitriou and Kordatos, 1993). This special occurrence is related to the mining and treatment of Pb-rich ores in the nearby mines, though it is difficult to define whether the layer investigated is a deposit developed as a byproduct of the ore processing or the conscious recycling of litharge slags during plaster manufacture.

As we will discuss, the reported occurrences bear two important consequences: (1) there is evidence of the appropriate use of hydraulic mortars and pozzolanic reactions in the Mediterranean area well before the Roman world; and (2) there is no firm evidence of the structural use of lime mortars in constructions (*i.e.* as a binder among masonry units) before the second half of the 1st millennium BC, except maybe the very early use of hot bitumen, discussed earlier, within the Mesopotamian context.

Outside the Mediterranean world, there is hardly any evidence for the use of lime plaster prior to the late 2nd millennium BC (Carran *et al.*, 2012).

No evidence is reported on the early use of plaster or mortar in the pueblo cultures of the US south west or in the South American Andean world, where the Inca's massive constructions of polygonal stones were assembled by smoothing the edges to make close contacts (Hyslop, 2014). Limited use of lime mortar, sometimes mixed with bitumen, has been reported at Machu Picchu (1450–1550 AD), though direct accounts and analyses are scarce (Prescott, 1847). In Central America the use of lime plaster is reported in Preclassical and Classical Central Mexico, and from the Classical Maya period (Hyman, 1970; Magaloni *et al.*, 1995; Abrams and Freter, 1996; Hansen *et al.*, 1997; Barba *et al.*, 2009). Some of the earlier examples are the plastered platforms at Cuello, Belize (1100–600 BC) and Nakbè, Guatemala (900–600 BC; Villaseor Alonso, 2009). In Mayan monuments the use of lime is structural, although spectacular use was made of lime as substrate to paintings and frieze decoration. Magnesian lime plasters were also extensively used (Villaseor and Price, 2008). An early use of pumices as lightweight aggregate in poured lime mortar is reported for the roof slab of Building Y in the El Tajin site, the sacred city of the Totonac people (850–1100 BC; Rivera-Villareal and Krayer, 1996).

Interestingly, there is ample evidence that in Mesoamerica the lime plaster was prepared by mixing different kinds of organic materials such as honey or juice extracted from a number of local plants such as cactus (nopal juice, *Opuntia ficus indica*; Littmann, 1957, 1960; Laws, 1962; Magaloni *et al.*, 1995). The type and extent of the effect of the organic molecules on the putty is still debated (Chandra *et al.*, 1998; Bensted, 1999; Rodriguez-Navarro *et al.*, 2017). On one hand there are claims that the organics destabilize portlandite chemically and release Ca ions, thus promoting pozzolanic reactions with impurity phases. On the other hand the physical effect of packing is emphasized, with the claim that the organics help the development of a more homogeneous microstructure, with smaller portlandite crystals, and ultimately a greater mechanical strength. The debate is open.

In China, there are accounts of the early use of lime during the Shang (1700–1027 BC) and the Zhou (1046–771 BC) Dynasties, the lime being mixed with sand and loess for wall facing and flooring. The palace site of Xianyang, Shaanxi province (350 BC) had the floors prepared with a mixture of lime, stone and pig's blood (Yang *et al.*, 2009) to give a dark red colour. The mortar formulation comprised lime, loess and sand; it became known as 'tabia' and became the regular material in use during the Eastern Tsin Dynasty (317–420 AD; Yang *et al.*, 2009). Lime-based mortars were used extensively in ancient architecture, with examples reported in the Qiantan River dam, Dutifulness Monument, Sticky Rice Bridge and others, completed mainly before the Ming (AD 1368–1644) and the Qing dynasties (AD 1644–1911; Huang, 2003; Yu and Chen, 2004; Zeng *et al.*, 2008). According to written sources (Song and KaiWu, 1982) lime-plaster production followed a traditional standard formula mixing lime and sticky rice solution up to 15–20 wt.%, and today the plaster is reported to be still in very good condition. Sticky rice is a type of rice grown in southeast and east Asia and it is composed mainly of amylopectin. Again, it seems that the organics influence the calcite dimensions and growth kinetics during carbonation (Zeng *et al.*, 2008). A variety of organic substances has been reported to have been used in traditional Chinese mortar recipes (Fang *et al.*, 2014; Zhang *et al.*, 2014; Zhao *et al.*, 2015), used both in masonry and in wooden buildings (Rao *et al.*, 2015).

The greater Indian region had a very early start with the use of mud and bitumen mortars (referred to as 'vajralepa' mortar; Kumar, 1984; Reza, 2008) related to the Indus valley civilizations, much as in Mesopotamia. The vajralepa bitumen was made of wood-tar and was used to waterproof walls and structures in bath houses (Sengupta, 1971; Kumar, 1984; Carran *et al.*, 2012). There is evidence of the early use of lime plaster in the ovens and pits of the houses in Kalibangan, Rajasthan, dated to the proto-Harappan period (3500–2500 BC; Sharma and Sharma, 2003), although it came into more common use since the last centuries of the 1st millennium BC. The pure lime binder was often obtained from 'kankar' (lime nodules) found abundantly among river gravel. Kankar contains ~70% calcite, 30% clay and a fraction of sand with other impurities, so that appropriate calcination of kankar may confer some hydraulic property to the mortar. Impure mortars with a variable sand/lime ratio are found in Buddhist settlements (Nagarjuna-Konda, 225–325 AD) and in the city of Kausambi, Uttar Pradesh (35–350 AD; Ghosh, 1989). A number of more recent buildings offer evidence of the widespread use of fired bricks and mortars, especially in mosques, temples (Thirumalini *et al.*, 2015), monuments (Singh *et al.*, 2014) and tombs (Gulzar *et al.*, 2013). The Great Stupa at Sanchi, Madhya Pradesh (Fig. 11) is the oldest stone building in India (3rd century BC) and one of the largest fired-brick domes in the world.

The Charminar in Hyderabad (1591–1593 AD) was built with granite stones and mortar (Singh, 1993). Lime plasters were also used extensively for decorations on temples and other monuments. The investigation of the decorations applied on rock at the carved Ellora caves (6–11 century AD; Fig. 12) surprisingly showed that many of the plasters contained a substantial amount of kemp hurd (*Cannabis sativa*), so that the decorations provide one of the first examples of kemp-reinforced plaster (Singh *et al.*, 2015). Furthermore, India offers a long tradition of organics mixed with lime allegedly to increase carbonation and durability: curd, jaggery, bel pulp (the fruit *Aegel*

Figure 11. The Great Stupa at Sanchi (3rd century BC) is the oldest stone building in India and one of the largest fired-brick domes (photo by Nagarjun Kandukuru, CC-BY-SA-2.0).

Figure 12. The Ellora caves show a spectacular labyrinth of rock carvings and paintings. Much of the artwork is plastered with one of the earliest examples of kemp-loaded lime of the 6th century AD.

marmelos), lentils, black gram and oil of margosa (Chandra and Aavik, 1983; Singh *et al.*, 1990).

The change in the paste properties resulting from the addition of organic compounds has long been known, (as mentioned briefly above) *e.g.* for traditional recipes for lime plasters in ancient Mesoamerica, China and India. Egyptians, Minoans and Romans modified the lime mortars with Arabic gum, animal glue, fig's milk, egg yolk and many other organic substances in an attempt to improve the mechanical or working properties (Sickels, 1981). In the *De Architectura* Vitruvius described explicitly the mixing of lime with oil (*calx ex oleo subacta*) to improve waterproofing. The practice was passed down to the Middle Ages and Renaissance, especially in the case of quality mortars for decoration and in applications exposed to weathering agents. A large variety of organic and proteic substances was apparently tested, with mixed results: cereal dough, animal blood, fermented wine, beer, milk derivatives including cheese, bee's wax, lard and many others.

3.2. Hydraulic mortars and pozzolans: the success of the 'Vitruvian' recipes

Mixing of lime plaster with partially or totally reacting aggregate marks the development from lime-based aerial mortar to hydraulic mortars. If the quartz aggregate of the mortar (formally inert) is replaced in part or fully with reactive Si,Al-rich material (generally defined 'pozzolanic material') then the highly alkaline environment created in the lime-saturated water induces the dissolution of the silicate or aluminosilicate phases and the subsequent precipitation of insoluble Si-rich hydrous phases (Hobbs and Siddall, 2011; Fig. 13).

The reactive pozzolanic material may be natural or synthetic silica glass, volcanic ash, radiolarite or diatomaceous earth, phytoliths, ceramics, clay, metallurgical slags, or any other reactive aluminosilicate compound. The materials employed in the past as pozzolanic additions in hydraulic lime mortars are also used today as supplementary cementitious materials (SCM) in the formulation of modern binders (Lothenbach *et al.*, 2011; Snellings *et al.*, 2012). Such materials can be represented in the lime–silica–

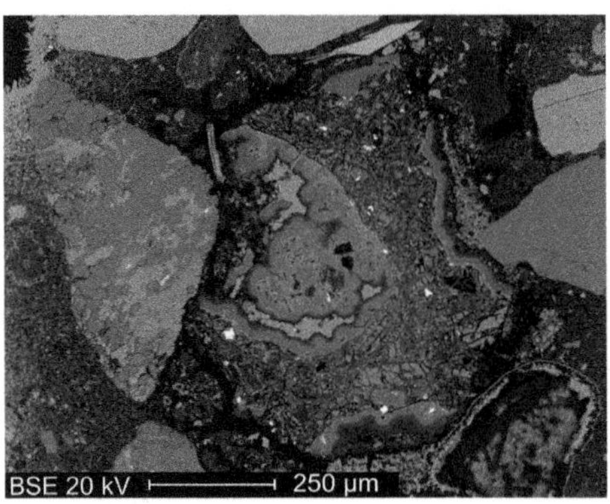

Figure 13. Reaction rim formed around a volcanic fragment in a lime matrix. The rim provides direct evidence of the pozzolanic reaction between the lime binder and the Si,Al-rich glass fragments inserted in the mix.

alumina ($CaO–SiO_2–Al_2O_3$) ternary diagram (Fig. 14). The reaction of lime with aluminosilicate phases is referred to as a 'pozzolanic reaction' or a 'hydraulic reaction'. When the presence of reactive aluminosilicate phases stimulates the pozzolanic reaction in the binder, then it is called a 'hydraulic' or 'pozzolanic' binder ('plaster', 'mortar') (Massazza, 1998). If natural volcanoclastic materials were not available, hydraulicity in the binder was often obtained by the use of crushed ceramics (pottery, bricks, tiles) together with traditional inert aggregates (Siddall, 2011). Fired ceramic is actually compositionally similar to volcanic materials, so the pozzolanic character of thermally activated clays and fired clay-based materials has been investigated thoroughly (Baronio and Binda, 1997; Böke *et al.*, 2006; Bakolas *et al.*, 2008; Zendri *et al.*, 2004; Fernandez *et al.*, 2011). The best ceramic-material yielding pozzolanic reaction is the very fine-grained kaolinite-rich pottery in which metakaolinite-type phases were formed by firing in the temperature range 600–900°C (Bellotto *et al.*, 1995). The use of clay minerals in the production of the ceramics as well as firing at higher temperatures induces the formation of less reactive phases, such as mullite (Gualtieri *et al.*, 1995, Gualtieri and Bellotto, 1998), and thus reduced pozzolanic reactivity.

One of the earliest instances of pyrotechnologically produced lime mortar mixed with volcanic ash occurs at Aşıklı Höyük, Turkey (Hauptmann and Yalcin, 2001). The pozzolanic reaction observed in the Pre-Pottery Neolithic B (PPNB) Aşıklı Höyük

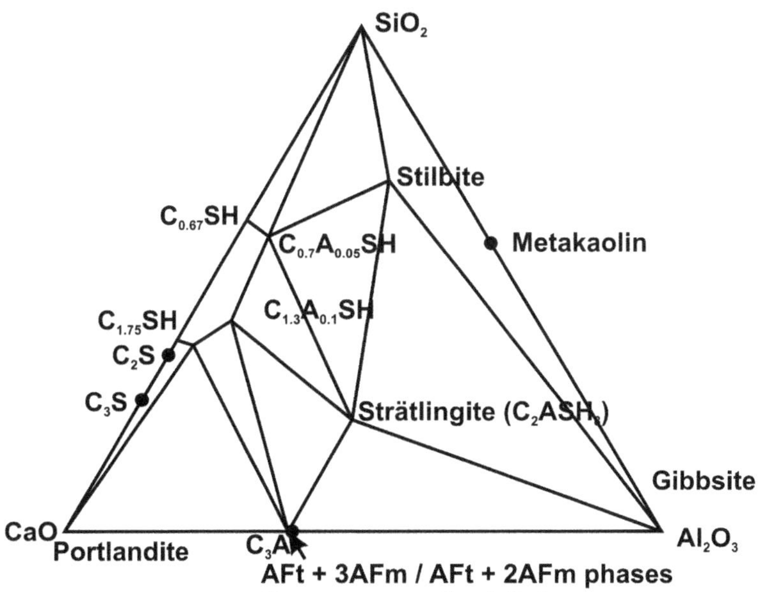

Figure 14. Simplified lime-silica-alumina diagram showing the clinker phases and some of the phases formed during cement hydration and pozzolanic reactions. Modified from Lothenbach *et al.* (2011).

plaster (8th millennium BC), however, is probably accidental and there is no evidence of systematic use of hydraulic materials in this early period. The possibility indeed exists that the silica phytoliths and other siliceous minerals derived from plant ashes could indeed have induced pozzolanic-type reactions in other cases of primitive plaster technology.

As discussed above, hydraulic binders based on pozzolanic reactions developed in the Greek and Aegean world in the 2nd millennium BC (Maravelaki-Kalaitzaki *et al.*, 2003; Moropoulou *et al.*, 2005; Theodoridou *et al.*, 2013). They became the technological basis for the highly successful binders used in later Roman architecture (Siddall, 2000). Between the Aegean tradition and the standardized adoption of the hydraulic mortar technology by Romans some thousand years later there is a fuzzy period of use and diffusion of these materials, possibly linked to cultural diffusion from the Eastern Mediterranean along the African coasts during the first Iron Age. Important clues to these pre-Roman developments are the hydraulic plasters of Tell es-Safi/Gath, Israel (Regev *et al.*, 2010a) and the carbon-containing mortars (ash mortars) used in Punic cisterns of Tunisia (Lancaster, 2012). It is again important to note that the hydraulic mortars in all these examples are limited to flooring and cistern plastering, mainly with a waterproofing function. Consistently, the analysis of the excavated structures and the mortars from a residential quarter in the Punic-Roman area of Palermo, dated firmly to before the final conquest of the city by the Romans during the First Punic War (254 BC), show only wall and floor plasters, with crushed pottery present in two of the samples investigated (Montana *et al.*, 2016).

The crucial evolution in the architectural use of hydraulic mortars occurred in a relatively short period in the first half of the 2nd century BC: the mortar is finally employed to strengthen masonry structures in both public and military buildings. It is the first fully recorded structural use of binders in architecture, in which the binder intimately links the units of the masonry and actively contributes to the mechanical strength of the composite structure. Marcus Porcius Cato (*De agri cultura liber*, dated approximately to 160 BC following the Oxford Classical Dictionary) described the systematic use of lime-based mortars as a binder in foundations, walls and complex architectural components of buildings (Greco, 2011). Following the critical chronological revision by Mogetta (2015), with which the present authors largely agree, the oldest known remaining example of concrete (*caementa*) in Rome is the *Porticus Metelli*, dated to ~130 BC. Most of the other concrete building traditionally dated to the late 3rd century or the early 2nd century BC (see table 2 in Mogetta, 2015) should be postdated to the last two decades of the 2nd century BC. Furthermore, there are recent indications that pozzolanic mortars were used to pour the foundations of fortification walls in wet areas far from Rome at very early periods (Aquileia: Bonetto *et al.*, 2016; Ravenna: Costa *et al.*, 2000). Apparently, Roman military engineers already knew that volcanic pozzolan performs better than crushed pottery in marshy or salt-water environments. If confirmed, such occurrences of pozzolanic reactions are even older than the examples in Rome, suggesting an early and rapid diffusion of the technology involving hydraulic reactions based on volcanoclastic material.

Three major innovations and/or technological optimization in the use of the binders were introduced rapidly and systematically at the beginning of the 2nd century BC or shortly thereafter (Lechtmann and Hobbs, 1987; Siddall, 2000; Adam, 2005; Hobbs and Siddall, 2011; Brandon *et al.*, 2014):

(1) The use of mortars to strengthen important architectural structures: for the first time binders are not limited to plastering and waterproofing.

(2) The systematic use of crushed ceramics (pottery, bricks, tiles) in terrestrial structures that required resistance to water. This technology, certainly derived from the Greek world, was carefully optimized in the so called *opus signinum*, widely used in cisterns, aqueducts, fountains and baths (Siddall, 2011).

(3) The exclusive use of highly pozzolanic volcanoclastic material in hydraulic mortars used to build structures in contact with sea water (Brandon *et al.*, 2014).

Not surprisingly, it was in late Republican times (2nd–1st centuries BC) that the architectural capabilities of Roman engineers, previously limited by the robust but cumbersome properties of ashlar (the Servian walls of the 6th century BC were still built in *opus quadratum*), started to produce efficient, solid and durable buildings rapidly complying with the formula later encoded by Marcus Vitruvius Pollio: "*Haec autem ita fieri debent, ut habeatur ratio firmitatis, utilitatis, venustatis*" (*De Architectura*, 1.3.2; 29–23 BC). As the only architectural book surviving from classical antiquity, the Vitruvian statements and prescriptions are considered the primary source of information concerning Roman and Greek building techniques. Vitruvius' *De Architectura* met with immense favour and reputation from the Renaissance to the end of the 19th century, becoming the conceptual reference for most Western architecture of the period. In his work, Vitruvius not only illustrated building techniques and architectural theories, but described at great length the nature and preparation of the materials, including mortars. As an example, Vitruvius in his Book II indicates clearly that the proportion of lime and sand must be increased if poor quality or marine sand is used as aggregate. His descriptions find neat experimental evidence in a number of recent investigations of Roman materials (Oleson *et al.*, 2006; Gotti *et al.*, 2008; Secco *et al.*, 2018), where Vitruvius' recipes for binder formulation and preparation found precise confirmation.

During the last part of Republican times, Roman engineers crucially improved the Aegean-derived technology of hydraulic binders by using slaked lime in a mixture with local high-alkali volcanic ash, first drawn from the banks of the river Tevere, and then from the volcanic sands found near Naples, at Pozzuoli; hence the name *pozzolan*, still in use today. Recent studies of the evolution of Roman mortars from Republican through Imperial times reveal specific changes in the composition and use of the mortar components. The mineral components of the ash outcrops within and around ancient Rome (especially the Pozzolane Rosse ignimbrite: Jackson *et al.*, 2007; Marra *et al.*, 2016) show that specific zeolite-rich tuffs with highly reactive properties were carefully selected from the very beginning in order to produce exceptionally hard and durable mortars. These important studies not only confirm the early chronology of

Roman mortars identified by macroscopic observations (Mogetta, 2015), but also confirm the incredibly detailed description of the materials that Vitruvius indicated as ideal for the preparation of quality mortars (black and red sands, or *harenae fossiciae*: *De Architectura* 2.4.1). Further, Vitruvius states that this pozzolanic material is the only one that is able to harden under water (*De Architectura* 2.6, hence the name 'hydraulic mortar'). The physicochemical and engineering characteristics of mortars used for the construction of Roman harbours have been investigated extensively in the frame of the ROMACONS project (Brandon *et al.*, 2005, 2014; Oleson *et al.*, 2006). Not only are the outstanding properties and durability confirmed, but the tuffs used in the mortars of the harbour of King Herod in Caesarea seem to have been transported all the way from the Bay of Naples. The Roman harbour builders knew very well that proper hardening of mortar at sea required good quality pozzolans, even at the cost of long-distance transportation.

The local occurrence of reactive volcanic sand (such as in Central Italy, or in the German Eifel area, known as Trass; Siddall, 2000), and the availability of recycled crushed pottery and bricks gave Roman engineers the appropriate technology to build flexible, innovative and durable architecture and ultimately set the basis for the efficient infrastructure of the Roman empire. In fact the Romans also developed the concept of lightweight robust structures by the use of innovative building techniques, such as the vaulting tubes, the use of appropriately shaped bricks, and the incorporation of lightweight pumiceous aggregates in the binder (Lancaster, 2005, 2015; Lancaster *et al.*, 2010, 2011. The arches of the Colosseum and the Pantheon dome are reported to have been made with such innovative materials and techniques (Lancaster, 2011).

Note that the word cement (*opus caementicium*) in ancient Roman times referred to the concrete masonry of monuments composed of cm-sized brick and tuff fragments (*caementa*) used as aggregates and which are bonded by hydraulic mortars with alkali-rich, calcium-alumino-silicate volcanic ash sands (Lechtman and Hobbs, 1987; Lamprecht, 1996). Only in recent times has the meaning changed to refer to modern clinker-based materials.

The ability of producing and using excellent mortars was lost slowly after the fall of the Roman Empire. The dome of the Hagia Sophia Basilica in Costantinople (532–537 AD) is probably the last great Roman building formed by the monolithic pouring of Roman pozzolanic mortar based on crushed ceramics (Livingston, 1970; Moropoulou *et al.*, 2002; Miriello *et al.*, 2017). Lime-based mortars continued to be used through the Middle Ages, though in many cases the mortars were of rather low quality, made of partially burned lime, sometimes mixed with charcoal and clays, and the lime putty was frequently unaged and poorly slaked. The careless preparation made them mostly porous and degradable (Furlan and Bissegger, 1975; Franzini *et al.*, 1999). Many of the Saxon, Norman and Longobardic materials are of this kind. The standardized production during the Roman Empire was thus followed by very local productions, mostly having only a small technological content, and it was only in specific and prestigious construction sites that high-quality binders were produced, as in the case of the Byzantine mosaics in Ravenna, or the Leaning Tower of Pisa (Franzini *et al.*, 2000).

In the latter case, the absence of volcanoclastic material and the presence of CSH products hints at the possible use of very reactive silica sources, such as diatomaceous earths. Most of the mortars used through the Middle Ages until the Renaissance to build cathedrals, fortresses and palaces are lime-based (Alvarez *et al.*, 2000; Franzini *et al.*, 2000; Pecchioni *et al.*, 2006). However, some binders show a limited hydraulicity (Moropoulou *et al.*, 2000; Elsen *et al.*, 2004) due to pozzolanic reactions. Although discontinuous, the tradition of making hydraulic mortars by the use of crushed ceramics persisted in several regions of the Roman Empire and beyond: it is called *cocciopesto* in Renaissance Italy, *horasan* in Turkey, *homra* in the Middle East, *surkhi* in India.

A peculiar kind of traditional material produced by clay activation is called *sarooj*. It was apparently used widely in ancient Persia and other areas around the Persian gulf, especially for cisterns and waterproof structures (Al-Rawas *et al.*, 1998, 2001). It is claimed that a number of ancient temples and buildings, such as the Elamite zigurrat at Chogha Zanbil, Iran (Fig. 7) were built partially using *sarooj*, although experimental evidence is scarce. The reported preparation of the material indicates that a mix of clays, dung and water was sun-dried into decimetre-sized blocks, which were then fire-heated for activation. The final product was ground and mixed with lime to produce the final binder (Masoumi *et al.*, 2015). *Sarooj* seems to be the earliest example in which thermal activation of clay was consciously and systematically employed to obtain hydraulic reactions.

As will be discussed below, the technology based on thermally activated clays is regarded as one of the future solutions for the replacement of modern Portland-based clinker with more sustainable products (Scrivener and Favier, 2015). As a matter of fact, our understanding of the pozzolanic reactions that occurred in ancient hydraulic binders allows modern interpretation of their mechanical durability (Moropoulou *et al.*, 2000; Charola *et al.*, 2005), and has fundamental implications for the development and technological control of modern cement alternatives to Portland clinkers (Mahasenan *et al.*, 2003; Schneider *et al.*, 2011). On a technical note, the protocols of separation and characterization of the binder fractions for reliable radiocarbon dating of ancient mortars require adequate understanding of the phases present and their reaction history (Lindroos *et al.*, 2011; Addis *et al.*, 2016).

The presence and extent of pozzolanic reactions in ancient mortars is commonly assessed by:

(1) measurement of the mechanical resistance to compression tests, which is substantially greater than those measured on traditional lime mortars,

(2) calculation of empirical parameters derived from bulk-chemistry measurements (*i.e.* hydraulicity index, cementation index),

(3) characterizing the nature and micro-textural features of the mineralogical phases present in the binder fraction of the composite, which are the result of pozzolanic reactions.

Testing the mechanical resistance is an indirect bulk measurement and leaves space for ambiguities in the interpretation of the mechanisms that lead to the development of the physical properties. The 'hydraulicity index' (HI = SiO_2 + Al_2O_3/CaO) or the 'cementation index' (CI = 2.8 SiO_2 + 1.1 Al_2O_3 + 0.7 Fe_2O_3/CaO + 1.4 MgO) are generally proportional to the amount of Si,Al-rich species in the mortar (Elsen *et al.*, 2012), though the presence of aluminosilicate phases is not sufficient to indicate that the pozzolanic reaction did actually take place to a sufficient extent in the binder, as they can be inert or very slowly reacting.

On the other hand, a thorough mineralogical characterization of the phases and of the textures resulting from the lime–pozzolan interaction yields direct evidence of the pozzolanic reaction. Adequate characterization, therefore, requires identification of reaction products (by X-ray powder diffraction (XRPD), FTIR, Raman spectroscopy, or other micro-analytical techniques) and/or micro-imaging of the reaction textures, such as the reaction rims formed around the pozzolanic particles through dissolution-recrystallization processes (Fig. 13; see for example: Maravelaki-Kalaitzaki *et al.*, 2003; Diekamp *et al.*, 2012). The limiting factor of the pozzolanic reaction rate in general is the dissolution kinetics of silicate phases. The issue is fundamental when assessing the reactivity of materials to be used as supplementary cementitious materials (SCM) in modern binder formulations (Malhotra and Mehta, 1996; Lothenbach *et al.*, 2011; Snellings *et al.*, 2012).

The hydrous phases formed by the pozzolanic reaction have a variable composition depending on the chemistry of the system (Fig. 14). They range from pure C-S-H phases in the lime-silica join (tobermoritic Ca,Si-hydrates; Richardson, 1999, 2004) to more complicated C-A-S-H phases as the Al activity increases (Hong and Glasser, 2002, L'Hôpital *et al.*, 2015, 2016). Zeolite-like phases are favoured in complex N-K-C-A-S-H environments, and their formation is reported to be enhanced by interaction with sea water (Jackson *et al.*, 2012, 2013, 2017). It is the interaction between sea water and the volcanic tuff that yielded tobermorite and zeolitic phases and produced the extremely hard and durable binders of Roman harbours.

3.3. Natural hydraulic lime (NHL) mortars

A particular type of hydraulic binder was developed in Europe in the 18th and 19th centuries by firing impure limestones containing clays in lime kilns; they are called natural hydraulic limes (NHL) or "Roman cements" (Weber *et al.*, 2012). In these natural hydraulic binders the reactive phases are not added as aggregate, but formed in the kiln through the reaction of the clays and the quicklime. They can be considered the precursors of modern Portland clinker. NHL were very popular for building construction and façade decoration in Central Europe in the 19th century (Callebaut *et al.*, 2000, 2001). They were, and are still, used extensively for restoration of historical buildings (Maravelaki-Kalaitzaki *et al.*, 2005; Gosselin *et al.*, 2009; Bras and Henriques, 2012) because of their good mechanical properties and compatibility with ancient materials.

A very early instance of production and use of NHL in the 14th century AD was identified during the investigation of the mortars and slags of the Gothic Obřany Castle, Moravia (Kropáč and Dolníček, 2013): the interpretation is that the lime binder incorporated high-temperature phases derived from the kiln lining and local metallurgical activities. Slags from ore processing were also observed in the mortars of the medieval archaeological complex at Montieri, Tuscany, Italy (Chiarelli *et al.*, 2015), which is located in an area of intense exploitation of argentiferous ores.

Both pozzolan-added mortars and NHL were used in the Czech Republic to build 14th century bridges, such as Charles Bridge in Prague and the Gothic bridge in Roudnice (Frankeová *et al.*, 2012). The slightly hydraulic character of the Narni bridge, in Central Italy (3rd century AD) is thought to be derived from the accidental firing of impure limestones, containing cherts (Cantisani *et al.*, 2002; Frankeová *et al.*, 2012). In extreme cases, the burning in lime kilns of chert-containing limestones in the presence of alkali lowering the melting point of silica yields the production of amazingly coloured silica glasses, which were found at the base of the furnace after firing (Artioli *et al.*, 2009).

Vitruvius recommended the use of very white and pure limestone for the production of lime in classical times. It appears that whenever the Vitruvian recipes were abandoned or not followed closely, the accidental use of impure limestones with various amounts of silica or aluminosilicate phases randomly produced binders with very variable degrees of hydraulicity. Sometimes the fine-grained size of the original material and the heterogeneity of the microtexture makes it very difficult to assess the degree of hydraulicity and/or deduce unambiguously the complex dynamic of the production process (Riccardi *et al.*, 2007).

The emerging picture is one of unstandardized binders and largely uncontrolled production processes from late antiquity until approximately the 18th century AD, when systematic experiments were carried out to produce optimized materials.

4. The development of modern cement materials and the hydration process

A significant change from traditional lime mortars was made by John Smeaton in England in 1756, when he was involved in the reconstruction of the Eddystone lighthouse (Blezard, 2003). As the story goes, he was driven by the need to develop a masonry construction durable in a marine environment and thus built using a binding lime mortar that did not dissolve in seawater. Among several attempts, he also departed from the Vitruvian recommendations of using pure white limestone. Using clay-rich carbonates of marly composition he obtained better hydraulic properties than lime. As discussed above, this class of materials may be defined as natural hydraulic limes and can be considered intermediate between slaked lime and modern Portland cement. Eventually the material that Smeaton selected for the lighthouse was a mortar prepared with equal proportions of local argillaceous limestones (blue Liassic limestones) and pozzolan from Civitavecchia, Italy, the closest stuff he could access resembling the ancient Roman material. Slightly improved mixtures of this kind were in use until the introduction of Portland cement.

In the first half of the 19th century the search for optimal hydraulic binders was pursued actively in several countries (Bentur, 2002; Blezard, 2003). In England

Smeaton was testing alternative formulations of hydraulic mortars, and J. Parker introduced and patented a so-called "Roman cement" (Patent by James Parker, of Northfleet, Kent, England, No. 2120, 1796), defined as a "cement or tarras to be used in aquatic and other buildings and stucco work". It was made by calcination of nodules of argillaceous limestone (known as septariae) and it produced a quick-setting cement. In France, Louis Vicat's experiments (Vicat, 1818) led to the preparation of hydraulic lime by calcination of a mixture of high-grade quicklime (produced by the chalk of the Upper Cretaceous carbonatic formation of the Paris Basin) and clay (Vicat, 1828). His formulation, called the "twice-kilned" process, met with considerable success and led his son Joseph Vicat to establish the well known Vicat Cement company. This is considered by many to be the predecessor of Portland cement.

A lot of patents were issued around the same time establishing plants in southern England, including the London area. The most famous one is that related to the three-stage process of Joseph Aspdin (patented in 1824), who named his product 'Portland cement', because, at that time, the Portland limestone had a reputation among builders for quality and durability, and he wanted to capture the similitude between his cement and Britain's favoured quarried stone. Portland cement was marketed as an improvement in the production of artificial stone. In one of the several plants established by Aspdin or his son, the temperatures were running high enough to produce partial or complete vitrification and crystallize alite, as shown by the retrospective analysis of the type of clinker material from Aspdin's kiln (Blezard, 2003). The 'clinker' is the reactive product formed by cooling from the high-temperature processing (1450°C) of the limestone and clay mixture within the kiln. The temperature is such that partial fusion occurs and the reacted molten material forms nodules that are partly crystalline and partly vitrified, the whole material being highly reactive with water. The four major phases making up the clinker are reported in Table 2, though the real situation is made more complicated by the existence of several polymorphs for each phase, and by compositional deviation from stoichiometry (Taylor, 1997). The basic chemical composition of Portland cement is shown in Fig. 14, where it is evident

Table 2. Main phase components of ordinary Portland clinkers.

Common phase name	Mineral name	Composition	Cement notation	Appx. wt %
Alite (tricalcium silicate)	hatrurite	Ca_3SiO_5	C3S	50–70
Belite (dicalcium silicate)	larnite	Ca_2SiO_4	C2S	15–30
Aluminate (tricalcium aluminate)	–	$Ca_3Al_2O_6$	C3A	5–10
Ferrite (tetracalcium aluminoferrite)	brownmillerite	$Ca_2(Al,Fe)_2O_5$	C4AF	5–15

that the composition of the two major phases (alite and belite) lies on the $CaO-SiO_2$ join as a result of the lime–silica reaction. The hydration products, composed mainly of calcium silica hydrates (C-S-H) and portlandite also lie in the same portion of the diagram.

Following Blezard (2003), the historical evolution of clinkers after the illustrated pioneering experiments moved from *proto*-Portland materials (*i.e.* Aspdin's original patent), which shows limited interaction between CaO and SiO_2 because of the limited temperature in the kiln, to *meso*-Portland materials, which are very heterogeneous materials showing some silica–lime interaction and containing mostly belite and some alite, and finally to *normal*-Portland cements as we know them today. The main characteristics of modern Portland clinkers are derived mostly from the use of rotary kilns in place of the traditional shaft kilns, a technical development that also allows continuous production in place of the batch process. The carefully-controlled initial formulation of calcareous and argillaceous components, together with the use of the rotary kiln that determines a long permanence at high temperature and continuous mixing, helps to: (1) minimize the amount of unreacted lime; (2) maximize the alite/belite ratio; and (3) obtain the appropriate crystal size of the mineral components, commonly in the range 10–40 μm. Modern clinkers contain about 60–65 wt.% of C3S, and <2 wt.% of unreacted free lime. The standard reaction properties of modern Portland clinker are essentially due to a high alite/belite ratio and to a careful control of the grinding of the clinker into a fine powder.

The very reactive Portland cement powder is mixed with water to produce a final hardened material through a series of complex reactions, the so-called 'hydration process', involving dissolution of the crystal phases, surface reactions, gel formation, precipitation of new phases and textural changes (Taylor, 1997; Gartner *et al.*, 2002; Bullard *et al.*, 2011). The different crystal phases present in the clinker have very different reactivities, C3A having the highest and C2S the lowest. In fact C3A is so reactive exothermally in water that it can cause unwanted rapid setting of the paste (flash set), with consequent loss of workability. Therefore calcium sulfate (commonly gypsum) is usually added to the clinker in the amount of ~4–8 wt.% as a set-controlling agent, because the sulfate ion retards the dissolution of the clinker phases, and promotes the formation of ettringite. The finely powdered mixture of clinker and gypsum is marketed as standard Portland cement (also called OPC = ordinary Portland cement), which is then mixed with water in the ratio water/cement >0.38, *i.e.* the least amount of water necessary for complete hydration of the cement phases. As in ancient binders (Table 1), when an aggregate material is added to the paste we have mortar (fine aggregate, such as sand with grain size <1 mm) or concrete (fine and coarse aggregate, such as coarse sands or gravel). They can both be considered to be cement composites, with the aggregate having the role of reducing the formation of fractures during the shrinkage that accompanies the hydration and hardening; for the same reason temper is added to pottery clays before firing. The aggregate may be composed of any type of loose or ground rock, and it is assumed not to react with the cement paste (hence it is referred to inappropriately as the 'inert' component), though this is not necessarily the case.

Upon mixing with water, the clinker phases react at different times to produce first a series of intermediate phases (the so-called AFt and AFm phases) such as ettringite ($Ca_6Al_2(SO_4)_3(OH)_{12}26H_2O$) or monosulfoaluminate ($Ca_4Al_2(SO_4)O_612H_2O$), and then the reaction proceeds to the final hydrous phases: portlandite and amorphous calcium silicate hydrate (C–S–H), which are the phases forming the interlocked gridwork of the material and producing high mechanical resistance. Figure 15 shows environmental scanning electron microscope (ESEM) evidence of the growth of ettringite and C–S–H at different times on the surface of the clinker grains. It is important to understand that it is the microstructure of the material, that is the final assemblage of crystalline and amorphous phases that governs the macroscopic and engineering properties of the material (Thomas *et al.*, 2011; Scrivener *et al.*, 2016). With time, portlandite will eventually convert into calcite through a carbonation reaction with atmospheric carbon dioxide, so that abundant calcite is found in old cements.

Reaction kinetics play a crucial role in determining the final microstructural properties. The completion of the hydration process may take days, months, or even years, depending on crystal size, defectivity and polymorphism of the phases, porosity of the paste, environmental conditions, *etc*. The system is so complex that reactions can be inhibited for a long time. It is often observed in old concrete that highly reactive species such as aluminates are still present years or decades after manufacturing (Secco *et al.*, 2014). Many of the practical problems in the modern cement industry derive from the lack of control of clinker composition, poor concrete design, or sloppy practice: even small changes may affect the hydration kinetics and determine substantial degradation of the engineering properties. Incidentally, the same problems, mostly due to human factors such as incompetence and fraud, were present and well recorded in Roman times (Oleson, 2011).

Strength development is related to the degree of hydration and to the speed of the process. In general, the slower the hydration kinetics, the higher the compressive strength developed. This is one of the differences observed between modern clinkers

Figure 15. (a) Prismatic crystals of ettringite growing amidst C-S-H felt during the hydration of a Portland cement, and (b) well developed platy crystals of Portlandite. Images obtained by ESEM. (Courtesy of D. Salvioni, Mapei S.p.a.).

and those produced in the early part of the last century. Modern cement is required to develop high resistance to compression (>40 MPa) within a few days of emplacement, whereas early 1900s cements developed high strengths (in the range of 20–30 MPa) over much longer periods, because of the greater belite content. This has some consequences also in terms of durability. High-compression resistance combined with tensional resistance are, of course, obtained by steel reinforced concrete (SRC), which is a composite in which the concrete is poured and solidified around a metal skeleton. The French-Swiss architect, Le Corbusier, was one of the first to understand and exploit the properties of SRC in modern architecture. He had discovered the use of reinforced concrete very early in Paris from one of the pioneers of its use, architect Auguste Perret, and then employed it to realize his architectural vision (Fig. 16). He later wrote: "Reinforced concrete provided me with incredible resources, and variety, and a passionate plasticity in which by themselves my structures will be rhythm of a palace, and a Pompeiian tranquility" (Letter to Auguste Perret, 1915).

Figure 16. Reinforced concrete structures by Charles-Édouard Jeanneret (Le Corbusier): (a) The convent of Sainte Marie de La Tourette, France (1953–1960) (photo by Alexandre Norman, French Wikipedia, CC-BY-SA-3.0); (b) The palace of Assembly, Chandigarh, India (1952–1961) (photo by English Wikipedia, CC-BY-SA-2.0).

Nowadays only ~35% of the cements produced industrially for the global market are OPC, and most of the materials are special composite formulations (Odler, 2000; Chatterjee, 2002), where special components or SCM are added to enhance specific physicochemical or mechanical properties, such as resistance to alkali or sulfates, thermal resistance, or low-heat emission. The main added constituents are: (1) latent hydraulic components (granulated blast furnace slags, class C fly ash) that have self-cementing properties that need to be activated by OPC; (2) pozzolanic aluminosilicate components (class F fly ash, silica fume, metakaolin, pozzolan) that have no self-cementing properties and need to be activated by portlandite; and (3) non-reactive or poorly reactive components that modify the grindability of the clinker or the rheology of the paste.

Because of the low market cost of cement, the materials added to the clinker are large-volume but low-cost byproducts of other industrial activities. Therefore on one hand the production of cement represents a virtuous route to recycle large quantities of waste material (*e.g.* slags from the metallurgical industry, fly ash from coal plants); on the other hand the market for special cements will need alternative materials should the global changes in metal and energy production decrease the availability of SCM materials in the future.

One of the most important properties of modern cement pastes and concrete composites is fluidity: it ensures the possibility of transportation, pumping, levelling concrete to fit modern architectural requirements, and also proper consolidation by grouting. The rheological and working properties of the cement and concrete are so important that they are now controlled invariably by the use of chemical admixtures (water-soluble organic polymers) that allow good fluidity of the paste with a smaller water/cement ratio. The net result is also a lower porosity of the paste and a greater mechanical strength: ultra-high performance concrete and high-strength concrete are made this way, they can reach compression strengths in excess of 150 MPa. The modern generations of plasticizers, ensuring fluidity at low water content are called water-reducing admixtures (Chandra, 2002; Edmeades and Hewlett, 2003). Besides rheology, organic and inorganic additives are also used to control specific properties of the mortars and cements, such as air entrainment (for better resistance to freeze and thaw cycles), acceleration or retardation of the setting (to control the time of progressive increase in mechanical resistance; Cheung *et al.*, 2011) and even resistance to chemical attack. Virtually all modern concrete formulations include polymeric admixtures to optimize performances.

How do we investigate cement materials? At the industrial level, most quality controls are done by optical microscopy, X-ray fluorescence (XRF) and XRPD, with the quantitative phase analysis by XRPD offering a number of crucial advantages over bulk chemical analysis by XRF. Cement producers are among the most active users of advanced on-line testing protocols and instrumentation using automatic XRPD, including full-profile refinement of all clinker phases (Bellotto and Signes-Frehel, 1998; Manias *et al.*, 2000; De la Torre and Aranda, 2003; Bequette and Dhanjal, 2011; Snellings, 2016).

At the research level, the cement system is complex enough that one experimental technique alone is hardly sufficient to unveil the fine interplay between all chemical reactions acting simultaneously during hydration. Furthermore, the system is extremely sensitive to any environmental perturbation, especially dehydration, so that sample preparation without artefacts is a crucial part of any experiment. Virtually all experimental techniques have been employed to characterize and understand the behaviour and evolution of cement systems. In this chapter we will mention only some of the most recent developments providing a better insight into fundamental mechanisms of cement hydration processes. Time-resolved XRPD is one of the most powerful techniques for following the evolution of the cement/water system during hydration (Dalconi *et al.*, 2008; Hesse *et al.*, 2011; Valentini *et al.*, 2015). It allows appropriate quantification of the phases and evaluation of reaction kinetics (Fig. 17). If reaction enthalpies are sufficiently well known (Matschei *et al.*, 2007; Damidot *et al.*, 2011), then it is possible to back-calculate the measured curves from isothermal calorimetry and discriminate the underlying reactions (Hesse *et al.*, 2011; Jansen *et al.*, 2012; Valentini, 2013).

The use of synchrotron radiation, of course, allows access to much faster reaction times with respect to laboratory instruments (Merlini *et al.*, 2007, 2008; Snellings *et al.*, 2010), down to the millisecond timescale (Schlegel *et al.*, 2012). Synchrotrons offer the advantage of highly brilliant and coherent radiation, so that experimental techniques can be designed flexibly to suit complex samples (Aranda, 2016). Simultaneous and combined investigations are thus possible only at large scale-facilities. Concerning cements, ultra-high resolution imaging (Harutyunyan *et al.*, 2009; Monteiro *et al.*, 2011) and sub-micron computed tomography (Artioli *et al.*, 2010; Parisatto *et al.*, 2015) provide considerable advances in understanding the 3D evolution of the microstructure

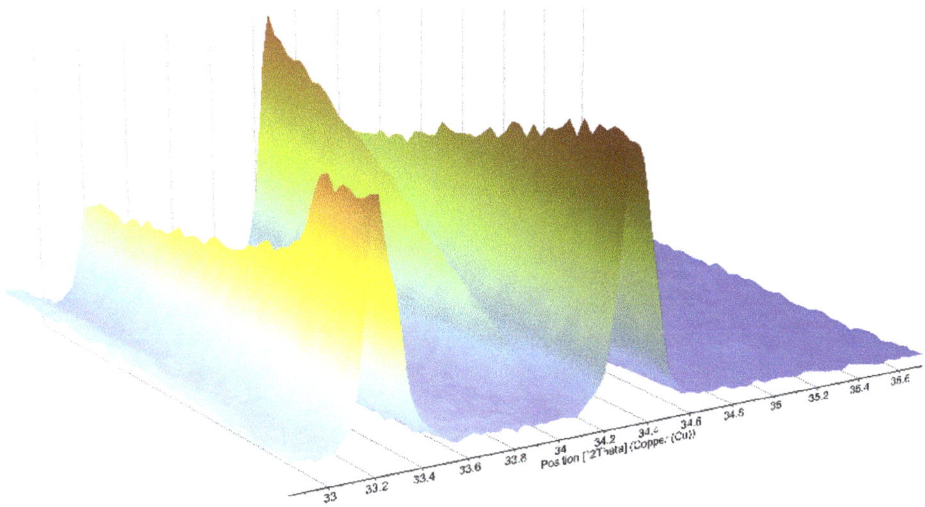

Figure 17. Time resolved XRPD patterns of a cement paste during the hydration process.

of the material in a totally non-invasive mode (Fig. 18). Furthermore, the innovative combination afforded by diffraction-enhanced tomographic imaging proved to be an extremely powerful tool for the investigation of complex materials (Voltolini *et al.*, 2013), and allowed, for the first time, direct imaging of nucleation processes (Artioli *et al.*, 2014a,b).

As a further link between the cement world and mineralogy, there is only one known natural occurrence of the calcium silicate phases (hatrurite, larnite) that are commonly produced in cement kilns. It is from a geological formation called the Hatrurim Formation, and it is exposed in various areas around the Dead Sea in Israel (Hatrurim Junction, near Arad), Palestine and Jordan (Maqarin). It is interpreted to occur as the result of pyrometamorphism of marls and limestones derived from gas or bitumen ignition in the underlying hydrocarbon-loaded formation (Burg *et al.*, 1991; Gilat, 1998; Sokol *et al.*, 2007). The natural combustion produced this unique rock formation of thermally metamorphosed 'cement zones' which are very close analogues of industrial cements. In the Hatrurim Formation the calcium silicate phases occur together with a number of exotic high-temperature mineral phases (shulamitite, bentorite, gazeevite, stracherite, zoharite, zuktamrurite, and others), most of them found only in this unique type locality.

The Maqarin site, located along the Yarmouk river near the Jordan–Syria border, has been investigated extensively because the subsequent interaction with groundwater caused the formation of hyperalkaline waters and C–S–H phases in the fractures of the HT-metamorphosed rocks. Such unique features were studied in detail in order to understand the long-term leaching and cation mobility in the only known natural analogue of a modern radioactive-waste repository. The investigation project was called The Maqarin Natural Analogue Project (Khoury *et al.*, 1992; Alexander and Smellie, 1998; Alexander and Blaser, 2002). This is another example showing the interplay of knowledge between natural and synthetic systems. This knowledge encompasses mineralogy, geology, hydrology, crystal chemistry, geochemistry and a number of other disciplines, and it is fundamental when tackling problems at a large or global scale.

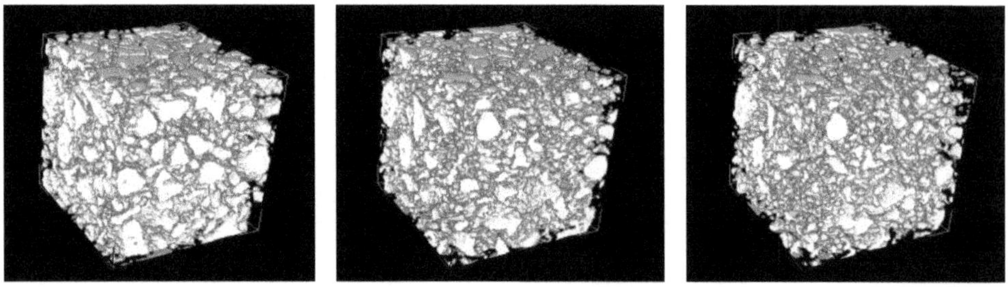

Figure 18. Virtual modelling of the microstructural evolution of cement can be compared directly to the experimental results obtained by computed micro-tomography (μ-CT).

4.1. Binders of the future

Modern cement formulations based on Portland clinkers are coming under serious criticism because of environmental issues. Cement is one of the industrial materials embodying the least amount of energy per unit volume, if compared, for example, with polymers, metals, and technical ceramics (Fig. 19). However, at present, modern society requires huge amounts of cement for infrastructures: the world production in 2016 was estimated at 4.2×10^9 tons/year (www.statista.com/statistics/219343/cement-production-worldwide/), higher than oil and gas, and second only to water. This is considered a problem because of the large emission of CO_2 during cement production (Worrell *et al.*, 2001; Barcelo *et al.*, 2014; Gartner and Hirao, 2015), estimated at ~1 ton of CO_2 emitted per ton of OPC. The emission of CO_2 derives from the fuel used in the kiln, grinding, transportation, but above all from the de-carbonation of limestone.

Increasing global concerns about sustainability and environmental compatibility (see the Intergovernmental Panel on Climate Change (IPCC) website: www.ipcc.ch/) stimulate active research towards more eco-friendly choices of materials (Ashby,

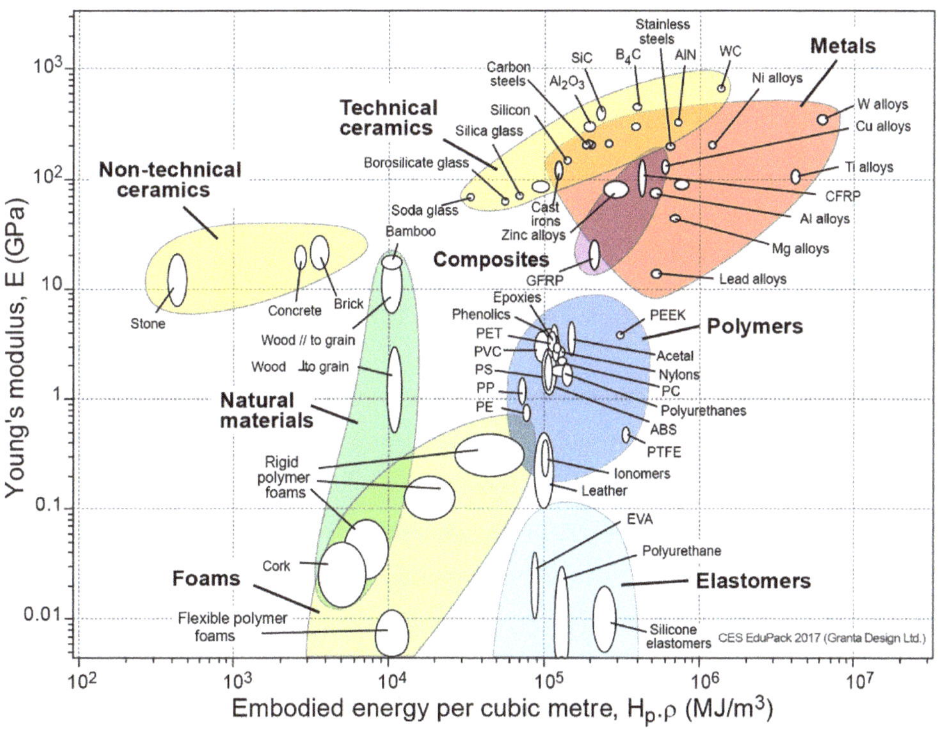

Figure 19. Young's Modulus *vs.* embodied energy per unit volume. (Chart created using CES EduPack, 2017, Granta Design Ltd. http://teachingresources.grantadesign.com/Charts-overview).

2013). The life-cycle assessment of building materials and methods is an active field of research, and concepts such as the 'transcendent quality' of building materials (Mora, 2007) and that of 'life-cycle thinking' (Buyle *et al.*, 2013) are finally entering the lexicon of construction planning. Comparative assessments of the life and energy cycles of traditional and alternative cements and mortars are on their way (Weil *et al.*, 2009; Chen *et al.*, 2010; Jiang *et al.*, 2014; Ouellet-Plamondon and Habert, 2015). Alkali-activated materials and fly-ash/slag-based geopolymers are being tested actively in the formulation of clinker-free binders (Shi *et al.*, 2006; Provis and Van Deventer, 2009, 2014; Pacheco-Torgal *et al.*, 2015). Thermally activated clays are also proposed to be a sustainable solution in the long time-span (Scrivener and Favier, 2015). It is accepted widely that the infrastructure and housing needs of modern society cannot avoid being dependent on cement-type materials (Schneider *et al.*, 2011), because the ratio between engineering properties and cost is more advantageous than for any other available material. Present trends and efforts are therefore clearly directed towards sustainability of construction materials (Liew *et al.*, 2017), which is the key word for future research and planning: despite the optimistic claims of the cement industry (Damtoft *et al.*, 2008) cement is certainly one of the materials in need of substantial improvement (Shi *et al.*, 2011; Flatt *et al.*, 2012; Kurtis, 2015).

5. Conservation of binders in architecture – Binders for conservation

Conservation of historical buildings and ancient architecture is a very intricate task. Sometimes the technical problems of conservation of the materials are not even the major difficulty: most of the problems are commonly connected with the management and decision-making of what to preserve and why. Economic, social, cultural and organizational constraints ought to be resolved before technical plans are developed (Aygen, 2013; Forsyth, 2007a. Are we sure we can distinguish the fuzzy boundaries of preservation, conservation, restoration, reconstruction, consolidation or stabilization? Following the indications of the ICOMOS Venice Charter (1964, International Charter for the Conservation and Restoration of Monuments and Sites, www.icomos.org/charters/venice_e.pdf), preservation involves the minimal repair and maintenance of remains in their existing state, possibly avoiding further degradation. Restoration involves the removal of subsequent additions in order to return to the pristine state of the building. In such a case archaeological and scientific support (dating the materials for example) are fundamental in understanding the different construction phases. Reconstruction also involves returning a building to an earlier state, but involves introducing new/old materials to complete or stabilize the structures. Therefore compatibility and reversibility are at stake, and the fundamental concept of authenticity should be pondered carefully (ICOMOS Nara Charter on Authenticity, 1994, www.icomos.org/charters/nara-e.pdf). Conservation may involve one or more of the previous interventions, as well as the adaptation of buildings to new uses. Present emphasis of engineering interventions is towards building retrofitting, *i.e.* adding new technology to old structures in order to comply with modern legislation or lifestyle standards, especially concerning seismic aspects (safety), energetics (environment) and comfort. The attitude to building preservation has shifted constantly in the past, and we can be sure that conservation philosophy (Earl and Saint, 2015) in the future will be different from that of today.

Leaving to managers and planners the general problems, which are clearly beyond the purpose of this review, there are two important points that should be kept in mind. The first one concerns the sustainability of any architectural or structural intervention: this includes a reasonable balance between resources and objectives, and then adequate planning aimed at monitoring the success and durability of the intervention. Monitoring and continuous maintenance are often neglected, mainly because of sheer costs, and this frequently jeopardizes, in the long run, the results of many technical interventions, even if they were carried out with state-of-the-art protocols. The second point concerns the planning of conservation at very early stages of the intervention plan, *i.e.* as soon as the building/site is earmarked for attention or, as in the case of archaeological excavations, even before the excavation has started. Present trends contemplate parallel planning of archaeological works and conservation (Agnew and Bridgland, 2006; Sullivan and Mackay, 2013; Pedelì and Pulga, 2013).

No matter what the intervention is going to be, materials are important. Prior to any preservation or conservation action, a considerable effort to identify the materials originally used and to assess the constructive techniques employed should be made, along with a careful analysis of compatibility with new materials and a critical evaluation of stability and permanence of restoration works to be undertaken. In conservation repair work, it is therefore imperative that an understanding of the structure and the materials be gained before specifying and undertaking any work (Forsyth, 2008). For ancient lime-based architectures this is particularly important, because the state of the structure depends on the materials employed, the quality of the application, the environmental conditions, the continuous or discontinuous use (abandonment) of the structure, *etc.*

The conservation process should encompass: (1) extensive diagnostics at various scales, including the characterization of materials; (2) modelling and structure analysis; (3) selection and testing of materials and architectural components; and (4) the intervention plan. The last should comply as much as possible with the guidelines of the ICOMOS Venice charter and the Nara document on authenticity.

Each structure must be treated as a specific case, involving a particular combination of material properties, architectural character and environmental parameters. Reaching a compromise between the structural intervention and the integrity/authenticity of the original structure is the most challenging task.

The choice of mortars and plasters to be used in conservation is important in relation to the three major factors governing the performance of historic fabric: porosity, flexibility and strength. Any change in these parameters will affect seriously the stability and durability of the structure. Therefore, the nature and properties of the materials employed must be controlled carefully in terms of compatibility with the existing materials (Van Balen *et al.*, 2005). If the intervention is extensive, reversibility

Figure 20. Brick masonry of an old factory converted to a modern building. The OPC cement used to repoint the joints is deleterious in terms of rigidity and salt leaching.

is hardly possible. Extensive work has been done to characterize and optimize repair mortars and grouts (*e.g.* Maurenbrecher, 2004; Faria *et al.*, 2008; Schueremans *et al.*, 2011). Compatibility between mortars and repaired stone materials has been addressed specifically (Isebaert *et al.*, 2014). While in the past Portland cement has been used variously to strengthen historical structures, it is now widely accepted that the use of OPC or formulations using OPC are detrimental to ancient structures in terms of mechanical and chemical incompatibility. OPC has two major adverse effects. The first is that the mechanical behaviour in terms of compressive strength or Young's modulus is much greater than that of lime mortars, thus creating rigid insertions in the fabric and reducing flexibility. The second is that OPC contains sulfate ions that are inevitably leached into the structure and redeposited as salts, usually at the surface, causing well known effects of surface deterioration (detachments, spalling, efflorescence, *etc.*). Sulfate salts are also deleterious to bricks and stone units of masonry (Fig. 20). For these reasons OPC has been virtually banned from all interventions to pre-industrial age buildings and ancient lime-based architecture. From all evidence, historical lime and hydraulic mortars proved to be less performing but more durable than 20th century reinforced concrete. Actually, the restoration and preservation of last century historical buildings based on Portland concrete is turning to be a diffuse and urgent problem, particularly where it is exposed and subject to damage from reinforcement corrosion (Forsyth, 2007b; Macdonald, 2008).

Acknowledgments

Small parts of the text have been adapted from chapter 3.2 of Artioli (2010).

References

Abrams, E.M. and Freter, A. (1996) A Late Classic lime-plaster kiln from the Maya centre of Copan, Honduras. *Antiquity*, **70**, 422–428.

Adam, J.P. (2005) *Roman Building: Materials and Techniques*. Routledge, London.

Addis, A., Secco, M., Preto, N., Marzaioli, F., Passariello, I., Brogiolo, G.P., Chavarria Arnau, A., Artioli, G. and Terrasi, F. (2016) New strategies for radiocarbon dating of mortars: Multi-step purification of the lime binder. Pp. 665–672 in: *Proceedings of the 4th Historic Mortars Conference – HMC 2016* (I. Papayianni M. Stefanidou and V. Pachta, editors). Aristotle University of Thessaloniki, Greece.

Affonso, M.T.C. and Freiberg, E.P. (2001) Neolithic lime plasters and pozzolanic reactions. Are they occasional occurrences? Pp. 9–13 in: *Lux Orientis. Archaeologie zwishen Asien and Europa* (R.M. Boehmer and J. Maran, editors). Festschrift fur Havald Hauptmann zum 65 Geburstag. Verlag Marie Leidorf GmbH. Rahden/Westfalia, Germany.

Agnew, N. and Bridgland, J. (editors) (2006) *Of the Past, for the Future: Integrating Archaeology and Conservation*. Proceedings of the Conservation Theme at the 5th World Archaeological Congress, Washington, DC, 22–26 June 2003. Getty Publications, Los Angeles, California, USA.

Alexander, W.R. and Blaser, P.C. (editors) (2002) *The use of technical natural analogues in radioactive waste disposal*. Nagra Unpublished Project Report, Nagra, Wettingen, Switzerland.

Alexander, W.R. and Smellie, J.A.T. (1998) *Maqarin natural analogue project: synthesis report on Phases I, II and III*. Nagra Unpublished Project Report, Nagra, Wettingen, Switzerland.

Al-Rawas, A.A., Hago, A.W., Corcoran, T.C. and Al-Ghafri, K.M. (1998) Properties of Omani artificial pozzolana (sarooj). *Applied Clay Science*, **13**, 275–292.

Al-Rawas, A.A., Hago, A.W., Al-Lawati, D. and Al-Battashi, A. (2001) The Omani artificial pozzolans (sarooj). *Cement, Concrete and Aggregates*, **23**, 19–26.

Alvarez, J.I., Navarro, I., Martın, A. and Casado, P.G. (2000) A study of the ancient mortars in the north tower of Pamplona's San Cernin church. *Cement and Concrete Research*, **30**, 1413–1419.

Ambers, J. (1987) Stable carbon isotope ratios and their relevance to the determination of accurate radiocarbon dates for lime mortars. *Journal of Archaeological Science*, **14**, 569–576.

Aranda, M.A. (2016) Recent studies of cements and concretes by synchrotron radiation crystallographic and cognate methods. *Crystallography Reviews*, **22**, 150–196.

Arandigoyen, M., Bicer-Simsir, B., Alvarez, J.I. and Lange, D.A. (2006) Variation of microstructure with carbonation in lime and blended pastes. *Applied Surface Science*. **252**, 7562–7571.

Arkun, B.H. (2003) *Neolithic plasters in the Near East: Çatalhöyük Building 5, a case study*. Master's Thesis, University of Pennsylvania.

Artioli, G. (2010) *Scientific Methods and Cultural Heritage: An Introduction to the Application of Materials Science to Archaeometry and Conservation Science*. Oxford University Press.

Artioli, G. and Secco, M. (2016) Modern and ancient masonry: Nature and role of the binder. Pp. 3–10 in: *Brick and Block Masonry – Trends, Innovations and Challenges* (C. Modena, F. Da Porto and M.R. Valluzzi, editors). Proceedings of the 16th International Brick and Block Masonry Conference, Padova, 26-30 June 2016. Taylor & Francis Group, London.

Artioli, G., Nicola, C., Montana, G., Angelini, I., Nodari, L. and Russo, U. (2009) The blue enamels in the baroque decorations of the churches of Palermo, Sicily: Fe^{2+}-coloured glasses from lime kilns. *Archaeometry*, **51**, 197–213.

Artioli, G., Cerulli, T., Cruciani, G., Dalconi, M.C., Ferrari, G., Parisatto, M., Rack, A. and Tucoulou, R. (2010) X-ray diffraction microtomography (XRD-CT), a novel tool for non-invasive mapping of phase development in cement materials. *Analytical and ioanalytical Chemistry*, **397**, 2131–2136.

Artioli, G., Valentini, L., Dalconi, M.C., Parisatto, M., Voltolini, M., Russo, V. and Ferrari, G. (2014a) Imaging of nano-seeded nucleation in cement pastes by X-ray diffraction tomography. *International Journal of Materials Research*, **105**, 628–631.

Artioli, G., Valentini, L., Voltolini, M., Dalconi, M.C., Ferrari, G. and Russo, V. (2014b) Direct imaging of nucleation mechanisms by synchrotron diffraction micro-tomography: superplasticizer-induced change of C–S–H nucleation in cement. *Crystal Growth & Design*, **15**, 20–23.

Ashby, M.F. (2013) *Materials and the Environment: Eco-informed Material Choice*. Elsevier, Amsterdam and Butterworth-Heinemann, Oxford.

Avrami, E., Guillaud, H. and Hardy M. (editors) (2008) *Terra Literature Review. An Overview of Research in Earthen Architecture Conservation*. The Getty Conservation Institute, Los Angeles.

Aygen, Z. (2013) *International Heritage and Historic Building Conservation: Saving the World's Past* (Vol. 3). Routledge, London.

Barba, L., Blancas, J., Manzanilla, L.R., Ortiz, A., Barca, D., Crisci, G.M., Miriello, D. and Pecci, A. (2009) Provenance of the limestone used in Teotihuacan (Mexico): A methodological approach. *Archaeometry*, **51**, 525–545.

Barcelo, L., Kline, J., Walenta, G. and Gartner, E. (2014) Cement and carbon emissions. *Materials and Structures*, **47**, 1055–1065.

Baronio, G. and Binda, L. (1997) Study of the pozzolanicity of some bricks and clays. *Construction and Building Materials*, **11**, 41–46.

Bakolas, A., Aggelakopoulou, E. and Moropoulou, A. (2008) Evaluation of pozzolanic activity and physico-mechanical characteristics in ceramic powder-lime pastes. *Journal of Thermal Analysis and Calorimetry*, **92**, 345–351.

Barnes, P. and Bensted, J. (2002) *Structure and Performance of Cements*. CRC Press, Boca Raton, Florida, USA.

Bar-Yosef, O. (1986) The walls of Jericho: An alternative interpretation. *Current Anthropology*, **27**, 157–162.

Bailey, S.W. (1988) *Hydrous Phyllosilicates (Exclusive of Micas)* (S.W. Bailey, editor). Reviews in Mineralogy, **19**. Mineralogical Society of America. Chantilly, Virginia, USA.

Bellotto, M. and Signes-Frehel, M. (1998) The Role of Powder X-ray Diffraction in the Cement Industry. European Powder Diffraction Conference. *Materials Science Forum*, Vols. **278–281**.

Bellotto, M., Gualtieri, A., Artioli, G. and Clark, S.M. (1995) Kinetic study of the kaolinite-mullite reaction sequence. Part I: kaolinite dehydroxylation. *Physics and Chemistry of Minerals*, **22**, 207–217.

Bensted, J. (1999) A discussion of the paper "Use of cactus in mortars and concrete" by S. Chandra, L. Eklund, and R.R. Villarreal. – Authors'reply. *Cement and Concrete Research*, **29**, 967–969.

Bentur, A. (2002) Cementitious materials – Nine millennia and a new century: Past, present, and future. American Society of Civil Engineers, 150th Anniversary Paper. *Journal of Materials in Civil Engineering*, JAN/FEB 2002, 2–22.

Bequette, A. and Dhanjal, S. (2011) Continuous X-ray diffraction at the Buzzi Unicem festus plant. Pp. 1–10 in: *Cement Industry Technical Conference*, 2011 IEEE-IAS/PCA 53rd. IEEE.

Biton, R., Goren, Y. and Goring-Morris, A.N. (2014) Ceramics in the Levantine Pre-Pottery Neolithic B: evidence from Kfar HaHoresh, Israel. *Journal of Archaeological Science*, **41**, 740–748.

Blauer-Bohm, C. and Jagers, E. (1997) Analysis and recognition of dolomitic lime mortars. Pp. 223–235 in: *Roman Wall Painting: Materials, Techniques, Analysis and Conservation* (H. Bearat, M. Fuchs, M. Maggetti and D. Paunier, editors). Proceedings of the International Workshop, Fribourg, 7–9 March 1996. Institute of Mineralogy and Petrography, Fribourg, Germany.

Blezard, R.G. (2003) The history of calcareous cements. Pp. 1–19 in: *Lea's Chemistry of Cement and Concrete* (P.C. Hewlett, editor). Butterworth-Heinemann, Oxford, UK.

Böke, H., Akkurt, S., İpekoğlu, B. and Uğurlu, E. (2006) Characteristics of brick used as aggregate in historic brick-lime mortars and plasters. *Cement and Concrete Research*, **36**, 1115–1122.

Bonetto, J., Artioli, G., Secco, M. and Addis, A. (2016) The use of the pozzolanic ashes in the big yards of Cisalpine Gaul during the Roman Republican Age: The cases of Aquileia and Ravenna. Pp. 29–44 in: Proceedings of the 5th International Workshop on the Archaeology of Roman Construction (S. Camporeale, J. DeLaine and A. Pizzo, editors). Oxford, 11-12 April 2015. "*Arqueología de la Construcción V. Man-made materials, engineering and infrastructure*". CSIC, Madrid.

Bonsall, C., Radovanović, I., Roksandic, M., Cook, G.T., Higham, T. and Pickard, C. (2008) Dating burial practices and architecture at Lepenski Vir. Pp. 175–204 in: *The Iron Gates in Prehistory: New Perspectives* (C. Bonsall, I. Radovanović and V. Boroneanţ, editors). Archaeopress, Oxford, UK.

Borić, D. (2002) The Lepenski Vir conundrum: reinterpretation of the Mesolithic and Neolithic sequences in the Danube Gorges. *Antiquity*, **76**, 1026–1039.

Brandon, C., Hohlfelder, R.L., Oleson, J.P. and Stern, C. (2005) The Roman Maritime Concrete Study

(ROMACONS): the harbour of Chersonisos in Crete and its Italian connection. *Méditerranée. Revue géographique des pays méditerranéens/Journal of Mediterranean Geography*, **104**, 25–29.

Brandon, C.J., Hohlfelder, R.L., Jackson, M.D. and Oleson, J.P. (2014) *Building for eternity: the history and technology of Roman concrete engineering in the sea*. Oxbow Books.

Bras, A. and Henriques, F.M. (2012) Natural hydraulic lime based grouts–The selection of grout injection parameters for masonry consolidation. *Construction and Building Materials*, **26**, 135–144.

Brown, G. and Brindley, G.W. (editors) (1980) *Crystal Structures of Clay Minerals and their X-ray Identification*. Monograph **5**. Mineralogical Society of Great Britain & Ireland, London.

Bullard, J.W., Jennings, H.M., Livingston, R.A., Nonat, A., Scherer, G.W., Schweitzer, J.S., Scrivener, K.L. and Thomas, J.J. (2011) Mechanisms of cement hydration. *Cement and Concrete Research*, **41**, 1208–1223.

Burg, A., Starinsky, A., Bartov, Y. and Kolodny, Y. (1991) Geology of the Hatrurim formation ("mottled zone") in the Hatrurim basin. *Israel Journal of Earth Sciences*, **40**, 107–124.

Buyle, M., Braet, J. and Audenaert, A. (2013) Life cycle assessment in the construction sector: A review. *Renewable and Sustainable Energy Reviews*, **26**, 379–388.

Callebaut, K., Elsen, J., Van Balen, K. and Viaene, W. (2000) Historical and scientific study of hydraulic mortars from the 19th century. Pp. 125–132 in: Proceedings of the International RILEM-workshop "*Historic Mortars: Characteristics and Tests*", Paisley, UK.

Callebaut, K., Elsen, J., Van Balen, K. and Viaene, W. (2001) Nineteenth century hydraulic restoration mortars in the Saint Michael's Church (Leuven, Belgium): Natural hydraulic lime or cement? *Cement and Concrete Research*, **31**, 397–403.

Cameron, M.A.S., Jones, R.E. and Philippakis, S.E. (1977) Scientific analyses of Minoan fresco samples from Knossos. *Annual of the British School at Athens*, **72**, 121–184.

Cantisani, E., Cecchi, A., Chiaverini, I., Fratini, F., Manganelli Del Fá, C., Pecchioni, E. and Rescic, S. (2002) The binder of the Roman concrete of the Ponte di Augusto at Narni (Italy). *Periodico di Mineralogia*, **71**, 113–123.

Carran, D., Hughes, J., Leslie, A. and Kennedy, C. (2012) A short history of the use of lime as a building material beyond Europe and North America. *International Journal of Architectural Heritage*, **6**, 117–146.

Cazalla, O., Rodriguez-Navarro, C., Sebastian, E., Cultrone, G. and Torre, M.J. (2000) Aging of lime putty: effects on traditional lime mortar carbonation. *Journal of the American Ceramic Society*, **83**, 1070–1076.

Chandra, S. (2002) Properties of concrete with mineral and chemical admixtures. Pp. 140–185 in: *Structure and Performance of Cements*. 2nd Edition. (J. Bensted and P. Barnes, editors). Spon Press, London – New York.

Chandra, S. and Aavik, J. (1983) Influence of black gram (natural organic material) addition as admixture in cement mortar and concrete. *Cement and Concrete Research*, **13**, 423–430.

Chandra, S., Eklund, L. and Villarreal, R.R. (1998) Use of cactus in mortars and concrete. *Cement and Concrete Research*, **28**, 41–51.

Charola, A.E., Rodrigues, P.F., McGhie, A.R. and Henriques, F.M. (2005) Pozzolanic components in lime mortars: Correlating behaviour, composition and microstructure. *Restoration of Buildings and Monuments*, **11**, 111–118.

Chatterjee, A.K. (2002) Special cements. Pp. 186–236 in: *Structure and Performance of Cements*. 2nd Edition (J Bensted and P. Barnes, editors). Spon Press, London, New York.

Chen, C., Habert, G., Bouzidi, Y., Jullien, A. and Ventura, A. (2010) LCA allocation procedure used as an incitative method for waste recycling: An application to mineral additions in concrete. *Resources, Conservation and Recycling*, **54**, 1231–1240.

Cheung, J., Jeknavorian, A., Roberts, L. and Silva, D. (2011) Impact of admixtures on the hydration kinetics of Portland cement. *Cement and Concrete Research*, **41**, 1289–1309.

Chiarelli, N., Miriello, D., Bianchi, G., Fichera, G., Giamello, M. and Memmi Turbanti, I. (2015) Characterisation of ancient mortars from the S. Niccol archaeological complex in Montieri (Tuscany–Italy). *Construction and Building Materials*, **96**, 442–460.

Chu, V., Regev, L., Weiner, S. and Boaretto, E. (2008) Differentiating between anthropogenic calcite in plaster, ash and natural calcite using infrared spectroscopy: implications in archaeology. *Journal of Archaeological Science*, **35**, 905–911.

Collepardi, M., Collepardi, S. and Troli, R. (2007) *Concrete Mix Design*. Grafiche Tintoretto, Lancenigo, Italy.

Costa, U., Gotti, E. and Tognon, G. (2000) Nota tecnica: malte prelevate da mura antiche dallo scavo della Banca Popolare di Ravenna. Pp. 25–28 in: Manzelli V. "Le mura di Ravenna repubblicana" (L. Quilici, and S. Quilici Gigli, editors). *"Fortificazioni antiche in Italia – Età Repubblicana"*. L'Erma di Bretschneider, Roma.

Coulson, W. and Wilkie, N.C. (1986) Ptolemaic and Roman kilns in the Western Nile delta. *Bulletin of the American Schools of Oriental Research*, **263**, 61–75.

Cultrone, G., Sebastián, E., Elert, K., De la Torre, M.J., Cazalla, O. and Rodriguez–Navarro, C. (2004) Influence of mineralogy and firing temperature on the porosity of bricks. *Journal of the European Ceramic Society*, **24**, 547–564.

Dalconi, M.C., Favero, M. and Artioli, G. (2008) In-situ XRPD of hydrating cement with lab instrument: reflection vs. transmission measurements. Proc. EPDIC-11, Warsaw, Poland, 18–22 September 2008. *Zeitschrift für Kristallographie, Proceedings*, **1**, 155–161.

Damidot, D., Lothenbach, B., Herfort, D. and Glasser, F.P. (2011) Thermodynamics and cement science. *Cement and Concrete Research*, **41**, 679–695.

Damtoft, J.S., Lukasik, J., Herfort, D., Sorrentino, D. and Gartner, E.M. (2008) Sustainable development and climate change initiatives. *Cement and Concrete Research*, **38**, 115–127.

Davidson, T.E. and McKerrell, H. (1976) Pottery analysis and Halaf Period trade in the Khabur Headwaters Region. *Iraq*, **38**, 45–56.

Day, K. (2003) *Concrete Mix Design, Quality Control and Specification* (with CD ROM). CRC Press, Boca Raton, Florida, USA.

De la Torre, A.G. and Aranda, M. G. (2003) Accuracy in Rietveld quantitative phase analysis of Portland cements. *Journal of Applied Crystallography*, **36**, 1169–1176.

Diekamp, A., Stalder, R., Konzett, J. and Mirwald, P.W. (2012) Lime mortar with natural hydraulic components: characterisation of reaction rims with FTIR imaging in ATR-mode. Pp. 105–113 in: *Historic Mortars*. Springer, Dordrecht, The Netherlands.

Dix, B. (1982) The manufacture of lime and its uses in the Western Roman provinces. *Oxford Journal of Archaeology*, **1**, 331–346.

Earl, J. and Saint, A. (2015) *Building Conservation Philosophy*. Routledge, London.

Edmeades, R.M. and Hewlett, P.C. (2003) Cement admixtures. Pp. 841–906 in: *Lea's Chemistry of Cement and Concrete* (P.C. Hewlett, editor). Butterworth-Heinemann, Oxford, UK.

El Ali, A., Barbin, V., Calas, G., Cervelle, B., Ramseyer, K. and Bouroulec, J. (1993) Mn^{2+}-activated luminescence in dolomite, calcite and magnesite: quantitative determination of manganese and site distribution by EPR and CL spectroscopy. *Chemical Geology*, **104**, 189–202.

Elsen, J. (2004) Characterisation of binder related particles (lime lumps) in historic lime mortars. *Proceedings of the 8th International Congress on Applied Mineralogy* (ICAM2004), Sao Paulo, 347–349.

Elsen, J. (2006) Microscopy of historic mortars – A review. *Cement and Concrete Research*, **36**, 1416–1424.

Elsen, J., Brutsaert, A., Deckers, M. and Brulet, R. (2004) Microscopical study of ancient mortars from Tournai (Belgium). *Materials Characterization*, **53**, 289–294.

Elsen, J., Van Balen, K. and Mertens, G. (2012) Hydraulicity in historic lime mortars: a review. Pp. 125–139 in: *Historic Mortars*. Springer, Dordrecht, The Netherlands.

Fang, S.Q., Zhang, H., Zhang, B.J. and Zheng, Y. (2014) The identification of organic additives in traditional lime mortar. *Journal of Cultural Heritage*, **15**, 144–150.

Faria, P., Henriques, F. and Rato, V. (2008) Comparative evaluation of lime mortars for architectural conservation. *Journal of Cultural Heritage*, **9**, 338–346.

Fernandez, R., Martirena, F. and Scrivener, K.L. (2011) The origin of the pozzolanic activity of calcined clay minerals: A comparison between kaolinite, illite and montmorillonite. *Cement and Concrete Research*, **41**, 113–122.

Flatt, R.J., Roussel, N. and Cheeseman, C.R. (2012) Concrete: An eco material that needs to be improved. *Journal of the European Ceramic Society*, **32**, 2787–2798

Forsyth, M. (editor) (2007a) *Understanding Historic Building Conservation*. John Wiley & Sons, New York.

Forsyth, M. (editor) (2007b) *Structures and Construction in Historic Building Conservation*. John Wiley & Sons, New York.

Forsyth, M. (editor) (2008) *Materials and Skills for Historic Building Conservation.* John Wiley & Sons.

Frankeová, D., Slížková, Z. and Drdácký, M., (2012) Characteristics of mortars from ancient bridges. Pp. 165–174 in: *Historic Mortars.* Springer, Dordrecht, The Netherlands.

Franzini, M., Leoni, L., Lezzerini, M. and Sartori, F. (1999) On the binder of some ancient mortars. *Mineralogy and Petrology*, **67**, 59–69.

Franzini, M., Leoni, L. and Lezzerini, M. (2000) A procedure for determining the chemical composition of binder and aggregate in ancient mortars: its application to mortars from some medieval buildings in Pisa. *Journal of Cultural Heritage*, **1**, 365–373.

Franzini, M., Leoni, L., Lezzerini, M. and Sartori, F. (2000) The mortar of the "Leaning Tower" of Pisa: the product of a medieval technique for preparing high-strength mortars. *European Journal of Mineralogy*, **12**, 1151–1163.

Fratini, F., Pecchioni, E., Rovero, L. and Tonietti, U. (2011) The earth in the architecture of the historical centre of Lamezia Terme (Italy): characterization for restoration. *Applied Clay Science*, **53**, 509–516.

Frierman, J.D. (1971) Lime burning as the precursor of fired ceramics. *Israel Exploration Journal*, **21**, 212–216.

Friesem, D.E., Wattez, J. and Onfray, M. (2017) Earth construction materials. Pp. 99–112 in: *Archaeological Soil and Sediment Micromorphology* (C. Nicosia and G. Stoops, editors). Wiley Blackwell, Hoboken, New Jersey, USA.

Furlan, V. and Bissegger, P. (1975) Les mortiers anciens – Historie et essais d'analyse scientifiques. *Zeitschrift fur Schweizerische Archaeologie und Kunstgeschichte*, **32**, 166–178.

Garfinkel, Y. (1987) Burnt lime products and social implications in the Pre-Pottery Neolithic B villages of the Near East. *Paléorient*, **13**, 69–76.

Gartner, E. and Hirao, H. (2015) A review of alternative approaches to the reduction of CO_2 emissions associated with the manufacture of the binder phase in concrete. *Cement and Concrete Research*, **78**, 126–142.

Gartner, E.M., Young, J.F., Damidot, D.A. and Jawed, I. (2002) Hydration of Portland cement. Pp. 57–113 in: *Structure and Performance of Cements.* 2nd Edition (J. Bensted and P. Barnes, editors). Spon Press, London, New York.

Ghosh, A. (editor) (1989) *An Encyclopaedia of Indian Archaeology.* Brill, Amsterdam.

Gibbs, K. (2015) Pottery invention and innovation in East Asia and the Near East. *Cambridge Archaeological Journal*, **25**, 339–351.

Gilat, A. (1998) Hydrothermal activity and hydro-explosions as a cause of natural combustion and pyrolysis of bituminous rocks; the case of Pliocene metamorphism in Israel (Hatrurim Formation). *Current Research – Geological Survey of Israel*, **11**, 96–102.

Goldberg, P. and Bar-Yosef, O. (1998) Site formation processes in Kebara and Hayonim caves and their significance in Levantine prehistoric caves. Pp. 107–126 in: *Neanderthals and Modern Humans in Western Asia* (T. Akazawa, K. Aoki and O. Bar-Yosef, editors). Springer-Verlag, Berlin-New York.

Goren, Y., Goring-Morris, A.N. and Segal, I. (2001) The technology of skull modelling in the Pre-Pottery Neolithic B (PPNB): regional variability, the relation of technology and iconography and their archaeological implications. *Journal of Archaeological Science*, **28**, 671–690.

Gosselin, C., Vergès-Belmin, V., Royer, A. and Martinet, G. (2009) Natural cement and monumental restoration. *Materials and Structures*, **42**, 749–763.

Gotti, E., Oleson, J.P., Bottalico, L., Brandon, C., Cucitore, R. and Hohlfelder, R.L. (2008) A comparison of the chemical and engineering characteristics of ancient Roman hydraulic concrete with a modern reproduction of Vitruvian hydraulic concrete. *Archaeometry*, **50**, 576–590.

Gourdin, W.H. and Kingery, W.D. (1975) The beginning of pyrotechnology: Neolithic and Egyptian limeplaster. *Journal of Field Archaeology*, **2**, 133–150.

Greco, A.V. (2011) *Virtutes materiae. Il contributo delle fonti latine nello studio di malte, intonaci e rivestimenti nel mondo romano.* Sandhi, Ortacesus.

Griffin, P.S., Grissom, C.A. and Rollefson, G.O. (1998) Three late eighth millennium plastered faces from 'Ain Ghazal, Jordan. *Paléorient*, **24**, 59–70.

Gualtieri, A., Bellotto, M., Artioli, G. and Clark, S.M. (1995) Kinetic study of the kaolinite-mullite reaction sequence. Part II: mullite formation. *Physics and Chemistry of Minerals*, **22**, 215–222.

Gualtieri, A. and Bellotto, M. (1998) Modelling the structure of the metastable phases in the reaction sequence kaolinite-mullite by X-ray scattering experiments. *Physics and Chemistry of Minerals*, **25**, 442–452.

Gulzar, S., Chaudhry, M.N. and Burg, J.P. (2013) Chemical and mineralogical characterization of old mortars from Jahangir tomb, Lahore-Pakistan. *Asian Journal of Chemistry*, **25**, 133.

Hansen, E.F., Rodriguez-Navarro, C. and Hansen, R.D. (1997) Incipient Maya burntlime technology: Characterization and chronological variations in Preclassic plaster, stucco and mortar at Nakbe, Guatemala. Pp. 207–216 in: *Materials Issues in Art and Archaeology V*. (P.B. Vandiver, J.R. Druzik, J.F. Merkel and J. Stewart, editors). Materials Research Society, Pittsburgh, Pennsylvania, USA.

Harutyunyan, V.S., Kirchheim, A.P., Monteiro, P.J.M., Aivazyan, A.P. and Fischer, P. (2009) Investigation of early growth of calcium hydroxide crystals in cement solution by soft X-ray transmission microscopy. *Journal of Materials Science*, **44**, 962–969.

Hauptmann, A. and Yalcin, Ü. (2001) Lime plaster, cement and the first puzzolanic reaction. *Paléorient*, **26**, 61–68.

Heimann, R.B., Maggetti, M., Heiman, G. and Maggetti, J. (2010) *Ancient and Historical Ceramics: Materials, Technology, Art and Culinary Traditions*. Schweizerbart Science Publishers, Berlin.

Herodotus. English translation by A.D. Godley. Harvard University Press, Cambridge, Massachusetts, USA; W. Heinemann, London. 1920.

Hesse, C., Goetz-Neunhoeffer, F. and Neubauer, J. (2011) A new approach in quantitative in-situ XRD of cement pastes: Correlation of heat flow curves with early hydration reactions. *Cement and Concrete Research*, **41**, 123–128.

Hewlett, P. (2003) *Lea's Chemistry of Cement and Concrete*. Butterworth-Heinemann.

Hobbs, L.W. and Siddall, R. (2011) Cementitious materials of the ancient world. Pp. 35–60 in: *Building Roma Aeterna* (Å. Ringbom and R.L. Hohlfelder, editors) (Proceedings). The Finnish Society of Sciences and Letters, Helsinki.

Hodder, I. (2006) *The Leopard's Tale: Revealing the Mysteries of Çatalhöyük*. Thames & Hudson, London.

Hodder, I. (2012) *Entangled – An Archaeology of the Relationship between Humans and Things*. Wiley-Blackwell.

Hong, S.Y. and Glasser, F.P. (2002) Alkali sorption by CSH and CASH gels: Part II. Role of alumina. *Cement and Concrete Research*, **32**, 1101–1111.

Huang, K.Z. (2003) Study and conservation of cultural architecture materials. *Southeast Cult.*, **173**, 93–96.

Hughes, J.J., Leslie, A.B. and Callebaut, K. (2001) The petrography of lime inclusions in historic lime based mortars. *Proceedings of the 8th Euroseminar on Microscopy Applied to Building Materials*, 359–364.

Huisman, D.J and Milek, K.B. (2017) Turf as construction material. Pp. 113–136 in: *Archaeological Soil and Sediment Micromorphology* (C. Nicosia and G. Stoops, editors). Wiley Blackwell, Hoboken, New Jersey, USA.

Hyman, D.S. (1970) *Precolumbian cements: A study of the calcareous cements in Prehispanic Mesoamerican building construction*. Ph.D. Dissertation, Johns Hopkins University, Baltimore, Maryland, USA.

Hyslop, J. (2014) *Inka Settlement Planning*. University of Texas Press.

Isebaert, A., Van Parys, L. and Cnudde, V. (2014) Composition and compatibility requirements of mineral repair mortars for stone – A review. *Construction and Building Materials*, **59**, 39–50.

Jackson, M.D., Marra, F., Deocampo, D., Vella, A., Kosso, C. and Hay, R. (2007) Geological observations of excavated sand (harenae fossiciae) used as fine aggregate in ancient Roman pozzolanic mortars. *Journal of Roman Archaeology*, **20**, 1–30.

Jackson, M.D., Vola, G., Všianský, D., Oleson, J.P., Scheetz, B.E., Brandon, C. and Hohlfelder, R.L. (2012) Cement microstructures and durability in ancient Roman seawater concretes. Pp. 49–76 in: *Historic Mortars*. Springer, Dordrecht, The Netherlands.

Jackson, M.D., Chae, S.R., Mulcahy, S.R., Meral, C., Taylor, R., Li, P., Emwas, A.-H., Moon, J., Yoon, S., Vola, G., Wenk, H.R. and Monteiro, P.J.M. (2013) Unlocking the secrets of Al-tobermorite in Roman seawater concrete. *American Mineralogist*, **98**, 1669–1687.

Jackson, M.D., Mulcahy, S.R., Chen, H., Li, Y., Li, Q., Cappelletti, P. and Wenk, H.R. (2017) Phillipsite and Al-tobermorite mineral cements produced through low-temperature water-rock reactions in Roman marine concrete. *American Mineralogist*, **102**, 1435–1450.

Jansen, D., Goetz-Neunhoeffer, F., Lothenbach, B. and Neubauer, J. (2012) The early hydration of Ordinary Portland Cement (OPC): An approach comparing measured heat flow with calculated heat flow from QXRD. *Cement and Concrete Research*, **42**, 134–138.

Jiang, M., Chen, X., Rajabipour, F. and Hendrickson, C.T. (2014) Comparative life cycle assessment of conventional, glass powder, and alkali-activated slag concrete and mortar. *Journal of Infrastructure Systems*, **20**, 04014020.

Karkanas, P. (2007) Identification of lime plaster in prehistory using petrographic methods: A review and reconsideration of the data on the basis of experimental and case studies. *Geoarchaeology*, **22**, 775–796.

Karkanas, P. and Goldberg, P. (2007) Micromorphology of sediments: Deciphering archaeological context. *Israel Journal of Earth Sciences*, **56**, 63–71.

Karkanas, P. and Efstratiou, N. (2009) Floor sequences in Neolithic Makri, Greece: micromorphology reveals cycles of renovation. *Antiquity*, **83**, 955–967.

Karkanas, P. and Stratouli, G. (2008) Neolithic lime plastered floors in Drakaina Cave, Kephalonia Island, Western Greece: Evidence of the significance of the site. *Annual of the British School at Athens*, **103**, 27–41.

Kenoyer, J.M. (1991) The Indus Valley Tradition of Pakistan and Western India. *Journal of World Prehistory*, **5**, 331–385.

Kenyon, K.M. (1981) *Excavations at Jericho III. The Architecture and Stratigraphy of the Tell*. British School of Archaeology in Jerusalem, London, 540 pp.

Khan, A. and Lemmen, C. (2013) Bricks and urbanism in the Indus Civilization. *arXiv preprint* arXiv:1303.1426.

Khoury, H.N., Salameh, E., Clark, I.D., Fritz, P., Bajjali, W., Milodowski, A.E., Cave, M.R. and Alexander, W.R. (1992) A natural analogue of high pH cement pore waters from the Maqarin area of northern Jordan. I: introduction to the site. *Journal of Geochemical Exploration*, **46**, 117–132.

Kingery, D.W., Vandiver, P.B. and Prickett, M. (1988) The beginnings of pyrotechnology, part II: Production and use of lime and gypsum plaster in the Pre-Pottery Neolithic Near East. *Journal of Field Archaeology*, **15**, 219–243.

Kingery, W.D., Vandiver, P.B. and Noy, T. (1992) 8,500-year-old sculpted plaster head from Jericho (Israel). *Materials Research Society Bulletin*, **17**, 46–52.

Kopelson, E. (1996) *Analysis and consolidation of architectural plasters from Çatalhöyük, Turkey*. Master's Thesis, University of Pennsylvania, USA.

Kosednar-Legenstein, B., Dietzel, M., Leis, A. and Stingl, K. (2008) Stable carbon and oxygen isotope investigation in historical lime mortar and plaster–Results from field and experimental study. *Applied Geochemistry*, **23**, 2425–2437.

Koui, M. and Ftikos, C. (1998) The ancient Kamirian water storage tank: A proof of concrete technology and durability for three millenniums. *Materials and Structures*, **31**, 623–627.

Kozlowski, S.K. and Kempisty, A. (1990) Architecture of the Pre-Pottery Neolithic settlement in Nemrik, Iraq. *World Archaeology*, **21**, 348–362.

Kropáč, K. and Dolníček, Z. (2013) Non-metallurgical Slags in the Masonry of Obřany Castle in the Czech Republic: Evidence for the Local Production of Hydraulic Lime in the 14th Century? *Geoarchaeology*, **28**, 544–556.

Kumar, K. (1984) The evidence of white-wash, plaster and pigment on North Indian sculpture with special reference to Sarnath. *Artibus Asiae*, **45**, 199–206.

Kurtis, K.E. (2015) Innovations in cement-based materials: Addressing sustainability in structural and infrastructure applications. *Materials Research Society Bulletin*, **40**, 1102–1109.

Kuzmin, Y.V. (2006) Chronology of the earliest pottery in East Asia: progress and pitfalls. *Antiquity*, **80**, 362–371.

L'Hôpital, E., Lothenbach, B., Le Saout, G., Kulik, D. and Scrivener, K. (2015) Incorporation of aluminium in calcium-silicate-hydrates. *Cement and Concrete Research*, **75**, 91–103.

L'Hôpital, E., Lothenbach, B., Scrivener, K. and Kulik, D.A. (2016) Alkali uptake in calcium alumina silicate hydrate (CASH). *Cement and Concrete Research*, **85**, 122–136.

Lamprecht, H.O. (1996) *Opus caementitium: Bautechnik der Römer*. Beton-Verlag, Germany.

Lancaster, L.C. (2005) *Concrete Vaulted Construction in Imperial Rome: innovations in Context*. Cambridge University Press, Cambridge, Cambridge, UK.

Lancaster, L.C. (2011) The use of lightweight concrete in Rome, Cilicia, and Tunisia. Pp. 60–72 in: *Building Roma Aeterna* (Å. Ringbom and R.L. Hohlfelder, editors). (Proceedings) The Finnish Society of Sciences and Letters, Helsinki.

Lancaster, L.C. (2012) Ash mortar and vaulting tubes: agricultural production and the building industry in North Africa. Pp. 145–160 in: *Arqueología de la construcción III: Los procesos constructivos en el mundo romano: la economía de las obras*. (S. Camporeale, H. Dessales and A. Pizzo, editors). Instituto de Arqueología de Mérida, Madrid.

Lancaster, L.C. (2015) *Innovative Vaulting in the Architecture of the Roman Empire: 1st to 4th Centuries CE*. Cambridge University Press, Cambridge, UK.

Lancaster, L.C., Sottili, G., Marra, F. and Ventura, G. (2010) Provenancing of lightweight volcanic stones used in ancient Roman concrete vaulting: Evidence from Turkey and Tunisia. *Archaeometry*, **52**, 949–961.

Lancaster, L.C., Sottili, G., Marra, F. and Ventura, G. (2011) Provenancing of lightweight volcanic stones used in ancient Roman concrete vaulting: evidence from Rome. *Archaeometry*, **53**, 707–727.

Laws, W.D. (1962) An investigation of temple plasters from Tikal, Guatemala, with evidence of the use by the ancient Maya of plant extracts in plaster making. *Wrightia*, **2**, 5217–5228.

Lechtman, H.K. and Hobbs, L.W. (1987) Roman concrete and the roman architectural revolution, ceramics and civilization. *High-Technology Ceramics–Past, Present, and Future*, **3**, 81–128.

Leslie, A.B. and Hughes, J.J. (2002) Binder microstructure in lime mortars: implications for the interpretation of analysis results. *Quarterly Journal of Engineering Geology and Hydrogeology*, **35**, 257–263.

Liew, K.M., Sojobi, A.O. and Zhang, L.W. (2017) Green concrete: Prospects and challenges. *Construction and Building Materials*, **156**, 1063–1095.

Lindroos, A., Heinemeier, J., Ringbom, Å., Brock, F., Sonck-Koota, P., Pehkonen, M. and Suksi, J. (2011) Problems in radiocarbon dating of Roman pozzolana mortars. Pp. 214–230 in: *Building Roma Aeterna*. (Å. Ringbom and R.L. Hohlfelder, editors). (Proceedings) The Finnish Society of Sciences and Letters, Helsinki.

Lindqvist, J.E. and Sandström, M. (2000) Quantitative analysis of historical mortars using optical microscopy. *Materials and Structures*, **33**, 612–617.

Littmann, E.R. (1957) Ancient Mesoamerican mortars, plasters, and stuccos: Comalcalco, part I. *American Antiquity*, **23**, 135–140.

Littmann, E.R. (1960) Ancient Mesoamerican mortars, plasters, and stuccos: The use of bark extracts in lime plasters. *American Antiquity*, **25**, 593–597.

Livingston, R.A. (1970) Materials analysis of the masonry of the Hagia Sophia Basilica, Istanbul. *WIT Transactions on the Built Environment*, **4**, 15–31.

Lothenbach, B., Scrivener, K. and Hooton, R.D. (2011) Supplementary cementitious materials. *Cement and Concrete Research*, **41**, 1244–1256.

Macdonald, S. (editor) (2008) *Concrete: Building Pathology*. John Wiley & Sons, Hoboken, New Jersey, USA.

Machel, H.G., Mason, R.A., Mariano, A.N. and Mucci, A. (1991) Causes and emission of luminescence in calcite and dolomite. In: *Luminescence Microscopy: Quantitative and Qualitative Aspects*. (C.E. Barker and O.C. Kopp, editors). Special Publication, SC25. SEPM, Tulsa, Oklahoma, USA.

Mackay, E.J.H. (1938) *Further Excavations at Mohenjo-Daro*. Government of India Press, New Delhi, 162 pp.

Magaloni, D., Pancella, R., Fruh, Y., Canetas, J. and Castano, V. (1995) Studies on the Mayan mortars technique. Pp. 483–489 in: *Materials Issues in Art and Archaeology IV* (P.B. Vandiver, J.R. Druzik, J.L.G. Madrid, I.C. Freestone and G.S. Wheeler, editors). Materials Research Society, Pittsburgh, Pennsylvania, USA.

Mahasenan, N., Smith, S., Humphreys, K. and Kaya, Y. (2003) The cement industry and global climate change: current and potential future cement industry CO_2 emissions. Pp. 995–1000 in: *Greenhouse Gas Control Technologies-6th International Conference* (Vol. 2). Elsevier, Amsterdam.

Malhotra, V.M. and Mehta, P.K. (1996) *Pozzolanic and Cementitious Materials*. Vol. 1. Taylor & Francis, Abingdon, UK.

Malinowski, R. and Garfinkel, Y. (1991) Prehistory of concrete. *Concrete International*, **13**, 62–68.

Manias, C., Retallack, D. and Madsen, I. (2000) XRD for On-line Analysis and Control. *World Cement*, Feb 2000.

Maravelaki-Kalaitzaki, P., Bakolas, A. and Moropoulou, A. (2003) Physico-chemical study of Cretan ancient mortars. *Cement and Concrete Research*, **33**, 651–661.

Maravelaki-Kalaitzaki, P., Bakolas, A., Karatasios, I. and Kilikoglou, V. (2005) Hydraulic lime mortars for the restoration of historic masonry in Crete. *Cement and Concrete Research*, **35**, 1577–1586.

Marra, F., d'Ambrosio, E., Gaeta, M. and Mattei, M. (2016) Petrochemical identification and insights on chronological employment of the volcanic aggregates used in ancient roman mortars. *Archaeometry*, **58**, 177–200.

Masoumi, M.M., Banakar H. and Boroomand, B. (2015) Review of ancient Persian lime mortar "Sarooj". *Malaysian Journal of Civil Engineering*, **27**, 94–109.

Massazza, F. (1998) Pozzolana and pozzolanic cements. Pp. 471–635 in: *Lea's Chemistry of Cement and Concrete.* 4th Edition (P.C. Hewlett, editor). Butterworth-Heimann, Oxford, UK.

Matschei, T., Lothenbach, B. and Glasser, F.P. (2007) Thermodynamic properties of Portland cement hydrates in the system $CaO-Al_2O_3-SiO_2-CaSO_4-CaCO_3-H_2O$. *Cement and Concrete Research*, **37**, 1379–1410.

Maurenbrecher, A.P. (2004) Mortars for repair of traditional masonry. *Practice Periodical on Structural Design and Construction*, **9**, 62–65.

McCown, D.E. (1952) Excavations at Nippur, 1948–50. *Journal of Near Eastern Studies*, **11**, 169–176.

Merlini, M., Artioli, G., Meneghini, C., Cerulli, T., Bravo, A. and Cella, F. (2007) The early hydration and the set of Portland cements: in situ X-ray powder diffraction studies. *Powder Diffraction*, **22**, 201–208.

Merlini, M., Artioli, G., Cerulli, T., Cella, F. and Bravo, A. (2008) Tricalcium aluminate hydration in additivated systems. A crystallographic study by SR–XRPD. *Cement and Concrete Research*, **38**, 477–486.

Milevski, I., Khalaily, H., Getzov, N. and Hershkovitz, I. (2008) The plastered skulls and other PPNB finds from Yiftahel, Lower Galilee (Israel). *Paléorient*, **34**, 37–46.

Minke, G. (2012) *Building with Earth: Design and Technology of a Sustainable Architecture.* Walter de Gruyter, Boston, Massachusetts, USA.

Miriello, D., Barba, L., Blancas, J., Bloise, A., Cappa, M., Cura, M., De Angelis, D., De Luca, R., Pecci, A., Taranto, M. and Yavuz, H.B. (2017) New compositional data on ancient mortars from Hagia Sophia (Istanbul, Turkey). *Archaeological and Anthropological Sciences*, **9**, 499–514.

Mogetta, M. (2015) A new date for concrete in Rome. *The Journal of Roman Studies*, **105**, 1–40.

Montana, G., Randazzo, L., Cerniglia, M.R., Aleo Nero, C. and Spatafora, F. (2016) Production technology of Early-Hellenistic lime-based mortars originating from a Punic-Roman residential area in Palermo (Sicily). *International Journal of Conservation Science*, **7**, 797–812.

Monteiro, P.J.M., Mancio, M., Kirchheim, A.P., Chae, R., Ha, J., Fischer, P. and Tyliszczak, T. (2011) Soft X-ray microscopy of green cements. In: *AIP Conference Proceedings* (Vol. **1365**, No. 1, pp. 351–356). American Institute of Physics, College Park, Maryland, USA.

Moorey, P.R.S. (1999) *Ancient Mesopotamian Materials and Industries: The Archaeological Evidence.* Eisenbrauns, Warsaw, Indiana, USA.

Mora, E.P. (2007) Life cycle, sustainability and the transcendent quality of building materials. *Building and Environment*, **42**, 1329–1334.

Moropoulou, A., Bakolas, A. and Bisbikou, K. (2000) Investigation of the technology of historic mortars. *Journal of Cultural Heritage*, **1**, 45–58.

Moropoulou, A., Cakmak, A.S., Biscontin, G., Bakolas, A. and Zendri, E. (2002) Advanced Byzantine cement based composites resisting earthquake stresses: the crushed brick and lime mortars of Justinian's Hagia Sophia. *Construction and Building Materials*, **16**, 543–552.

Moropoulou, A., Bakolas, A. and Anagnostopoulou, S. (2005) Composite materials in ancient structures. *Cement and Concrete Composites*, **27**, 295–300.

Oates, D. (1961) The excavations at Nimrud (Kalu), 1960. *Iraq*, **23**, 1–14.

Oates, D. (1990) Innovations in mud-bricks: decorative and structural techniques in ancient Mesopotamia. *World Archaeology*, **21**, 388–406.

Oates, J.A.H. (1998) *Lime and Limestone: Chemistry and Technology, Production and Uses.* Wiley–VCH, Weinheim, Germany.

Odler, I. (editor) (2000) *Special Inorganic Cements*. Spon Press, London, New York.

Oleson, J.P. (2011) Harena sine calce: Building disasters, incompetent architects, and construction fraud in ancient Rome. Pp. 9–27 in: *Building Roma Aeterna*. (Å. Ringbom and R.L. Hohlfelder, editors). (Proceedings) The Finnish Society of Sciences and Letters, Helsinki.

Oleson, J.P., Bottalico, L., Brandon, C., Cucitore, R., Gotti, E. and Hohlfelder, R.L. (2006) Reproducing a Roman maritime structure with Vitruvian pozzolanic concrete. *Journal of Roman Archaeology*, **19**, 31–52.

Oppenheim, L.A. (1965) On royal gardens in Mesopotamia. *Journal of Near Eastern Studies*, **24**, 328–333.

Organ, R.M. (1961) The conservation of cuneiform tablets. *The British Museum Quarterly*, **23**, 52–58.

Ouellet-Plamondon, C. and Habert, G. (2015) Life cycle assessment (LCA) of alkali-activated cements and concretes. Pp. 663-686 in: *Handbook of Alkali-activated Cements, Mortars and Concretes*. (F. Pacheco-Torgal, J. Labrincha, C. Leonelli, A. Palomo and P. Chindaprasit, editors). Elsevier, Amsterdam.

Pacheco-Torgal, F., Labrincha, J., Leonelli, C., Palomo, A. and Chindaprasit, P. (editors) (2015) *Handbook of Alkali-activated Cements, Mortars and Concretes*. Elsevier, Amsterdam.

Papadimitriou, G. and Kordatos, J. (1993) The brown waterproofing plaster of the ancient cisterns in Laurion and its weathering and degradation. Pp. 277–284 in: *ASMOSIA (Association for the Study of Marble and Other Stones used in Antiquity) Proc. 3rd International Conference*.

Parisatto M., Dalconi M.C., Valentini L., Artioli G., Rack A., Tucoulou R., Cruciani G. and Ferrari G. (2015) Examining microstructural evolution of Portland cements by in situ synchrotron micro-tomography. *Journal of Materials Science*, **50**, 1805–1817.

Pecchioni, E., Fratini, F. and Cantisani, E. (2006) The ancient mortars, an attestation of the material culture: the case of Florence. *Periodico di Mineralogia*, **75**, 255–262.

Pecchioni, E., Fratini, F. and Cantisani, E. (2014) *Atlas of the Ancient Mortars in Thin Section under the Optical Microscope*. Kermes quaderni, Nardini Editore, Florence, Italy.

Pedelì, C. and Pulga, S. (2013) *Conservation Practices in Archaeological Excavations: Principles and Methods*. The Getty Conservation Institute, Los Angeles, California, USA.

Pesce, G.L., Ball, R.J., Quarta, G. and Calcagnile, L. (2012) Identification, extraction, and preparation of reliable lime samples for 14C dating of plasters and mortars with the "Pure Lime Lumps" technique. *Radiocarbon*, **54**, 933–942.

Poduska, K.M., Regev, L., Berna, F., Mintz, E., Milevski, I., Khalaily, H., Weiner, S. and Boaretto, E. (2012) Plaster characterization at the PPNB site of Yiftahel (Israel) including the use of 14 C: implications for plaster production, preservation, and dating. *Radiocarbon*, **54**, 887–896.

Potts, D.T. (2014) Elamite Monumentality and Architectural Scale. Lessons from Susa and Choga Zanbil. Pp. 23–38 in: *Approaching Monumentality in Archaeology* (J.F. Osborne, editor). IEMA Proceedings, Vol. 3. State University of New York Press.

Prescott, W.H. (1847) *The Conquest of Peru*. Harper & Brothers (reprinted 2007, Digital Antiquaria).

Provis, J.L. and Van Deventer, J.S.J. (editors) (2009) *Geopolymers: Structures, Processing, Properties and Industrial Applications*. Elsevier, Amsterdam.

Provis, J.L and Van Deventer, J.S.J (editors) (2014) *Alkali activated materials*. State-of-the-Art Report, RILEM TC 224-AAM, Springer, Berlin.

Radovanović, I. (1996) *The Iron Gates Mesolithic*. International Monographs in Prehistory. Ann Arbor, Michigan, USA.

Radovanović, I. (2000) Houses and burials at Lepenski Vir. *European Journal of Archaeology*, **3**, 330–349.

Rao, H., Li, B., Yang, Y., Ma, Q. and Wang, C. (2015) Proteomic identification of organic additives in the mortars of ancient Chinese wooden buildings. *Analytical Methods*, **7**, 143–149.

Redman, C. (1999) *Human Impact on Ancient Environments*. University of Arizona Press, Tucson, Arizona, USA.

Regev, L., Zukerman, A., Hitchcock, L., Maeir, A.M., Weiner, S. and Boaretto, E. (2010a) Iron age hydraulic plaster from Tell es-Safi/Gath, Israel. *Journal of Archaeological Science*, **37**, 3000–3009.

Regev, L., Poduska, K.M., Addadi, L., Weiner, S. and Boaretto, E. (2010b) Distinguishing between calcites formed by different mechanisms using infrared spectrometry: archaeological applications. *Journal of Archaeological Science*, **37**, 3022–3029.

Reza, H. (2008) Bengal Gupta Viharas: Did such a phenomenon exist? *The International Journal of*

Interdisciplinary Social Sciences, **3**, 211–216.

Riccardi, M.P., Lezzerini, M., Car, F., Franzini, M. and Messiga, B. (2007) Microtextural and microchemical studies of hydraulic ancient mortars: two analytical approaches to understand pre-industrial technology processes. *Journal of Cultural Heritage*, **8**, 350–360.

Richardson, I.G. (1999) The nature of CSH in hardened cements. *Cement and Concrete Research*, **29**, 1131–1147.

Richardson, I.G. (2004) Tobermorite/jennite-and tobermorite/calcium hydroxide-based models for the structure of CSH: applicability to hardened pastes of tricalcium silicate, β-dicalcium silicate, Portland cement, and blends of Portland cement with blast-furnace slag, metakaolin, or silica fume. *Cement and Concrete Research*, **34**, 1733–1777.

Rivera-Villarreal, R. and Krayer, S. (1996) Ancient structural concrete in Mesoamerica. *Concrete International*, **18**, 67–70.

Robson, E. (1996) Building with bricks and mortar. In: *Houses and Households in Ancient Mesopotamia* (K. Veenhof, editor). Leiden, The Netherlands.

Rodriguez-Navarro, C., Hansen, E. and Ginell, W.S. (1998) Calcium hydroxide crystal evolution upon aging of lime putty. *Journal of the American Ceramic Society*, **81**, 3032–3034.

Rodriguez-Navarro, C., Ruiz-Agudo, E., Luque, A., Rodriguez-Navarro, A.B. and Ortega-Huertas, M. (2009) Thermal decomposition of calcite: mechanisms of formation and textural evolution of CaO nanocrystals. *American Mineralogist*, **94**, 578–593.

Rodriguez-Navarro, C., Ruiz-Agudo, E., Burgos-Cara, A., Elert, K. and Hansen, E.F. (2017) Crystallization and colloidal stabilization of $Ca(OH)_2$ in the presence of nopal juice (*Opuntia ficus indica*): Implications in architectural heritage conservation. *Langmuir*, **33**, 10936–10950.

Rollefson, G.O. and Kohler-Rollefson, I. (1992) Early Neolithic exploitation patterns in the Levant: Cultural impact on the environment. *Population and Environment: A Journal of Interdisciplinary Studies*, **13**, 243–254.

Ruiz-Agudo, E. and Rodriguez-Navarro, C. (2009) Microstructure and rheology of lime putty. *Langmuir*, **26**, 3868–3877.

Sanizadeh, S.K. (2008) Novel hydraulic structures and water management in Iran: a historical perspective. *Options Mediterraneennes*, **83**, 26–43.

Sauvage, M. (2011) Construction work in Mesopotamia in the time of the third Dynasty of Ur: Archaeological and textual evidence. Pp. 40–50 in: *TerrAsia 2011* (H. Wang, S. Kim, H. Guillaud and D. Gandreau, editors). Proceedings of the 2011 International Conference on Earthen Architecture in Asia, Mokpo, Terrakorea.

Schlegel, M.C., Sarfraz, A., Müller, U., Panne, U. and Emmerling, F. (2012) First seconds in a building's life – In situ synchrotron X-ray diffraction study of cement hydration on the millisecond timescale. *Angewandte Chemie International Edition*, **51**, 4993–4996.

Schmandt-Besserat, D. (1977) The earliest uses of clay in Syria. *Expedition*, **19**, 28.

Schneider, M., Romer, M., Tschudin, M. and Bolio, H. (2011) Sustainable cement production–present and future. *Cement and Concrete Research*, **41**, 642–650.

Schueremans, L., Cizer, Ö., Janssens, E., Serré, G. and Van Balen, K. (2011) Characterization of repair mortars for the assessment of their compatibility in restoration projects: research and practice. *Construction and Building Materials*, **25**, 4338–4350.

Scrivener, K. and Favier, A. (2015) *Calcined clays for sustainable concrete*. RILEM series, Springer, Berlin.

Scrivener, K., Snellings, R. and Lothenbach, B. (editors) (2016) *A Practical Guide to Microstructural Analysis of Cementitious Materials*. CRC Press, Boca Raton, Florida, USA.

Secco, M., Peruzzo, L., Palasse, L., Artioli, G., Viani, A. and Gualtieri, A.F. (2014) Crystal chemistry of clinker relicts from aged cementitious materials. *Journal of Applied Crystallography*, **47**, 1626–1637.

Secco, M., Dilaria, S., Addis, A., Bonetto, J., Artioli, G. and Salvadori, M. (2018) The evolution of the Vitruvian Recipes over 500 years of floor-making techniques: The case studies of the Domus delle Bestie Ferite and the Domus di Tito Macro (Aquileia, Italy). *Archaeometry*, **60**, 185–206.

Sengupta, R. (1971) Influence of certain Harappan architectural features on some texts of early-historic period. *The Indian Journal of History of Science*, **6**, 23–26.

Shahack-Gross, R., Albert, R.M., Gilboa, A., Nagar-Hilman, O., Sharon, I. and Weiner, S. (2005) Geoarchaeology in an urban context: the uses of space in a Phoenician monumental building at Tel Dor (Israel). *Journal of Archaeological Science*, **32**, 1417–1431.

Sharma, D.P. and Sharma, M. (2003) *Panorama of Harappan civilization*. New Delhi.

Shaw, J. (1973) *Minoan Architecture, Materials and Techniques*. Annuario, NS XXXIII, Roma.

Shi, C., Roy, D. and Krivenko, P. (2006) *Alkali-activated Cements and Concretes*. CRC Press, Boca Raton, Florida, USA.

Shi, C., Jiménez, A.F. and Palomo, A. (2011) New cements for the 21st century: the pursuit of an alternative to Portland cement. *Cement and Concrete Research*, **41**, 750–763.

Sickels L.-B. (1981) Organic additives in mortars. *Edinburgh Architecture Research*, **8**. 7–20.

Siddall, R. (2000) The use of volcaniclastic material in Roman hydraulic concretes: A brief review. Pp. 339–344 in: *Archaeology of Geological Catastrophes* (W.J. McGuire, D.R. Griffiths, P.L. Hancock and I.S. Stewart, editors). Special Publication, **171**, The Geological Society, London.

Siddall, R. (2011) From kitchen to bathhouse: the use of waste ceramics as pozzolanic additives in Roman mortars. Pp. 152–168 in: *Building Roma Aeterna* (Å. Ringbom and R.L. Hohlfelder, editors). (Proceedings) The Finnish Society of Sciences and Letters, Helsinki.

Singh, M. (1993) Analysis and characterization of Charminar lime plaster. *Current Science-Bangalore*, **64**, 760–764.

Singh, M., Jain, K.K. and Singh, T. (1990) Some studies on the suitability of lime mortars based on ancient Indian formulations for restoration purposes. Pp. 151–158 in: *Superfici dell'architettura: le finiture* (G. Biscontin and S. Volpin, editors). Atti del convegno di studi, Bressanone, 26–29 June 1990. Libreria Progetto Editore, Italy.

Singh, M., Waghmare, S. and Kumar, S.V. (2014) Characterization of lime plasters used in 16th century Mughal monument. *Journal of Archaeological Science*, **42**, 430–434.

Singh, M., Kumar, S.V. and Waghmare, S.A. (2015) Characterization of 6–11th century AD decorative lime plasters of rock cut caves of Ellora. *Construction and Building Materials*, **98**, 156–170.

Snellings, R. (2016) X-ray powder diffraction applied to cement. Pp. 107–176 in: *A Practical Guide to Microstructural Analysis of Cementitious Materials* (K. Scrivener, R. Snellings and B. Lothenbach, editors). CRC Press, Boca Raton, Florida, USA.

Snellings, R., Mertens, G., Cizer, Ö. and Elsen, J. (2010) Early age hydration and pozzolanic reaction in natural zeolite blended cements: Reaction kinetics and products by in situ synchrotron X-ray powder diffraction. *Cement and Concrete Research*, **40**, 1704–1713.

Snellings, R., Mertens, G. and Elsen, J. (2012) Supplementary cementitious materials. Pp. 211–278 in: *Applied Mineralogy of Cement & Concrete* (M.A.T.M. Broekmans and H. Pöllmann, editors). Reviews in Mineralogy and Geochemistry, **74**, Mineralogical Society of America and Geochemical Society, Chantilly, Virginia, USA.

Sokol, E.V., Novikov, I.S., Vapnik, Y. and Sharygin, V.V. (2007) Gas fire from mud volcanoes as a trigger for the appearance of high-temperature pyrometamorphic rocks of the Hatrurim Formation (Dead Sea area). In *Doklady Earth Sciences* (Vol. **413**, No. 2, pp. 474–480). Nauka/Interperiodica.

Song, Y.X. and KaiWu, T.G. (1982) *The exploitation of the works of Nature*. First written in 1587. Times and Culture Publishing Co., Taibei, China.

Srejović, D. (1972) *Europe's First Monumental Sculpture: New Discoveries at Lepenski Vir*. Thames & Hudson, London.

Srejović, D. (1981) *Lepenski Vir: Menschenbilder einer frühen europäischen Kultur*. Verlag Philipp von Zabern, Mainz am Rhein, Germany.

Staubach, S. (2013) *Clay: The History and Evolution of Humankind's Relationship with Earth's Most Primal Element*. Penguin Group, New York.

Stoops, G., Canti, M.G. and Kapur, S. (2017) Calcareous mortars, plasters, and floors. Pp. 189–199 in: *Archaeological Soil and Sediment Micromorphology* (C. Nicosia and G. Stoops, editors). Wiley Blackwell, Hoboken, New Jersey, USA.

Sui Pheng, L. (2001) Construction of dwellings and structures in ancient China. *Structural Survey*, **19**, 262–274.

Sullivan, S. and Mackay, R. (editors) (2012) *Archaeological sites: conservation and management*. Vol. 5. Getty

Publications, Los Angeles, California, USA.

Taylor, H.F. (1997) *Cement Chemistry*. Thomas Telford, London.

Theodoridou, M., Ioannou, I. and Philokyprou, M. (2013) New evidence of early use of artificial pozzolanic material in mortars. *Journal of Archaeological Science*, **40**, 3263–3269.

Thickett, D., Odlyha, M. and Ling, D. (2002) An improved firing treatment for cuneiform tablets. *Studies in Conservation*, **47**, 1–11.

Thirumalini, S., Ravi, R., Sekar, S.K. and Nambirajan, M. (2015) Knowing from the past – Ingredients and technology of ancient mortar used in Vadakumnathan temple, Tirussur, Kerala, India. *Journal of Building Engineering*, **4**, 101–112.

Thomas, J.J., Biernacki, J.J., Bullard, J.W., Bishnoi, S., Dolado, J.S., Scherer, G.W. and Luttge, A. (2011) Modeling and simulation of cement hydration kinetics and microstructure development. *Cement and Concrete Research*, **41**, 1257–1278.

Thuesen, I. and Gwozdz, R. (1982) Lime plaster in Neolithic Hama, Syria. A preliminary report. *Paléorient*, **8**, 99–103.

Valentini, L. (2013) RieCalc: Quantitative phase analysis of hydrating cement pastes. *Journal of Applied Crystallography*, **46**, 1899–1902.

Valentini, L., Dalconi, M.C., Favero, M., Artioli, G. and Ferrari, G. (2015) In situ XRD measurement and quantitative analysis of hydrating cement: Implications for sulfate incorporation in C–S–H. *Journal of the American Ceramic Society*, **98**, 1259–1264.

Van Balen, K. (2005) Carbonation reaction of lime, kinetics at ambient temperature. *Cement and Concrete Research*, **35**, 647– 657.

Van Balen, K., Papayianni, I., Van Hees, R., Binda, L. and Waldum, A. (2005) Introduction to requirements for and functions and properties of repair mortars. *Materials and Structures*, **38**, 781–785.

van Strydonck, M.J., Dupas, M. and Keppens, E. (1989) Isotopic fractionation of oxygen and carbon in lime mortar under natural environmental conditions. *Radiocarbon*, **31**, 610–618.

Vandiver, P.B., Soffer, O., Klima, B. and Svoboda, J. (1989) Origins of ceramic technology at Dolní Věstonice, Czechoslovakia. *Science*, **246**, 1002–1008.

Vatandoust, R., Dehkordi, M.H., Abdollahi, P. and Madani, F.S. (2011) Development of Earthen Architecture in Iran. Pp. 27–39 in: *TerrAsia 2011* (H. Wang, S. Kim, H. Guillaud and D. Gandreau, editors). Proceedings of the 2011 International Conference on Earthen Architecture in Asia, Mokpo, Terrakorea.

Vicat, L.J. (1818) *Recherches expérimentales sur les chaux de construction, les bétons et les mortiers ordinaires*. Goujon, Paris.

Vicat, L.J. (1828) *Mortier et Ciment calcaires*. Paris.

Villaseor Alonso, M.I. (2009) *Lowland Maya lime plaster technology: A diachronic approach*. Ph.D. Dissertation. University College London.

Villaseor, I. and Price, C.A. (2008) Technology and decay of magnesian lime plasters: the sculptures of the funerary crypt of Palenque, Mexico. *Journal of Archaeological Science*, **35**, 1030–1039.

Voltolini, M., Dalconi, M.C., Artioli, G., Parisatto, M., Valentini, L., Russo, V., Bonnin, A. and Tucoulou, R. (2013) Understanding cement hydration at the microscale: new opportunities from pencil-beam synchrotron X-ray diffraction tomography. *Journal of Applied Crystallography*, **46**, 142–152.

Walker, C.B.F. (1987) *Cuneiform*. University of California Press, Los Angeles, California, USA.

Weber, J., Bayer, K. and Pintér, F. (2012) Nineteenth century "Novel" building materials: Examples of various historic mortars under the microscope. Pp. 89–103 in: *Historic Mortars*. Springer, Dordrecht, The Netherlands.

Weil, M., Dombrowski, K. and Buchwald, A. (2009) Life-cycle analysis of geopolymers. Pp. 194–210 in: *Geopolymers: Structures, Processing, Properties and Industrial Applications* (J.L. Provis and J.S.J. Van Deventer, editors). Elsevier.

Wernecke, D.C. (2008) A burning question: Maya lime technology and the Maya forest. *Journal of Ethnobiology*, **28**, 200–210.

Williams, R. (2004) *Lime Kilns and Lime Burning*. Osprey Publishing, Oxford, UK.

Worrell, E., Price, L., Martin, N., Hendriks, C. and Meida, L.O. (2001) Carbon dioxide emissions from the global cement industry. *Annual Review of Energy and the Environment*, **26**, 303–329.

Wright, G.R.H. (1983) Puddled mud walling: An ancient survival in the Orient. *Mitteilungen der Deutschen Orientgesellschaft zu Berlin Berlin*, **115**, 9–14.

Wright, G.R.H. (2005) *Ancient Building Technology, Volume* **2**: Materials (2 Vols). Brill, Leiden, The Netherlands.

Wu, X., Zhang, C., Goldberg, P., Cohen, D., Pan, Y., Arpin, T. and Bar-Yosef, O. (2012) Early pottery at 20,000 years ago in Xianrendong Cave, China. *Science*, **336**, 1696–1700.

Yang, F., Zhang, B., Pan, C. and Zeng, Y. (2009) Traditional mortar represented by sticky rice lime mortar – One of the great inventions in ancient China. *Science in China Series E: Technological Sciences*, **52**, 1641–1647.

Yu, H.Y. and Chen, D. (2004) Protection and development of Qiantan River's dyke constructed in Ming and Qing dynasty as a tourism resource. *Zhejiang Hydrotechnology*, **134**, 9–10.

Zendri, E., Lucchini, V., Biscontin, G. and Morabito, Z.M. (2004) Interaction between clay and lime in "cocciopesto" mortars: a study by ^{29}Si MAS spectroscopy. *Applied Clay Science*, **25**, 1–7.

Zeng, Y.Y., Zhang, B.J. and Liang, X. (2008) A case study and mechanism investigation of typical mortars used on ancient architecture in China. *Thermochimica Acta*, **473**, 1–6.

Zhang, K., Zhang, H., Fang, S., Li, J., Zheng, Y. and Zhang, B. (2014) Textual and experimental studies on the compositions of traditional Chinese organic–inorganic mortars. *Archaeometry*, **56**(S1), 100–115.

Zhao, P., Jackson, M.D., Zhang, Y., Li, G., Monteiro, P.J. and Yang, L. (2015) Material characteristics of ancient Chinese lime binder and experimental reproductions with organic admixtures. *Construction and Building Materials*, **84**, 477–488.

Zoppi, A., Lofrumento, C., Ricci, M., Cantisani, E., Fratini, T. and Castellucci, E.M. (2012) A novel piece of Minoan art in Italy: the first spectroscopic study of the wall paintings from Phaistos. *Journal of Raman Spectroscopy*, **43**, 1663–1670.

EMU Notes in Mineralogy, Vol. 20 (2019), Chapter 5, 203–231

Mineralogy of slags: A key approach for our understanding of ancient copper smelting processes

David BOURGARIT

Centre de Recherche et de Restauration des Musées de France, Palais du Louvre, 14 quai François Mitterrand, 75001 Paris, France
and
CNRS-UMR 7055 Préhistoire et Technologie, maison Archéologie and ethnologie, René-Ginouvès (MAE), 21 Allée de l'Université - 92 023, Nanterre cedex, France
e-mail: david.bourgarit@culture.gouv.fr

Copper was the first metal to have been smelted (extracted from its ore) some seven thousands year ago in the ancient Near East. For most pre-industrial periods, the documentation of copper smelting *chaine operatoire* relies mainly on investigations by archaeometallurgists of the metallurgical waste recovered during archaeological excavations, namely the copper slags. Copper slags are mostly an assemblage of crystals of oxides (iron, manganese, *etc.*), olivine (fayalite, *etc.*) and/or pyroxenes embedded in a polymetallic more-or-less glassy matrix. The mineralogy of the slags is directly related to the initial charge and the working conditions prevailing in the pyrometallurgical reactor. This chapter aims to give an overview of how copper slag mineralogy is investigated and the type of information it yields in order to help our understanding of past metallurgies and societies.

1. Introduction

1.1. A short history of copper metallurgy

Copper and its alloys were among the earliest valuable and strategic metals used by humans together with gold and silver (Tylecote, 1992; Mille and Carozza, 2009). Copper alloys such as arsenical copper and then tin-copper – also called bronzes –gradually replaced lithic tools and weapons during the Bronze Age (~2000 to 800 BC in Europe), before being replaced in turn by iron and steel during the Iron Age (Tylecote, 1992). Copper alloys, however, were still being widely used afterwards for common ware. Day-to-day items such as vessels, candlesticks, *etc.* began to be mass produced in copper and alloys in Europe from the 13th Century AD onwards (Thomas *et al.*, 2013). From the 15th Century AD to the 19th, artillery manufacture meant that copper and particularly bronze were highly prized materials (Killick and Fenn, 2012). One particularly illustrative example of the strategic importance of bronze cannons, as a guarantee of a safe and sustainable shipping trade, is very thoroughly and nicely depicted for 14th–17th Century Venice (Avery, 2011). From the very beginnings of

DOI: 10.1180/EMU-notes.20.5

their exploitation up to the present time, copper and/or some of its alloys have been associated continuously with a high aesthetic value. Among the most famous examples are the so-called 'bronze' statues and statuettes that were produced widely during Antiquity, in medieval Asia, in Africa, as well as in Early modern Europe. By means of splendid inlaid basins, ewers, and candlesticks, the medieval Islamic World spread the shine of bronze and brass all around the world as well.

1.2. Copper smelting: from ore to metal

The most important sources of copper are sulfides, oxides and carbonates (Artioli, 2010). When exploited by humans, these minerals may be called ores (Killick, 2014). Copper ores need to be processed thermally and chemically – 'smelted', in order to extract the metallic copper. The processes of copper smelting belong to the field of extractive metallurgy. We will briefly explore the main processes in use during pre-industrial times in the next section. Native copper may also have been present in a number of copper deposits, notably in North America and may have been exploited occasionally by human groups. It must be emphasized that the working of native copper solely by hammering, as done notably at the very beginning of copper metallurgy, refers more to tacit know-how directly inherited from the lithic industry than to the acquisition of new skills and knowledge pertaining to metallurgy, that is the mastering of thermo-chemical and/or thermo-mechanical treatments (extraction of ore, refining, alloying, casting, annealing, *etc.*). Thus, the evidence of copper items within a given chrono-cultural context does not necessarily indicate the mastering of metallurgy (Bourgarit and Mille, 2007).

The aim of copper smelting is to recover metallic copper from its ore. This implies that in most cases two chemical transformations have taken place (Davenport *et al.*, 2002; Lossin, 2003). On the one hand, copper compounds have to be separated from the other species of minerals present in the ore, *i.e.* from gangue material. Iron and/or silica compounds are encountered very frequently in copper deposits (equation 1). On the other hand, copper has to be reduced, either from its oxides or from any other compounds including sulfides (equation 2). In practice, these two chemical events – which may take place at one or several different stages of the smelting process – are thermally activated: at relatively high temperatures for equation 1 (typically ~1200–1300°C) and at lower temperatures for equation 2 (~500–700°C). High temperatures in pre-industrial periods were achieved by burning wood and/or charcoal in more or less elaborate reactors (charcoal beds, bone fire, ceramic vases, built furnaces, *etc.* see Rehder, 2000). Depending on the process, the period and the ore being smelted, a large quantity of waste would have been produced. There is agreement among those studying ancient copper smelting processes that this waste should be referred to as 'slag'.

$$FeO + SiO_2 \rightleftharpoons \langle Fe{-}Si{-}O \rangle_{slag} \quad (1)$$

$$CuFeS_2 + \text{``O''} \rightarrow (Cu_xFe_yS_z)_{matte} + (1-x)Cu + (1-y)FeO + (1-z)SO_2 \quad (2)$$

1.3. Copper slags: the main material studied

1.3.1. Other sources exploited for the investigation of smelting processes

Most ancient copper-reduction processes are documented exclusively by historical sources, namely excavation of production sites. Written documents are sometimes referred to, but these may cause confusion because the main studies to date deal with prehistoric societies. The 16th Century AD description of a medieval copper smelting process in Bohemia is a good example (Agricola, 1556). Being one of the first technical treatises on the subject published in western Europe, it has long led scholars to misinterpret protohistoric processes (see for example the nice review by Metten, 2003). Ethnographic studies may also shed some light on ancient metallurgical processes (Iles and Childs, 2014) but parallels should be drawn with care. Few such studies have been undertaken so far for copper smelting (Zschoke and Preuschen, 1932; Lewin and Hauptmann, 1984; Herdits, 1993; Bisson, 2000; Anfinset, 2011). Archaeological assemblages commonly provide evidence of metallurgical structures and associated items such as furnaces, ceramic vases, tuyeres and metallurgical *chaine operatoires* indeed are based mainly on ceramics (Martinón-Torres and Rehren, 2014). More generally, archaeology aims to shed light on the evidence of the spatial organization of the activity (spatial distribution of structures and waste). However, most of the evidence usually records only a small part of the information related to the processes. Moreover, the informative potential of such archaeological material is often severely reduced because of poor conservation: the superstructure of furnaces is a good example of this, being systematically absent from the archaeological record. Therefore, copper slags represent the most interesting source of information with which to work.

1.3.2. Slag description

As noted above, most researchers working on ancient metallurgies have agreed to refer broadly to all waste produced during copper pyrometallurgical processes, including smelting operations, as 'copper slag' (Hauptmann, 2014). Bachmann (1982) was the first to set up a typology of ancient copper smelting slags which encompasses more or less all the types encountered at protohistoric to medieval sites (Fig. 1). This typology is still in wide use. In pre-industrial contexts, copper slags usually reveal complex compositions. Their macrostructure ranges from a very heterogeneous aggregate of unreacted minerals bound by a fused phase, up to almost perfectly homogeneous melt (Fig. 2). Due to rapid cooling, the melt generally consists of a fine grained mixture of different phases (Fig. 3). Crystalline phases are usually olivines, (fayalite, Fe_2SiO_4 or tephroite, Mn_2SiO_4), spinel (magnetite, Fe_3O_4 or hercynite, $FeAl_2O_4$) and pyroxene (*e.g.* hedenbergite, $CaFeSi_2O_6$). Cu-based inclusions are frequently present, either as sulfides or as metallic Cu, the latter often being oxidized as copper oxides, carbonates or chlorides. These phases grow in a developed silica- or manganese-based glass.

1.3.3. Slag representativeness

Depending on the production context and on the period, the quantity of slags recovered may vary from a few kilograms to thousands of tons. The first question for

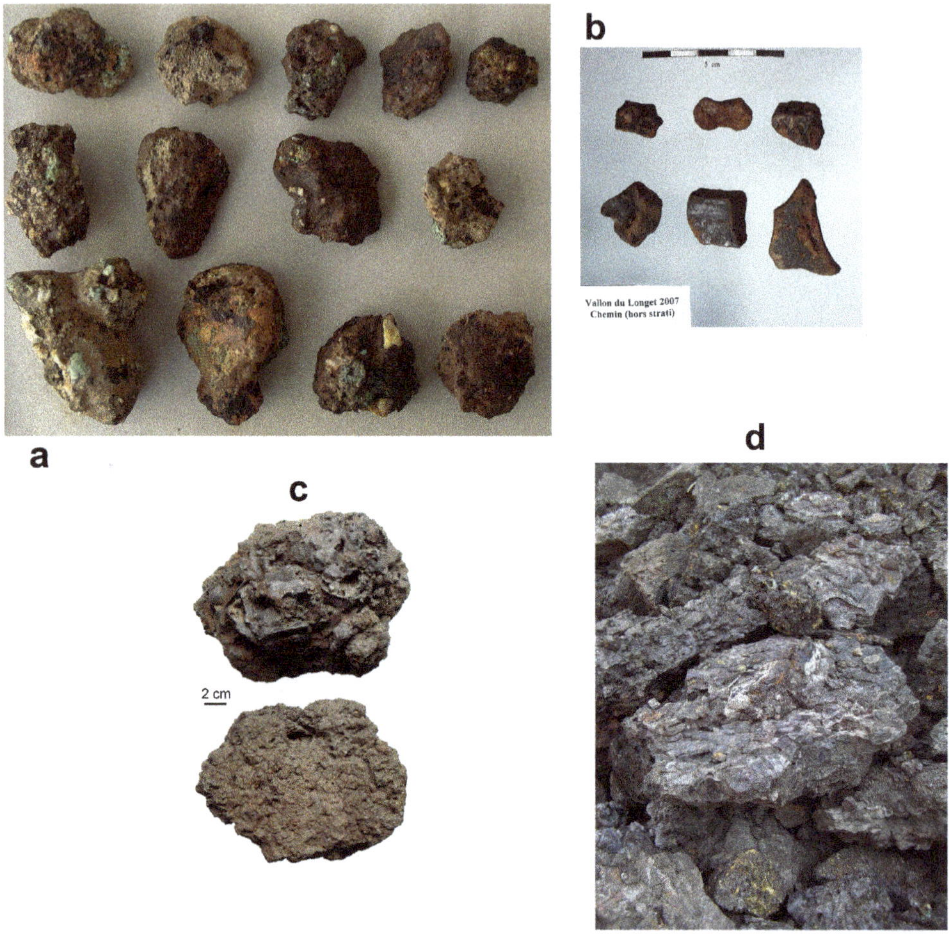

Figure 1. Main types of copper smelting slags recovered so far from pre-industrial sites. (a) Conglomerate slags from Chalcolithic La Capitelle du Broum, France (size $\approx$ 1 cm); (b) plate slags from Early Bronze Age Saint-Véran; (c) coarse slags from Late Bronze Age Luzerna, Trentino; and (d) tapped slags from Roman Skouriotissa, Cyprus.

archaeometallurgists is to decide what these copper slags represent with respect to the process as a whole and its possible evolution during the chronology of a site's occupation. Let us examine two extreme situations.

In the following: 'Chalcolithic' will be used to designate the period when copper smelting first appeared. At most Chalcolithic sites excavated so far tiny quantities of smelting debris have been recovered: a maximum of a few hundred small pieces weighing several kilograms at the most (Bourgarit, 2007). Several reasons may be put forward. First, Chalcolithic copper production is usually small and is devoted to small implements intended for a restricted circle of consumption (Mille and Carozza, 2009).

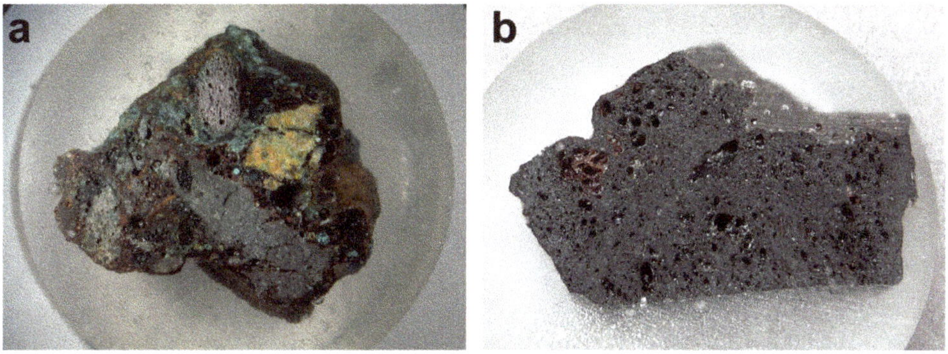

Figure 2. Two extreme cases of copper-slag macrostructures on bulk polished sections (daylight macrophotography). (a) Unreacted minerals in a conglomerate slag from Chalcolithic La Capitelle du Broum, and (b) homogeneous melt in a plate slag from Early Bronze Age Saint-Véran (size of image ≈ 2.5 cm).

Second, slags are usually too viscous for proper copper separation, thus requiring crushing to recover the entrapped copper prills. The resulting minute slag pieces or even fine slag sand can easily escape the archaeological record (Bourgarit and Mille, 2005). Whether crushed or not, slags may disappear during burial mechanisms including corrosion. Such mechanisms are usually well documented by archaeologists, whereas copper slags are usually quite chemically inert in most burial environments, although selective corrosion may occur at the rims (Carozza *et al.*, 2010). Third, slags may be recycled as seen in a number of slag-tempered ceramics. This phenomenon is nevertheless quite rare and easily quantifiable. It has also been proposed that slags may be reintroduced in the charge with fresh ore as catalyst. Such recycling should reappear in the archaeological record... as slags. Another possible reason for the small quantity of slags recovered is that smelting sites may be seasonal, *i.e.* ephemeral. This has been proposed for an Early Bronze Age site in the Alps (Bourgarit *et al.*, 2008), where the amount of slag (~100 kg) was estimated to have been produced over a period of 4–8 months. This does not apply to all sites. For example, the 500 slag items (<1 kg) recovered during one excavation campaign at a Chalcolithic site in southern France (Fig. 1) could not have been produced by more than one or two smelting runs (Bourgarit and Mille, 2005). Finally, when combined, these

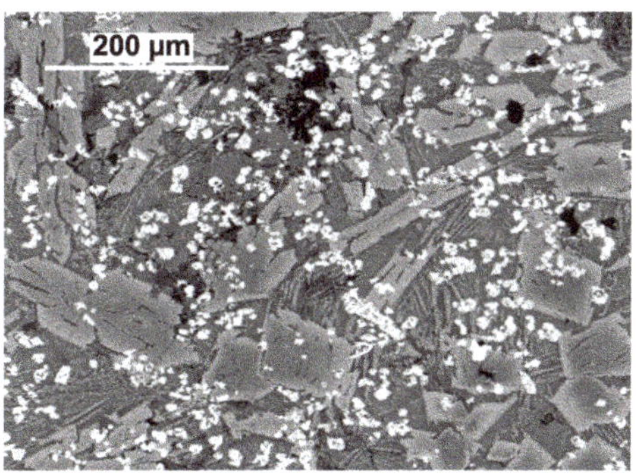

Figure 3. Typical copper-slag microstructure as seen by SEM (backscattered electrons) on a plate slag from Saint-Véran, French Alps: Mg-fayalite appears here as zoned polyhedra with a rim depleted in Mg; grains of magnetite (white) and dendritic augite-like clinopyroxens (dark grey) are also present.

hypotheses may explain only partly the minute quantities of slags recovered: opened some decades ago, the debate has not evolved much beyond this point (Craddock, 2007). One extreme hypothesis is that the earliest smelting operation used to work high-grade ore thus producing few slags, if any. 'Non-slagging' processes are evoked; the few slags recovered would thus stem from trial and error which would represent in the smelting process either a lower-grade ore or a less carefully beneficiated ore. Consequently, more gangue elements would need to be removed. Ultimately, the question remains: what do the few slags recovered represent? Whatever the explanation, the relative homogeneity of the slag corpus under study in terms of both typology and composition will guarantee its archaeological consistency, and thus its meaning. What meaning? In the extreme case of trial and error products, it may be assumed that the few slags recovered would represent the upper limit of the current skills and know-how.

Given the large number of slags, other issues may arise. For example, the huge Roman slag heaps in the Troodos mountains, Cyprus – the largest known so far in the ancient World – enclose millions of tons of slag, each lump of which weighing several hundreds of grams (Fig. 1). Such large slag deposits were produced centuries ago (in Roman times), probably over a large time span (several centuries). Due to possible reworking, displacements, *etc.*, the current structure of the heaps may not reflect the original one, thus rendering interpretation of stratigraphic dating difficult. Absolute dating of slags would be one solution to help sorting. Archaeomagnetism represents a promising route (Shaar *et al.*, 2011). Some preliminary work has been carried out by thermoluminescence on the quartzite of a slagged sherd of tuyere (Hauptmann and Wagner, 2007). Also, ^{14}C dating may be applied to entrapped charcoal pieces. However, absolute dating of slags is not common practice, notably because of the cost. Moreover, the chronological resolution may not always be sufficient. A marked variability of smelting residues from coeval workshops cannot be excluded (Humphris *et al.*, 2009). A large sample may appear as the best solution, but how large? In Cyprus, the 80 slags sampled from three different late Roman heaps revealed a large variability (Georgakopoulou and Kassianidou, 2013). Several hundred pieces were probably examined, requiring rapid and inexpensive selection methods. A number of such methods are applied routinely by archaeologists: typology, colour, density, *etc.* For the Cyprus story, all the Roman slags appear to belong to the same tapped-slag type and probably exhibit similar densities. Colour does not seem to be relevant either (Kassianidou, 2003). Handheld X-ray fluorescence may offer an intermediate solution, having been developed for analogous modern purposes (*e.g.* metal sorting prior to recycling).

2. A brief historiography of archaeometallurgical studies of copper smelting

2.1. When and where?

It is not within the scope of the present chapter to draw a comprehensive historiography; for a general view see Craddock (1995), Killick (2001, 2014). There are a few markers.

Theodore Wertime's expedition in 1968 in Afghanistan, Iran and Turkey marked the first large-scale attempt to survey ancient copper smelting sites – among other places involved in pyrotechnical activities (Arab and Rehren, 2004a,b). One of the main aims of the expedition was to document the very beginning of copper extractive metallurgy which was thought to have arisen in those areas under survey. For several reasons the expedition did not yield the expected results. Almost simultaneously, large-scale excavations were carried out in the Timna mining and metallurgical district of Israel (Rothenberg, 1990). These represent the first focused studies of ancient copper mining and smelting processes. A short time after, another big project was started in the nearby Faynan valley in Jordan (Hauptmann, 2007), leading to one of the most comprehensive studies on copper mining and smelting carried out to date. The large time span covered by the numerous sites excavated allowed the evolution of copper smelting to be established, from its very beginning (Chalcolithic, *i.e.* 4th millenium BC in this region) up to the early Islamic period. Simultaneously, the Eastern Alps were examined thoroughly, mainly for Bronze Age occupations (end of 3rd to 1st millenium BC, see Weisgerber, 2004). The British Isles have provided the field with major milestones as well, although the main interest here has been copper mining rather than smelting (Timberlake, 2003; O'Brien, 2004). The Iberian Peninsula has also yielded interesting data regarding the beginning of copper extractive metallurgy (Rovira and Montero-Ruiz, 2013). It is stressed that protohistoric periods (Bronze and Iron Ages) have been and are still being studied much more than the following periods including Roman and medieval times. The geographical distribution of copper smelting sites being excavated is also very uneven, the main efforts being directed to the Near and Middle East, the Alps and the Mediterranean. In comparison, studies in Eurasia and in eastern and northern Europe are rare. Africa, east Asia and the Americas are underexplored despite recent attempts notably in Japan, China, Mexico and Chile (Yoshikawa *et al.*, 2006; Maldonado and Rehren; 2009; Mei, 2009; Mille *et al.*, 2013; Zori *et al.*, 2013; Mei *et al.*, 2015; Figueroa *et al.*, 2018).

2.2. Why?

The aims of the copper-slag investigations have been evolving depending on the teams involved and archaeological contexts. Three goals may be targeted. One primary objective is to determine whether copper smelting has been carried out or not. In a number of cases this may not be straightforward; we will come to this in Section 4. The types of ore smelted including their provenance may also be an objective of slag investigation, particularly when smelting sites are far from the mines. This is notably the case at the beginning of copper metallurgy in Eneolithic Trentino (Cierny *et al.*, 2004) and Chalcolithic Faynan Valley (Hauptmann, 2007). Finally, the aim is to characterize the copper smelting processes. The main working conditions to be documented are the nature of the charge being smelted (*e.g.* deliberate mixture of ores, addition of fluxing agents, *etc.*), the thermal cycles, the atmosphere(s), and the duration of the different steps. This may start by understanding the nature of the physical and chemical mechanisms involved (reduction of copper, formation of the different com pounds, *etc.*).

As for most of the studies dealt with in this volume, these three 'technical' objectives are underlined by a number of issues raised by the archaeology and history. How is copper production organized (*e.g.* as a household activity performed in the domestic sphere or as more specialized systems)? What is the whole *chaine operatoire* including ore prospection and destination of final product? How are technological choices made, if at all? Is it possible to reconstruct a technological lineage? The smelting strategy is constrained by one or several factors including technological, social, economic, political as well as cultural. Moreover, the complexity of the processes may constitute an accurate record of innovation and/or inventions, *i.e.* of technological transfers between human societies. In the last two decades a number of Anglo-Saxon archaeologists and archaeometallurgists have monopolized the social anthropological questioning on copper smelting and more generally on ancient copper metallurgies (Budd and Taylor, 1995; Ottaway, 2001; Kienlin, 2014; Thornton, 2014; White and Hamilton, 2014). Ironmaking has generated a number of papers as well (Charlton *et al.*, 2010, 2013b). It is beyond the author's competence to comment on these works, and on the scepticism sometimes expressed about the Anglo-Saxon approaches (Killick and Fenn, 2012). Note also the interesting (French) work carried out on late prehistoric lithic and ceramic technologies (Rouillard *et al.*, 2007; Pelegrin, 2009; Roux, 2010, 2013).

2.3. Who?

Our understanding of all the ancient copper-production modes is a matter of interdisciplinary collaboration involving archaeologists, historians and a number of different specialists from natural sciences. That said, the researchers working on the smelting processes belong to a well defined discipline called archaeometallurgy (Rehren and Pernicka, 2008; Killick and Fenn, 2012; Killick, 2015). So far, most studies have been carried out by European teams. Though the USA has hosted some of the first archaeometallurgists (C.S. Smith, R. Maddin, J. Muhly), none of them has ever worked on copper smelting. American archaeometallurgists working on copper smelting are still few in number. China, Japan and some South and Central American countries have only recently entered the field.

Who are these archaeometallurgists? Mineralogists have always been very active in the field. This is easily understandable as the study of ancient copper reduction processes relies mainly on the investigation of complex mineral assemblages, *i.e.* copper slags. As far as metallurgical processes are involved, metallurgists have also naturally been involved since the very beginning. Archaeologists are also involved of course. Increasingly, the younger generation of investigators has benefited from a mixed background in both archaeology and material sciences. This may appear as a drawback because the basic knowledge in mineralogy and metallurgy is diluted, whereas the materials (slags) and processes to be examined are often very complex. Yet the high quality of a number of recent papers tends to invalidate this pessimistic view.

2.4. How?

Once the slags are sorted according to their typology (size, form, colour, texture) and density, their investigation follows more or less standard mineralogical protocols, from petrographic studies of thin sections to investigations of mounted thick sections and samplings, including elemental analysis and powder X-ray diffraction (Hauptmann, 2007, 2014). To our knowledge, only two attempts – both promoted by Gilberto Artioli – have been made so far to quantify the crystalline phases by Rietveld refinement of the diffraction diagrams (Bourgarit *et al.*, 2008; Addis *et al.*, 2016). Due to the complexity of the diagrams, the precision on quantification is often rather poor. Yet it is often adequate for understanding production modes. In the two aforementioned cases this quantification has proven to be determinant (see Section 4). Scanning electron microscopy (SEM) and microanalysis systems have gained importance in slag studies, not only in helping to identify mineral phases, but also in characterizing chemical compositions of amorphous phases and pointing out chemical segregations such as zoning. Yet, one may regret that SEM examinations and analysis are slowly replacing the less time and money-consuming petrographic observations. Less common analytical approaches such as Mössbauer spectrometry (Moesta *et al.*, 1984; Metten, 2003), Fe-K-edge X-ray absorption near edge structure (XANES) and high-wavelength resolution WDS electron microprobe (Burger *et al.*, 2010) have been carried out in order to quantify the Fe^{3+}/Fe^{2+} ratio in the bulk or in particular phases. Note that, to the author's knowledge, grouping of slags using statistical treatments on elemental composition as has been done for iron smelting (Charlton *et al.*, 2013a) has never been carried out before.

Metallurgical approaches and notably those dealing with modern slag systems mostly inspired by the steel industry (VDEh, 1995) are underrepresented in the archaeometallurgical literature. The best example of this may be seen in the scarcity in the archaeometallurgical literature of simple empirical methods for determining the physical and chemical properties of ancient slags such as melting temperature and viscosity (Lutz *et al.*, 1988). This brings us to thermodynamics and kinetics calculations. Theoretical approaches are used at various levels to assess some of the working conditions. The main working conditions are temperature, atmosphere, initial charge composition, slag viscosity and process duration. How these parameters are deduced from slag investigations will be detailed in the next section. Yet, the chemical systems and reactions involved in the smelting processes are often too complex for fully satisfactory thermodynamics and kinetics calculations because of the heterogeneity of ancient slags and because the systems are commonly far from thermodynamic equilibrium. The experimental simulation route may open complementary avenues. Various aims are targeted (Doonan and Dungworth, 2013). Small-scale model experiments carried out in laboratories focus on specific aspects of the chemical reactions and working conditions. As far as we know, few such experiments have been published to date (Neuninger *et al.*, 1970; Tylecote, 1974; Moesta, 1983; Rostoker *et al.*, 1989; Pollard *et al.*, 1991; Bourgarit *et al.*, 2004; Burger *et al.*, 2010). Full-scale models – often carried out in the field – are much more common and have been used since early on in the history of archaeometallurgy (Gowland, 1912; Coghlan, 1939/1940; Böhne, 1968; Tylecote and Merkel, 1985; Zwicker *et al.*, 1985; Merkel, 1990; Herdits, 1993; Ambert *et al.*, 1997; Woelk *et al.*, 1998; Rovira, 1999; Bourgarit *et al.*, 2002; Lorscheider *et al.*, 2003; Yoshikawa *et al.*, 2006; Timberlake, 2007; Catapotis *et al.*, 2008; Hanning *et al.*, 2010). They generally choose to reconstruct either part or the whole process with the aim of understanding the economic and social environment (production rates, organization of the workshop, human skills, *etc.*). The term 'experiential' has been proposed for experiments focusing on the role of the human in the process, while 'experimental' simulations are devoted more to the study of the scientific aspects (Jeffery, 2004). Depending on the rigour of the approach (problem, set-up, publication), the data stemming from these numerous simulations may be of variable quality.

3. Main working conditions documented by slag investigation

3.1. Temperature

In modern industrial processes, and especially in the steel industry, there are two types of basic methods for measuring melting characteristics of a slag, namely the pressure drop method and the slag atlas method (VDEh, 1995; Dash *et al.*, 2012). In the archaeometallurgical field, minimal temperature is estimated mainly by plotting bulk composition of major elements into phase diagrams (Fig. 4). As pointed out by Hauptmann (2014), because the charge often only melted gradually and sometimes only partially it is informative to follow the phase diagram by increasing the temperature. Yet it must be kept in mind that the slag we observe *post mortem* is the result of a decrease in temperature leading to complete solidification. Because minor elements are not taken into account, and because thermodynamic equilibrium is rarely reached these calculations yield only a rough approximation. The archaeometallurgical community is now well aware of that (see the very interesting discussion in May 2014 on the arch-metals web forum hosted at Oxford University (arch.oucs.ox.ac.uk/detail/95684/index.html) after a comment on a paper by Young and Poyner (2012). Note that a recent experimental study on lead smelting by the iron reduction process has shown that chalcopyrite, $CuFeS_2$, may play a key role in fostering the formation of fayalite, Fe_2SiO_4, at temperatures below the fayalite liquidus (Liu *et al.*, 2015). When present, chemical zoning

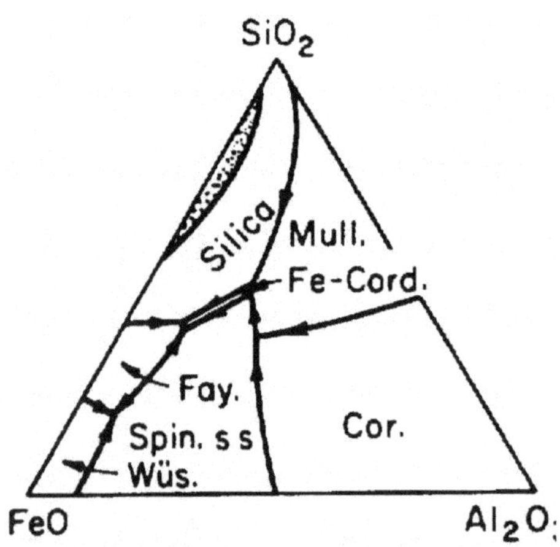

Figure 4. Ternary system $FeO-Al_2O_3-SiO_2$ showing different phases at thermodynamic equilibrium.

offers a complementary way to infer temperature. Moreover, not only minimal but also maximal temperatures may be estimated in that way. Zoning occurs quite often for olivine in the fayalite–forsterite system, with an enrichment of the forsteritic component at the rim (Fig. 3). Temperatures at different Bronze Age sites in the Alps were thus shown to range between 1200 and 1350°C (Fig. 5) (Bourgarit *et al.*, 2008; Burger, 2008; Addis *et al.*, 2016). The estimates are far from precise, but they are generally adequate for understanding ancient processes. The main question here is how were such temperatures reached? The different technological answers (wind-furnaces, hand- or feet-powered bellows, blowpipes, quality of insulation, *etc.*) actually lead to the same temperature range. To date archaeology has proven to be the best way to document the heating technology.

3.2. Atmosphere

In the usual working temperature range (1200–1350°C), copper smelting takes place within two main redox systems. First, the Cu system: Cu^{+} and/or Cu^{2+} have to be reduced to Cu. Second, depending on the gangue, the Fe or Mn system controls the formation of the slag, as frequently exemplified by the quartz, fayalite and magnetite (QFM) buffer equilibrium (equation 3). Both temperature and oxygen fugacity control these two systems. The latter is influenced mainly by the CO/CO_2 ratio prevailing in charcoal-powered smelting reactors. The main objective of archaeometallurgists when trying to infer the CO/CO_2 ratio is to reconstruct the height of the charcoal column in the reactor, and consequently the height of the reactor itself (Rehder, 1999). As a

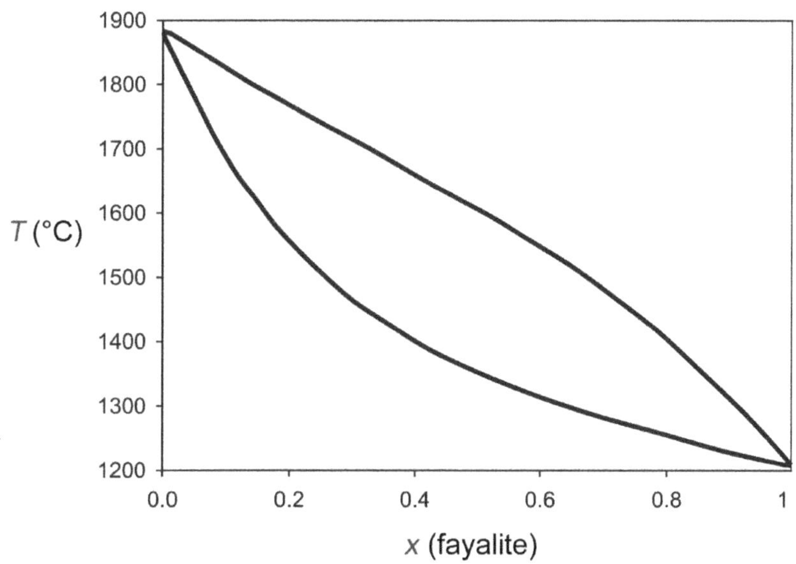

Figure 5. Quasi-binary phase diagram forsterite–fayalite $(1-x)Mg_2SiO_4 + xFe_2SiO_4$. (after Klimm and Ganschow, 2005).

reminder, furnace superstructures – if they ever existed – are systematically absent from the archaeological record.

Because most copper is removed from the slag, Fe or Mn oxides are the only available indicators of the redox conditions (although delafossite, $Cu^{+}Fe^{3+}O_2$, has been the main indicator of oxidizing conditions in early copper slags (Bourgarit, 2007). In most studies, the redox conditions are estimated by looking at particular equilibria such as the aforementioned quartz, fayalite and magnetite (QFM) buffer equilibrium (equation 3). Usually, the equilibrium conditions are related unequivocally to oxygen fugacity using T–f_{O_2} Ellingham diagrams (Fig. 6). Because equilibrium is rarely reached, a better approach is to quantify the different iron-bearing phases and deduce the Fe^{3+}/Fe^{2+} ratio in the bulk (Burger *et al.*, 2010). Mössbauer spectrometry is the best method for this. Remember that the high-temperature composition may have been altered significantly during cooling, although laboratory experiments have shown that the cooling rate has no influence (Burger, 2008). That said, the equilibrium approximation is actually not problematic when high precision is not required.

This may become problematic, however, if both sulfide and oxide copper compounds are in the smelted charge. In addition to the three gases CO, CO_2 and O_2, two other species have to been taken into consideration, namely S and SO_2 (Krismer *et al.*, 2013). Furthermore, experimental simulations and Mössbauer measurements (Burger *et al.*, 2010) have shown that the solid oxygen brought about by CuO exerts the main control

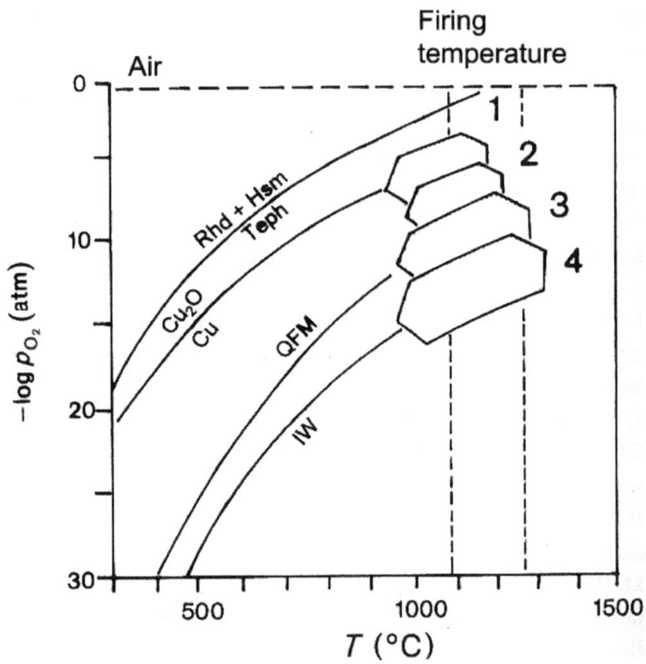

Figure 6. T-log f_{O_2} diagram for several redox systems including Cu_2O/Cu and quartz/fayalite/magnetite (QFM) (after Hauptmann, 2007).

on the reactions with respect to O_2. This is easily conceivable given the difference in molar volumes of these O-bearing compounds. At 1200°C, the molar volume of O_2 is $\sim 10^5$ cm^3; it is <20 cm^3 for CuO. In other words, in the same volume one may put almost 10^4 times more atoms of O if one chooses to insert CuO instead of O_2! For the experiments, the partial pressure of oxygen (pO_2) was forced in the system by a draft of CO/CO_2; the amount of solid oxygen/solid sulfur (O/S) was controlled by a mixture of malachite $CuCO_3.Cu(OH)_2$/chalcopyrite $CuFeS_2$. The main results are shown in Table 1. As an example, the same slag with a global amount of 35 wt.% Fe^{3+}could be obtained either under an imposed $pO_2 = 10^{-10}$ and a molar ratio O/S = 2.5, or at $pO_2 = 10^{-4}$ and O/S = 2. Thus, a small difference in the composition of the charge yields a change in equilibrium oxygen partial pressure of six orders of magnitude! If inferred solely from the oxidation state of Fe, the real redox conditions prevailing in the furnace can be incorrect.

$$3Fe_2^{2+}SiO_4 + O_2 <---> 2Fe^{2+}OFe_2^{3+}O_3 + 3SiO_2 \quad (3)$$

3.3. Charge composition

The example above shows how important it may be to document properly the initial composition of the charge. A model has thus been proposed to estimate the amount of oxidic *vs.* sulfidic copper in the initial charge (Burger *et al.*, 2010). Besides the control of redox conditions, the ideal charge composition should obey another constraint: to balance the quantity of those elements forming the slag. Following the example of Fe, the ideal slag composition is the fayalitic one, Fe_2SiO_4, *i.e.* two Fe for one Si. To discover whether copper smelters did manage to master the slag composition is a crucial issue for archaeologists and historians. This marks the beginning of productive extractive metallurgy. It is usually quite an easy question for archaeometallurgists to answer. Bulk chemical elemental analysis of slags is a quick way to distinguish between erratic and optimal compositions, as seen when comparing the silica-saturated Chalcolithic slags and the much more fluid Early Bronze Age fayalitic slags (Fig. 8).

Table 1. Global relative amount of Fe^{3+} (%) in laboratory experimental copper smelting slags according to the charge composition (chalcopyrite/malachite expressed as a molar O:S ratio) and the oxygen partial pressure in the atmosphere. Measurements by Mössbauer spectroscopy confirmed by Rietveld refinement.

	O:S					
Log (pO_2)	0	0.8	2	2.5	3	4
−3			63		>66	100
−4	0	7	35	60	>66	100
−7			24		>66	100
−10				36	>66	100

Remember, though, that the complex systems encountered in the slags often prevent the compositions from being plotted in relevant diagrams.

Behind this rather simple task of measuring an elemental composition, a more complex question hides: how was slag composition mastered? Theoretically, there are two main possibilities. On the one hand, the ore and its gangue may happen to yield a near-to-optimum composition which may just need some adjustments by careful beneficiation. To the author's knowledge, there is no archaeological evidence showing unequivocally that this was the only way to control the slag composition, although evidence showing the importance of ore beneficiation is abundant. On the other hand, adding of exogenous material may be carried out in the melt. At the Early Bronze Age Saint-Véran site, French Alps, numerous fragments of riebeckite ($Na_2Fe_3^{2+}Fe_2^{3+}(Si_8O_{22})(OH)_2$) have been found at one such smelting site. Because of both the distance from the riebeckite deposits and the size and quantity of the fragments recovered, this mineral is thought to have been brought there intentionally (Rostan and Malaterre, 1994) as a fluxing agent due to its silica content. Unpublished preliminary laboratory-scale experiments tend to show that the soda plays no significant role, although soda may promote removal of sulfur through the mechanism of *kernröstung* (Rosenqvist, 2004). Another means of fluxing is to use sacrificial ceramic tuyeres which have been shown to be applied intentionally to provide the necessary chemical input for early iron smelting in Jordan (Veldhuijzen, 2003). Contrary to the two previous examples, the large, exogenous Ca and Mg contents in the Chalcolithic slags of Roque Fenestre, Southern France (Bourgarit and Mille, 1997) were shown not to be deliberate additions. The resulting slag compositions were indeed erratic and far from any low-temperature eutectic. Yet Ca and Mg proved to follow systematically the dolomite stoichiometry. This has shown that dolomite was entering the melt, and therefore that the local furnaces – which are absent from the archaeological record – were more than simple holes in the ground: they did have rock walls (made out of the local rock, namely dolomite).

3.4. Process duration

Process duration is crucial for estimating the production rate. The duration depends mainly on two factors, namely the separation rate of the final product from the slag, and the slag cooling rate. To our knowledge the only attempt so far to quantify separation rates of the final product (mainly molten copper sulfides and metallic copper inclusions) from the slag was carried out by Addis *et al.* (2016). Therefore, the sinking velocity or buoyancy was calculated as a function of slag viscosity. Viscosity was assessed from the bulk chemical composition following a semi-empirical model (Lutz *et al.*, 1988). Other models and methods are available (VDEh, 1995). Depending on the type of slags recovered (see Section 4), minimum process duration periods of 2 h to >2 days were inferred.

Cooling temperatures can be assessed by the morphological texture of olivines and spinels (Donaldson, 1976) (see Fig. 9). Archaeometallurgical studies reporting on this are rare (Hess, 1998; Anguilano *et al.*, 2002; Manasse and Mellini, 2003; Artioli *et al.*,

2005; Burger, 2008; Addis *et al.*, 2016). At Saint-Véran, the slags exhibit an original columnar morphology of the fracture surface (Fig. 10). Whatever the thickness of the slag, a prismatic morphology is observed consistently for the first 4–6 mm of the section below the top surface. This feature has been interpreted as resulting from water quenching as part of an optimization of the production rate. The morphological texture of olivines and spinel do not testify to such rapid cooling, even near to the upper surface. Consequently, it is assumed that the quenching has been carried out after the formation of these crystals. Despite a high variability of the olivine morphology in the Saint-Véran slags, a cooling rate of ~50°C/hour may be a good approximation (Burger, 2008). Considering a 1200–1050°C solidification interval, the water quenching would have been performed some 3 h after the end of heating.

4. Three archaeological issues for copper smelting: cases studies

4.1. The appraisal of copper smelting: sulfides or oxides?

It has long been assumed that the first type of copper ores to have been smelted were oxides and carbonates, with sulfides being exploited much later. In other words, it has been stated that the appraisal of copper metallurgy was totally driven by a particular geological determinism: therefore metallurgy began where oxides and carbonates were to be found. Yet, an overview of Chalcolithic smelting evidence (Bourgarit, 2007) has proven that a number of extractive metallurgies started with sulfides, although the mineralogical evidence for sulfide smelting may not be straightforward in a copper slag (see in Section 4.3 the discussion on melting crucibles). Another argument against early copper sulfide smelting was that sulfides are far more complex to smelt than oxides. In copper sulfides, people meant mainly iron-bearing sulfides such as chalcopyrite $CuFeS_2$: there would be a double difficulty in separating copper from both Fe and S. Yet, the question was biased. First, Chalcolithic processes were not necessarily looking for productivity and thus for efficient separation. Remember the aforementioned 'non-slagging' early processes where the slags had to be crushed in order to recover the entrapped copper prills. Such an immature process obviously satisfied the small demand for copper at the beginning of copper metallurgy (Carozza and Mille, 2007). Second, Fe and S are actually promoting the recovery of Cu. On the one hand, the large affinity of Fe for Si facilitates the separation of Cu from its siliceous gangue by forming olivines and/or pyroxenes (equation 1). Therefore of course, proper thermodynamic conditions must be met. This was already the case during Chalcolithic periods, as evidenced by the presence of olivine and/or pyroxenes in the very first slags recovered so far. On the other hand, the large affinity of S for Cu has proven to promote the separation of Cu from the slag by gravity. The neo-formed copper sulfide droplets such as covellite (CuS) and chalcocite (Cu_2S) do indeed exhibit relatively low melting point and viscosity, together with high density (Hauptmann, 2003). Here again, mineralogical studies have revealed sulfidic phases in a number of Chalcolithic slags (see Krismer *et al.*, 2013 for the formation conditions of Cu_2S), thus demonstrating that copper sulfide ores were being smelted from an early period.

4.2. Multistep processes for copper sulfide smelting: from Late Bronze Age to Chalcolithic

We have seen in Section 3.2 how difficult it is to bring sufficient O into the system to remove S. Several models have been proposed so far to explain how chalcopyrite used to be desulfurized, and more generally how it was smelted. Models are based on archaeological and slag mineralogy investigations, as well as experimental simulations. Here are a few examples, starting from the more recent ones.

4.2.1. Chalcopyrite smelting during the Late Bronze Age in the Alps

During the Late Bronze Age in the Eastern Alps, *i.e.* >1000 years after the first attempts to smelt copper in this region, quite complex processes have been assumed (Metten, 2003) at sites producing massive quantities of metallic copper (or copper sulfides) out of chalcopyrite. Here, archaeologists have revealed systematically a complex set up of reactors including batteries of furnaces and possibly roasting beds (Cierny *et al.*, 2004; Goldenberg, 2004; Weisgerber, 2004). Roasting beds are large banks of open fire on which the ore is displayed in order to oxidize it in air at relatively low temperatures (500–700°C is an optimum, see Davenport *et al.*, 2002; Lossin, 2003). The roasting process has produced no waste in the archaeological record so far, thus no mineralogical study could be carried out on this first step. Note that the very existence of a roasting bed is still a matter of debate (Doonan *et al.*, 1996; Metten, 2003). The archaeological record encompasses different types of slags obviously stemming from the high-temperature furnaces. Mineralogical studies have helped understand why (see also the mineralogical study of a furnace by Moesta in Moesta and Schlick, 1990). The first thorough study of the three types of slag recovered at Acqua Fredda, Austrian Alps, namely coarse cake-like slags, thin plate slags and sand slag, has shown that all types are chemically and mineralogically very similar (Metten, 2003). Metten has thus proposed a one-step process, although a preliminary roasting operation and a second smelting step are not entirely excluded.

Another very interesting investigation on coeval slags derived mainly from Luzerna, Trentino, has distinguished between three slightly different types, namely coarse (Fig. 1), massive and flat (Addis *et al.*, 2016). The distinction is based on a number of criteria including density, bulk chemistry, amount and composition of sulfide inclusions and relative amounts of fayalite/magnetite/pyroxene/quartz. These types are associated by the author with three distinct, successive, high-temperature metallurgical steps where chalcopyrite is transformed progressively into copper-sulfide (matte) growing in purity. Each step has various durations (see Section 3.3).

4.2.2. Early Bronze Age and Chalcolithic bornite and fahlore smelting

Such a multi-step process may be confirmed by another example of copper-sulfide smelting. The Early Bronze Age mining and metallurgical district of Saint-Véran, French Hautes-Alpes, provides the earliest evidence so far in Europe of primary copper mass production. A production of some seven tons of copper per year has been estimated (Rostan *et al.*, 2002) dated from the beginning of the 2nd millennium BC, *i.e.*

at least half a millennium before the two aforementioned sites. As for the Late Bronze Age sites, productivity was certainly a main concern. It has been shown that the mining and smelting rates were comparable (Bourgarit *et al.*, 2008, 2010). In order to promote efficient and rapid recovery of the copper compounds, slag viscosity was maintained below a critical level by controlling the amount of magnetite (Fig. 7). Also, as mentioned in the previous section, water quenching has been inferred as shown by the prismatic morphology below the top surface of the flat slags (Fig. 10). Interestingly, no roasting beds have been recovered so far at Saint-Véran, and only one type of slag has been exhumed at the two metallurgical sites excavated, namely the flat-type (Fig. 1). Does this mean that smelting was performed by a single-step process? Unlike the two aforementioned Late Bronze Age sites, the exploited ore is not chalcopyrite $CuFeS_2$ but bornite Cu_5FeS_4. Much smaller amounts of S and Fe have to be removed. Hence, the starting product at Saint-Véran would be more or less the starting product of the third step at Luzerna, thus leading to similar flat slags.

Yet, early copper sulfide smelting does not systematically yield such homogenous ideal slags. More primitive types have been recovered in Chalcolithic southern France (Bourgarit and Mille, 2004) as well as during Early Bronze Age in northern Tyrol, Austria (Goldenberg, 1998; Martinek and Sydow, 2004; Goldenberg and Rieser, 2004; Höppner *et al.*, 2005). There, fahlores, *i.e.* solid solutions of tetrahedrite $Cu_{12}Sb_4S_{13}$ and tennantite $Cu_{12}As_4S_{13}$ have been exploited. How? Several hypotheses have been

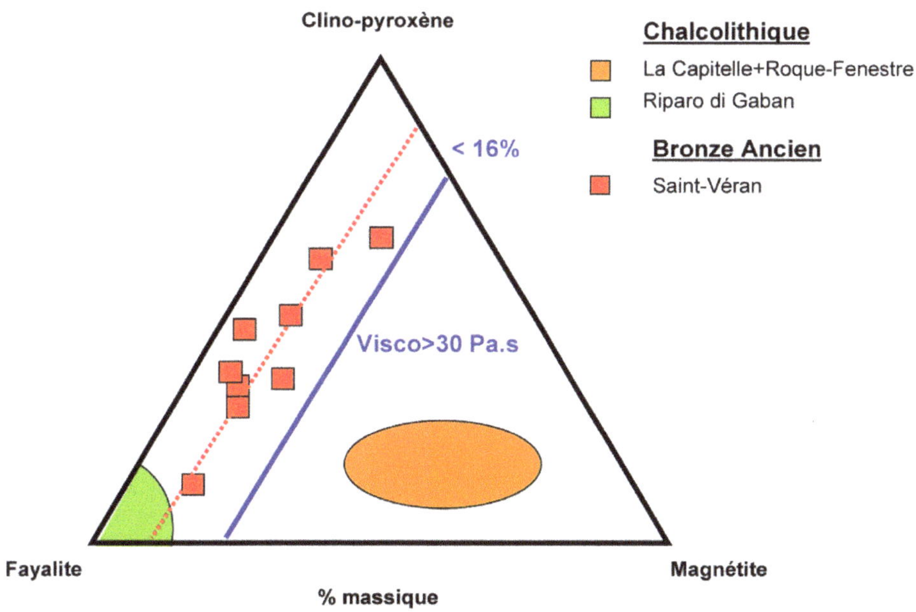

Figure. 7. Relative amounts of fayalite, pyroxenes and magnetite in the Saint-Véran slags (red squares) as deduced by Rietveld refinement of powder diffraction diagrams. 16 wt.% of magnetite is the modern upper limit beyond which slags are considered too viscous. As a comparison, chalcolithic slags from La Capitelle and Roque Fenestre (orange area) are much too rich in magnetite, thus leading to high viscosity.

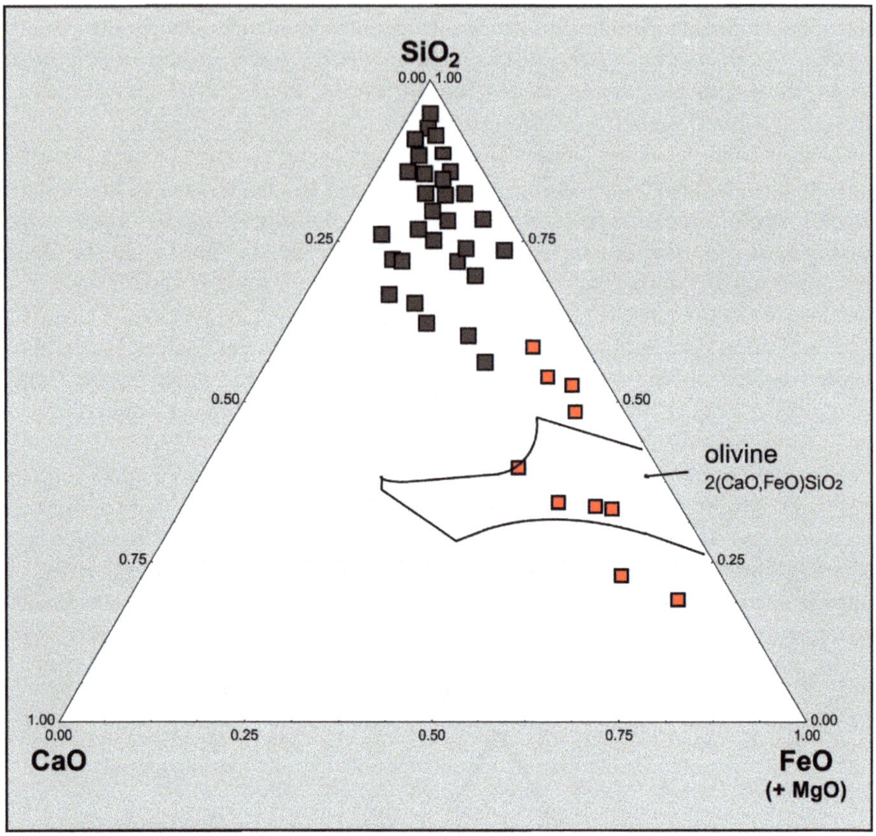

Figure 8. Bulk chemical composition (wt.%) of of Chalcolithic slags from La Capitelle du Broum (grey squares) and Early Bronze Age (EBA) slags from Saint-Véran (red squares) plotted on the ternary diagram FeO(+MgO)–CaO–SiO_2 (analysis by proton-induced X-ray emission of pellets of homogenized powder). The line shows the fayalite liquidus domain (after Osborn and Muan, 1960) elemental composition. Most EBA slag compositions fit within the fayalite domain, whereas chalcolithic slags are much too rich in silica.

formulated, including a two-step process finishing with matte conversion (Bourgarit and Mille, 2005) as suggested by matte pieces in the archaeological record. The most crucial issue remains the removal of S. Therefore, both low-temperature roasting-like (Burger *et al.*, 2011) and high-temperature smelting steps (Burger *et al.*, 2010) have been carried out on chalcopyrite. Similar investigations have been performed on tennantite at high temperature (Bourgarit *et al.*, 2003). The main conclusion is that under pre-industrial working conditions – *i.e.* without massive O_2 input – such as those prevailing in the modern matte conversion processes (Davenport *et al.*, 2002), it is virtually impossible to de-sulfurize entirely the copper sulfide in a simple manner. There are three ways. First, dead-roasting is carried out for several days or even weeks, as testified by 16th and 18th Century texts (Agricola, 1556; Maréchal, 1985). The

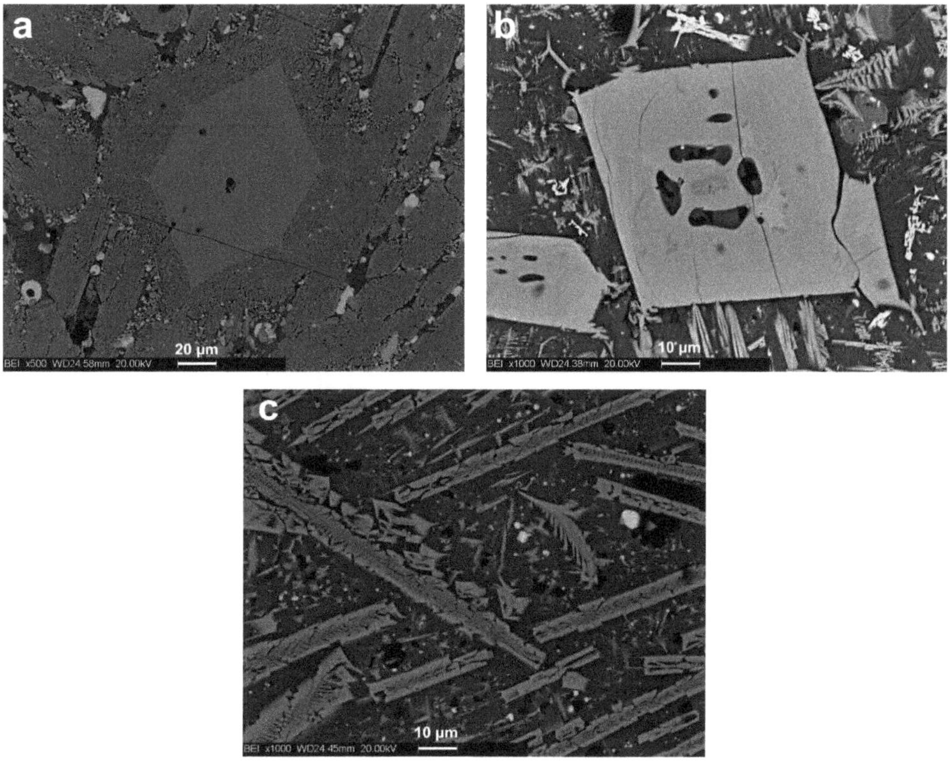

Figure 9. Different fayalite habits encountered at the Luzerna, Late Bronze Age Trentino smelting site, and corresponding cooling rates (after Addis *et al.*, 2016). (a) Prismatic olivine in coarse slags testifies to slow cooling rates (0.5°C/h), (b) Elongate hoppers in massive slags (40°C/h), and (c) chain olivine in flat slags indicate faster cooling rate (80–350°C/h).

resulting S-free ore is then smelted. Second, solid instead of gaseous oxygen is introduced in the system. This idea was raised first by Rostoker *et al.* (1989), confirmed by the present author's own experiments (Bourgarit *et al.*, 2003; Burger *et al.*, 2010) and by archaeological evidence (Bourgarit, 2007; Pelton *et al.*, 2015) showing that at a number of early sites mixed oxidic and sulfidic ores might have been smelted. Third, final product was not metallic copper. At all the sites dealing with chalcopyrite and bornite mentioned in this paragraph, the very nature of the final product is not known. Copper sulfide known as black copper might have been the primary product (Doonan, 1999). Where and how this black copper was converted has still to be found.

4.3. Smelting or melting?

As seen above, our understanding of copper-smelting processes and production modes has increased dramatically over recent decades, due mostly to thorough mineralogical and petrological studies and associated theoretical and experimental simulations. In such a scientific context, the apparently simple question of how to distinguish between

Figure 10. Side view of a slag fragment from Saint-Véran showing the typical prismatic layer 1–2 mm below the upper surface (total thickness ~10 mm)

melting and smelting should seem trivial. It is not. Smelting reactors at the beginning of copper metallurgy were rather rudimentary, being either mere holes in the ground or ceramic vessels (Craddock, 1999; Bourgarit, 2007). The type of metallurgy carried out in the latter has often been controversial (Tylecote, 1974). The ceramic materials used for both processes are indeed often the same, namely ordinary clay with no particular refractory properties. Moreover, smelting a copper ore may affect the ceramic vessel the same way melting of copper does. In other words, both processes may generate similar slaggy layers on the inner surface of the vessel.

To distinguish between melting and smelting activity is a crucial issue for archaeologists, especially at the beginnings of metallurgy. Each activity indeed refers to very specific knowledge and possible particular social, economic and political status. Thorough mineralogical investigations of Neolithic crucibles from Switzerland (Maggetti *et al.*, 1990) have concluded that chalcopyrite had been smelted therein. Yet, simpler optical observations of the same crucibles (Rehren, 2009) have shown that, because of their location in the slag and their texture, the $CuFeS_2$ inclusions could not be remnant of a copper ore. Instead, these inclusions have been formed by biochemical reactions after deposition of the crucible in silty sediments, as confirmed by the presence of particular aggregates of pyrite crystals. Rehren (2009) concluded by stressing the "strength of optical methods, common sense and an open mind when considering the seemingly firm and indisputable results obtained by advanced scientific instruments". In some instances, archaeological observations may even provide the unique decisive arguments. This was notably the case at another Neolithic settlement, in SW France, with the opposite conclusion (Carozza *et al.*, 1997; Carozza, 1998;

Bourgarit *et al.*, 2002). Here the slaggy material layering the common domestic ceramic-ware sherds was proved to stem from chalcopyrite smelting. But the proof was not brought by the mineralogical investigations. The unusual fragmentation of the vases, their relative thinness, and the systematic absence of ceramic reshaping – as usually seen on melting crucibles due to re-use (Queixalos *et al.*, 1987) – constituted the main arguments towards a smelting operation. The considerable fragmentation of slagged sherds has also been used elsewhere as an argument in support of smelting (Müller *et al.*, 2004), although in that case, the argument was supported by chemical evidence as well (*i.e.* the presence of gangue elements).

5. Conclusion

This review focuses on pre-industrial copper smelting. Yet other non-ferrous and ferrous metallurgies have been documented by using slag mineralogical investigations as well. The first studies of iron slags and related material started slightly before first copper slag studies (Morton and Wingrove, 1969, 1972). They led to a number of very interesting advances in our understanding of metal production modes, lineages and evolutions (Humphris and Rehren, 2013) which can be applied to any kind of metallurgy including copper extractive metallurgy. To date, much less attention has been paid to tin, lead and silver (Rehren *et al.*, 1999; Yener *et al.*, 2003; Ettler *et al.*, 2009; Chirikure *et al.*, 2010). Arsenic-rich compounds, speiss, aimed at the elaboration of copper-arsenic alloys has recently been discovered at Arisman, Central Iran, from the Early Bronze Age (Rehren *et al.*, 2012). The production process has been documented by the investigation of the associated slags (Boscher, 2016). Whatever the metallurgy, slags have long shown their importance for specialized research on ancient techniques. Archaeological collections are under construction in the British Isles (see National slag collection at http://hist-met.org/resources/national-slag-collection.html). Some museums – mainly German and Anglo Saxon – are keen to present these wastes together with more valuable/aesthetic representations of cultural heritage. Slags may even form an integral part contemporaneous artworks as seen notably in *Blackout* by Mike Kelley. The contribution of mineralogy and petrography to the study of ancient extractive metallurgies is not restricted to slag. Technical ceramics such as furnaces, vessels and tuyeres, provide invaluable materials for investigation also. Other ceramics related to metal melting and casting including melting crucibles, molds (Katona *et al.*, 2007) and cores (Castelle *et al.*, 2016) are also studied increasingly.

Back to ancient copper smelting, there is no doubting that mineralogical and petrographic slag studies represent a major archaeometallurgical approach. In the past two decades, these techniques have been combined increasingly closely with a number of other sciences, including the humanities, to deal successfully with what are often complex archaeological and/or historical problems. Although scientific investigations in the field are increasing both in terms of quantity and quality, a few directions for the future may be suggested. The mineralogical approach should be maintained at its current high level involving as much petrography as possible. Studies of slag and metallurgical technical ceramics would benefit greatly from this. Modern metallurgical tools should be brought into play. There should be more collaboration with those working on ancient iron smelting. Finally, a more systematic but careful use of relevant anthropological concepts and tools may help to extract from the slags some fundamental answers regarding technological changes between human groups.

Acknowledgments

Thanks are due to Andrew Lacey for the English editing.

References

Addis, A., Angelini, I., Nimis, P. and Artioli, G. (2016) Late Bronze Age copper smelting slags from Luserna (Trentino, Italy): interpretation of the metallurgical process. *Archaeometry*, **58**, 96–114.

Agricola, G. (1556) *De Re Metallica*. Dover, New York.

Ambert, P., Bourgarit, D., Mille, B. and Happ, J. (1997) Un nouveau point sur l'expérimentation métallurgique des cuivres gris du midi de la France. *Archéologie en Languedoc*, **21**, 45–50.

Anfinset, N. (2011) *Social and Technological Aspects of Mining, Smelting and Casting Copper. An Ethnoarchaeological Study from Nepal.* Der Anschnitt. Deutsches Bergbau-Museum Bochum, Bochum, Germany.

Anguilano, L., Angelini, I., Artioli, G., Moroni, M., Baumgarten, B. and Oberrauch, H. (2002) Smelting Slags from Copper and Bronze Age archaeological sites in Trentino and Alto Adige. Pp. 627–638 in: *Atti del II Congresso Nazionale di Archeometria, Bologna, 29 gennaio-1 febraio 2002* (C. D'Amico, editor). Patron, Bologna, Italy.

Arab, R. and Rehren, T. (2004a) The Wertime Pyrotechnological Expedition of 1968. *Journal of the Institute of Archaeometallurgical Studies (IAMS)*, **24**, 29–34.

Arab, R. and Rehren T. (2004b) The Pyrotechnological Expedition of 1968. Pp. 550–555 in: *Persiens antike Pracht, Bergbau-Handwerk-Archäologie* (T. Stöllner, R. Slotta and A. Vatandoust, editors). Deutsches Bergbau Museum, Bochum, Germany.

Artioli, G. (2010) *Scientific Methods and Cultural Heritage. An Introduction to the Application of Materials Science to Archaeometry and Conservation Science.* Oxford University Press, Oxford, New York.

Artioli, G., Angelini, I. and Pedrotti, A. (2005) La prima metallurgia in Trentino alla luce delle piu recenti indagini. *Proceedings of the Il sito fusorio della tarda età del Rame di Millan presso Bressanone dell'area alpina*, Bolzano, Italie, 15 juin 2005.

Avery, V. (2011) *Vulcan's Forge in Venus' City. The Story of Bronze in Venice, 1350-1650.* Oxford University Press, Oxford, UK, 700 pp.

Bachmann, H.G. (1982) Copper smelting slags from Cyprus?: Review and classification of analytical data. Pp. 143–151 in: *Early Metallurgy in Cyprus, 4000–500 B.C.* (J.D. Muhly, R. Maddin and V. Karageorghis, editors). The Foundation, Nicosia.

Bisson, M.S. (2000) Precolonial copper metallurgy: Sociopolitical context. Pp. 83–145 in: *Ancient African Metallurgy* (J. Vogel, editor). Altamira-Press, Walnut-Creek, California, USA.

Böhne, C. (1968) Über die Kupferverhüttung der Bronzezeit. *Archaeologia Austriaca*, **44**, 49–60.

Boscher, L. (2016) *Reconstructing the arsenical copper production process in early Bronze Age southwest Asia.* Unpublished PhD Thesis, University College London.

Bourgarit, D. (2007) Chacolithic copper smelting. Pp. 3–14 in: *Metals and Mines* (S. La Niece, D. Hook and P.T. Craddock, editors). Archetype Publications Ltd, London.

Bourgarit, D. and Mille, B. (1997) La métallurgie chalcolithique de Cabrières: confrontation des données expérimentales et archéologiques en laboratoire. *Archéologie en Languedoc*, **21**, 51–63.

Bourgarit, D. and Mille, B. (2005) Nouvelles données sur l'atelier métallurgique de la Capitelle du Broum (district minier de Cabrières, Hérault, France): la transformation de minerais de cuivre à base de sulfures se précise! Pp. 97–108 in: *Mémoire XXXVII de la Société Préhistorique Française, La première métallurgie en France et dans les pays limitrophes* (P. Ambert and J. Vacquer, editors). Société Préhistorique Française, Carcassonne, France.

Bourgarit, D. and Mille, B. (2007) Les premiers objets métalliques ont-ils été fabriqués par des métallurgistes? *Actualité Chimique*, **312–313**, 54–60.

Bourgarit, D., Mille, B., Burens, A. and Carozza, L. (2002) Smelting of chalcopyrite during chalcolithic times : some have done it in ceramic pots as vase-furnaces. Pp. 297–302 in: *Proceedings of the 33rd International Symposium on Archaeometry*. Vrije Universiteit, Amsterdam.

Bourgarit, D., Mille, B. Prange, M., Ambert, P. and Hauptmann, A. (2003) Chalcolithic fahlore smelting at Cabrières: Reconstruction of smelting processes by archaeometallurgical finds. Pp. 431–440 in: *Archaeometallurgy in Europe*, vol **1**. Associazione Italiana di Metallurgia.

Bourgarit, D., Wirth, E. and Mille, B. (2004) Fahlore smelting at chalcolithic time: Laboratory experiments for the reconstruction of an original copper extractive metallurgy. Poster presented at the 34th International Symposium on Archaeometry, Zaragoza, Spain, 3-7 May 2004.

Bourgarit, D., Rostan, P., Burger, E., Carozza, L. and Artioli, G. (2008) The beginning of copper mass production in the western Alps: the Saint Véran mining area reconsidered. *Historical Metallurgy*, **42**, 1–11.

Bourgarit, D., Rostan, P., Carozza, L., Mille, B. and Artioli, G. (2010) Vingt ans de recherches à Saint-Véran, Hautes Alpes: état des connaissances de l'activité de production de cuivre à l'âge du Bronze ancien. *Trabajos de Prehistoria* **67**, 269–285.

Budd, P. and Taylor, T. (1995) The faerie smith meets the bronze industry: magic versus science in the interpretation of prehistoric metal-making. *World Archaeology*, **27**, 133–143.

Burger, E. (2008) *Métallurgie extractive protohistorique du cuivre : Etude thermodynamique et cinétique des réactions chimiques de transformation de minerais de cuivre sulfurés en métal et caractérisation des procédés*. Unpublished PhD thesis, University Paris VI, Paris.

Burger, E., Bourgarit, D., Wattiaux, A. and Fialin, M. (2010) The reconstruction of the first copper smelting processes in Europe during the 4th to the 3rd millennium BC: Where does the oxygen come from? *Applied Physics A*, **100**, 713–724.

Burger, E., Bourgarit, D., Frotté, V. and Pilon, F. (2011) Kinetics of iron-copper sulphides oxidation in relation to protohistoric copper smelting. *Journal of Thermal Analysis and Calorimetry*, **103**, 249–256.

Carozza, L., Bourgarit, D. and Mille, B. (1997) L'habitat et l'atelier de métallurgiste chalcolithique d'Al Claus: analyse et interprétation des témoins d'activité métallurgique. *Archéologie en Languedoc*, **21**, 147–160.

Carozza, L. and Mille, B. (2007) Chalcolithique et complexification sociale : quelle place pour le métal dans la définition du processus de mutation des sociétés de la fin du Néolithique en France? Pp. 195–232 in: *Le Chalcolithique et la construction des inégalités, tome 1, le continent européen, séminaires du collège de France* (J. d. Guilaine, editor). Errance, Paris.

Carozza, L., Rostan, P., Bourgarit, D., Mille, B., Coquinot, Y., Burens, A. and Escanilla Artigas, N. (2010) Un site métallurgique du Bronze ancien dans le vallon du Longet à Molines-en-Queyras (Hautes-Alpes) : caractérisation du contexte archéologique et des déchets liés aux activités de métallurgie extractive. Pp. 261–281 in: *Archéologie de la Montagne Européenne, actes de la table ronde internationale de Gap, Archéologie de l'espace montagnard: confrontation d'expériences européennes, septembre/octobre 2008* editors). Errance, Paris.

Castelle, M., Coquinot, Y. and Bourgarit, D. (2016) Casting cores: an under-explored witness of the fabrication of early modern bronze statues. *Microchemical Journal*, **126**, 14121–131.

Catapotis, M., Pryce, O. and Bassiakos, Y. (2008) Preliminary results from an experimental study of perforated copper-smelting shaft furnaces from Chrysokamino (Eastern Crete). Pp. 113–121 in: *Proceedings of the Aegean Metallurgy in the Bronze Age, Proceedings of an International Symposium Held at the University of Crete, Rethymnon, Greece, on November 19–21, 2004*. Ta Pragmata Publications. Athens.

Charlton, M.F., Crew, P., Rehren, T. and Shennan, S.J. (2010) Explaining the evolution of ironmaking recipes – An example from northwest Wales. *Journal of Anthropological Archaeology*, **29**, 352–367.

Charlton, M.F., Crew, P., Rehren, T. and Shennan, S.J. (2013a) Measuring variation in iron smelting slags: an empirical evaluation of group-identification proceedures. Pp. 421–432 in: *The World of Iron* (J. Humphris and T. Rehren, editors). Archetype Publications, London.

Charlton, M. F., Shennan, S. J., Rehren, T. and Crew, P. (2013b) Evolutionary analysis of ironmaking slag. Pp. 288–295 in: *The World of Iron* (J. Humphris and T. Rehren, editors). Archetype Publications, London.

Chirikure, S., Heimann, R.B. and Killick, D. (2010) The technology of tin smelting in the Rooiberg Valley, Limpopo Province, South Africa, ca. 1650–1850 CE. *Journal of Archaeological Science*, **37**, 1656–1669.

Cierny, J., Marzatico, F., Perini, R. and Weisgerber, G. (2004) La riduzione del rame in località Acqua Fredda al Passo del Redebus (Trentino) nell'età del Bronzo Recente e Finale. Pp. 125–154 in: *Alpenkupfer - Rame delle Alpi* (G. Weisgerber and G. Goldenberg, editors). Deutsches Bergbau Museum, Bochum, Germany.

Coghlan, H.H. (1939/1940) Prehistoric copper and some experiments in smelting. *Transactions of the Newcomen Society*, **20**, 49–65.

Craddock, P.T. (1995) *Early Metal Mining and Production.* Edinburgh University press, Edinburgh, UK.

Craddock, P.T. (1999) Paradigms of metallurgical innovation in prehistoric Europe. Pp. 175–192 in: *The Beginnings of Metallurgy* (A. Hauptmann, E. Pernicka, T. Rehren and Ü. Yalçin, editors). Deutsches Bergbau Museum, Bochum, Germany.

Craddock, P. (2007) Evidences for the earliest smelting processes in Western Europe *Proceedings of the 2nd International conference* "Archaeometallurgy in Europe", 17-21 June 2007, Aquileia (CD ROM).

Dash, S., Mohanty, N., Mohanty, U.K. and Sarkar, S. (2012) A novel technique for determination of flow characteristics of blast furnace slag. *Open Journal of Metal*, **2**, 42–47.

Davenport, W.G., King, M., Schlesinger, M. and Biswas, A.K. (2002) *Extractive Metallurgy of Copper.* 4th edition. Elsevier, Oxford, UK.

Donaldson, C. H. (1976) An experimental investigation of olivine morphology. *Contributions to Mineralogy and Petrology*. **57**, 187–213.

Doonan, R.C.P. (1999) Copper production in the Eastern Alps during the Bronze Age: Technological change and the unintended consequences of social reorganization. Pp. 72–77 in: *Metals in Antiquity* (S.M.M. Young, P.A.M. Pollard, P. Budd and R.A. Ixer, editors). Archaeopress, Oxford, UK.

Doonan, R. and Dungworth, D. (2013) Experimental archaeometallurgy in perspective. Pp. 1–10 in: *Accidental and Experimental Archaeometallurgy* (D. Dungworth and R. Doonan, editors). Historical Metallurgy Society, London.

Doonan, R.C.P., Klemm, S., Ottaway, B.S., Sperl, G. and Weinek, H. (1996) The east alpine Bronze Age copper smelting process : evidence from the Ramsau Valley, Eisenerz, Austria. Pp. 17–22 in: *Archaeometry'94: Proceeedings of the 29th International Symposium on Archaeometry* (S. Demirci, A.M. Özer and G.D. Summers, editors). Tübitak, Turkey.

Ettler, V., Cervinka, R. and Johan, Z. (2009) Mineralogy of medieval slags from lead and silver smelting (Bohutín, Příbram District, Czech Republic): towards estimation of historical smelting conditions. *Archaeometry*, **51**, 987–1007.

Figueroa, V., Mille, B., Salazar, D., Berenguer, J., Menzies, A., Sapiains, P., Cifuentes, A. and Joly, D. (2018) A prehistoric copper production center identified at Collahausi, southern Tarapacá Altiplano (Chile). *Chungará (Arica)*, **50**(4). http://dx.doi.org/10.4067/S0717-73562018005001001.

Georgakopoulou, M. and Kassianidou, V. (2013) Archaeometallurgical finds and analytical results. Pp. 237–253 in: *Landscape and Interaction. The Troodos Archaeological and Environmental Survey Project, Cyprus* (M. Givenn, A.B. Knapp, J. Noller, L. Sollars and V. Kassianidou, editors), **1**. Oxbow, London.

Goldenberg, G. (1998) L'exploitation Du Cuivre dans Les Alpes Autrichiennes à l'Age Du Bronze. Pp. 9–24 in: *L'atelier Du Bronzier En Europe Du XXe Au VIIIe Siècle Avant Notre Ère* (C. Mordant, M. Pernot and V. Rychner, editors), Tome II, CTHS, Paris.

Goldenberg, G. (2004) Ein Verhüttungsplatz der mittleren Bronzezeit bei Jochberg (Nordtirol). Pp. 165–176 in: *Alpenkupfer – Rame delle Alpi* (G. Weisgerber and G. Goldenberg, editors). Deutsches Bergbau Museum, Bochum, Germany.

Goldenberg, G. and Rieser, B. (2004) Die Fahlerzlagerstätten von Schwaz/Brixlegg in Nordtirol - ein weiteres Zentrum urgeschichtlicher Kupferproduktion in den österreichischen Alpen. Pp. 37–52 in: *Alpenkupfer - Rame delle Alpi* (G. Weisgerber and G. Goldenberg, editors). Deutsches Bergbau Museum, Bochum, Germany.

Gowland, W. (1912) The metals in antiquity. *Journal of the Royal Anthropological Institute*, **42**, 235–287.

Hanning, E., Gauβ, R. and Goldenberg, G. (2010) Metal for Zambujal: Experimentally reconstructing a 5000-year-old technology. *Trabajos de Prehistoria*, **67**, 287–304.

Hauptmann, A. (2003) Rationales of liquefaction and metal separation in earliest copper smelting: basics for reconstructing Chalcolithic and Early Bronze Age smelting processes. Pp 459–468 in: *Proceedings of the 1st International Conference 'Archaeometallurgy in Europe'*, Milano, 24–26 september 2003, Associazione Italiana di Metallurgia. Volume **1**.

Hauptmann, A. (2007) *The Archaeometallurgy of Copper. Evidence from Faynan, Jordan*. Springer, Berlin.

Hauptmann, A. (2014) The investigation of archaeometallurgical slag. Pp. 91–105 in: *Archaeometallurgy in Global Perspective* (B.W. Roberts and C.P. Thornton, editors). Springer, New York.

Hauptmann, A. and Wagner, I. (2007) Prehistoric copper production at Timna: thermoluminescence (TL) dating and evidence form the East. Pp. 67–75 in: *Metals and Mines* (S. La Niece, D. Hook and P.T. Craddock, editors). Archetype Publications Ltd, London.

Herdits (1993) Zum Beginn experimentalarchäologischer Untersuchungen einer bronzezeitlichen Kupferverhüttungsanlage in Mühlbach, Salzburg. *Archaeologia Austriaca*, **77**, 31–38.

Hess, K. (1998) *Zur frühen Metallurgie am oberen Euphrat : Untersuchungen an archäometallurgischen Funden vom Arslantepe aus dem 4. und 3. Jahrtausend v. Chr.* Unpublished PhD, Universität Frankfurt/ Main, Frankfurt/Main.

Höppner, B., Bartelheim, M., Huijsmans, M., Krauss, R., Martinek, K.-P., Pernicka, E. and Schwab, R. (2005) Prehistoric copper production in the Inn Valley (Austria), and the earliest copper in Central Europe. *Archaeometry*, **47**, 293–315.

Humphris, J. and Rehren, T., editors (2013) *The World of Iron*. Archetype Publications, London.

Humphris, J., Martinon-Torres, M., Rehren, T. and Reid, A. (2009) Variability in single smelting episodes – a pilot study using iron slag from Uganda. *Journal of Archaeological Science*, **36**, 359–369.

Iles, L. and Childs, S.T. (2014) Ethnoarchaeological and historical methods. Pp. 193–215 in: *Archaeometallurgy in Global Perspective* (B.W. Roberts and C.P. Thornton, editors). Springer, New York.

Jeffery, D. (2004) Experiential and experimental archaeology with examples in iron processing. *Institute for Archaeo-Metallurgical Studies Newsletter*, **24**, 13–16.

Kassianidou, V. (2003) Archaeometallurgy: Data, analyses, and discussion. Pp. 214–227 in: *The Sydney Cyprus Survey Project. Social Approaches to Regional Archaeological Survey* (M. Given and A.B. Knapp, editors). The Cotsen Institute of Archaeology, University of California, Los Angeles.

Katona, I., Bourgarit, D., Thomas, N. and Bouquillon, A. (2007) From furnace to casting moulds: an exceptional 14th century copper-metallurgy workshop studied in the light of refractory ceramic materials. Pp. 161–167 in: *Archaeometric and Archaeological Approaches to Ceramics* (Y. Waksman, editor). Archaeopress, Oxford, UK.

Kienlin, T.L. (2014) Aspects of metalworking and society from the Black Sea to the Baltic Sea from the fifth to the second millennium BC. Pp. 447–472 in: *Archaeometallurgy in Global Perspective* (B.W. Roberts and C.P. Thornton, editors). Springer, New York.

Killick, D. (2001) Science, speculation, and the origins of extractive metallurgy. Pp. 483–492 in: *Handbook of Archaeological Sciences* (D.R. Brothwell and A.M. Pollard, editors). John Wiley & Sons, New York.

Killick, D. (2014) From ores to metals. Pp. 11–45 in: *Archaeometallurgy in Global Perspective* (B.W. Roberts and C.P. Thornton, editors). Springer, New York.

Killick, D. (2015) Archaeometallurgy as archaeology. *Proceedings of the Archaeometallurgy in Europe III, Proceedings of the 3rd International Conference, Deutsches Bergbau-Museum Bochum, June 29 – July 1, 2011*, 2015, Deutsches Bergbau-Museum Bochum (Der Anschnitt. Beiheft **26**).

Killick, D. and Fenn, T. (2012) Archaeometallurgy: The Study of Preindustrial Mining and Metallurgy. *Annual Review of Anthropology*, **41**, 559–575.

Klimm, D. and Ganshow, S. (2005) The control of iron oxidation state during FeO and olivine crystal growth. *Journal of Crystal Growth*, **275**, 849–854.

Krismer, M., Töchterle, U., Goldenberg, G., Tropper, P. and Vavtar, F. (2013) Mineralogical and petrological investigations of Early Bronze Age copper-smelting remains from the Kiechlberg (Tyrol, Austria). *Archaeometry*, **55**, 923–945.

Lewin, B. and Hauptmann, A. (1984) *Kodo-Zuroko. Illustrierte Abhandlung über die Verhüttung des Kupfers in*

Japan aus dem Jahre 1801. Veröffentlichungen aus dem deutschen Bergbau-Museum, 29th edition, Bochum, Germany.

Liu, S., Rehren, T., Chen, J., Xu, C., Venunan, P., Larreina-Garcia, D. and Martinón-Torres, M. (2015) Bullion production in imperial China and its significance for sulphide ore smelting worldwide. *Journal of Archaeological Science*, **55**, 151–165.

Lorscheider, F., Mass, A. and Steiniger, D. (2003) Frühe Kupferproduktion – archäologischer Befund und Experiment: Versuche zur Fahlerzverhüttung in einem einzigen Ofengang. Pp. 301–308 in: *Man and Mining* (G. Weisgerber, T. Stöllner *et al.*, editors). Bochum, Germany.

Lossin, A. (2003) Copper. In: *Ullmann's Encyclopedia of Industrial Chemistry*, **9**. Wiley-VCH Verlag GmbH & Co. KGaA, Weinheim, Germany.

Lutz, C., Bachmann, H.-G. and Thiemann, U. (1988) Viscosities of slags from non-ferrous metal smelting: Experimental determination and calculation. *Erzmetall*, **41**, 394–398.

Maggetti, M., Baumgartner, D. and Galetti, G. (1990) Mineralogical and chemical studies on Swiss Neolithic crucibles. Pp. 95–104 in: *Archaeometry'90* (E. Pernicka and G.A. Wagner, editors). Birkhäuser, Basel, Switzerland.

Maldonado, B. and Rehren, T. (2009) Early copper smelting at Itziparátzico, Mexico. *Journal of Archaeological Science*, **36**, 1998–2006.

Manasse, A. and Mellini, (2003) M. Diachronic overview of non-ferrous metallurgic slags from southern Tuscany (Italy). Pp. 507–515 in: *Proceedings of the 1st Internationla conference 'Archaeometallurgy in Europe'*, Milano, 24–26 Sept 2003, Associazione Italiana di Metallurgia. Volume **1**.

Maréchal, J.R. (1985) Methods of ore roasting and the furnaces used. Pp. 29–42 in: *Furnaces and Smelting Technology in Antiquity* (P.T. Craddock and M.J. Hughes, editors). British Museum, London.

Martinek, K.P. and Sydow, W. (2004) Frühbronzezeitliche Kupfermetallurgie im Unterinntal (Nordtirol) Pp. 199–211 in: *Alpenkupfer – Rame delle Alpi* (G. Weisgerber and G. Goldenberg, editors). Deutsches Bergbau Museum, Bochum, Germany.

Martinón-Torres, M. and Rehren, T. (2014) Technical ceramics. Pp. 107–131 in: *Archaeometallurgy in Global Perspective* (B.W. Roberts and C.P. Thornton, editors). Springer, New York.

Mei, J. (2009) Early metallurgy in China: Some challenging issues in current studies. Pp. 9–16 in: *Metallurgy and Civilisation* (J. Mei and T. Rehren, editors). Archetype Publications, London.

Mei, J., Wang, P., Chen, K., Wang, L., Wang, Y. and Liu, Y. (2015) Archaeometallurgical studies in China: some recent developments and challenging issues. *Journal of Archaeological Science*, **56**, 221–232.

Merkel, J.F. (1990) Experimental reconstruction of Bronze Age copper smelting based on archaeological evidence from Timna. Pp. 78–122 in: *The Ancient Metallurgy of Copper* (B. Rothenberg, editor). Institute for Archaeo-metallurgical studies, London.

Metten, B. (2003) Beitrag zur spätbronzezeitlichen Kupfermetallurgie im Trentino (Südalpen) im Vergleich mit anderen prähistorischen Kupferschlacken aus dem Alpenraum. *Metalla*, **10 1/2**, 1–122.

Mille, B. and Carozza, L. (2009) Moving into the metal ages: The social importance of metal at the end of the Neolithic period in France. Pp. 143–171 in: *Metals and Society, studies in honour of Barbara S. Ottaway* (T.L. Kienlin and B.W. Roberts, editors). Habelt, Bonn, Germany.

Mille, B., Salazar, D., Bourgarit, D., Figueroa, V., Perlès, C., Berenguer, J. and Burens-Carozza, A. (2013) Emergence of large scale copper production during the Early Bronze Age in Saint-Véran (France) and in Prehispanic Northern Chile: a comparative research program. *The Crucible* (Newsletter of the Historical Metallurgical Society), pp. 8–9.

Moesta, H. (1983) *Erze und Metall – Ihre Kulturgeschichte im Experiment*. Springer, Berlin.

Moesta, H. and Schlick, G. (1990) The furnace of Mitterberg: An oxidizing Bronze Age copper process. *Bulletin of the Metals Museum*, **14**, 5–16.

Moesta, H., Schnau-Roth, G. and Wagner, H.G. (1984) Mössbauer-Studien Zu Bronzezeitlichen Kufperhütten-Prozessen, I. Schlacken, Besonders Der Ostalpinen Kupfertechnik. *Berliner Beiträge Zur Archäometrie*, **9**, 95–112.

Morton, H. and Wingrove, J. (1969) Constitution of bloomery slags. Part I: Roman. *Journal of Iron and Steel Institute*, **207**, 1556–1564.

Morton, H. and Wingrove, J. (1972) Constitution of bloomery slags. Part II: Medieval. *Journal of Iron and Steel*

Institute, **210**, 478–488.

Müller, R., Rehren, T. and Rovira, S. (2004) Almizaraque and the early copper metallurgy of Southeast Spain: new data. *Madrider Mitteilungen*, **45**, 33–56.

Neuninger, H., Preuschen, E. and Pittioni, R. (1970) Versuch einer Fahlerzverhüttung nach dem alten Arbeitsgang auf der Insel Luzon. *Archaeologia Austriaca*, **47**, 87–90.

O'Brien, W. (2004) *Ross Island: Mining, metal and society in early Ireland* Department of Archaeology, National University of Ireland, Galway, Ireland.

Osborn, E.F. and Muan, A. (1960) *Phase Equilibrium Diagrams of Oxide Systems*. American Ceramic Society, Columbus, Ohio, USA.

Ottaway, B.S. (2001) Innovation, production ansd specialization in early prehistoric copper metallurgy. *European Journal of Archaeology*, **4**, 87–112.

Pelegrin, J. (2009) Cognition and the emergence of language: a contribution from lithic technology. Pp. 95–107 in: *Cognitive Archaeology and Human Evolution* (S. de Beaune, F. Coolidge and T. Wynn, editors). Cambridge University Press, Cambridge, UK.

Pelton, A., Stamatakis, M.G., Kelepertzis, E. and Panagou, T. (2015) The origin and archaeometallurgy of a mixed sulphide ore for copper production on the island of Kea, Aegean Sea, Greece. *Archaeometry*, **57**, 318–343.

Pollard, A.M., Thomas, R.G. and Williams, P.A. (1991) Some experiments concerning the smelting of arsenical copper. Pp. 169–174 in: *Proceedings of 'Archaeological sciences 1989'*, Oxbow Monograph **9**, Oxford, UK.

Queixalos, I., Menu, M. and Mohen, J.-P. (1987) Creusets pour La Fonte des Alliages à Base de Cuivre Du Bronze Final Au Fort-Harrouard à Sorel-Moussel (Eure-et-Loir). *Bulletin de La Société Préhistorique Française*, **84**, 23–30.

Rehder, J.E. (1999) High temperature technology in antiquity. A Sourcebook on the design and operation of ancient furnaces. Pp. 305–315 in: *The Beginnings of Metallurgy* (A. Hauptmann, E. Pernicka, T. Rehren and Ü. Yalçin, editors). Deutsches Bergbau Museum, Bochum, Germany.

Rehder, J. E. (2000) *The Mastery and Uses of Fire in Antiquity*. McGill-Queen's Univeristy Press, Montreal and Kingston, Canada.

Rehren, T. (2009) From mine to microbe – the Neolithic copper melting crucibles from Switzerland. Pp. 155–162 in: *From Mine to Microscope* (A. Shortland, I. Freestone and T. Rehren, editors). Oxbow, Oxford, UK.

Rehren, T. and Pernicka, E. (2008) Coins, artefacts and isotopesùArchaeometallurgy and archaeometry. *Archaeometry*, **50**, 232–248.

Rehren, T., Schneider, J. and Bartels, C. (1999) Medieval lead-silver smelting in the Siegerland, West Germany. *Historical Metallurgy*, **33**, 73–84.

Rehren, T., Boscher, L. and Pernicka, E. (2012) Large scale smelting of speiss and arsenical copper at Early Bronze Age Arisman, Iran. *Journal of Archaeological Science*, **39**, 1717–1727.

Rosenqvist, T. (2004) *Principles of Extractive Metallurgy (Second edition)*. Tapir Academic Press, Trondheim, Norway.

Rostan, P. and Malaterre, J.-F. (1994) Les Clausis-Pinilière: Ensemble Minier, Saint-Véran. *Bilan Scientifique SRA PACA, Marseille*, 34–35.

Rostan, P., Rossi, M. and Gattiglia, A. (2002) Approche économique et industrielle du complexe minier et métallurgique de Saint-Véran (Hautes-Alpes) dans le contexte de l'Age du Bronze des Alpes du Sud. *Proceedings of the Les Alpes dans L'Antiquité*, Tende, France, 2002, Société Valdôtaine de Préhistoire et d'Archéologie.

Rostoker, W., Pigott, V.C. and Dvorak, J.R. (1989) Direct reduction to copper metal by oxide-sulfide mineral interaction. *Archeomaterials* **3**, 69–87.

Rothenberg, B. (1990) *The Ancient Metallurgy of Copper*. Institute of Archaeology, London.

Rouillard, P., Perlès, C. and Grimaud, E., editors (2007) *Mobilités, Immobilismes. L'emprunt et Son Refus*. De Boccard, Paris.

Roux, V. (2010) Technological Innovations and Developmental Trajectories: Social Factors as Evolutionary Forces. Pp. 217–233 in: *Innovations in Cultural systems. Contributions from Evolutionary Anthropology*

(M.J. O'Brien and S.J. Shennan, editors). MIT Press, Cambridge, Massachusetts, USA.

Roux, V. (2013) Spreading of innovative technical traits and cumulative technical evolution: continuity or discontinuity? *Journal of Archaeological Method and Theory*, **20**, 312–330.

Rovira, S. (1999) Una propuesta metodologica para el estudio de La metalurgia prehistorica?: El caso de Gorny en la region de Kargaly (Orenburg, Rusia). *Trabajos de Prehistoria*, **56**, 85–113.

Rovira, S. and Montero-Ruiz, I. (2013) Iberia: Technological development of prehistoric metallurgy. Pp. 231–239 in: *Metal Matters. Innovative Technologies and Social Change in Prehistory and Antiquity* (S. Burmeister, S. Hansen, M. Michael and N. Müller-Scheeβel, editors). Leidorf, Rahden/Westf., Germany.

Shaar, R., Ben-Yosef, E., Ron, H., Tauxe, L., Agnon, A. and Kessel, R. (2011) Geomagnetic field intensity: How high can it get? How fast can it change? Constraints from Iron Age copper slag. *Earth and Planetary Science Letters*, **301**, 297–306.

Thomas, N., Bourgarit, D., Verbeek, M., Plumier, J. and Asmus, B. (2013) Commerce et techniques métallurgiques: les laitons mosans dans le marché européen au Moyen Âge (XIIIe-XVIe siècles). Pp. 169–182 in: *L'archéologie au laboratoire* (S. Thiébault and P. d. Depaepe, editors). La Découverte, Paris.

Thornton, C.P. (2014) The emergence of complex metallurgy on the Iranian Plateau. Pp. 665–696 in: *Archaeometallurgy in Global Perspective* (B.W. Roberts and C.P. Thornton, editors). Springer, New York.

Timberlake, S. (2003) Early Mining research in Britain: The developments of the last ten years. Pp. 21–41 in: *Mining and Metal Production through the Ages* (P. Craddock and J. Lang, editors). British Museum Press, London.

Timberlake, S. (2007) The use of experimental archaeology/archaeometallurgy for the understanding and reconstruction of Early Bronze Age mining and smelting technologies. Pp. 27–36 in: *Metals and Mines* (S. La Niece, D. Hook and P.T. Craddock, editors). Archetype Publications Ltd, London.

Tylecote, R.F. (1974) Can copper be smelted in a crucible? *Journal of the Historical Metallurgy Society*, **8**, p. 54.

Tylecote, R.F. (1992) *A History of Metallurgy*. Brookfield : The Institute of Materials, London.

Tylecote, R.F. and Merkel, J.F. (1985) Experimental smelting techniques?: Achievements and future. Pp. 3–20 in: *Furnaces and Smelting Technology in Antiquity* (P.T. Craddock and M.J. Hughes, editors). British Museum, London.

VDEh, Verein Deutscher Eisenhüttenleute (1995) *Slag Atlas*. Verlag Stahleisen GmbH, Düsseldorf, Germany.

Veldhuijzen, X. (2003) Technical ceramics in early iron smelting. The role of ceramics in the early first millennium BC iron production at Tell Hammeh (az-Zarqa), Jordan. Pp. 295–302 in *Understanding People through their Pottery; Proceedings of the 7th European Meeting on Ancient Ceramics (EMAC '03)*, Instituto Português de Arqueologia, Lisboa.

Weisgerber, G. (2004) Schmelzanlagen früher Kupfergewinnung - ein Blick über den Alpen. Pp. 15–36 in: *Alpenkupfer - Rame delle Alpi* (G. Weisgerber and G. GoldenberPg, editors). Deutsches Bergbau Museum, Bochum, Germany.

White, J.C. and Hamilton, E.G. (2014) The transmission of early bronze technology to Thailand: New perspectives. Pp. 805–852 in: *Archaeometallurgy in Global Perspective* (B.W. Roberts and C.P. Thornton, editors). Springer, New York.

Woelk, G., Gelhoit, P. and Bunk, W. (1998) Reconstruction and processing of a Bronze Age copper-reduction furnace. *Der Anschnitt*, **8**, 263–277.

Yener, K.A., Adriaens, A., Earl, A. and Özbal, H. (2003) Analyses of metalliferous residues, crucibles fragments, experimental smelts, and ores from Kestel Tin Mine and the tin processing site of Göltepe, Turkey. Pp. 181–196 in: *Mining and Metal Production through the Ages* (P. Craddock and J. Lang, editors). British Museum Press, London.

Yoshikawa, R., Motomura, Y., Nakanishi, T. and Izawa, E. (2006) Smelting of oxidized copper ore and slag from the Naganobori Mine in the early 8th century. *Proceedings of the Beginning of the use of Metals and Alloys (BUMA) VI.*

Young, T. and Poyner, D. (2012) Two medieval bloomery sites in Shropshire: the adoption of water power for iron smelting. *Historical Metallurgy*, **46**, 78–97.

Zori, C., Tropper, P. and Scott, D. (2013) Copper production in late prehispanic northern Chile. *Journal of Archaeological Science*, **40**, 1165–1175.

Zschoke, K. and Preuschen, E. (1932) Über das Herdschmelzen der arsenikalischen Kupfererze von Mancayan auf der Insel Luzon (Manila). In: *Das Urzeitliche Bergbaugebiet von Mühlbachbischofshofen* (K. Zschoke and E. Preuschen, editors). Die anthropologischen Gesellschafat, Vienna.

Zwicker, U., Greiner, H., Hofman, K.H. and Reithinger, M. (1985) Smelting, refining and alloying of copper and copper alloys in crucible furnaces during prehistoric up to Roman times. Pp. 103–116 in: *Furnaces and Smelting Technology in Antiquity* (P.T. Craddock and M.J. Hughes, editors). British Museum, London.

EMU Notes in Mineralogy, Vol. 20 (2019), Chapter 6, 233–281

The struggle between thermodynamics and kinetics: Phase evolution of ancient and historical ceramics

ROBERT B. HEIMANN[1] and MARINO MAGGETTI[2]

[1]*Am Stadtpark 2A, D-02826 Görlitz, Germany*
robert.heimann@ocean-gate.de
[2]*University of Fribourg, Department of Geosciences, Earth Sciences, Chemin du Musée 6, CH-1700 Fribourg, Switzerland*
marino.maggetti@unifr.ch

This contribution is dedicated to the memory of Professor Ursula Martius Franklin, a true pioneer of archaeometric research, who passed away at her home in Toronto on July 22, 2016, at the age of 94.

Making ceramics by firing of clay is essentially a reversal of the natural weathering process of rocks. Millennia ago, potters invented simple pyrotechnologies to recombine the chemical compounds once separated by weathering in order to obtain what is more or less a rock-like product shaped and decorated according to need and preference. Whereas Nature reconsolidates clays by long-term diagenetic or metamorphic transformation processes, potters exploit a 'short-cut' of these processes that affects the state of equilibrium of the system being transformed thermally. This 'short-cut' is thought to be akin to the development of mineral-reaction textures resulting from disequilibria established during rapidly heated pyrometamorphic events (Grapes, 2006) involving contact aureoles or reactions with xenoliths. In contrast to most naturally consolidated clays, the solidified rock-like ceramic material inherits non-equilibrium and statistical states best described as 'frozen-in'. The more or less high temperatures applied to clays during ceramic firing result in a distinct state of sintering that is dependent on the firing temperature, the duration of firing, the firing atmosphere, and the composition and grain-size distribution of the clay. Hence, the salient properties of the ceramics have to be assessed in a temperature-time-composition space. Owing to the variability of clay composition, the mineralogical processes during thermal transformation of clay minerals can be very complex, not least because most reactions occur far removed from thermodynamic equilibrium and hence are kinetically controlled; that is, they are time- and temperature-dependent. Indeed, kinetics imposes constraints on thermodynamics by retarding reaction rates because of low temperatures, large temperature gradients present in primitive pottery kilns, short reaction times, inhomogeneously distributed reaction partners, and varying redox conditions triggered, for example, by ingress of air during reducing firing cycles. In the context of ceramic technological development over time, the role and development of pottery technology within complex societies is discussed. The close relationship between pottery development and changes in life/societal organization appears to be a major driver in this endeavour.

In this chapter, the phase evolution of some typical ancient and historical ceramics will be traced using ceramic phase diagrams, *i.e.* chemographical expressions of Goldschmidt's mineralogical phase rule. In particular, the systems $CaO-Al_2O_3-SiO_2$

DOI: 10.1180/EMU-notes.20.6

(in which most ancient low- to medium-fired ceramics can be accommodated), $K_2O-Al_2O_3-SiO_2$ (applicable to high-fired Chinese stoneware and European hard-paste porcelain) and $Na_2O-CaO-(Al_2O_3)-SiO_2$ (typical of some ancient Egyptian and Mesopotamian alkaline glazes and French soft-paste porcelain) are discussed.

1. Introduction

The dawn of ceramic technology is irretrivably lost in the distant past, concealed from us by the impenetrable veil of ignorance and time. The history of ceramics as part of every-day life of man presumably had its origin at a time when late Neolithic hunter-gatherer bands turned into agrarian societies, some 12,000 years ago. There is, however, growing evidence that much earlier some Late Palaeolithic or Early Mesolithic genius observed that a soft clay lump would harden in a campfire, and may have realised that this novel property could be exploited to fashion cooking pots and storage containers to hold liquids.

During the past decades, our knowledge of the origin of ceramics technology has been pushed back by archaeological research well into the Upper Palaeolithic/Mesolithic period. Arguably, among the first objects fashioned from clay were maternal goddess images such as the famous Upper Palaeolithic 'Venus of Dolni Věstonice', Moravia, and fragments of animal and human figurines dating around 28,000 BP (Klima, 1962). Close to the end of the Mesolithic period (13,000–12,000 BP), hunter-gatherer communities who roamed what today is Japan, discovered ceramic technology independently (Nakamura *et al.*, 2001), but this time applied it to manufacture the rope-patterned ceramic vessels of the Jōmōn culture, among the world's oldest pots (Chard, 1974; Klein, 1980). Very recently, in a cave in southern China even earlier remains of ceramic technology became known, dated to between 18,300 and 15,430 cal BP (cal BP = calibrated date before present) (Boaretto *et al.*, 2009). In the African realm, pottery appeared in sub-Saharan West Africa (Mali) as early as in the middle of the tenth millennium cal BCE (cal BCE = calibrated date before the common era) (Huysecom *et al.*, 2009). Because ceramic shards are generally well preserved in all but the most acidic soils they are of overriding importance to archaeology to date and distinguish prehistoric cultures by the unique and enduring physical and stylistic features of their characteristic pottery.

There is a close relationship between technological progress, and life and work organization within a society (Heimann, 1991). The evolution of ceramic technology commonly reflects this interdependent relationship. Roman Terra Sigillata, for example, was a ceramic product highly cherished because of its consistent properties, durability, easy availability, practicality, and artistic expression. Hence, many consider this medium–high-fired earthenware the 'Meissen' ware of its time. At this point, it may be appropriate to reflect on ceramic technology progress in general, and to classify any technological development by two descriptive terms: holistic and prescriptive. 'Holistic technologies' are associated essentially with crafts and artistry, prescriptive ones with industry. During much of its history, the process of making pottery was, with a few notable exceptions, a typical holistic process involving a single, stepwise approximation towards the final object whereby the potter, starting with a conscious selection of appropriate raw materials, was required to master the whole succession of steps to produce the pot. In contrast to this, 'prescriptive technologies' involve the division of labour whereby the total work process is subdivided into rather simple unit-procedural steps that represent autonomous skills and thus draw on different groups of workers. Consequently, a considerable degree of abstraction and a solid technical understanding are necessary to integrate the individual unit processes by appropriate measures of work organization and control. The making of Roman Terra Sigillata constitutes an important early pre-industrial example of this tendency, during which labour was organized and subdivided (Roberts, 1997): different groups of workers mined clay, prepared the ceramic masses, made moulds, formed the vessels using these moulds, gathered fire wood, stoked the kilns, inspected, sorted, crated and shipped the finished pots.

The invention of ceramics was an immensely important and far-reaching paradigmatic shift in human evolution. It was the first conscious attempt of humans to alter their given natural material environment. This attempt provides a prime example of a trend observed generally in technology development. At the formative stage of any society, all the utilized technology is holistic. Only after accumulation of extensive practical knowledge and theoretical understanding, can generalization and abstraction occur. Only then can a prescriptive process emerge that in time achieves the necessary level of standardization and organization (Franklin, 1980). In this regard, the production of Roman Terra Sigillata was an important turning point in ceramic development as the holistic process of making pottery was replaced by a novel prescriptive technology that relied on process-determined division of labour, using the combined skills of many individuals. It should be emphasized that the presence and utilization of machinery is not a prerequisite for this development. Instead, the essential characteristics and indispensable requirement for a prescriptive process is 'control' that includes being in command of material resources including capital, information and people. Information includes not only technical knowledge but also the tools of control such as regulations, co-ordination, supervision, and not least, enforcement.

As will be described below in detail, clay-based ceramics are the result of an attempt to turn weathered remnants of rock back into artificial rock-like products by the action of heat on earthy raw materials (Heimann and Franklin, 1979; Cultrone *et al.*, 2001; Maggetti, 2001). This development was thought to have been triggered by the Neolithic transition from hunter-gatherer to agrarian societies, a long-championed but actually outdated view (Jordan and Zvelebil, 2009). Indeed, the manufacturing of pottery typically implies some form of sedentary lifestyle, because pottery is inherently breakable and thus generally useless to hunter-gatherers who are constantly on the move. Firing of clay results in a product with a mineralogical composition different from that of the original natural material. All thermal reactions that transform the raw clay material into the final ceramic object proceed towards some equilibrium state. True thermodynamic equilibrium, however, will never be completely attained because of the relatively low firing temperature and the short firing time applied by the potter. Consequently, we have to focus our attention on the kinetics, *i.e.* the dynamic nature of the firing process. Much research in the field has shown that a fired ceramic body is characterized by 'frozen-in' and thus aborted phase transitions that deviate strongly from true, *i.e.* thermodynamic equilibrium. However, this non-equilibrium state provides the ceramic analyst with important clues as to how, when and where the object under investigation was produced (Heimann, 1982).

Modelling the composition of fired ceramics requires the use of standards based on the expected equilibrium composition, if the limiting factors imposed by reaction kinetics are taken into account. Thermodynamics, as important a guidepost as it may be, can tell only which reactions among the minerals in a clay are allowed to occur at a given temperature and kiln atmosphere. Kinetics, however, constrains the thermodynamic 'licence' to proceed towards equilibrium by retarding reaction rates because of low temperatures, large temperature gradients in primitive pottery kilns, short reaction times, non-uniform grain-size distributions, inhomogeneously distributed reaction partners and varying redox conditions. Thermodynamics will thus control the reactions at higher temperatures and for longer reaction times under equilibrium conditions whereas kinetics will exercise control at lower temperatures and for shorter times, resulting in non-equilibrium states.

Chemically speaking, nearly all ancient low- to medium-fired earthenware can be contained in the component space of the ternary oxide system $CaO-Al_2O_3-SiO_2$. High-fired stoneware and porcelains can be accommodated in the ternary oxide system $K_2O-Al_2O_3-SiO_2$. Iron oxide, magnesia and other elements are always present in smaller quantities, and can generally be summed with geochemically similar elements. Whereas this suggests high chemical complexity, the number of phases formed during firing is surprisingly small, and their stoichiometry is comparatively simple. Condensed three-component phase diagrams, therefore, are invaluable sources of information, even though they have been derived under equilibrium conditions. Knowledge of the ceramic phase diagrams provides the key to understand, on the one hand, the relations between mineralogical and chemical composition of the precursor clays, the maximum firing temperature reached, and the firing atmosphere utilized. On the other hand, important information can be obtained about the firing time, and the microstructure, fabric, colour and eventually the mechanical and thermal properties of the ceramic product under investigation (Heimann, 1989).

2. Ceramic phase diagrams

2.1. Derivation of phase diagrams

In mineralogy and materials science, it is an established procedure to express the composition of an assembly of different phases using a 'phase diagram'. Depending on the number of components present, there exist unary, binary, ternary and multinary phase diagrams. Although a natural clay contains many oxide components owing to its complex mineralogical composition, the following brief discussion of phase diagrams will be limited to binary and ternary cases.

Phase diagrams are chemographical representations of the famous Gibbs' phase rule (Gibbs, 1878) that relates the number of phases *[Ph]*, the number of components *C*, and the number of degrees of freedom *df* of a (closed) system by the deceptively simple equation

$$[Ph] + df = C + 2 \qquad (1)$$

Components are defined as those simple oxides that combine to constitute the phases present at equilibrium, *i.e.* those portions of a system that are physically homogeneous and mechanically separable. The number of degrees of freedom is the number of variables that can be altered without changing the number of phases present. The number '2' in equation 1 is applicable when both temperature and pressure have to be considered. In systems of importance for ceramics, however, the vapour pressure is very low over a wide range of temperatures. The pressure variable can thus be ignored safely. Such systems with a reduced degree of freedom are called 'condensed' systems, and equation 1 reduces to

$$[Ph] + df = C + 1 \qquad (2)$$

This equation is also called the 'mineralogical phase rule' (Goldschmidt, 1911). Systems with $C = 2$ are known as binary systems, *e.g.* $Al_2O_3-SiO_2$, systems with $C = 3$ are known as ternary systems, *e.g.* $CaO-Al_2O_3-SiO_2$. In the remainder of the current chapter, (crystalline) mineral phases, *[Ph]*, will be denoted by square brackets to distinguish them from components, *C*. This is done to avoid confusion between components and phases.

2.1.1. Binary systems

Instead of oxide components such as CaO and SiO_2, mineral phases or other chemical molecules and compounds can be displayed in binary or ternary phase diagrams. Figure 1a shows the simple binary system anorthite–diopside at ambient pressure. This particular system plays a crucial role in describing the composition and texture of basaltic rocks, *i.e.* of effusive igneous rocks with 45–52 wt.% SiO_2 (Fig. 2). The composition is given on the abscissa, and the percentage of diopside increases towards the right. The ordinate depicts the temperature; the melting points of the pure compounds are 1550°C (anorthite) and 1390°C (diopside), respectively. With increasing admixture of diopside to anorthite, and anorthite to diopside, respectively, the melting point of the mixture decreases according to Raoult-van't Hoff's law that describes quantitatively the melting point depression of each end-member composition as

$$\Delta T = \frac{RT_{\mathrm{A}}^2}{\Delta H_{\mathrm{f}}} \cdot x_{\mathrm{B}} \qquad (3)$$

whereby ΔT = change in melting temperature, T_A = the melting point of pure component A, x_B = mole fraction of B in the melt, R = the universal gas constant and ΔH_f = molar heat of fusion (Wachtman, 1969).

The melting points for all anorthite–diopside ratios lie on the so-called liquidus curves that intersect at the eutectic point e_{An-Di}. At any point along the liquidus curves two phases exist: either solid anorthite + melt or solid diopside + melt. Consequently, the degree of freedom (*df*) of the system is one, *i.e.* one variable: either the composition or the temperature can be varied without changing the number of phases *[Ph]* = 2, to

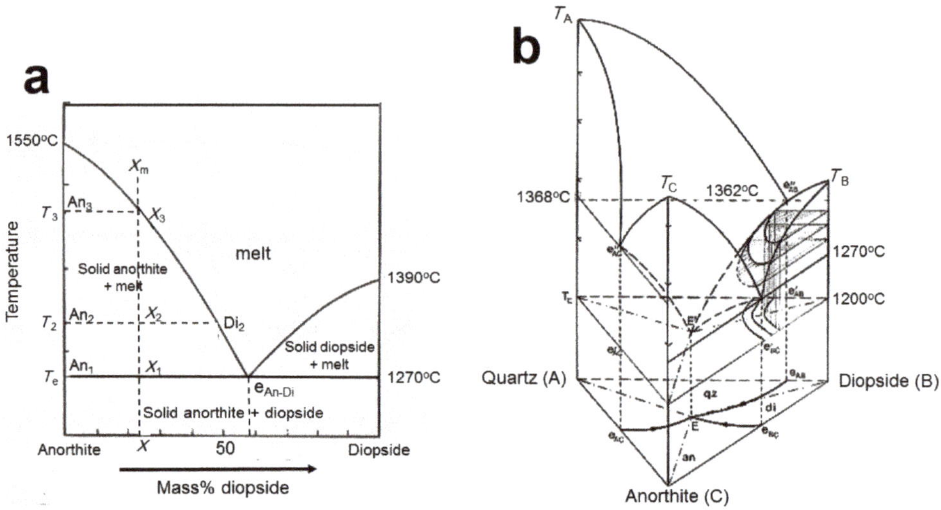

Figure 1. (a) Binary phase diagram anorthite–diopside without intermediate compound or solid solution (after Bowen, 1915). (b) Ternary phase diagram anorthite–diopside–quartz (after Clark *et al.*, 1962; Osborn and Tait, 1952), showing (i) the complex curved liquidus surface (top), (ii) the planar isothermal solidus surfaces (middle) tangent to the ternary eutectic point E′ at 1200°C, and (iii) the planar subsolidus surface (bottom) at ambient temperature. The binary eutectics at 1362°C (quartz–diopside), 1368°C (quartz–anorthite) and 1270°C (anorthite–diopside) are also shown. On the right, as an example, isotherms have been projected downwards from the liquidus surface onto the solidus surface. Both images were taken from Heimann (1989).

fulfill equation 2 for this univariant equilibrium. At the eutectic point (from Greek: *eu* easy, well; *tektein* to melt) e_{An-Di} at 1270°C, the two solid phases anorthite and diopside are in equilibrium with the melt. Hence, the number of coexisting phases, *[Ph]* is three, and $df = 0$ as any change of temperature or composition will displace the system from point e_{An-Di}. This is called an invariant equilibrium.

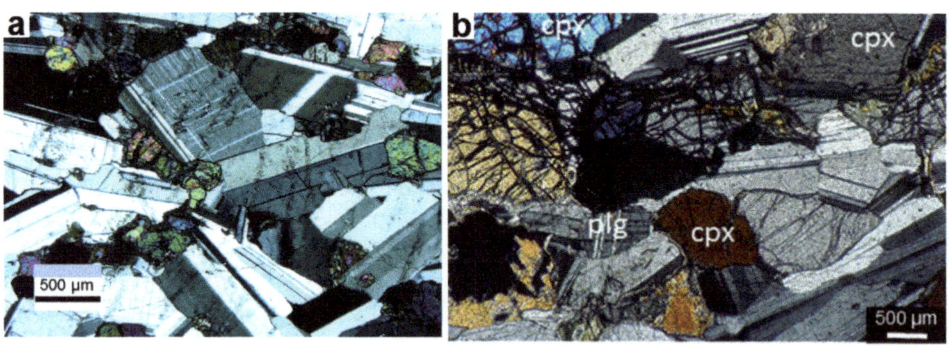

Figure 2. (a) Thin section of diabase with porphyritic or interstitial texture. (b) Thin section of gabbro with ophitic texture, showing large crystals of clinopyroxene (cpx) next to plagioclase (plg). Images by David Walker (http://microscopy-uk.org.uk).

The construction of phase diagrams usually starts from a homogeneous melt, *i.e.* at a temperature well above the liquidus. During cooling of this melt, the onset of crystallization is observed and the corresponding temperatures recorded. In the binary anorthite–diopside system, the overall composition of the melt X_m may be 77 mass% anorthite, 23 mass% diopside and is represented by an isopleth, the vertical dashed line in Fig. 1b. At the starting point X_m, above temperature T_3, the melt starts to cool. At X_3, the temperature of the melt equals T_3. A small amount of solid anorthite appears because the melt composition has now reached the liquidus curve. On further cooling, anorthite will continue to solidify while the composition of the remaining melt in equilibrium with solid anorthite follows the curve X_3–e_{An-Di} moving towards smaller amounts of the anorthite component because solid anorthite is being removed by solidification.

The amount of melt coexisting with solid anorthite at each temperature can be estimated by applying the so-called lever rule (Fig. 1a). For example, at the temperature T_2 solid anorthite An_2 is in equilibrium with a melt of composition Di_2 (52 mass% anorthite, 48 mass% diopside) in the ratio of the line lengths $X_2Di_2/An_2X_2 = 0.92$. Further cooling results in solidification of still more anorthite until at the eutectic temperature, T_e, the eutectic point, e_{An-Di} is reached. At this point (42 mass% anorthite, 58 mass% diopside) the temperature will remain constant owing to the release of the latent heat of crystallization, until all traces of melt have solidified into a mixture of solid anorthite and solid diopside. Frequently, a typical eutectic microstructure develops that is commonly lamellar (layered) due to the reduced diffusion distances in the solid state.

Whereas the size and the shape of the crystals formed during cooling of a melt depend heavily on the rate of cooling, their compositions depend on where the starting melt composition is located in the phase diagram with respect to the eutectic point. If the starting composition of the melt (X_m) is to the left of the eutectic point (hypoeutectic) (Fig. 1a), the first phase to appear during cooling is anorthite. As shown in Fig. 2a, large crystals of anorthitic plagioclase dominate the microstructure of typical porphyritic basalts. Plagioclase crystals continue to crystallize until at the eutectic temperature the remaining melt solidifies, forming a fine-grained aggregate of pyroxene + anorthite that fills the interstitial spaces among plagioclase crystals. In contrast, when starting from a melt composition to the right of the eutectic point (hypereutectic), pyroxene is first to crystallize in large crystals until on reaching the eutectic temperature a fine-grained mixture of pyroxene and plagioclase crystallizes. Such so-called ophitic texture is characteristic for gabbros, *i.e.* intrusive igneous rocks with 45–52 wt.% SiO_2, a typical composition of clinopyroxene and plagioclase, with minor contributions of olivine and amphibole (Fig. 2b).

Whereas the basaltic rocks shown in Fig. 2 appear to have crystallized under near-thermodynamic equilibrium, for ceramics the situation is strikingly different. Here the rates of heating and cooling are very fast compared to rocks such as diabase or gabbro formed from a crystallizing melt within the Earth's crust. Hence, crystals in ceramics will generally be very small, and there is frequently a large proportion of amorphous, *i.e.* glassy material present. Moreover, during ceramic firing, the liquidus will not be approached by cooling a melt from high temperatures but instead by heating a solid mixture of the components. Then, in our example, a mixture of solid anorthite and diopside of composition X will lead at the temperature T_e to the formation of a partial melt of eutectic composition e_{An-Di}. According to the lever rule, the ratio of solid to melt is then $X_1e_{An-Di}/An_1X_1 = (58-23)/23 = 1.52$. With increasing temperature, the amount of melt increases until at temperature T_3 all traces of solid will have disappeared. Then the composition of the melt is X.

Because potters do not desire to melt completely the mineral mixture called clay during ceramic firing, the comparatively low firing temperatures and short firing times will never allow thermodynamic equilibrium conditions under which a ceramic phase diagram may be obtained experimentally to be reached. Hence, thermodynamics will be overruled by kinetics, and application of ceramic phase diagrams to the situation prevailing in ancient pottery kilns must abide by the constraints imposed by kinetic retardation of the potential multitude of mineral reactions as discussed above.

2.1.2. Ternary systems

Another level of complexity is attained when a third component is added to a binary phase diagram. A ternary diagram is then obtained that can be depicted in a compositional triangle (Fig. 1b, Fig. 3a). However, in this case the temperature cannot be plotted directly but can be described only by isotherms that are projections down the temperature axis perpendicular to the triangle. The liquidus surface describes the melting points of all given compositions whereby the points T_A, T_B and T_C are the melting points of the pure components A (quartz), B (diopside), and C (anorthite). If, for example, component B is added to pure A, the melting temperature of this mixture decreases along the liquidus line $T_A-e''_{AB}$. At the binary eutectic point e''_{AB}, the lowest melting temperature (1362°C) of all compositions in the binary subsystem A–B is reached. Adding component C eventually leads to the lowest possible melting temperature, T_E, of the ternary system, the ternary eutectic point E′ at 1200°C in the centre of the triangle A–B–C (Fig. 1b).

The second surface shown in Fig. 1b is the isothermal solidus surface tangent to the ternary eutectic point E′. This surface determines the compositions of the solid phases in equilibrium with the ternary eutectic melt. Below this surface a liquid phase cannot occur. The third surface shown is the projection of the 3D-phase diagram down to ambient temperature, the isothermal subsolidus surface A–B–C. Indicated on this surface are the stability fields of the phases [A], [B] and [C], separated by the curved phase boundary lines $e_{AC}-E$, $e_{BC}-E$ and $e_{AB}-E$, as well as the conodes A–E, B–E and C–E. In actual phase diagrams, the isotherms shown in Fig. 1b as (curved) planes parallel to the solidus surface and intersecting the liquidus surface $T_B-e''_{BC}-E'-e''_{AB}$ are projected downwards onto the solidus surface in much the same way as elevation contours on a topographical map. The 3D-ternary phase diagram shown in Fig. 1b depicts the actual system quartz (A)–diopside (B)–anorthite (C) with the melting points T_{qz} = 1723°C, T_{di} = 1390°C, T_{an} = 1550°C; the binary eutectics at 1362°C

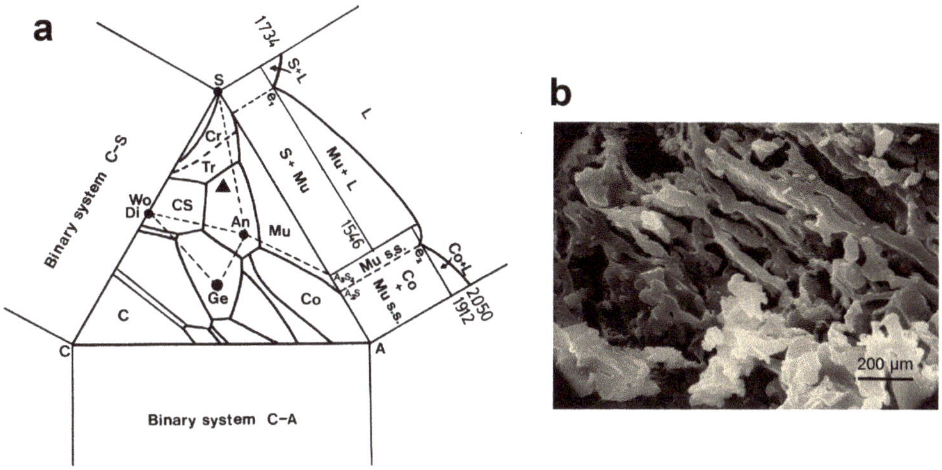

Figure 3. (a) The ternary system CaO (C)–Al_2O_3 (A)–SiO_2 (S) surrounded by the three binary systems CaO–SiO_2 (C–S, calcium silicates), CaO–Al_2O_3 (C–A, calcium aluminates) and SiO_2–Al_2O_3 (S–A). The binary system SiO_2–Al_2O_3 is shown explicitly and reveals two eutectic points, SiO_2–mullite [Mu]–liquid [L] at e_1 (T_e = 1546°C) and mullite–corundum [Co]–liquid at e_2 (T_e = 1912°C). Within the triangle depicting the ternary system, the phases have been denoted as follows: An: anorthite, C: calcium oxide, Co: corundum, Cr: cristobalite, CS: (Wo) wollastonite, Ge: gehlenite, Mu: mullite, and Tr: tridymite. Thick lines denote phase boundaries; dashed lines are representative of compatibility triangles (after Rankin and Wright, 1915). (b) Calcareous illitic clay of the composition indicated by the black triangle in the ternary phase diagram, experimentally fired at 1010°C under oxidizing conditions (Heimann *et al.*, 1980). With permission by cfi Ber. DKG.

(quartz–diopside), 1270°C (diopside–anorthite), 1368°C (quartz–anorthite); and the ternary eutectic at 1200°C. The composition of the ternary eutectic is 30 mass% quartz, 33 mass% diopside and 37 mass% anorthite (Osborn and Tait, 1952; Clark *et al.*, 1962). The ternary phase diagram discussed above refers to high-fired calcareous illitic clays with the four components SiO_2, Al_2O_3, CaO and MgO but fails to incorporate an important fifth component such as Fe_2O_3. Hence, even more complex phase relations have to be considered in appropriate multicomponent (multinary) systems.

Each ternary system can be deconvolved into its three constituent binary systems as shown in Fig. 3a. The binary boundary systems of the ternary system CaO (C)–Al_2O_3 (A)–SiO_2 (S) have been collapsed onto the plane of the ternary system. The black triangle in the compatibility triangle wollastonite [Wo]–anorthite [An]–silica [S] depicts the composition of a typical calcareous illitic clay reportedly used by Roman potters during the first few centuries CE to produce Terra Sigillata at *Tabernae* (today's Rheinzabern, Palatinate, Germany) (Heimann and Maggetti, 2014).

Figure 3b is a scanning electron microscopy image of this calcareous illitic clay, with a chemical composition shown in Table 1, which was fired in air at 1010°C. Intensive vitrification is apparent (Maniatis and Tite, 1975). The fabric of the pottery has been dissolved into sheets of vitrified material that are arranged parallel to the direction of compression during extrusive forming or throwing on a potter's wheel. X-ray diffraction analysis (Heimann *et al.*, 1980) revealed that crystalline quartz, anorthite-rich plagioclase and diopside are the main constituents. This is exactly the equilibrium composition to be expected from the location of the clay composition in the compatibility triangle [silica = quartz]–[anorthite]–[diopside] shown in Fig. 3a. Minor amounts of hematite and sanidine (high-temperature alkali feldspar) account for the iron and alkali contents, respectively, of the raw clay.

In Fig. 3a the conode or tie line connecting the compositions of the conjugate phases wollastonite [Wo] and anorthite [An] (dashed line) is partitioned by the intersection with the phase boundary CS [calcium silicate]–An [anorthite], approximately in the ratio Wo/An = 58/42, corresponding to the position of the eutectic point shown in Fig. 1a. The formation of diopside would be expected because the starting clays contain 2.7 mass% MgO. However, the representation of [Di] as an individual phase would require a quaternary phase diagram that cannot be rendered in two dimensions.

The discussion above of the construction, evaluation and interpretation of binary and ternary phase diagrams considered only very simple and hence, straightforward cases. Situations in reality are more complex, involving formation of multinary compounds from the melt, occurrence of miscibility gaps among solid solutions, spinodal demixing, peritectic systems with incongruent melting, solid solutions, redox equilibria, construction of crystallization paths during freezing of a melt, and liquid immiscibility conditions. To obtain information on these peculiarities, the reader is referred to the pertinent literature (see Levin *et al.*, 1964–1990; Kingery *et al.*, 1976; Rosenberger, 1979; West, 1982; The American Ceramic Society, 1992–1996; Heimann, 2010). With the advent of modern high-speed computers, a vast number of software programs and thermodynamic modelling approaches is available, sometimes as freeware distributed *via* the internet (Schultz and Chang, 1985; Andersen *et al.*, 1993; Chatterjee *et al.*, 1998). Kattner (1997) provided a useful guide to these programs and their capabilities. Older but pedagogically excellent information on the laws of stability and coexistence of minerals including application of Schreinemaker analysis to multicomponent systems of geological relevance can be obtained from Niggli (1954).

2.2. Micro- or local equilibria

In contrast to the straightforward cases discussed above in which global equilibrium is regarded as extending over the entire spatial dimension of the system, the concept of micro- or local equilibria applies to ceramic systems under generally non-equilibrium conditions. These local equilibria are established whenever individual mineral grains are in close contact with each other. Their differing chemical potentials, μ_i, will become the driving force for heterogeneous mineral reactions that will approach, but never completely reach, chemical equilibrium. Ideally, these mineral reactions will proceed until thermodynamic equilibrium has been attained, defined by $dG = 0 = -s \cdot dT + \Sigma\mu_i dn_i$, whereby s is the molar entropy, μ_i the chemical potentials of components i, and n_i are the mole numbers (concentrations) of the components i.

The firing process to which clays are subjected results in a distinct state of sintering, dependent on firing temperature, firing time, composition of the raw material and kiln atmosphere. It can be expected, therefore, that the relationship between chemical and mechanical properties of the ceramic product and the multitude of predictor variables is exceptionally complex, and in many cases impossible to unravel. Fortunately, there are several ways to overcome this problem. One involves retrospective estimation of firing conditions by assessing temperature-dependent properties of the particular clay. Thus, the properties of the ceramic body, such as phase content, colour, density, hardness, strength, toughness and magnetic properties will be compared to those of a series of similar or, ideally, identical clays fired at various temperatures under controlled conditions in the laboratory. These series form a system of reference samples with a characteristic, temperature-dependent mineralogical phase composition against which the composition of the actual ceramics can be matched (Heimann and Franklin, 1979). Analytical tools to determine the mineralogical phase composition are X-ray diffraction (XRD), optical and electron microscopy, vibrational spectroscopy such as infrared (IR) or Raman, and a host of other modern surface analytical techniques (see, for example, Kingery and Vandiver, 1986; Kockelmann *et al.*, 2001).

Because the firing of a ceramic body is a short-term event, the time during which the object was subjected to the maximum firing temperature is rarely sufficient to attain a global equilibrium situation ($dG = 0$) with respect to the phase transitions initiated by the thermal treatment. In particular, a calcareous clay fired for only a few minutes to a maximum temperature below 800–850°C in a typical bonfire retains its calcite content completely even though the thermodynamic decomposition temperature of calcite has been exceeded (Maggetti *et al.*, 2011). This again witnesses how kinetic constraints override thermodynamic laws.

Likewise, kinetic retardation must be taken into account in thermal decomposition reactions of primary clay minerals such as illite or kaolinite, and non-plastic minerals such as quartz as well as the formation of high-temperature minerals such as gehlenite, diopside or mullite. As discussed above, the ceramic phase diagrams generally have been constructed from information gathered during slow cooling of silicate melts, in which the formation of new phases from the melt is accelerated because the mass-transfer coefficients are much larger in a liquid than in a solid. In fired clays, however, the temperature is usually not sufficiently high to melt the raw material completely and is not desired either. Even if partially melted phases existed in the ceramic body as in high-fired stoneware or porcelain, or in the low-to medium-fired slips of Mesopotamian Tell Halaf and Cretan Kamares wares or Roman Terra Sigillata, the considerable viscosity of these silica-rich melts largely prevented the crystallization of equilibrium phases. Hence, the state of the slips remains more or less glassy.

The ceramic body, therefore, is left in a state of distinct non-equilibrium. At lower firing temperatures, it is composed of a succession of micro- or local equilibria that are confined to the reactive interfaces between mineral grains in an inhomogeneous clay matrix (Maggetti, 1986). This situation is demonstrated in Fig. 4. On the left an aggregate of two calcite crystals, one quartz grain and one illite flake is shown schematically. The five existing microequilibria are denoted 1 to 5, and their loci in the

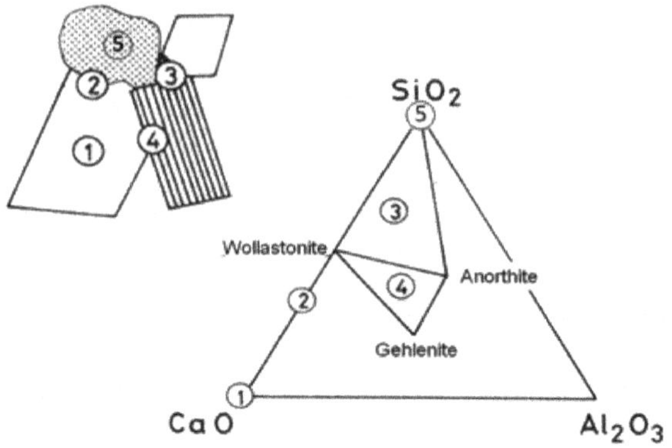

Figure 4. Schematic representation of microequilibria in a typical calcareous clay fired under oxidizing conditions (Maggetti, 1986). Left: white = calcite, stippled = quartz, laminated = illite. Right: Location of typical microequilibria 1 to 5 in the ternary phase diagram $CaO-Al_2O_3-SiO_2$. For explanation, see text.

ternary phase diagram $CaO-Al_2O_3-SiO_2$ are shown on the right. This diagram is particularly important for the calcareous illitic clays widely used by potters in antiquity, and hence is discussed in more detail here. Calcite converts on heating to CaO, and thus obtains position 1 at the CaO apex of the phase diagram. At the boundary between calcite and quartz crystals, two compatible phases of the binary subsystem $CaO-SiO_2$ form, for example, wollastonite, $Ca_3(Si_3O_9)$ and rankinite, $Ca_3Si_2O_7$. The approximate positions of the contact zones are indicated at 2. The microequilibrium 3 is stable at the three-grain boundary quartz–calcite–illite, and the position of the reaction product is somewhere in the subsystem wollastonite–anorthite–quartz, depending on the relative proportion of the components: the exact position depends on the relative proportions of the original minerals. Microequilibrium 4 between calcite and illite contains less SiO_2 component than microequilibrium 3, and is therefore located somewhere in the subsystem wollastonite–anorthite–gehlenite. Finally, the quartz will be located at the SiO_2 apex 5.

The existence of microequilibria gives the ceramic analyst an important tool at hand to perform mineralogical analyses of grain-boundary melts, reaction rims and pores using optical and electron-optical techniques (Heimann *et al.*, 1980). The results obtained often lead to sound conclusions about the crystallization paths towards equilibrium (see Section 4.3.4.).

In conclusion, the useful concept of microequilibrium explains why thermodynamically incompatible mineral phases such as gehlenite and quartz can occur side-by-side in the ceramic body obtained from a calcareous paste. This is caused by the inherent inhomogeneity (graininess) of the clay, and the low temperatures and short firing times applied during firing. These conditions are not sufficient to attain thermodynamic equilibrium with respect to the global system. This situation reflects again the universal struggle between thermodynamic and kinetic principles in that there exists the thermodynamic ‘license’ to proceed with a certain phase-transformation reaction, but the low maximum firing temperature, steep heating rate and short reaction time impose a kinetic constraint resulting in retarding the expected reactions. Thermodynamics will control the reactions at high temperatures and for long reaction times; kinetics will exercise control at lower temperatures and shorter times. This occurs because large amounts of activation energy and configurational entropy, $-s{\cdot}dT$, respectively are required to nucleate a new phase. Because these driving forces are not available at the relatively low temperatures employed by an ancient potter, metastable 'frozen-in' equilibria dominate the ceramic microstructure as evidenced, for example, in Neolithic ware.

In any case, the formation of gehlenite from calcareous clays by grain-boundary reactions (Maggetti and Heimann, 1979) is surprising because, from a thermodynamical point of view, the formation of an Al-rich pyroxene such as fassaite, $Ca(Mg,Al,Fe^{3+})(Si,Al)_2O_6$ (Tilley, 1938) would be expected (Dondi *et al.*, 1998; Cultrone *et al.*, 2001). However, in this case another example of kinetics overriding thermodynamics is brought to bear because, according to Goldsmith's 'ease of crystallization' principle (Goldsmith, 1953), a silicate phase with Al only in fourfold coordination is energetically favoured over phases containing Al in mixed four- and sixfold coordinations, at least in the absence of hydroxyl ions. Because in the crystal structure of gehlenite all Al ions are in tetrahedral coordination (Louisnathan, 1971), this phase will form preferentially. Similarly, the preferential formation of mullite with a larger $(AlO_4)/(AlO_6)$ ratio than structurally similar sillimanite (Rehak *et al.*, 1998) could be explained by Goldsmith's rule.

3. Oxidizing *vs.* reducing firing – the role of the oxygen partial pressure

Figure 5 summarizes the disappearance of several typical minerals that make up calcareous illitic clay and the appearance of new phases that occur under oxidizing (left) and reducing (right) conditions. The phases were identified by XRD.

The onset of formation of diopside and/or gehlenite in oxidizing fired calcareous illitic clays at ~800°C leads to a profound change in the fabric of the ceramic body as

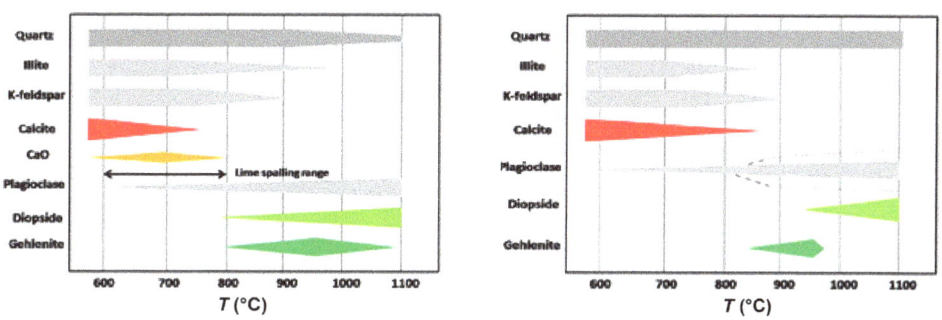

Figure 5. Experimentally determined phase stability in calcareous illitic clay fired under oxidizing (left) (Maggetti, 1982) and reducing (right) (Letsch and Noll, 1983) conditions. Under oxidizing conditions, illite is found to be stable to higher temperatures while decarbonization of calcite is impeded under reducing conditions. Onset of formation of diopside occurs at a lower temperature in the presence of high oxygen partial pressure.

partial melts are formed, the amount of which increases with increasing firing temperature. In contrast to this, calcareous ceramics that do not contain diopside are thought to be fired much below 800°C, typical of the overwhelming majority of Neolithic pottery. In this case, the thermal decomposition of calcite, which starts at 600°C and finishes at ~850°C, is not completed yet (Rodriguez-Navarro *et al.*, 2009) or these wares were fired for only very short times that were insufficient to decompose larger calcite grains into reactive CaO that would give rise to diopside and gehlenite formation (Maggetti *et al.*, 2011). Nevertheless, the occurrence of diopside can, with caution, be considered an important indicator of the (maximum) firing temperature attained, and thus is at least a rudimentary marker of the technological skill of the potter. The same clay fired under reducing conditions shows a delayed onset of both diopside and gehlenite formation owing to the fact that the stability of calcite is extended by at least 100°C under reducing conditions, *i.e.* in the presence of CO_2 in the kiln atmosphere formed by the water–gas equilibrium $CO + H_2O \leftrightarrow CO_2 + H_2$ (Escardino *et al.*, 2006). In contrast, phase evolution studies on Mg-rich calcareous illitic archaeological (?) clays from Troy, Turkey, fired under reducing conditions revealed that the onset of formation of the Mg-rich phase åkermanite ($Ca_2MgSi_2O_7$) instead of the Al-rich phase gehlenite ($Ca_2Al[AlSiO_7]$) occurs at ~750°C, together with anorthite and sanidine, whereas diopside starts to form only at temperatures ~850°C, *i.e.* ~100°C lower as indicated in Fig. 5 (right) (Görres *et al.*, 2000). These results were inferred from high-resolution Rietveld refinement corroborated by transmission electron microscopy (TEM), a technique capable of detecting those minute mineral neoformations that elude detection by XRD. These results show the importance of considering the chemical composition of the clays, the firing atmosphere as well as the technique employed, if determination of ancient firing temperatures is the goal. Unfortunately, studies employing such advanced analytical techniques to solve technological riddles of the ceramic past are rare.

Gehlenite formed during ceramic firing at 800–850°C (Fig. 5, left) is metastable and reacts at higher temperatures (>1050°C) with silica released during decomposition of metakaolinite or illite to anorthite + wollastonite (or in the presence of MgO to diopside, $CaMgSi_2O_6$) according to

$$Ca_2Al[AlSiO_7] + 2\ SiO_2 \rightarrow CaAl_2Si_2O_8 + CaSiO_3 \qquad (4)$$

The so-called 'gehlenite problem' is another example of the intricate interplay between thermodynamic principles and kinetic reality. The formation of gehlenite outside its field of thermodynamic stability, and its occurrence in and disappearance from calcareous illitic clays has long been considered an analytical challenge as a thermodynamic problem (Heimann and Franklin, 1979). However, the solution to this problem has been found in the realm of reaction kinetics. Roman Terra Sigillata, produced in Switzerland, does not show gehlenite, whereas Terra Helvetica, a specific Swiss subtype of the former, contains significant amounts of this mineral (Maggetti and Küpfer, 1978; Maggetti and Heimann, 1979). Because both wares were produced from similar clays and fired in the stability range of gehlenite (Fig. 5, left), it has to be assumed that well processed clays with a narrow and small grain-size distribution and, in particular, absence of larger calcite grains, were utilized to make Terra Sigillata, whereas Terra Helvetica was made from less-processed clays with coarser calcite grains. Evidently, the reaction rate of gehlenite formation is a function of the primary calcite grain sizes which, in turn, are influenced by the degree of processing of the clay; a clear measure of the technical skill of the potter.

The mineralogical changes of a non-calcareous iron-rich illitic clay under oxidizing (top, dark bars) and reducing (bottom, light bars) conditions are shown in Fig. 6a. The main differences lie in the extended stability range of illite to somewhat higher temperatures and the earlier onset of mullite formation under oxidizing conditions. Because in contrast to calcareous clays, reaction products, *i.e.* mullite, will be formed only at much higher temperatures beyond 1000°C, the fabric of non-calcareous clays fired below that temperature threshold corresponds, in principle, to that of low-fired calcareous clays. Hence, for firing temperatures below 1000°C, attempts to relate the fabric of a non-calcareous ceramic body to its firing temperature are doomed in practice (Heimann and Franklin, 1979).

Varying redox conditions during the ceramic firing process account for the formation of coloured iron oxide pigments. This property provided the ancient potters with a versatile tool to adorn their objects with contrasting black and red surfaces, as shown impressively by the black-figured technique of the Corinthian potters invented around 700 BCE as well as the red-figured technique of the Attic potters discovered around 530 BCE (Heimann and Maggetti, 2014). In a pottery kiln, the prevailing redox conditions can be adjusted and manipulated *via* the Boudouard equilibrium $CO_2 + C \leftrightarrow 2CO$ that fixes the oxygen

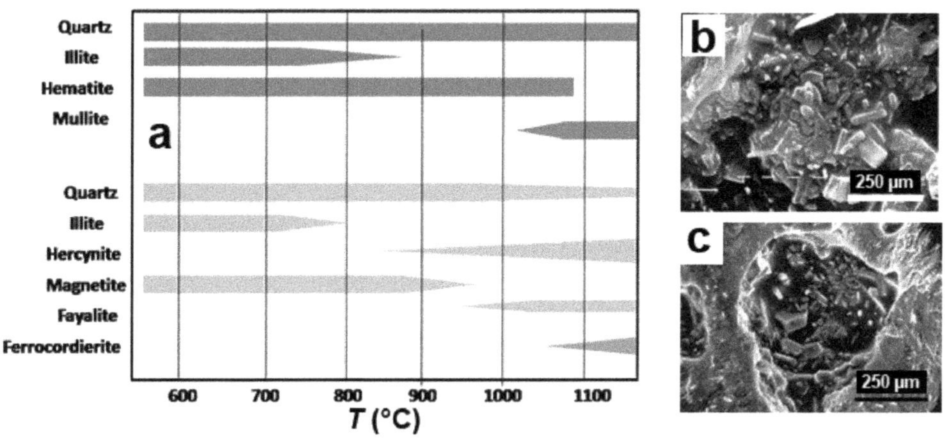

Figure 6. (a) Experimentally (XRD) determined phase stability in a non-calcareous iron-rich illitic clay under oxidizing (top) and reducing (bottom) firing conditions. Under oxidizing conditions, illite is found to be stable to somewhat higher temperatures. Onset of formation of mullite occurs at lower temperature in the presence of high oxygen partial pressure (after Heimann *et al.*, 1980, Letsch and Noll, 1983). Neo-formation of anorthite-rich plagioclase (b) and ferrocordierite (c) are observed; these occur as a result of experimental firing of a non-calcareous iron-rich illitic clay under strongly reducing conditions at 1035°C (Heimann *et al.*, 1980). With permission by cfi Ber. DKG.

fugacity in such a way that ferric ions (Fe^{3+}) in red hematite are reduced to a spinel-type phase (magnetite, black) containing ferrous ions (Fe^{2+}). However, the system follows the Boudouard equilibrium only if airtight conditions prevail and gas-free fuels such as charcoal or coke are used, conditions that were never fully realized in ancient pottery kilns (Letsch and Noll, 1983). If the clay body releases oxygen and/or carbon dioxide during firing, the Boudouard equilibrium will be disturbed briefly but restored quickly by consuming additional carbon from the fuel. Any water released reacts with carbon monoxide until the water–gas equilibrium $CO + H_2O \leftrightarrow CO_2 + H_2$ is achieved; that will be catalysed in the presence of magnetite.

Because ancient pottery kilns required frequent refuelling, opening of the stoking channel and the presence of cracks in the kiln walls allowing uncontrolled ingress of air, and other changes in the air supply further disturbed the Boudouard equilibrium and created transitional states between fully and partially oxidizing firing. As a consequence, many 'reduction-fired' ceramics not only contain spinel phases such as magnetite or hercynite but also elemental carbon as black pigment phases. A telltale sign of such uncontrolled air ingress is the notorious misfired 'red-stained' ware.

During oxidizing firing, hematite is stable throughout the entire temperature range up to 1490°C, where it reaches its dissociation pressure of 1 bar. In a reducing atmosphere, hematite is quantitatively converted to magnetite already below ~600°C. In fired (illitic) clays, magnetite is stable until 850–900°C, but converts through reaction with alumina released by decomposing illite to hercynite ($FeAl_2O_4$). Further temperature increase shifts the equilibrium towards the formation of Fe_xO (wüstite) that converts to metallic iron at 730°C. Furthermore, the formation of fayalite (Fe_2SiO_4) under ideal thermodynamically controlled conditions ought to commence at 350°C. However, at this low temperature fayalite formation is kinetically constrained by the lack of reactive silica that will be available only at much higher temperatures through decomposition of the clay minerals. Consequently, fayalite may occur in measurable amounts only in ceramics fired under reducing conditions to at least 950°C as shown in Fig. 6a.

A useful tool to assess the degree of reduction obtained in ancient pottery kilns has been found in determining the magnetic (mass) susceptibility, χ, of fired clay. This physical property is a dimensionless proportionality constant that indicates the degree of magnetization of a material in response to an applied magnetic field. Figure 7 shows the real component of the magnetic susceptibility χ_g as a function of the firing temperature and the oxygen fugacity of the firing atmosphere of an experimentally fired calcareous illitic clay with a composition shown in Table 1. The oxygen fugacities established in an electrically heated furnace were controlled by solid-state buffer mixtures (Mn_2O_3/Mn_3O_4: weakly reducing with $f_{O_2} > 10^{-4}$ atm; Ni/NiO and Fe_2O_3/Fe_3O_4: strongly reducing with $f_{O_2} < 10^{-4}$ atm). Clay samples heated in air were used as a control (f_{O_2} = 0.2 atm). At a standard firing temperature of 1000°C, the oxygen fugacities achievable with the solid-state buffers are as follows: $\log f_{O_2} = -0.56$ atm for Mn_2O_3/Mn_3O_4; $\log f_{O_2} = -5.39$ for Fe_2O_3/Fe_3O_4; and $\log f_{O_2} = -10.22$ for Ni/NiO (Huebner, 1971).

The average magnetic mass susceptibilities of iron minerals to be expected during ceramic firing increase from hematite, $(2.5–3.7)\cdot 10^{-7}$ m^3/kg to fayalite,

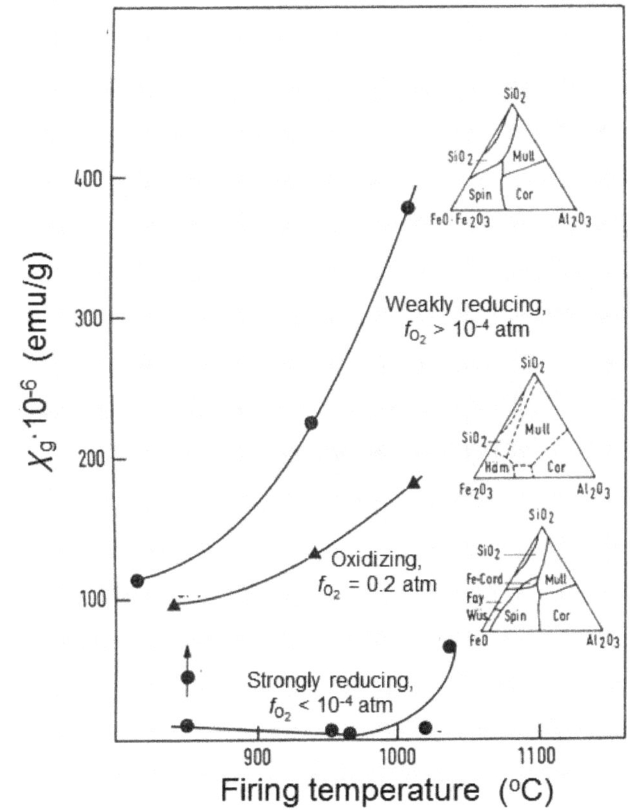

Figure. 7. Real component of the magnetic susceptibility of calcareous illitic clay fired under reducing and oxidizing conditions (Heimann *et al.*, 1980). The ternary diagrams show the phases stable at the oxygen fugacities selected for the firing experiments. Cor: cordierite, Fay: fayalite, Fe-Cord: ferrocordierite, Häm: hematite, Mull: mullite, Spin: spinel (magnetite–hercynite solid solutions), Wüs: wüstite. With permission by cfi Ber. DKG.

$(0.5-13)\cdot10^{-7}$ m^3/kg to wüstite, $(10-100)\cdot10^{-7}$ m^3/kg to magnetite, $(2.5-8)\cdot10^{-4}$ m^3/kg (Powell *et al.*, 2002). Consequently, as is evident from Fig. 7, the magnetic susceptibilities of the samples fired under weakly reducing conditions are much greater than those of the

Table 1. Chemical composition of an archaeological clay from Otterbach, Jockgrim, Palatinate, Germany in mass% fired at 1000°C (Heimann *et al.*, 1980). With permission by cfi Ber. DKG.

SiO_2	61.7	MgO	2.7	P_2O_5	0.1
TiO_2	0.8	CaO	7.0	Ba (ppm)	514
Al_2O_3	19.3	Na_2O	0.8	Cr (ppm)	134
Fe_2O_3	5.5	K_2O	3.5	Rb (ppm)	197

samples fired under oxidizing conditions, owing to the strongly ferrimagnetic nature of magnetite. The low magnetic susceptibilities of samples fired in strongly reducing atmosphere are, at first glance, rather surprising. However, an explanation for this phenomenon can be found in the fact that the reduction of iron ions continued until all were transformed to Fe_xO. These wüstites are, in contrast to the ferrimagnetic magnetite, antiferromagnetic with a tendency to suppress the susceptibility. Hence, it causes only a weak paramagnetic response because it does not possess a spontaneous macroscopic magnetic momentum. Moreover, as Fe_xO acts as a strong mineralizer it promotes the formation of partial melts at mineral grain-boundaries from which iron aluminium silicates such as the cyclosilicate ferrocordierite (sekaninaite, $(Fe^{2+},Mg)_2Al_4Si_5O_{18}$) or the iron silicate fayalite $(Fe^{2+})_2SiO_4$ can crystallize during cooling. Indeed, the increase in the magnetic susceptibility of the strongly reduced samples beyond 1000°C shown in Fig. 7 may be related to the formation of ferrocordierite (Heimann *et al.*, 1980). It can be further surmized that a large proportion of Fe^{2+} ions generated during strongly reducing firing will remain dispersed in the glassy phase and thus cannot establish any cooperative magnetic phenomena, again accounting for the comparatively low magnetic susceptibilities measured (Fig. 7).

In summary, the onset of the formation of a partial melt phase at much lower temperature in calcareous clays compared to non-calcareous clays was the technological reason why prehistoric potters searched for clay deposits rich in calcium (and alkali) ions. Such clays could be processed easily in primitive wood-fired kilns with a bare minimum of sophistication to yield dense, mechanically durable cooking pots, bowls, strainers, plates and cups of all sorts that could well fulfil their function in the ancient kitchens and dining rooms. On the downside, highly calcareous clays have a narrow sintering interval that requires considerable temperature control (see Fig. 18). Consequently, a pronounced risk of overfiring exists as evidenced by the high proportion of misfired wasters littering ancient pottery production sites. In addition, there was always the risk of 'lime blowing', that is, spalling of the pot by rehydration of CaO to portlandite, $Ca(OH)_2$, and subsequent carbonation of this phase to re-form calcite. Hence, such calcareous clays, when used for cooking pots must be fired well below the decomposition temperature of calcite that is known to start beyond 600°C in oxidizing atmosphere but beyond 800°C in a reducing atmosphere (Letsch and Noll, 1983). However, these low firing temperatures lead to mechanically inferior wares. To remedy this problem, cooking pots fashioned from highly calcareous clays were frequently reinforced with coarse quartz or rock grains as temper (Kilikoglou *et al.*, 1995, 1998).

Skilful manipulation of the redox equilibria of iron oxides during oxidizing and reducing firing enabled ancient potters to play with the entire colour palette from jet black to flaming red, and hence, to create aesthetically pleasing and highly treasured ceramic objects for domestic use and trade over considerable distances within the ancient world.

4. Case studies

In the following sections, we will discuss several examples of the application of ceramic phase diagrams to the technological interpretation of ancient and historical ceramics. We will proceed from low-fired pottery of old Mesopotamia, Egypt, Minoan Crete and Greece to medium high-fired earthenware such as Terra Sigillata to high-fired ceramics such as ancient Chinese and Thai stoneware, and European hard-paste and soft-paste porcelains. We have chosen this approach to show that the first group is subject to complexities introduced by kinetic constraints, whereas the second and third groups conform approximately to the thermodynamic principles developed above. Most of the information provided below has been adapted from recent work by the authors (Heimann and Maggetti, 2014). Basic references to the ceramic development and singular references to major overview publications for the period considered as well as much more information on the beauty and striking variability of ancient and historical ceramics, supported by a host of colour photographs, can be extracted from this book.

4.1. Low-fired earthenware

4.1.1. Neolithic Mesopotamian wares

Utilitarian pottery appeared around 6500 BCE in a region extending from Mesopotamia to Anatolia, the eastern arc of the Fertile Crescent. The oldest Mesopotamian wares were fired at temperatures as low as 600–700°C as estimated for bonfired Proto-Hassuna ware. Owing to the composition of clays won from the beds of the Euphrates and Tigris rivers, the clays utilized in Mesopotamian pottery were very rich in calcium carbonate (Fig. 8). The range of fluvio-lacustrine Quaternary clay composition comprises 20–25 mass% quartz, 30–40 mass% carbonate minerals (calcite, dolomite), 30–35 mass% clay minerals, 3–6 mass% iron oxides and hydroxides, and 2–5 mass% gypsum and halite (Mustafa, 2011). The average clay content (30–35 mass%) consists approximately of 25 mass% illite, 30 mass% illite-smectite mixed-layer mineral, 20 mass% palygorskite, 15 mass% kaolinite and 10% chlorite (Kadum and Hussain, 2011). The overall chemical compositions of typical clays from the Mesopotamian plain (Nahrawan, 65 km east of Baghdad; Seleucia, 30 km south of Baghdad) are shown in Fig. 8 (open circle and star symbols).

Because the sintering temperatures of the highly calcareous illitic-smectitic clays used were <1000°C, the resultant pottery was prone to overfiring with associate slumping owing to their small sintering interval. The composition of the calcareous river silt of Mesopotamia is in stark contrast to that of the rather lime-poor Nile silt of Egypt that dominated the composition of Egyptian ceramics independent of time and locality (Fig. 9). The low firing temperature applied to the Mesopotamian pottery rendered the ceramic body porous. This initial drawback was remedied by coating the outside of the pot with a clay slip as early as the 5th millennium BCE (Figs 8b, 10b). Soon after 2000 BCE, the Mesopotamian potters invented lead oxide glazing to render their pots impervious to water. The lead oxide glaze fused at relatively low temperature and lent itself easily to colouring with metal oxides such as copper and manganese

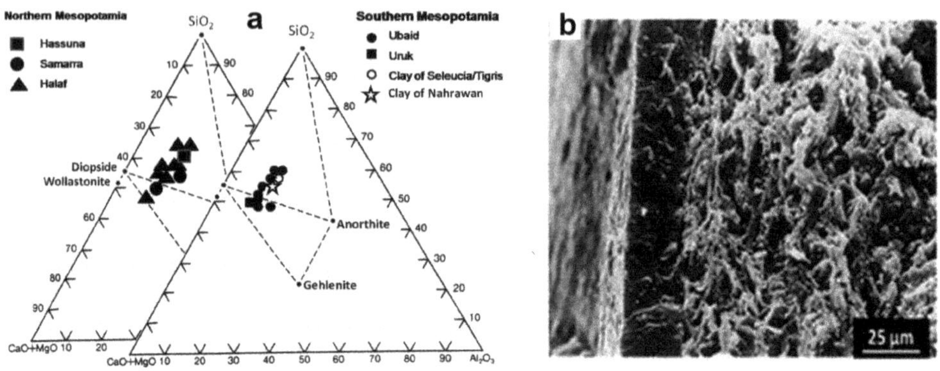

Figure 8. (a) Chemical composition of the Neolithic-Chalcolithic wares and clays from northern (Hassuna, Samarra, Halaf; left) and southern (Ubaid, Uruk; right) Mesopotamia shown in the triangle (CaO+MgO)−Al_2O_3−SiO_2. An open circle and a star (Noll, 1984, 1991; Jabbori, 1980) depict, respectively, the clays from Seleucia and Nahrawan. (b) Vitreous black engobe (left) on the porous ceramic body (right) of pottery from Tell Halaf, Mesopotamia, 5th millennium BCE (Noll and Heimann, 2016). With permission by Schweizerbart Science Publishers.

oxides, producing rich green and black colour palettes. This ingenious discovery by Mesopotamian potters led to widespread application of lead glaze that adhered particularly well to almost any fired clay surface and thus constituted an important technological advancement.

Characteristic of all locally and temporally different wares from northern (Hassuna, Samarra, Halaf) and southern (Ubaid, Uruk) Mesopotamia is their chemical

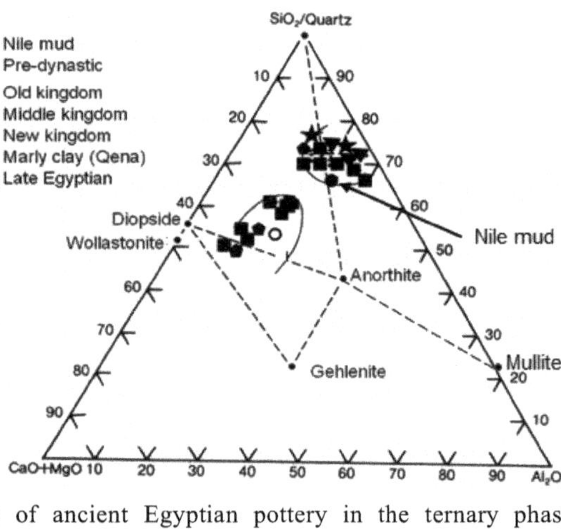

Figure 9. Composition of ancient Egyptian pottery in the ternary phase diagram (CaO+MgO)−Al_2O_3−SiO_2. Dashed lines show the three compositional triangles diopside−anorthite−quartz, diopside−anorthite−gehlenite, and quartz−anorthite−mullite (in Heimann and Maggetti, 2014). Data from Noll (1984). The composition of Nile mud was obtained from Hängst (1979).

homogeneity, owing to the homogeneity of the river-derived clay raw material. This chemical homogeneity of the ceramic body of the northern and southern Mesopotamian wares is shown in Fig. 8a. In the ternary $(CaO+MgO)-Al_2O_3-SiO_2$ phase diagram, the compositions of Tigris clays (open circle and star, right triangle) and the resulting ceramic wares are depicted. The clay compositions sit squarely in the compositional triangle diopside–anorthite–quartz whereas the composition of the ceramics deviates toward larger amounts of lime, presumably the result of inclusions of larger grains of calcite/dolomite in the ceramic mass owing to insufficient processing of clay that otherwise would have removed these inclusions (Noll, 1984). The content of alumina, however, is constant at ~10–12 mass% thus confirming the homogeneity of the clay source. The large CaO content of some of the ceramics shifts the phase composition into the triangle diopside–anorthite–gehlenite. Superposition of the two ternary diagrams confirms that northern and southern Mesopotamian wares are congruent, meaning that their raw materials can be traced back to the same riverine source.

Note that the ternary diagrams shown in Figs. 8 and 9 are oversimplified in that CaO and MgO were combined to only one component. Strictly speaking, two compositional triangles wollastonite–anorthite–gehlenite and diopside–anorthite–gehlenite would then exist, the first of which is not stable in the presence of larger amounts of MgO.

The history of painted ceramics produced by the iron reduction technique started with the northern Mesopotamian Hassuna and Samarra cultures in the 6th millennium BCE and reached a technologically remarkably high development stage during the Halaf period in the first half of the 5th millennium BCE (Fig. 8a, left triangle). While Proto-Hassuna ware was probably fired at a low 600–700°C for a short time as suggested by the extent of the inner black cores of the ceramic shards, Hassuna pottery was fired at higher temperatures in updraught pottery kilns dated to the second half of the 7th millennium BCE.

Still higher temperatures estimated at between 850 and 1050°C were attained by potters of the Samarran and Tell Halaf cultures of the first half of the 5th millennium BCE. Their kilns produced bowls and jars painted with geometric patterns in reddish-brown colours, featuring designs attesting to skilful manipulation of the valence state of iron oxides during either reducing or oxidizing firing (Fig. 10a). Halaf potters, in particular, decorated their products with a densely sintered, vitreous *engobe* (slip) that not only sealed the surface but also provided an aesthetically pleasing decorative effect (Figs 8b, 10b). This required the careful preparation of the clay suspension by efficient elutriation to obtain a homogeneous and smooth surface of the paint layer with an appealing gloss after burnishing in the leather-hard state. Despite its technological superiority, this glossy engobe technique slipped into oblivion until it was rediscovered millennia later by Middle Minoan potters who created the famous eggshell Kamares ware (Fig. 11b). From this early highlight the glossy engobe or slip technique was carried on through time: Mycenaean ware, Corinthian black-on-light and Attic red-on-black, and Roman Terra Sigillata wares all show these shiny surface layers, formed by partial vitrification during reducing (black layer) or oxidizing (red layer) firing of fine-grained, potassium- and iron-rich fusible illitic clays.

Figure 10. Iron oxide-painted earthenware pottery. (a) Samarra, Mesopotamia, 6500–6000 BCE, Reg. No. 1924,0416.8. AN 00310643.© The Trustees of the British Museum, London. (b) Ceramic cup (left) and bowl (right) from Tell Halaf with geometric cross-hatched decoration and black *engobe* around the carinated rim ($6^{th}/5^{th}$ millennium BCE). Reg.No.VA 12468 (height 6.1 cm, diameter 8.4 cm), VA 12402 (height 8.4 cm, diameter 15.8 cm). © Staatliche Museen zu Berlin, Photo: Olaf M. Teßmer.

4.1.2. Egyptian Neolithic pottery

In Egypt, relatively lime-poor Nile silt dominated the composition of ceramics independent of time and locality. The 5th millennium BCE Badarian ware is among the finest pottery ever produced in the prehistoric Near East. Their vessels excel with extreme thinness, and are well fired and highly polished. A lustrous surface gloss emphasizes the colours of black, brown and red, the latter being used frequently with a black upper body. This black colour was produced by 'smoking', *i.e.* oriented deposition of carbon flakes on the highly burnished upper rim of the so-called 'black-

Figure 11. (a) Slip-painted earthenware jar. Naqada II period, Predynastic Egypt *ca.* 3300 BCE. Height 29.5 cm, diameter 22.5 cm. Reg.No. 1901,1012.2. AN 00076453. © The Trustees of the British Museum, London. (b) Middle Minoan IIB (1800–1700 BCE) 'egg-shell' Kamares-style cup (restored with gypsum), Height 5.7 cm. Palaiokastro, Crete. Reg.No. 1907,0119.296. AN 00756559. © The Trustees of the British Museum, London.

topped ware' (C-black technique). To add contrast, a decorative palm-leaf pattern was polished onto the air-dried clay body prior to smoking that subsequently would stand out by virtue of its silver-grey gloss over the dull black, unpolished background. This gave the vessels the appearance of high-priced and hence, prestigious metal objects.

During the Predynastic Naqada period, earthenware pottery was produced from marly clays (Noll, 1991; Noll and Heimann, 2016), slip-painted with geometric patterns as well as animal, plant and boat decorations. The Naqada I period somewhat overlapped the Badari era and eventually succeeded it with red-grounded pottery made from Nile mud, sometimes cold-painted white with gypsum. During the Naqada II stage of Predynastic Egypt, additional colours and colour combinations appeared, often disguising the red-grounded ceramic body with a light-coloured slip (Fig. 11a). This slip served as a painting ground for geometric ornaments and images of human figures and animals executed in red-brown and black colours by using iron and manganese oxide pigments instead of smoking that became a less frequently applied decoration technique.

Production of fine pottery in the Badarian tradition continued during the 3rd to 6th dynasties of the Old Kingdom (2650–2150 BCE). Red-slipped bowls of 'Meidum' type were produced from marl or silty Nile clay with a thick red slip on the interior and exterior surfaces. Such 'Meidum' bowls were highly polished and elegantly executed, with a rather large diameter, rounded bottom, and flared rim (Fig. 12b).

From a chemical point of view, all of these ceramic bodies were remarkably homogeneous, with raw clay compositions closely associated with that of lime-poor Nile silt, the composition of which is nearly constant over long distances. As shown in Fig. 9, the composition of Nile silt (normalized for loss on ignition) is ~68 mass% SiO_2, 24 mass% $Al_2O_3 + Fe_2O_3$, and 8 mass% $CaO + MgO$ (Hängst, 1979). Hamdan *et al.* (2014) subdivided the composition of Nile silt in two types, a high-silica

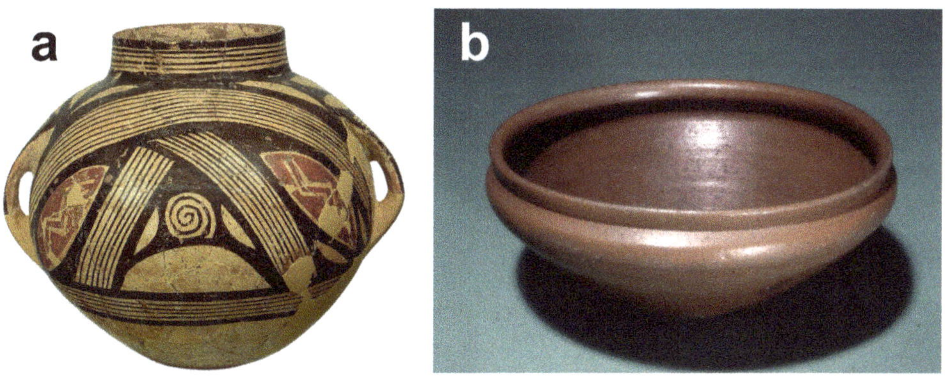

Figure 12. (a) Late Neolithic (5300–4800 BCE) spherical vase from Dimini, Thessaly. Reg. No. 5922 of the National Museum Athens, Greece. © Hellenic Ministry of Culture and Sports/Archaeological Receipts Fund. (b) Red-slipped pottery bowl of 'Meidum' type. El Badari, Upper Egypt. Old Kingdom (2650–2150 BCE). Diameter 21.6 cm. Reg. No. 1925,1012.11.AN 00424005. © The Trustees of the British Museum, London.

(~73 mass%), high-alumina (~18 mass%) and low-iron (~7 mass%) type of granitic origin (White Nile, central Africa) and a low-silica (~58 mass%), low-alumina (~11 mass%) and high-iron (~12 mass%) type of basaltic origin (Blue Nile, Ethiopia).

The composition of ceramics made from silica- and alumina-rich Nile mud straddles the tie line quartz–anorthite. Moreover, a substantial number of pottery analyses are located in the triangle quartz–anorthite–mullite, attesting to decreased CaO and increased alumina contents. However, this is not to mean that mullite had actually been formed as a firing product because the temperatures to which the clay was subjected were too low to overcome the activation energy barrier required to nucleate mullite. Indeed, the onset of mullite formation is known to require temperatures well above 1000°C, temperatures that could be achieved only much later in improved natural or forced draught kilns.

The fact that the Nile silt composition is lowest in SiO_2 compared to the analyses of the Egyptian wares is almost certainly related to the fact that fine quartz sand was used as an intentionally added temper during production of the pottery.

The New Kingdom wares produced from very lime-rich marly clays of Qena (Dendara) and El-Ballas form a clearly separated group in Fig. 9 that extends from the low-calcareous compositional triangle diopside–quartz–anorthite to the calcareous triangle diopside–anorthite–gehlenite.

4.1.3. Aegean Neolithic pottery

Ceramics of the Neolithic sites of Thessalian Sesklo (Wijnen, 1981) and Dimini in northern Greece are two early highlights of the European ceramic development. These ceramics are considered a crucial fulcrum of the expansion of the Neolithic ceramic tradition from Anatolia to the remainder of Europe.

Close to the ancient archaeological sites Sesklo and Dimini, three types of clay occur that all were used at one time or another to produce pottery in Neolithic times (Schneider *et al.*, 1990; Table 2). These raw materials comprise: (1) white illitic clays with low iron content and very poor in calcium and chromium, formed by weathering of mica schists and gneisses; (2) red tertiary clays with high iron, low calcium and relatively high chromium and nickel contents; as well as (3) alluvial clays high in chromium but with rather variable calcium contents. The composition of clay type 1 matches closely that of non-calcareous locally produced red-and-white painted pottery.

Table 2. Compositions of white illitic (1) and red tertiary (2) clays found near the Sesklo site (Schneider *et al.*, 1990). Oxides are given in mass%, trace elements in ppm. Distinguishing elemental contents are shown in italics.

Type	SiO_2	TiO_2	Al_2O_3	Fe_2O_3	MnO	MgO	CaO	Na_2O	K_2O	Cr	Ni	Rb	Y
1	72.6	0.44	17.9	*2.20*	0.03	1.09	*0.54*	1.86	3.31	*51*	11	179	*13*
2	66.0	0.80	16.6	*6.54*	0.12	2.18	*1.95*	3.37	2.32	*164*	*76*	96	*51*

Various pottery styles could be distinguished from each other by their different calcium, chromium and yttrium contents (Heimann and Maggetti, 2014). Chromium contents are probably related to the presence of the mineral serpentine which is typical of ophiolitic rocks that are common throughout the mountains surrounding the Thessalian plains to the east, south and west. Only the northern parts of the plains, covered by Pleistocene terraces, yield clays that are low in chromium but have differing yttrium contents. Hence, coarse ceramic ware of the Tsangli-Larissa phase (Dimini I) found in the north could be subdivided by their yttrium content. The differences in the yttrium content of the local clays and the pottery excavated are thought to be related to geochemical differences in the metamorphic rocks located nearby. Likewise, clays and pottery from the northern site of Soufli are low in yttrium but rich in chromium (>250 ppm), a fact that attests to the alluvial nature of the clay type 3, being derived from the mountains to the east and west.

The Late Neolithic painted Classic Dimini IV wares, derived from type 3 alluvial clays form a narrow compositional group rich in calcium (>6%) and chromium (>250 ppm), indicating a high degree of standardization. Indeed, this ware is one of the earliest examples of considerable pottery specialization in the European Neolithic. In contrast to this, the Incised Classic Dimini ware spreads out widely in a Ca–Cr plot, indicating that pottery of this particular style was produced at many sites outside Dimini, and was exchanged and distributed widely. As suggested by Schneider *et al.* (1990), the ancient potters might have realised that a light-coloured ceramic body was more important as a background for a visually contrasting, red-painted decoration than for incised patterns. Hence, there was no need to standardize the calcium content of the clay destined to produce incised pottery.

In conclusion, from chemical analyses of the pottery and relating their results by multivariate clustering, a much more complex picture emerges than previously assumed by classic archaeologists. According to this new view, Neolithic pottery production in Thessaly was already surprisingly standardized, realised by selection of specific raw materials, an improved kiln technology that may have achieved temperatures as high as 1000°C for the most advanced pottery, as well as abundance and diversity of decorating pigments and their application techniques. Hence, the transition from rather crude, low-fired Middle Neolithic Sesklo III pottery to Late Neolithic Tsangli-Larissa (Dimini I) fine decorated ware was accompanied by impressive and intensive technological and stylistic innovation (Schneider *et al.*, 1991). Subsequently, during the Late Neolithic to Early Helladic periods Thessalian potters used calcareous clays to brighten the appearance of their pots. Calcareous clays provide both artistic and technological benefits. Aesthetically, on these light surfaces, several black and red pigments would create strikingly beautiful contrasting colourations (Fig. 12a). Technologically, the combination of calcareous ceramic bodies and illitic slips was very fortuitous as the engobe surface layers adhered well to the body owing to the small gradient in the coefficients of thermal expansion. In addition, lower firing temperatures were required to obtain dense ceramics owing to the fluxing properties of CaO.

4.1.4. Bronze Age Cretan (Minoan) pottery

Whereas Neolithic Aegean pottery was produced using free-forming techniques such as coiling, slab building and paddling, only in ~2000 BCE did Cretan potters regularly use a fast-spinning potter's wheel, at this time a radical advance in ceramic technology that enabled the artistic expression of clays in a much more sophisticated manner (Williams, 1997). The use of the potter's wheel is thought to have spread from Crete to mainland Greece around 1800 BCE, and from there to other parts of Europe.

Similar advances also pertain to the selection of raw materials. The examples of Early to Late Minoan pottery (Fig. 13) show clearly a development trend reflecting the desire of the ancient potters to boost the quality of their products by beneficiation of clay raw materials even though the still rather primitive kiln technology (Heimann and Maggetti, 2014) frequently counteracted their endeavours.

Chemical analyses reveal that the composition of Cretan pottery changed with time in response to technological developments and possibly aesthetic requirements. Figure 13a depicts the phase triangle $CaO-Al_2O_3-SiO_2$ with the binary compounds wollastonite ($CaSiO_3$, Wo) and mullite ($3Al_2O_3{\cdot}2SiO_2$, Mu), as well as the ternary compounds anorthite ($CaAl_2Si_2O_8$, An) and gehlenite ($Ca_2Al_2SiO_7$, Ge).

The open circles refer to Neolithic pottery made from low-calcareous clays found at Phaistos in southern Crete, the solid circles to pottery from very calcareous clays found at Knossos in northern Crete. The Phaistos samples are situated in a compositional triangle formed by the conodes quartz–anorthite, anorthite–mullite and mullite–quartz. Thus, the equilibrium composition should consist of the mineral assembly quartz + mullite + anorthite (glass) similar to that of stoneware or porcelain. However, low-firing temperatures, presumably close to or below 600°C, prevented the attainment of this equilibrium composition. In addition, because small amounts of residual primary dolomite occur in the shards, and dolomite decomposes at 550°C in oxidizing ceramic firing, its preservation in the ceramic body also points to very low firing temperatures. The Neolithic Knossos ceramic bodies are highly calcareous, and occupy

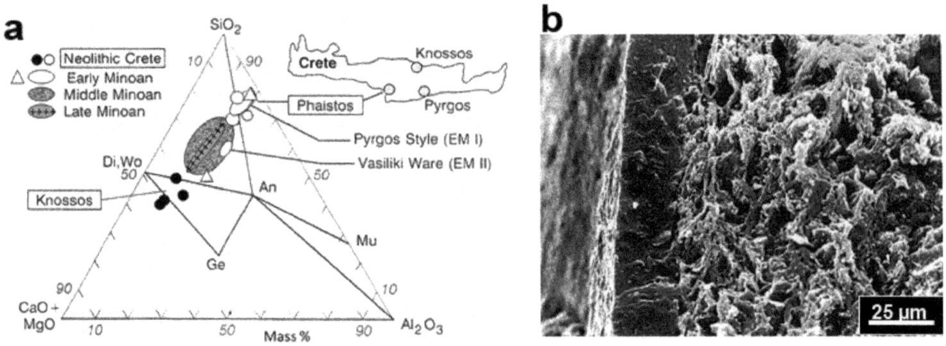

Figure 13. (a) Composition of Neolithic Crete and Bronze Age Minoan ceramics in the ternary phase diagram $CaO-Al_2O_3-SiO_2$ (Heimann, 1989; data after Noll, 1982). (b) Scanning electron microscopy image of the black, highly vitrified engobe (left) on a porous ceramic body of Middle Minoan Kamares ware (Noll, 1982; in Heimann and Maggetti 2014). With permission by Schweizerbart Science Publishers.

the compositional triangle anorthite–gehlenite–wollastonite. Low firing temperatures, however, prevented the formation of equilibrium high-temperature phases such as gehlenite and diopside. Large amounts of pristine calcite grains are still present in the shards. Here, the solidification has been attained solely by shrinkage due to thermally induced dehydration of the clay matrix and the concomitant increase of packing density of the clay particles, with only marginal mediating action of true sintering processes. In this context it should be noted that the apparent absence of gehlenite might be related to either thermal reactions beyond 950°C, forming anorthite during unintentional kiln temperature excursions or chemical decomposition forming a sequence of zeolites or calcium carbonate during burial in soils saturated with carbon dioxide and/or humic acids (Heimann and Maggetti, 1981). Figure 14 shows an example of scalenohedral calcite crystallites formed within a pore space of a Late Minoan pottery shard and attributed to the decomposition during burial of initially formed gehlenite (Noll, 1982).

The characteristics of ceramics changed considerably during the Early Minoan (EM) period. Although Pyrgos-style ware of Late EM I (*ca.* 2600 BCE) still shows many features akin to those of the Neolithic Phaistos pottery the compositional field of which it partly shares (open ellipses), sintering of the clay matrix attests to an improvement of the technology through higher firing temperatures. The Vasiliki ware of EM II (*ca.* 2000 BCE) became more lime-rich with a comparatively narrow distribution of the alumina content. The presence of small amounts of the equilibrium phase diopside points to a firing temperature >850°C. Early Minoan potters may have realised that it was more advantageous to use calcareous clays rather than non-calcareous ones as the former resulted in a denser fabric at moderate firing temperatures. On the other hand, highly calcareous clays, similar to those from which the Neolithic Knossos pottery was made, yielded pottery prone to lime spalling because of the presence of residual free CaO from the thermal decomposition of calcite. Only if the firing temperatures exceeded 850 to 900°C and were maintained for sufficient time, could free CaO combine with alumina and silica released by the thermal decomposition of clay minerals to form high-temperature so-called 'index' minerals such as diopside, gehlenite and anorthite.

Middle Minoan (2160/1970–1600 BCE) potters improved both the quality of the raw clay materials and the firing technology. Their high-quality Kamares ware (mottled

Figure 14. Calcite crystallites within a pore of a Late Minoan ceramic shard presumably formed by decomposition of gehlenite during burial (Noll, 1982). With permission by Schweizerbart Science Publishers.

grey ellipse in Fig. 13a) is now squarely located in the compositional triangle quartz–anorthite–wollastonite, the 'ceramic triangle'. Kamares pottery generally contains less alumina than Early Minoan ware. Diopside is now abundantly present, calcite is usually absent, but gehlenite is still only rarely visible in X-ray diffractograms. Hence, all these features point to a maximum firing temperature in the range 900 to 950°C. Later Kamares ware may have been fired at still higher temperatures around 1050°C, because in the white parts of the painted decoration proto-enstatite ($MgSiO_3$; Smith, 1959) formed by thermal decomposition of talc ($Mg_s[(OH)_2Si_4O_{10}]$), applied to the leather-hard body prior to firing (Noll, 1982) (Fig. 11b). Hence, the occurrence of proto-enstatite constitutes an important temperature marker for this particular type of pottery. The onset of formation of enstatite from talc in reducingly fired Mg-rich calcareous illitic clays has been determined to be ~850°C (Görres *et al.*, 2000). Figure 13b shows in an oblique view the interface between densely sintered engobe and a porous ceramic body of Middle Minoan Kamares ware.

Finally, during the Late Minoan period (1600/1580–1070 BCE), the processing and firing technologies of the pottery were further improved. The variation ellipse (light grey with crosses in Fig. 13) has narrowed considerably pointing to a very effective compositional control of the alumina/silica ratio, presumably by standardizing the ratio of plastic clay to non-plastic inclusions. This even suggests that the Late Minoan potters had learned to blend clays from different sources with different lime contents and workability to achieve a consistent product. In addition, efficient elutriation of the raw clay materials presumably removed coarse constituents such as calcite grains. Ceramics lower in lime and richer in silica shown in the variation ellipse originate from Mochlos, Gournia and Pseira along the northern coast of Crete, whereas the samples richer in lime and lower in silica were found at Knossos, Haghia Triada (Belfiore *et al.*, 2007) and Phaistos. The estimated firing temperature was ~1050°C (Heimann and Franklin, 1979) as judged by the high proportion of closed near spherical pores, and the highly sintered and slightly vitrified ceramic matrix.

4.2. Medium high-fired earthenware

The prototypical Roman medium high-fired earthenware is known as Terra Sigillata which was produced in virtually all parts of the Roman Empire, from Italy to Spain to Gaul to Britain to Syria. Owing to its ubiquity, it has been dubbed the Roman *'leitfossil'* (type fossil; Bemmann, 1993). Indeed, Roman pottery has been exported from Italy since the 4th century BCE, first as so-called Campanian Ware, a black-surfaced pottery type, followed in the mid-1st century BCE by Terra Sigillata, a red slipped ware produced initially in Arretium (modern Arezzo) (Fig. 15a). Terra Sigillata pottery workshops flourished between the 1st century BCE and about the 4th century CE. This mass-produced pottery was highly standardized and their forms and types (Dragendorff, 1895, 1896; Déchelette, 1904; Knorr, 1919) were limited to a few practical forms for easy stacking and transport. The ware was produced from calcareous illitic clays that were collected carefully, processed thoroughly by elutriation and settling and fired between 950 and 1050°C in special kilns. In these kilns, the ware was prevented from reacting with hot-flame gases by

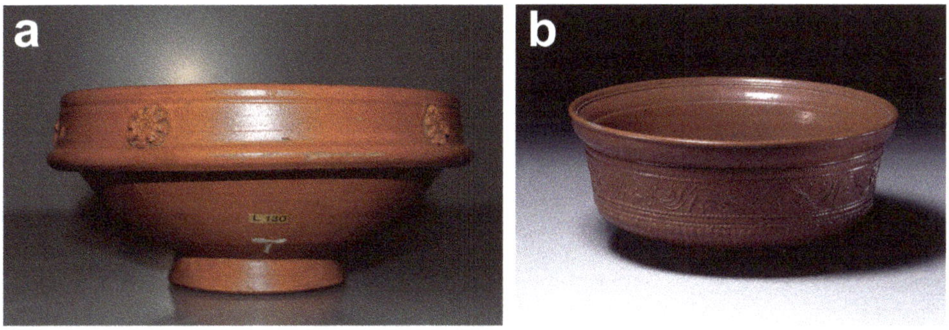

Figure 15. (a) Italian Terra Sigillata bowl (type Conspectus 34.1) from Arretium, Italy (30–100 CE). Workshop of C. Jul. Firmicus. Reg. No. 1814,0704.1558. AN 00894802. © The Trustees of the British Museum, London. (b)Terra Sigillata picture bowl (type Dragendorff 29) from La Graufesenque, Southern Gaul. (20-40 CE). Workshop of Bassus and Coelus. Height 10 cm, diameter 22.5 cm. Reg. No. 1869,0205.5. AN 00469677. © The Trustees of the British Museum, London.

channelling the latter through terracotta pipes away from the product. The finest clay fraction (<2 μm), obtained as the last fraction during settling with a large amount of illite, was used as a slip (engobe) into which the leather-hard vessels were dipped prior to firing. Roman potters also used slips with greater viscosity to trail on decorations by the so-called barbotine technique. Plain undecorated pottery vessels for domestic consumption were formed on a potter's wheel. In contrast to this, the production of the celebrated picture bowls catering to the refined taste of the affluent portion of Roman society involved pressing or throwing wet clay into an unglazed ceramic mould adorned with impression of stamps or '*sigilla*' from which the ware obtained its descriptive name (Fig. 15b).

During oxidizing firing, the potassium- and iron-rich illitic slip transformed into a partially vitrified, highly glossy and bright red, hematite-containing surface coating. Firing under reducing conditions and subsequent 'smoking' of burnished ware produced shiny black surfaces by incorporation of elemental carbon (soot). This type of pottery is known as 'Terra Nigra'. However, true black Terra Sigillata was produced from calcareous clay coated with a slip made from non-calcareous iron-rich illitic clay that after reducing firing turned black owing to the formation of magnetite below 950°C and an additional hercynite phase at firing temperatures exceeding 950°C (Maggetti and Galetti, 1986; Fig. 6a). A subsequent oxidizing firing step turned the porous ceramic body red again due to formation of hematite whereas the dense surface layer maintained its glossy black appearance. This was precisely the technology applied by Attic potters to make their celebrated black-on-red ware.

Figure 16 shows that the ceramic bodies of Terra Sigillata vary within narrow compositional bounds whereby the ware from Italian Arezzo is slightly more siliceous and less aluminous than that of the southern Gaulish workshops of La Graufesenque.

Much effort has been expended on the determination of the provenance, assignment of workshops, trade pattern and production techniques of Roman pottery. This endeavour has been aided by the use of ceramic reference groups (Schneider and

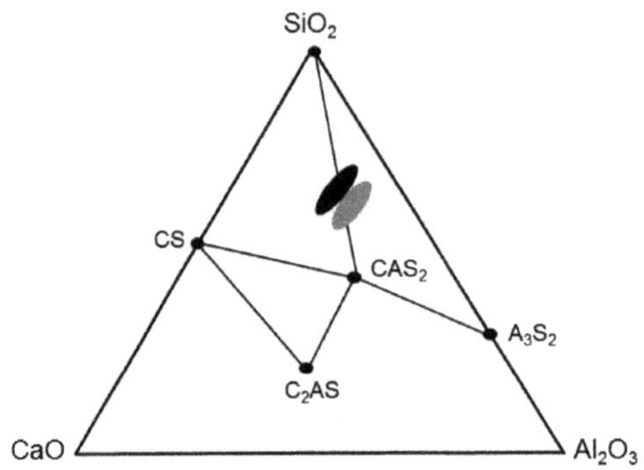

Figure 16. Compositional scatter of Terra Sigillata bodies from Arezzo, Etruria, Italy (black area, 256 analyses) and La Graufesenque, southern Gaul, France (grey area, 139 analyses). Symbols are as in Fig. 20.

Hoffmann, 1990) that are characterized by a unique combination of elements and their concentrations, thus providing a geochemical fingerprint to distinguish among workshops and provenance. For example, Fig. 17 shows the fit of Terra Sigillata, definitely produced by the Roman potter Lucius Gellius, with the reference groups 'Arretium' (Arezzo) and 'Lugdunum' (Lyon) in a bivariate TiO_2/MgO plot. Apparently, the potter worked in both places even though the spread of the compositional pattern suggest that pottery produced by Gellius (open circles) may have been altered by selective dissolution of alkali and alkaline earth elements during burial thus increasing the relative concentration of more environmentally stable elements such as titanium, aluminium and iron. On the one hand, samples with greater TiO_2, and generally also elevated phosphorus levels, may actually belong to the bulk of the Arezzo and Lyon populations. On the other hand, the high Mg concentration combined with lower TiO_2 levels cannot be explained easily by alteration during burial.

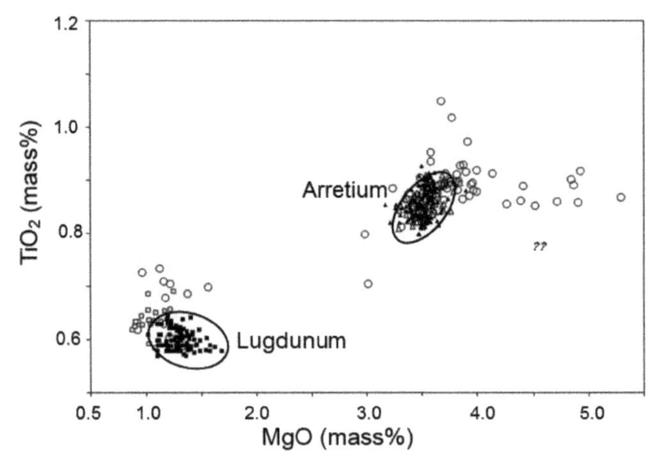

Figure 17. Bivariate plot of the titania and magnesia concentrations of Terra Sigillata produced by the potter Lucius Gellius (open circles) and of the ceramic reference groups for Terra Sigillata from Arretium (Arezzo) and Lugdunum (Lyon) (variation ellipses). From Schneider and Daszkiewicz (2006), where the exact meaning of "*??*" is also explained.

Instead, this feature may indicate the existence of a different workshop or production centre within the Lyon and Arezzo regions (Schneider and Daszkiewicz, 2006).

As mentioned above, almost all Roman Terra Sigillata has been manufactured from calcareous illitic clays. Indeed, such clays were used nearly exclusively for Italian Terra Sigillata from Arezzo and Padana, and provincial ware from southern (La Graufesenque, Banassac, Montans), central (Lezoux, Lyon) and eastern (Blickweiler, Chèmery-Faulquemont, Rheinzabern, Trier) Gaul (Picon *et al.*, 1971). The use of calcareous clays has several technological advantages (Tite *et al.*, 1982). First, a calcareous clay body has a significantly higher thermal expansion coefficient of $(4.5-7.0)\times 10^{-6}$ °C^{-1} than that of comparable non-calcareous clay bodies with $(2-3.5)\times 10^{-6}$ °C^{-1}. Hence, its thermal expansion coefficient will match more closely that of the partially vitrified slip that has a thermal expansion coefficient in the range of $(5-10)\times 10^{-6}$ °C^{-1}. Consequently, there will be a smaller risk of slip-crazing or even delamination during cooling. Second, calcareous clays produce bodies with greater rigidity and compressive strength owing to the ready formation of crystalline mineral phases (Peters and Iberg, 1978). The resulting pottery is, therefore, tougher and more resilient when used as domestic tableware and, in particular, for use by the Roman military. The inevitable breakage of earthenware encountered during transport over long distances on rocky roads required continuous replacement. Hence, each legion had an associated contingent of potters.

The sintering and, eventually, vitrification processes of the calcareous illitic clays used by Roman potters in their Terra Sigillata workshops can be described quantitatively by defining two stages during which the microstructural changes are brought about by different mechanisms. Stage 1 between ambient temperature and 900°C is governed by solid-state dehydroxylation of clay minerals, leading to a close-to-linear mass loss that causes shrinkage-related changes of the fabric imposed by the formation process of the ceramic body. In stage 2 beyond 900°C, the reactions associated with mass loss have ceased to occur, and microstructural changes occur by changes in the pore volume that is, sintering *sensu strictu* caused by sealing of open pores. The densification obtained during these two stages can be expressed by the volume quotients V_{900}/V_0 and V_T/V_{900} as

$$\frac{V_{900}}{V_0} = \left(1 - \frac{\Delta m}{m}\right)\frac{\rho_{900} - R_{900}}{\rho_0 - R_0}\cdot\frac{\rho_0 R_0}{\rho_{900} R_{900}} \tag{5}$$

$$\frac{V_T}{V_{900}} = \frac{\rho_T - R_T}{\rho_{900} - R_{900}}\cdot\frac{\rho_{900} R_{900}}{\rho_T R_T} \tag{6}$$

whereby m is mass of clay, Δm is the mass loss between ambient temperature and 900°C, V_0 pore volume of the dried ceramic body. Furthermore, V_{900} and V_T are the pore volumes at 900°C and $T > 900$°C, respectively; ρ_0 is the specific gravity of the dried ceramic body, ρ_{900} and ρ_T are the specific gravities at 900°C and $T > 900$°C, respectively; R_o is the volumetric density of the dried ceramic body, and R_{900} and R_T are the volumetric densities at 900°C and $T > 900$°C. Plotting V_T/V_{900} against firing temperature yields an S-shaped curve (Fig. 18) whereby $V_T/V_{900} = 0.8$ is defined as the temperature of the onset of densification and $V_T/V_{900} = 0.5$ to the mid-point sintering temperature (Heimann, 1978/79).

The steep negative slope of the sintering curve of a calcareous clay suggests an extremely narrow firing interval of only 15–25°C as suggested by the small difference between the onset of densification ($V_T/V_{900} = 0.8$ at 1015°C) and the mid-point of sintering ($V_T/V_{900} = 0.5$ at 1035°C). Hence, such clays require very strict control of the firing temperature in order to avoid undesirable slumping and warping of the ware caused by rapid formation of grain boundary melts with subsequent destabilization of the vessel. The rise of the sintering curve beyond 1150°C can be explained by bloating, that is, swelling caused by the expansion and subsequent expulsion of air trapped in closed pores (Maniatis and Tite, 1975).

After a period of considerable economic success, the factories at Italian Arretium began to decline both in their technological and aesthetic aspects in the 1st century CE and hence, gave way to the ascent of factories in the Roman province of Gaul. However, the economic importance of the latter also decreased at the end of the 2nd century CE and their place, in turn, was taken by pottery factories in Roman North Africa, principally in the areas of modern Tunisia and Libya (Leptis Magna).

To demonstrate essential technological aspects of Gaulish Terra Sigillata wares, a comparison is shown (Fig. 19) of the densification of the ceramic body of eastern Gaulish Terra Sigillata from Tabernae (today's Rheinzabern) with those of other provincial Roman potteries from southern (La Graufesenque, G) and eastern Gaul (Chémery-Faulquemont, C and Blickweiler, B). The lower $SiO_2/(Al_2O_3+K_2O)$ (quartz/illite+chlorite) ratios, in particular larger amounts of fluxing K_2O, of both La Graufesenque and Blickweiler Terra Sigillata, lead to greater densification compared to wares and moulds from Rheinzabern as well as Chémery-Faulquemont. The fact that densification of Blickweiler ware exceeds that of ware from La Graufesenque despite

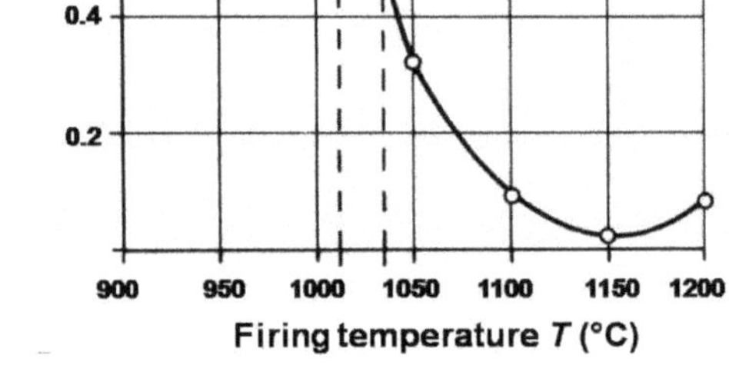

Figure 18. Typical sintering curve of calcareous illitic clay fired in air (Heimann, 1978/79). V_{900} and V_T are the pore volumes at 900°C and at $T > 900$°C, respectively.

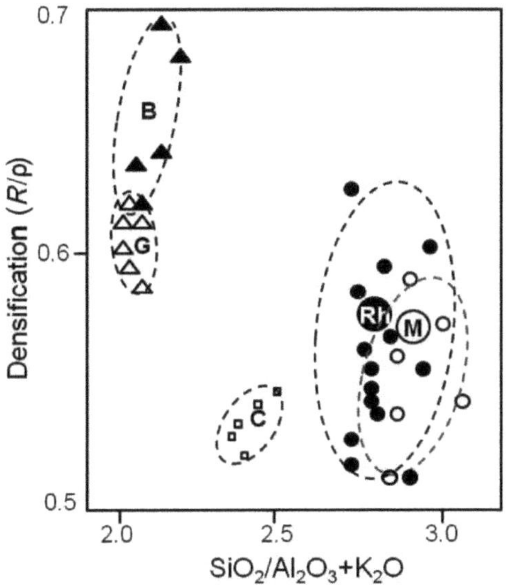

Figure 19. Density coefficient (densification) R/ρ of eastern (Rheinzabern Rh, M; Chémery-Faulquemont, C; Blickweiler, B) and southern (La Graufesenque, G) Gaulish Terra Sigillata as a function of the quartz/clay ratio, expressed as the ratio $SiO_2/(Al_2O_3+K_2O)$. R refers to raw density, *i.e.* density with pores; ρ refers to specific gravity, *i.e.* density without pores. The large circles Rh and M denote the averages of Rheinzabern ware and moulds, respectively (Heimann, 1982).

almost identical quartz/clay ratios suggests somewhat higher firing temperatures of the former, which were due to improved kiln technology.

4.3. High-fired stoneware and porcelain

Firing of non-calcareous illitic and kaolinitic clays at temperatures >1200°C has been achieved over time in ever more sophisticated ceramic kilns and has resulted in stoneware pottery with increased mechanical and chemical stability, much reduced porosity, and smooth shiny surfaces owing to the presence of wood ash-derived and, later, salt and feldspathic glazes.

Stoneware can be considered the historical precursor of true hard-paste porcelain from which it is distinguished by its generally coloured nature, its lack of translucency and some remaining porosity. It had already been produced in copious amounts in China during the late Shang dynasty (~1700–1028 BCE) and reached its acme during the Han (202 BCE–220 CE) and Yuan (1279–1368 CE) dynasties. Many of the early stoneware products were coated with a greenish celadon glaze thought to imitate highly priced jade. Important stoneware traditions existed in Japan, Korea and the Indochinese realm including Thailand, Laos and Vietnam. At the beginning of the 14th century CE, a special development occurred in the Rhenish part of Germany (Frechen, Cologne and Siegburg) based on high-refractory clays that reached its zenith between about 1440 and 1620 CE. The products were blue-greyish plates, bowls and drinking vessels, salt-glazed to improve their appeal and to provide added smoothness to aid easy cleaning.

4.3.1. Chinese stoneware (proto-porcelain)

The clay raw materials landscape of China is characterized by two main types of clay that are separated by a distinct geographical dividing line, the Nanshan–Qinling (N–Q) divide or kaolinite-porcelain stone (*petuntse*) line, also known as clay-rich/quartz-rich division line. The clays used for high-temperature ceramics located north of this line are rich in true clay minerals derived from sedimentary strata. These highly refractory clays contain ~40 mass% alumina, mostly fixed in kaolinitic clay minerals. By contrast, micaceous clays found south of the divide have formed mostly from weathered feldspar of igneous, predominantly granitic, rocks and contain comparatively small amounts of clay minerals but large amounts of fine quartz and secondary potassium mica (sericite) instead. Consequently, their alumina content is <20 mass%. White-firing, *i.e.* iron-poor versions of these rocks are referred to as 'porcelain or china stone' (*petuntse*) and constitute the mainstay of 'true' Chinese porcelain produced at the important kiln site of Jingdezhen, Jiangxi province from the 10th century CE onwards and reaching its perfection during the Song, Ming and Qing dynasties.

Typical proto-porcelain, *i.e.* high-fired stoneware (Yin *et al.*, 2011) shows a composition of the ceramic body of 75–80 mass% SiO_2, 15–18 mass% Al_2O_3, ~2 mass% K_2O, ~2 mass% Fe_2O_3, and 1–1.5 mass% each of CaO, Na_2O, MgO and TiO_2. This composition is essentially that of true porcelain, and is consistent with the mineralogy of the clays occurring south of the N–Q divide which are dominated by micaceous porcelain stone (*petuntse*). The shiny greenish glaze of the proto-porcelain greenware is composed of 60–70 mass% SiO_2, ~15 mass% Al_2O_3, 10–20 mass% CaO, <3 mass% MgO, 2–3 mass% K_2O and 2–4 mass% Fe_2O_3. The glazing process was based primarily on eutectic-melt formation during firing, during which lime-rich material reacted with the surface of the ceramic body to form a vitreous glaze with a composition determined by the firing temperature, and possibly the duration of firing (Yin *et al.*, 2011). The source of the lime-rich glazing material was washed wood ash, high in CaO and low in K_2O. The difference in composition between the ceramic body (B) and the applied glaze (G) is shown in Fig. 20.

The exact compositions of the glazes were automatically tuned by a temperature-controlled mechanism through selective absorption of ceramic material into the melting glaze. The lime-rich compound interacted with the material of the ceramic body to form a calcium aluminium silicate glaze at an assumed firing temperature of 1200–1250°C. Figure 21a shows a glazed stoneware dish produced during the Yuan dynasty at the Imperial Jiaotanxia Guan kiln at Hangzhou, Zhejiang province.

4.3.2. *Chinese porcelain*

Arguably, true high-fired porcelain ware with a white translucent body originated during the Sui (581–617 CE) and Tang (618–906 CE) periods even though some scholars maintain that it was already produced in Zhejiang province during the Eastern Han (25–220 CE) period. During the Song (960–1279 CE) and Mongol Yuan (1279–1368 CE) dynasties, Chinese ceramic workmanship reached one of its most brilliant phases with the famous monochrome Longquan celadon porcelaneous stoneware. However, it was not until the late 14th century CE that the techniques of painting porcelain in polychrome underglaze and overglaze colours were developed and brought to technical perfection during the Ming (1368–1644 CE) and Qing (1644–1912 CE) dynasties.

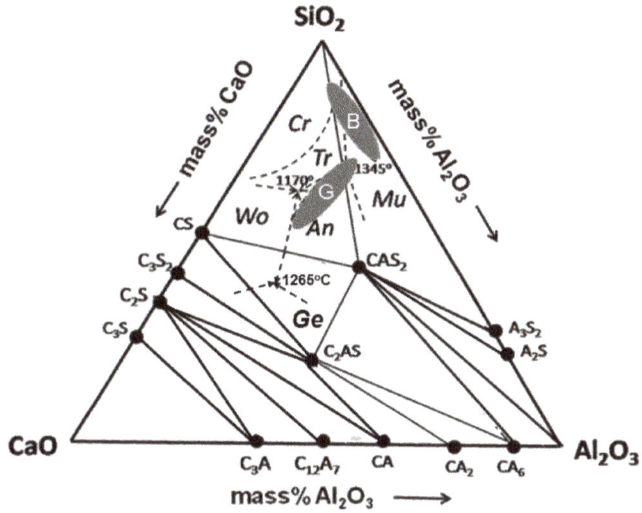

Figure 20. Compositional scatter ellipses of the body (B) and the glaze (G) of Chinese proto-porcelain from the Deqing site of Zhejiang province (Yin *et al.*, 2011), located in the 'ceramic triangle' $CaO-Al_2O_3-SiO_2$. Three important triple points are indicated at the intersection of the phase boundaries (dashed lines). An – anorthite, CAS_2; Cr – cristobalite, SiO_2; Ge – gehlenite, C_2AS; Mu – mullite, A_3S_2 to A_2S; Tr – tridymite, SiO_2; Wo – wollastonite, CS. CS is the composition of wollastonite in cement chemical notation, meaning $CaOSiO_2$. The binary and ternary phases are denoted according to the cement chemical notation, C = CaO, A = Al_2O_3, S = SiO_2.

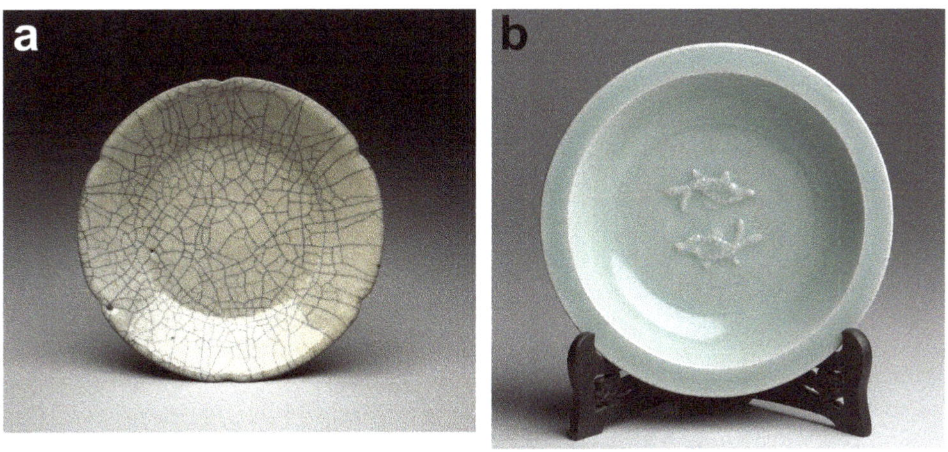

Figure 21. (a) Guan stoneware dish with craquelling celadon glaze. Yuan dynasty (1279–1368). Hangzhou, Zhejiang province. Height 3 cm, diameter 16.4 cm. Reg. No. PDF.69. AN 1051096. © The Trustees of the British Museum. London. (b) Longquan celadon-glazed porcelaneous dish with twin fish decoration. Southern Song dynasty (12th to 13th centuries CE). Diameter 22 cm. Reg. No. 43.1. © Minneapolis Institute of Arts, The John R. Van Derlip Fund.

By its nature, ancient Chinese porcelain was soft-paste porcelain, produced from kaolinite (china clay), *petuntse* or china stone ('pottery stone'), the historic term for a wide range of micaceous or feldspathic rocks, and fine quartz sand. It was fortuitous that Nature provided a ready-to-use mix of these minerals at Jingdezhen, Jiangxi Province in southern China. This enabled Chinese potters to process the natural raw materials at temperatures between 1250 and 1300°C, without any modification other than cleaning and aging of the clay mix. As a result, Jingdezhen became the main production hub for Imperial porcelain during the Southern Song dynasty (1127–1279 CE). Porcelain produced at Jingdezhen was highly cherished as it was said to be as white as jade, as bright as a mirror, and as thin as paper (Schroeder and Erickson, 2014).

Today, Chinese scholars use the term 'porcelain' in a broad sense that includes fine, hard dusky and grey micaceous mineral-based materials fired to a vitrified state such that they emit a musical note when struck. In contrast to this, their Western colleagues prefer the term 'porcelain' in a narrower sense for fine, hard, white and translucent pottery. They also distinguish between hard-paste (*pâte dure*) calcareous (early Meissen, Vienna, some Sèvres) and feldspathic (also called triaxial) porcelains, produced from kaolin (china clay), micaceous feldspar (also called *petuntse*, china stone, porcelain stone) and quartz, and soft-paste (*pâte tendre*) porcelain, made from ground fused glass and mixed with clay or some other ingredients (Section 4.3.5.). Whereas most Far Eastern porcelains were fired only once below 1320°C with a likely average of 1280°C, Western hard-paste porcelains such as Meissen ware (Section 4.3.4.) were treated to higher temperatures, often in excess of 1400°C, fired twice (biscuit and glaze firing), and covered with a clear, transparent feldspathic glaze. Application of overglaze colours would require a third firing.

The ceramic body of Longquan porcelaneous celadon ware consists of a mixture of finely ground *petuntse* (china stone), kaolinitic clay and some red-firing clay that would contrast admirably with the blue-green jade-like glaze (Wood and Rastelli, 2014). Firing to temperatures of 1240–1260°C causes the clay to break down completely and transform to clusters of small 2:1 mullite needles and some silica glass. The feldspar component of the china stone will be converted to larger prismatic 3:2 mullite crystals and more glass into which quartz crystals dissolve (Fig. 22a). This means that there are two generations of mullite with different origins, compositions and physical appearance (Tite *et al.*, 2012; see also Section 4.3.3.). On cooling, residual quartz grains undergo a displacive phase transformation at 573°C from high quartz to low quartz, accompanied by shrinking, causing the glass adhering to the quartz grains to crack. As more and more quartz dissolves in the silicate glass during prolonged firing, the viscosity of the glass increases and provides strength to the ceramic body, that is, resistance to slumping. The mullite needles, acting as efficient reinforcement, augment this structural strengthening.

The longstanding enigma that is the translucent, soft and in its ethereal beauty unique lustre of the glaze of Longquan celadon ware (Fig. 21b) was solved eventually through the work of Kingery and Vandiver (1986). Figure 22b shows schematically a cross-

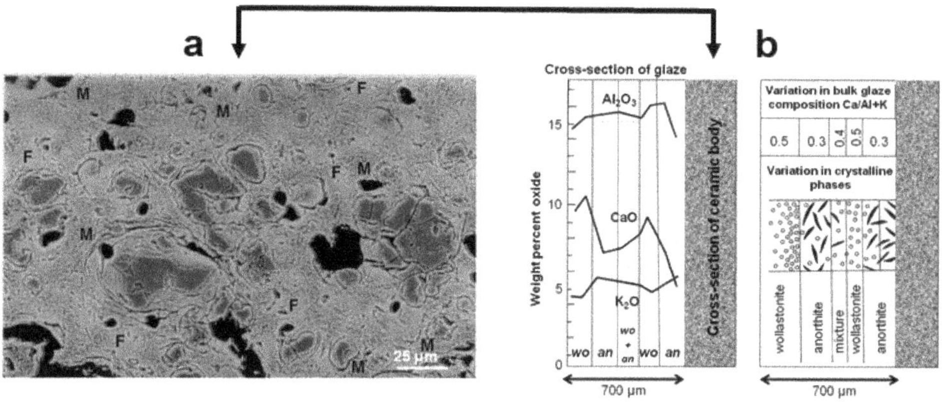

Figure 22. (a) Backscattered electron image of Longquan celadon ware body, etched for 3 min in 2% hydrofluoric acid. Quartz grains appear as dark grey, surrounded by rims of glass and cracks related to the displacive high–low phase transformation of quartz. There are clusters of filaments consisting of secondary mullite (M) as well as glassy pools containing elongated mullite crystals (F) in a continuous matrix (Tite *et al.*, 2012). Image courtesy of Prof. Michael Tite, Oxford, UK. (b) Variation in composition across the glaze of a Longquan celadon shard, showing a structure consisting of alternate zones comprising wollastonite and anorthite crystallites. Modified and redrawn after Kingery and Vandiver (1986).

sectional view of the glaze, characterized by local variation in composition on a scale of <100 μm. The left panel of Fig. 22b depicts the uneven distribution of Al_2O_3, CaO, and K_2O across the glaze that leads to precipitation of nano- to micro-sized needle-shaped crystallites of anorthite in places where the alumina content is high and the lime content is low. Likewise, wollastonite crystallites precipitate in regions high in CaO and low in alumina. The resulting zoning shown in Fig. 22b is at the heart of the optical performance of the glaze. Both wollastonite and anorthite can form from the viscous calcium aluminium silicate glass if mixing is restricted by high viscosity and the bulk glass composition is similar to that of the crystalline precipitates. Hence, in the $CaO-Al_2O_3-SiO_2$ phase diagram shown in Figs 3 and 20, the composition of the crystallites varies along the Wo–An join.

Longquan celadon ware was fired in hill-climbing (dragon) kilns of considerable lengths, up to 70 m or more. In the largest of these kilns, up to 100,000 ceramic pieces could be fired in a single firing cycle at temperatures as high as 1240–1260°C (N. Wood, pers. comm. 2012). During the Yuan dynasty, further advances in firing technology allowed very large dishes and jars to be made for export purposes in great quantities. However, their quality, and in particular the optical properties of their glazes, never reached those of the Song wares.

4.3.3. Thai stoneware

As early as the 10th century CE, potters living in what today is Thailand developed independently a ceramic tradition that was increasingly influenced by the technology of their northern neighbours in China, but without achieving the technological competency of their Chinese potter colleagues. Nevertheless, the wealth and prosperity of the state of Sukhothai (1238– ~1438 CE) depended to a large extent on the manufacture and export of high-quality stoneware that was hugely popular and widely traded throughout the southeast Asian realm. During the Sukhothai reign, glazed stoneware known as '*sangkhalok*' was produced in large quantities at the kilns of Sukhothai and Si Satchanalai but also at other kiln sites in northern Thailand including Kalong, Phan, Phayao and San Kamphaeng (Shaw, 2008, 2009).

The clays used by the potters of northern Thailand were kaolinitic, and the mineralogical phase composition of the stoneware body consisted of 9–14 mass% mullite, 15–26 mass% quartz and a balance of a glassy phase (Kilb, 1979; Heimann, 1989). This composition is located in the compositional triangle mullite—K-feldspar–quartz of the ternary system K_2O–Al_2O_3–SiO_2 (Fig. 23a). The relatively high proportion of fluxing K_2O leads to partial melting during firing at 1200°C or beyond. The ternary eutectic point is at ~10 mass% K_2O, 11 mass% Al_2O_3 and 79 mass% SiO_2, at the junction of the stability fields of mullite, K-feldspar and quartz. The associated ternary eutectic temperature is 985°C, *i.e.* firing of the clay at this temperature produces partial grain-boundary melts of the composition of the ternary eutectic. Indeed, the chemical composition of the Si Satchanalai ware shown in Fig. 23a is close to that of the ternary eutectic composition.

At this point, it is necessary to describe the thermal reactions of kaolinite in more detail. Firing of a kaolinitic clay leads to a structural collapse between 500 and 600°C with the formation of a more or less amorphous dehydroxylated 'metakaolinite' phase. This phase transforms at >900°C to an Al–Si spinel (AS_2, $Al_2O_3{\cdot}2SiO_2$). This spinel, in turn, releases silica as cristobalite when heated beyond 1100°C to form Al-rich primary platy mullite (2:1 mullite, $2Al_2O_3{\cdot}SiO_2$). At even higher temperatures, the amount of melt present in the ceramic body increases, and eventually Si-rich secondary needle-

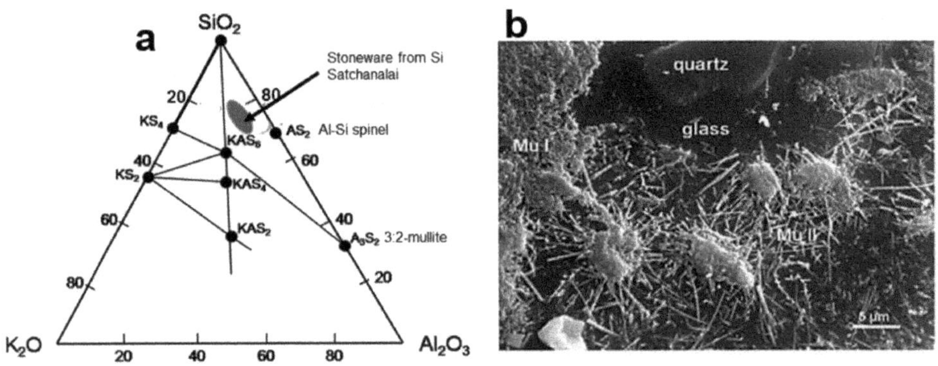

Figure 23. (a) Compositional scatter ellipse of stoneware from Si Satchanalai, Thailand, in the ternary phase diagram K_2O–Al_2O_3–SiO_2 (Heimann and Maggetti, 2014). (b) Scanning electron microscopy image of the ceramic body of stoneware from Si Satchanalai showing primary platy mullite (Mu I, left) and secondary needle mullite (Mu II, centre) as well as a quartz grain dissolving in glass (top) (Kilb, 1979). The firing temperature of the stoneware was estimated to have been between 1200 and 1250°C (Kilb, 1979; Heimann, 1989). Image courtesy Dr. Lothar Kilb.

like mullite (3:2 mullite, $3Al_2O_3 \cdot 2SiO_2$) crystallizes directly from the melt (Schüller, 1964; Iqbal and Lee, 2000). Frequently, primary mullite provides a growth site for this crystallization process because of structural similarity between these two phases that form a solid solution series as shown in the binary $SiO_2-Al_2O_3$ phase diagram of Fig. 3a. The two limiting compositions of the mullite solid solution, *i.e.* platy mullite (Mu I) and needle mullite (Mu II) are shown in the SEM micrograph of the ceramic body of stoneware from Si Satchanalai, together with a quartz grain dissolving in glass (Fig. 23b). These two types of mullite can be distinguished from each other by careful XRD analysis supported by Rietveld refinement (Heimann, 2016).

4.3.4. Hard-paste porcelain of Meissen

The fascinating history of the discovery and technological development of European hard-paste porcelain of Meissen type has been described, for example, by Walcha (1981) and Heimann and Maggetti (2014). This unique ceramic ware is special as it is fired at such high temperatures that the mineral reactions both at grain boundaries and in the bulk are being controlled by nearly true thermodynamic relationships. Compared to East Asian porcelain, Meissen hard-paste porcelain had an entirely different development as it contains much less free quartz but more mullite. It also shows a substantially denser microstructure compared to Chinese and Japanese porcelains, related to its higher firing temperature that enabled quartz to be dissolved efficiently in the glassy phase. These facts induce drastically increased nucleation and growth of mullite crystals, and promote enhanced sealing of pores to yield a denser microstructure. In contrast to Chinese porcelain that is fired in a once-through cycle, Meissen porcelain is normally fired twice. The first (biscuit) firing occurs between 900 and 950°C. After cooling, the white biscuit porcelain can be decorated by painting (underglaze decoration). After dipping in the glaze slurry and appropriate drying, the second firing (glost or glaze firing) is performed between 1200 and ~1500°C. After cooling, the ceramic can be painted with enamel colours including gold and fired a third time at 700 to 850°C to fix the enamel to the transparent glaze (overglaze decoration) Fig. 24a).

As shown in Fig. 25a, the chemical composition of the typical raw materials mix of 'triaxial' porcelain (50% kaolinite, 25% quartz, 25% (K-) feldspar) is located in the phase field of mullite at point 'A' (69.3 mass% SiO_2, 26.2 mass% Al_2O_3, 4.5 mass% K_2O). As the liquidus temperature (melting point) of this composition is beyond 1700°C it will never be reached during ceramic firing as the typical firing temperature of contemporary Meissen porcelain is between 1370 and 1470°C. Firing at and cooling from this temperature causes crystallization of mullite along the crystallization path A–B (dashed line) formed by connecting the point corresponding to the mullite composition through point 'A'. This crystallization path meets the phase boundary mullite–tridymite at 1280°C at point 'B'. Further crystallization occurs when the system follows the phase boundary along the path B–C, until at 985°C the ternary eutectic K-feldspar–tridymite–mullite is reached, at which point any liquid phase ceases to exist.

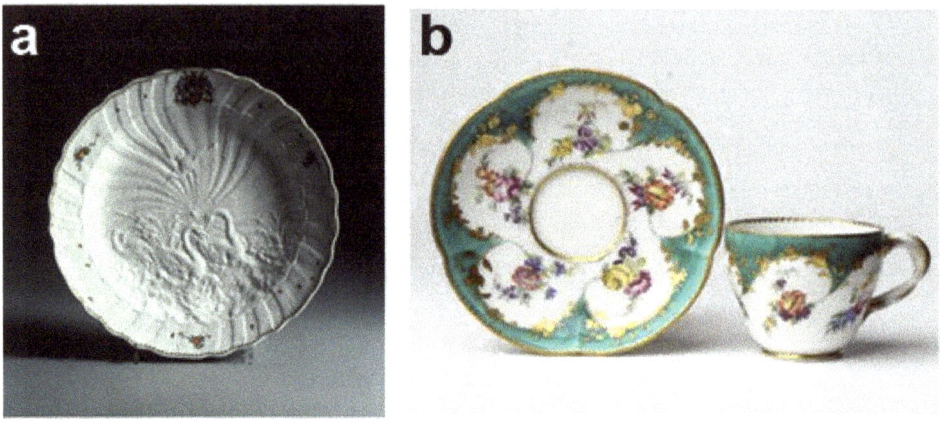

Figure 24. (a) Large bowl of the Swan service with overglaze decoration; Meissen (1737–1742). Diameter 43 cm. Inv. No. PE1427. © Staatliche Kunstsammlungen Dresden. (b) Soft-paste porcelain saucer and cup, painted with overglaze colours. Sèvres (1751–1757). Reg. No. C.1418A-1919 and C.1418-1919. © Victoria and Albert Museum, London.

Crystallization during cooling of the near-eutectic melt from high temperature leads first to formation of mullite, reaching a maximum amount of 27.4 mass% at 1280°C.

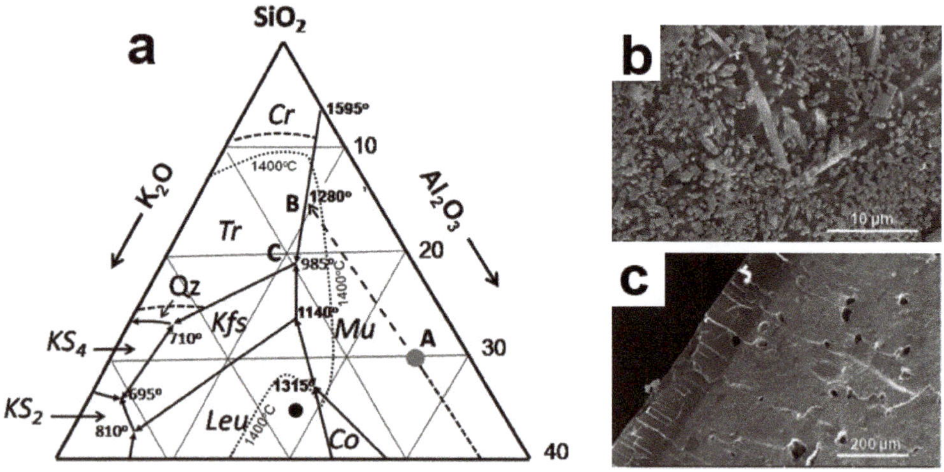

Figure 25. (a) Phase boundaries and coexisting phases in the SiO_2-rich section of the ternary system $K_2O-Al_2O_3-SiO_2$ (after Schairer and Bowen, 1947). The path A–B–C (dashed line) characterizes the crystallization history of a typical Meissen porcelain body. Co – corundum, Cr – cristobalite, Kfs – potassium feldspar, KS_2 – potassium disilicate, KS_4 – potassium tetrasilicate, Leu – leucite, Mu – mullite, Qz – quartz, Tr – tridymite. The ternary eutectic is at the triple point Kfs–Mu–Tr with the melting temperature of 985°C. The dotted line shows the isotherm at 1400°C. (b, c) Microstructure of historical Meissen porcelain from 1774. Mullite crystals in a glassy matrix (b) and contact zone between glaze and the ceramic body below (c). Images courtesy of Dr Bernd Ullrich, TU Bergakademie Freiberg, Germany.

Below this temperature, tridymite crystallization starts, reaching 22.9 mass% at 985°C besides 47.4 mass% melt phase, and an additional amount of 2.3 mass% mullite. During further cooling to ambient temperature, crystallization of the remaining melt yields 26.6 mass% K-feldspar and an additional amount of 20.8 mass% tridymite. Hence, the thermodynamic equilibrium composition of the final porcelain body ought to be 29.7 (27.4 + 2.3) mass% mullite + 43.7 (22.9 + 20.8) mass% tridymite + 26.6 mass% K-feldspar (Salmang *et al.*, 2007). In practice, however, the microstructure of a typical porcelain body is very different. It consists of ~30 mass% mullite and 70 mass% $K_2O-Al_2O_3-SiO_2$ glass (Fig. 25b,c) because the highly viscous melt thus formed freezes without crystallization. This is precisely the composition found in historical (see for example Ullrich and Ballmair, 2002; Ullrich *et al.*, 2010) and contemporary Meissen porcelains.

Figure 26 shows the reaction scheme leading to the microstructure of porcelain (see also Iqbal and Lee, 1999, 2000; Lee and Iqbal, 2001).

4.3.5. *Soft-paste porcelain*

The earliest attempts by European potters to replicate Chinese porcelains date back to the 16th century CE when soft-paste (frit) porcelain bodies, so-called Medici

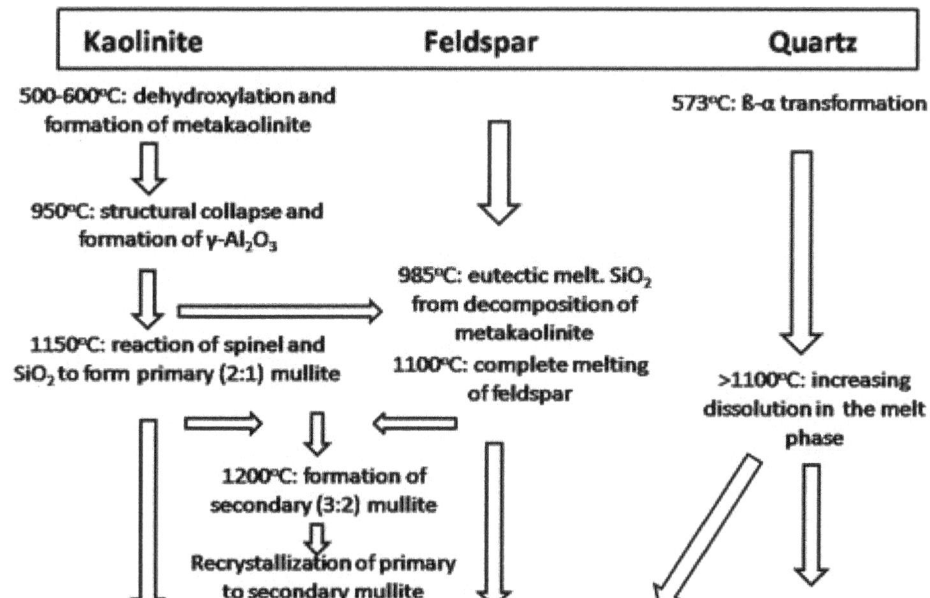

Figure 26. Simplified reaction scheme of formation of the microstructure of triaxial porcelain (adapted from Schüller, 1982; in Heimann and Maggetti (2014).

'porcelain', were crafted in Florence under the patronage of Grand Duke Francesco I de Medici. Medici 'porcelain' was manufactured by mixing a vitreous alkali silicate frit with white clay, feldspar and quartz. This 'surrogate' porcelain was re-invented some hundred years later in France and England, using traditional Near Eastern quartz-glass (frit)-clay compositions to copy East Asian porcelain. In France, this soft-paste (frit) porcelain is called *pâte tendre*.

In contrast to hard-paste porcelain, firing of soft-paste porcelain proceeds at considerably lower temperatures (biscuit firing at ~1100°C, glaze firing at ~1000°C). However, the decisive difference between hard- and soft-paste porcelain is not the firing temperature but the different sintering behaviour. The feldspar compound in hard-paste porcelain forms, with alumina and silica released from the kaolinite, a eutectic melt that is quenched during cooling and thus remains in the vitreous state without crystallizing (Zanelli *et al.*, 2011). Hence, hard-paste porcelain contains up to 70 mass% of glass as opposed to the more crystalline microstructure of soft-paste porcelain, with only ~30 mass% of glass. This renders the latter more prone to mechanical destruction because it possesses lower hardness, abrasion resistance, bending strength and thermal shock resistance.

French soft-paste porcelains resemble Chinese porcelains in terms of translucency and whiteness. These soft-paste porcelains were produced in the 18th century CE in French manufactures at Saint-Cloud, Chantilly and Vincennes/Sèvres (fig. 24b). The frit used at the Vincennes/Sèvres pottery was a complex mix of sand, soda, calcined alum, calcined gypsum, saltpetre and rock salt, fired at the hottest part of the kilns to obtain a glassy substance. Subsequently, this frit was crushed, milled with chalk and marl, and finally shaped, mostly by moulding. The green ware was biscuit-fired at ~1100°C for 4 days, then glazed and fired at ~1000°C for use another day. The bodies of Sèvres porcelain consist of relict quartz and newly formed crystalline phases such as tridymite and wollastonite, all embedded in a glassy phase.

The simple ternary phase diagram Na_2O-SiO_2-CaO (Fig. 27a) is well suited to study equilibrium phase relationships of French soft-paste porcelain. The chemical analyses of porcelain bodies plot in the compatibility triangle connecting the phases [β-CS; β-wollastonite]–[NC_3S_6; devitrite]–[SiO_2; tridymite/cristobalite]. This is the equilibrium phase association for French soft-paste porcelains. It should be remembered that the simple three-component system shown in Fig. 27 does not consider K_2O and several other oxides that have a small but measurable impact on phase relationships, as well as on invariant and divariant equilibria. However, the simplified system can be used as a reasonably good approximation to reality. The region of interest shows two invariant points: (1) a ternary peritectic point at a temperature of 1035°C with the four equilibrium (coexisting) phases [liquid L]–[β–CS]–[NC_3S_6]–[SiO_2] and (2) a ternary eutectic point at a temperature of 725°C with the four equilibrium phases [liquid L]–[NS_2]–[NC_3S_6]–[SiO_2].

The 1100°C isothermal section of this phase diagram illustrates the equilibrium composition attained during firing of French soft-paste porcelain bodies for five days at this temperature (Fig. 27b). According to theory, the phase assembly would be

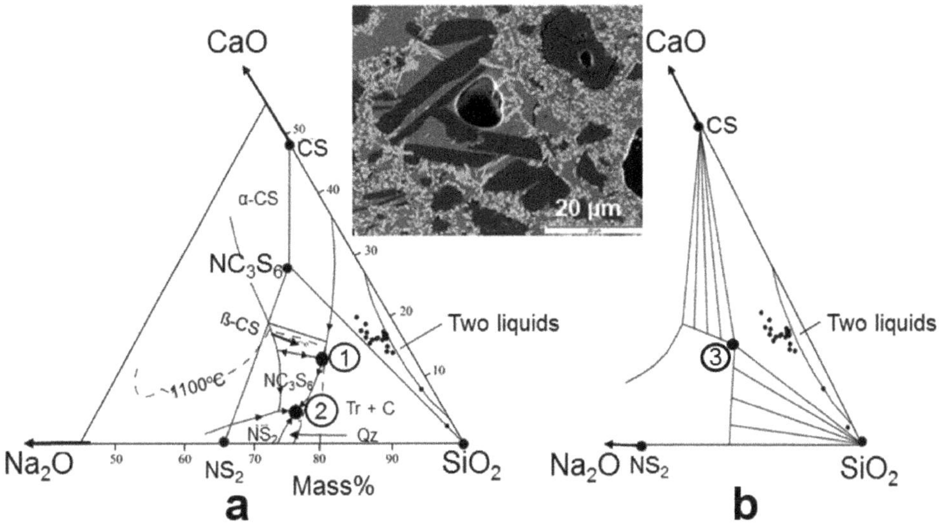

Figure 27. (a) Phase boundaries and coexisting phases in the high SiO_2 corner of the ternary system Na_2O–SiO_2–CaO (redrawn after Morey and Bowen, 1925). C – cristobalite, SiO_2; CS – wollastonite, $CaO{\cdot}SiO_2$; NS_2 – water glass, sodium disilicate, $Na_2O{\cdot}2SiO_2$; NC_3S_6 – devitrite, $Na_2O{\cdot}3CaO{\cdot}6SiO_2$; Qz – quartz, SiO_2; Tr – tridymite, SiO_2. **1** ternary peritectic point (1035°C), **2** ternary eutectic point (725°C). Dots = 17 analyses of French soft-paste porcelain (Brongniart, 1844, Kingery and Smith, 1985, Kingery and Vandiver, 1986). (b) Isothermal section at 1100°C of the phase diagram shown on the left. The composition of 17 French soft paste porcelains are shown in the compositional triangle [β-wollastonite, CS] + [SiO_2 phase (cristobalite or tridymite)] + [liquid] of composition **3**. The backscattered electron image shows large subidiomorphic tridymite laths and tiny wollastonite grains as crystallization products from the 1100°C melt (now glassy matrix), contrasting with a relict α-quartz particle (upper right corner), surrounded by the characteristic crack due to the crystallographic β/α-inversions during cooling (Heimann and Maggetti, 2014).

expected to consist of [β-wollastonite] + [tridymite/cristobalite] + [liquid of composition 3]. However, when the kiln cools, the composition of the liquid follows the curve to the peritectic point 1, at which point simultaneously [β-wollastonite] and [tridymite/cristobalite] crystallize. The peritectic reaction [L] + [β-CS] → [NC_3S_6] + [SiO_2] starts at 1035°C and consumes all remaining liquid, leading to the final equilibrium phase association [β-wollastonite] + [devitrite] + [tridymite/cristobalite].

The microstructure of typical Sèvres soft-paste porcelain (Fig. 27, centre) shows quartz grains, partially or completely transformed to tridymite/cristobalite, wollastonite and glass. Although the equilibrium phases tridymite and wollastonite are present as expected from the phase diagram, devitrite is conspicuously absent. Devitrite is sometimes found as an unwanted devitrification product of modern soda-lime-silica sheet glass, but is readily resorbed in the glass matrix after prolonged heating to high temperatures (Aboud *et al.*, 2005), in accord with the findings above. The surviving large amount of a non-equilibrium glassy phase is due to the presence of alumina that inhibits devitrification.

5. Conclusion

Mineralogical investigations of ancient and historical ceramics provide a sensitive probe to gauge the technical competence of potters as well as the overall state of technological development of a particular society. The study of the microstructure of ancient and historical ceramics provides many ways to determine how effectively craftspeople responded to the challenge of turning natural raw material 'clay' into the finished product 'pottery'. Plots of chemical composition of the original material as well as the mineralogical phase composition of the resulting product in ceramic phase diagrams show clearly how well the potter succeeded in the intuitive quest towards a stable phase equilibrium that would result in dense wares. As it turns out, for all Neolithic and most Bronze Age ceramics this quest was clearly not high on the agenda of the ancient potters, as their major goal was to produce porous vessel bodies for cooking pots, water containers to keep the water fresh and cool, or fermentation containers. Surface densification by burnishing vessels in the leather-hard state and, later, glazing would provide significant incentive for technological development toward surficially denser pottery without necessarily relying on higher firing temperatures. Indeed, at the rather low firing temperatures applied, most mineral-grain boundary reactions would have occurred far away from thermodynamic equilibrium, and hence were kinetically controlled. Kinetics exerts constraints on thermodynamics by retarding reaction rates because of low temperatures, large temperature gradients present in primitive pottery kilns, short reaction times, inhomogeneously distributed reaction partners, and varying redox conditions. Observation of mineral phases formed during ceramic firing such as diopside, gehlenite, anorthite, sanidine or mullite can tell the ceramics analyst how well the starting clay was processed and how efficiently high heat could be generated in ever more sophisticated pottery kilns; eventually leading to very high fired, dense and translucent porcelain.

The microstructural analyses of ancient pottery, assisted by ceramic phase diagrams, show impressively how the intellectual environment, *i.e.* the technological knowledge of a particular ancient society filled the requirements of the cultural environment (Kolb, 1982), *i.e.* individual needs by optimum exploitation and manipulation of raw materials extracted from the natural environment (Stimmell *et al.*, 1982; Heimann, 1989).

References

Aboud, T.K., Stoch, L. and Sroda, M. (2005) Quartz crystallization in soda-lime-silica glass. *Optica Applicata*, **35**, 829–836.

Andersen, D.J., Lindsley, D.H. and Davidson, P.M. (1993) QUILF: a Pascal program to assess equilibria among Fe-Mg-Mn-Ti oxides, pyroxenes, olivine, and quartz. *Computers and Geosciences*, **19**, 1333–1350.

Belfiore, C.M., Day, P.M., Hein, A., Kilikoglou, V., La Rosa, V., Mazzoleni, P. and Pezzino, A. (2007) Petrographic and chemical characterisation of pottery production of the Later Minoan I kiln at Haghia Triada, Crete. *Archaeometry*, **49**, 621–653.

Bemmann, H. (1993) Terra Sigillata – Tafelgeschirr der Römer und "Leitfossil" der Archäologie. In: *Die Römer in der Pellenz*, **2** (K. Schäfer, editor).

Boaretto, E., Wu, X., Yuan, J., Bar-Yosef, O., Chu, V., Pan, Y., Liu, K., Cohen, D., Jiao, T., Li, S., Gu, H.,

Goldberg, P. and Weiner, S. (2009) Radiocarbon dating of charcoal and bone collagen associated with early pottery at Yuchayan Cave, Hunan Province, China. *Proceedings of the National Academy of Sciences of the USA*, **106**, 9595–9600.

Bowen, N.L. (1915) The crystallization of haplobasaltic, haplodioritic and related magmas. *American Journal of Science*, **40**, 161–185.

Brongniart, A. (1844) *Traité des Arts Céramiques ou des Poteries considérées dans leur Histoire, leur Pratique et leur Théorie.* 3 vols and 1 atlas, Paris: Béchet Jeune; 2nd edn reviewed, corrected and augmented with notes and additions by A. Salvetat, Paris: Béchet Jeune, 1854; 3rd edn with notes and additions by A. Salvetat, 1877; facsimile of the 3rd edn. Paris: Dessain and Tolra, 1977.

Chard, C.S. (1974) *Northeast Asia in Prehistory*. University of Wisconsin Press, Madison, Wisconsin, USA.

Chatterjee, N.D., Krüger, R., Haller, G. and Olbricht, W. (1998) The Bayesian approach to an internally consistent thermodynamic database: theory, database, and generation of phase diagrams. *Contributions to Mineralogy and Petrology*, **133**, 149–168.

Clark, S.P., Schairer, J.F. and de Neufville, J. (1962) Phase relations in the system $CaMgSi_2O_6$–$CaAl_2Si_2O_8$–SiO_2 at low and high pressure. *Carnegie Institute of Washington, Year Book*, **61**, 59–68.

Cultrone, G., Rodriguez-Navarro, C., Sebastian, E., Cazalla, O. and De la Torre, M.J. (2001) Carbonate and silicate phase reactions during ceramic firing. *European Journal of Mineralogy*, **13**, 621–634.

Déchelette, J. (1904) *Les Vases céramiques ornés de la Gaul romains (Narbonnaise, Aquitaine et Lyonnaise).* A. Picard and fils, Paris.

Dondi, M., Ercolani, G., Fabbri, B., and Marsigli, M. (1998) An approach to the chemistry of pyroxenes formed during the firing of Ca-rich silicate ceramics. *Clay Minerals*, **33**, 443–452.

Dragendorff, H. (1895/96) Terra Sigillata. Ein Beitrag zur Geschichte der griechischen und römischen Keramik. *Bonner Jahrbücher*, **96/97**, 18–155.

Escardino, A., García-Ten, J. and Soriano, M. (2006) Study of calcite decomposition during wall tile firing. *Qualicer* 2006, 265–278.

Franklin, U.M. (1980) *The potential of technical studies*. Paper presented at the Symposium on the Bronze Age of China. New York, June 2-3, 1980.

Gibbs, J.W. (1878) On the equilibrium of heterogeneous substances. *Transactions of the Connecticut Academy of Arts and Science*, **3**, 108–248, 343–524.

Goldschmidt, V.M. (1911) Anwendung der Phasenregel auf Silikatgesteine. *Zeitschrift für Elektrochemie*, **17**, 686–689.

Goldsmith, J.R. (1953) A "simplexity principle" and its relation to "ease" of crystallization. *Bulletin of the Geological Society of America*, **64**, 439–451.

Görres, M., Evangelakakis, C., Kroll, H. and Kohl, V. (2000) The application of mineralogical techniques to the characterization of archaeological ceramic finds. Pp. 985–988 in: *Applied Mineralogy; In Research, Economy, Technology, Ecology and Culture* (D. Rammlmair, editor). Balkema, Rotterdam, The Netherlands.

Grapes, R. (2006): *Pyrometamorphism*. Springer, Berlin.

Hamdan, M.A., Martinez, S.M., Garcia Vallès, M.T., Nogués, J.M., Hassan, F.A., Flower, R.J., Aly, M.H., Senussi, A. and Ebrahim, E.S. (2014) Ancient Egyptian pottery from the subsurface floodplain of the Saqqara-Memphis area: its mineralogical and geochemical implications. *Archaeometry*, **56**, 987–1008.

Hängst, K. (1979) *Mineralogische Untersuchungen an altägyptischer Keramik der 18. und 19. Dynastie.* Unpublished Masters thesis, Universität Köln, Germany.

Heimann, R.B. (1978/79) Mineralogische Vorgänge beim Brennen von Keramik und Archäothermometrie. *Acta Praehistorica and Archaeologica*, **9/10**, 79–102.

Heimann, R.B. (1982) Firing technologies and their possible assessment by modern analytical methods. Pp. 89–96 in: *Archaeological Ceramics* (J.S. Olin and A.D. Franklin, editors). Smithsonian Institution Press, Washington, D.C.

Heimann, R.B. (1989) Assessing the technology of ancient pottery: The use of ceramic phase diagrams. *Archeomaterials*, **3**, 123–148.

Heimann, R.B. (1991) Technological progress and market penetration of advanced ceramics in Canada. *Bulletin of the American Ceramic Society*, **70(7)**, 1120–1124.

Heimann, R.B. (2010) *Classic and Advanced Ceramics. From Fundamentals to Applications*, Wiley-VCH Verlag GmbH, Weinheim, Germany.

Heimann, R.B. (2016) X-ray powder diffraction. Chapter 19 in: *Oxford Handbook of Archaeological Ceramic Analysis (*A.M.W. Hunt, editor). Oxford University Press, Oxford, UK, 768 pp.

Heimann, R.B. and Franklin, U.M. (1979) Archaeo-thermometry: The assessment of firing temperatures of ancient ceramics. *Journal of the International Institute of Conservation, Canadian Group (IICCG)*, **4(2)**, 23–45.

Heimann, R.B. and Maggetti, M. (1981) Experiments on simulated burial of calcareous terra sigillata (mineralogical change). Preliminary results. Pp. 163–177 in: *Scientific Studies in Ancient Ceramics*, occasional paper no. 19. British Museum Research Laboratory, London.

Heimann, R.B. and Maggetti, M. (2014) *Ancient and Historical Ceramics. Materials, Technology, Art, and Culinary Traditions.* Schweizerbart Science Publishers, Stuttgart, Germany.

Heimann, R.B., Maggetti, M. and Einfalt, H.C. (1980) Zum Verhalten des Eisens beim Brennen eines kalkhaltigen illitischen Tons unter reduzierenden Bedingungen. *Berichte der Deutschen Keramischen Gesellschaft*, **57**, 145–152.

Huebner, J.S. (1971) Buffering techniques for hydrostatic systems at elevated temperatures. Pp. 123–127 in: *Research Techniques for High Pressure and High Temperatures* (G.C. Ulmer, editor). Springer, Berlin, Heidelberg, New York.

Huysecom, E., Rasse, M., Lespez, L., Neumann, K., Fahmy, A., Ballouche, A., Ozainne, S., Maggetti, M., Tribolo, Ch. and Soriano, S. (2009) The emergence of pottery in Africa during the tenth millennium cal BC: new evidence from Ounjougou (Mali). *Antiquity*, **83**, 905–912.

Iqbal, Y. and Lee, W.E. (1999) Fired porcelain microstructures revisited. *Journal of the American Ceramic Society*, **81**, 3584–3590.

Iqbal, Y., and Lee, W.E. (2000) Microstructural evolution in triaxial porcelain. *Journal of the American Ceramic Society*, **83**, 3121–3127.

Jabbori, S. (1980) Report on geological investigation of clay deposits for brick industry in Nahrawan area with reserve estimation. *GEOSURV, Internal Rept.* No. 1039.

Jordan, P. and Zvelebil, M. (editors) (2009) *Ceramics before Farming: The Dispersal of Pottery Among Prehistoric Eurasian Hunter-gatherers*. University College London Institute of Archaeology, Publication **53** and Left Coast Press, Walnut Creek, California, USA.

Kadum, L.S. and Hussain, A.I. (2011) Origin and mineralogy of construction materials of gravel and sand in Tikrit area. *Tikrit University Journal for Humanities*, **18(8)**, 33–46.

Kattner, U.R. (1997) The thermodynamic modeling of multicomponent phase equilibria. *JOM*, **49(12)**, 14–19.

Kilb, L. (1979) *Gefügeanalytische Untersuchungen historischer steinzeugartiger Werkstoffe zur Behandlung archäometrischer Fragestellungen.* Unpublished Ph.D. dissertation. Technical University Clausthal-Zellerfeld, Germany.

Kilikoglou, V., Vekinis, G. and Maniatis, Y. (1995) Toughening of ceramic earthenware by quartz inclusions: an ancient art revisited. *Acta Metallurgica et Materialia*, **43**, 2959–2965.

Kilikoglou, V., Vekinis, G., Maniatis, Y. and Day, P.M. (1998) Mechanical performance of quartz-tempered ceramics. Part I: Strength and toughness. *Archaeometry*, **40**, 261–279.

Kingery, W.D. and Smith, D. (1985) The development of European soft-paste (frit) porcelain. Pp. 273–292 in: *Ancient Technology to Modern Science*. Ceramics and Civilization, vol. I, The American Ceramic Society, Columbus, Ohio, USA.

Kingery, W.D. and Vandiver, P.B. (1986) *Ceramic Masterpieces. Art, Structure, and Technology*. The Free Press, New York.

Kingery, W.D., Bowen, H.K. and Uhlmann, D.R. (1976) *Introduction to Ceramics*. 2nd edition John Wiley & Sons, New York, Chichester, Brisbane, Toronto, Singapore.

Klein, R. (1980) Later Pleistocene hunters. Pp. 90–91 in: *The Cambridge Encyclopedia of Archaeology* (A. Sherratt, editor). Prentice Hall of Canada Ltd., Scarborough, Ontario, Canada.

Klima, B. (1962) The first ground plan of an Upper Palaeolithic loess settlement in Middle Europe and its meaning. Pp. 193–210 in: *Courses Towards Urban Life* (R.J. Braidswood and G.R. Willey, editors). Chicago University Press, Chicago.

Knorr, R. (1919) *Töpfer und Fabriken verzierter Terra-Sigillata des 1. Jahrhunderts.* W. Kohlhammer, Stuttgart.

Kockelmann, W., Kirfel, A. and Hähnel, E. (2001) Non-destructive phase analysis of archaeological ceramics using TOF neutron diffraction. *Journal of Archaeological Science*, **28**, 213–222.

Kolb, C.C. (1982) Ceramic technology and problems and prospects of provenience in specific ceramics from Mexico and Afghanistan. Pp. 193–208 in: *Archaeological Ceramics* (J.S. Olin and A.D. Franklin, editors). Smithsonian Institution Press, Washington, D.C.

Lee, W.E. and Iqbal, Y. (2001) Influence of mixing on mullite formation in porcelain. *Journal of the European Ceramic Society*, **21**, 2583–2586.

Letsch, J. and Noll, W. (1983) Phase formation in several ceramic subsystems at 600°C–1000°C as a function of oxygen fugacity. *Ceramic Forum International/Berichte der Deutschen Keramischen Gesellschaft*, **60**, 259–267.

Levin, E.M., Robbins, C.R. and McMurdie, H.F. (1964–1990) *Phase Diagrams for Ceramists.* Vol. I – VIII. The American Ceramic Society, Westerville, Ohio, USA.

Louisnathan, S.J. (1971) Refinement of the crystal structure of a natural gehlenite, $Ca_2Al(Al,Si)_2O_7$. *The Canadian Mineralogist*, **10**, 822–837.

Maggetti, M. (1982) Phase analysis and its significance for technology and origin. Pp. 121–133 in: *Archaeological Ceramics* (J.S. Olin and A.D. Franklin, editors). Smithsonian Institution, Washington, D.C.

Maggetti, M. (1986) Majolika aus Mexiko-ein archäometrisches Fallbeispiel. *Fortschritte der Mineralogie*, **64**, 87–103.

Maggetti, M. (2001) Chemical analyses of ancient ceramics: What for? *CHIMIA International Journal of Chemistry*, **55**, 923–930.

Maggetti, M. and Galetti, G. (1986) Chemischer Herkunftsnachweis der "Schwarzen Sigillata" vom Magdalensberg. *Magdalensberg-Grabungsbericht*, **15**, 391–431.

Maggetti, M. and Heimann, R.B. (1979) Bildung und Stabilität von Gehlenit in römischer Feinkeramik. *Schweizer Mineralogische und Petrographische Mitteilungen*, **59**, 413–417.

Maggetti, M. and Küpfer, T. (1978) Composition of the Terra Sigillata from La Péniche (Vidy, Lausanne, Switzerland). *Archaeometry*, **20**, 183–188.

Maggetti, M., Neururer, Ch. and Ramseyer, D. (2011) Temperature evolution inside a pot during experimental surface (bonfire) firing. *Applied Clay Science*, **53**, 500–508.

Maniatis, Y. and Tite, M.S. (1975) A scanning electron microscopic examination of the bloating of fired clay. *Transactions of the British Ceramic Society*, **74**, 229–232.

Morey, G.W. and Bowen, N.L. (1925) The system sodium metasilicate–calcium metasilicate–silica. *Journal of the Society of Glass Technology*, **9**, 232–233.

Mustafa, M.M. (2011) Mineral resources and industrial deposits in the Mesopotamian plain. *Iraqi Bulletin of Geology and Min*eralogy, Special issue **4**, 105–118.

Nakamura, T., Taniguchi, Y., Tsuji, S. and Oda, H. (2001) Radiocarbon dating of charred residues on earliest pottery in Japan. *Radiocarbon*, **43**, 1129–1138.

Niggli, P. (1954) *Rocks and Mineral Deposits.* Pp. 347–427. Freeman & Company, San Francisco, California, USA.

Noll, W. (1982) Mineralogie und Technik der Keramiken Altkretas. *Neues Jahrbuch der Mineralogie, Abhandlungen,.* **143(2)**, 150–199.

Noll, W. (1984) Antike Keramik und ihre Dekoration in naturwissenschaftlicher Sicht. Pp. 42–53 in: *International Vase Symposium, Amsterdam.* Allard Pierson Series, Vol 5.

Noll, W. (1991) *Alte Keramiken und ihre Pigmente.* E. Schweizerbart, Stuttgart. 334 pp.

Noll, W. and Heimann, R.B. (2016) *Ancient Old World Pottery. Materials, Technology, and Decoration.* Schweizerbart Science Publishers, Stuttgart, Germany. 311 pp.

Osborn, E.F. and Tait, D.B. (1952) The system diopside–forsterite–anorthite. *American Journal of Science*, Bowen volume, 413–433.

Peters, T. and Iberg, R. (1978) Mineralogical changes during firing of calcium-rich brick clays. *Bulletin of the American Ceramic Society*, **57**, 503–509.

Picon, M., Vichy, M. and Meille, E. (1971) Composition of Lezoux, Lyon and Arezzo Samian ware. *Archaeometry*, **13**, 191–208.

Powell, A.J., McDonnell, J.G., Batt, C.M. and Vernon, R.W. (2002) An assessment of the magnetic response of an iron-smelting site. *Archaeometry*, **44**, 651–665.
Rankin, G.A. and Wright, F.E. (1915) The ternary system $CaO–Al_2O_3–SiO_2$. *American Journal of Science*, **39**, 1–79.
Rehak, P., Kunath-Fandrel, G., Losso, P., Hildmann, B., Schneider, H. and Jäger, C. (1998) Study of the Al coordination in mullite with varying Al:Si ratio by ^{27}Al NMR spectroscopy and X-ray diffraction. *American Mineralogist*, **83**, 1266–1276.
Roberts, P. (1997) Mass-production of Roman finewares. Pp. 188–193 in: *Pottery in the Making. World Ceramic Traditions* (I. Freestone and D. Gaimster, editors), British Museum Press. London.
Rodriguez-Navarro, C., Ruiz-Agudo, E., Luque, A., Rodriguez-Navarro, A. and Ortega-Huerto, M. (2009) Thermal decomposition of calcite: Mechanisms of formation and textural evolution of CaO nanocrystals. *American Mineralogist*, **94**, 578–593.
Rosenberger, F. (1979) *Fundamentals of Crystal Growth I*. Springer, Berlin.
Salmang, H., Scholze, H. and Telle, R. (2007) *Keramik*. 7th edition. Springer, Berlin, Heidelberg. 1148 pp.
Schairer, J.F. and Bowen, N.L. (1947) Melting relations in the systems Na_2O-Al_2O_3-SiO_2 and K_2O-Al_2O_3-SiO_2. *American Journal of Science*, **245**, 193–204.
Schneider, G. and Daszkiewicz, M. (2006) Chemical analysis of Italian Sigillata from Italy and from the northern provinces Pp. 537–543 in: *Old Pottery in a New Century – Innovating Perspectives on Roman Pottery Studies* (D. Malfitano, J. Poblome and J. Lund, editors). *Atti del Convegno Internazionale di Studi*, Catania, 22-24 Aprile.
Schneider, G. and Hoffmann, B. (1990) Chemische Zusammensetzung italischer Sigillata. Pp. 27–38 in: *Conspectus formarum terrae sigillatae italico modo confactae*. Materialien zur römisch-germanischen Keramik (E. Ettlinger *et al.*, editors), Bd. 10. Bonn, Germany.
Schneider, G., Knoll, H., Gallis, K. and Demoule, J.P. (1990) Production and circulation of Neolithic Thessalian pottery: Chemical and mineralogical analyses. In: *La Thessalie, Quinze années de recherchés archeologiques, 1975-1990*. Actes du Colloque International, Lyon, France.
Schneider, G., Knoll, H., Gallis, K. and Demoule, J.P. (1991) Transition entre les cultures néolithiques des Sesklo et de Dimini: Recherches minèralogiques, chimiques et technologiques sur les ceramiques et les argiles. *Bulletin de Correspondence Héllenique* (BCH), **115**, 1–64.
Schroeder, P.A. and Erickson, G. (2014) Kaolin: From ancient porcelain to nanocomposites. *Elements*, **10**, 177–182.
Schüller, K.H. (1964) Reactions between mullite and glassy phase in porcelains. *Transactions of the British Ceramic Society*, **63**, 103–117.
Schüller, K.H. (1982) The development of microstructures in silicate ceramics. 1. Whitewares. *Journal of Educational Modules in Materials Science and Engineering*, **4**, 529–565.
Schultz, A. and Chang, Y.A. (1985) Computer graphics for ternary phase diagrams. *Journal of Metals*, November, 10–13.
Shaw, J.C. (2008) *Northern Thai Ceramics*. 2nd edition. Duangphorn Kemasingki, Chiang Mai, Thailand.
Shaw, J.C. (2009) *Thai Ceramics*. Citylife Publications, Chiang Mai, Thailand.
Smith, J.V. (1959) The crystal structure of proto-enstatite, $MgSiO_3$. *Acta Crystallographica*, **12**, 515–519.
Stimmell, C., Heimann, R.B. and Hancock, R.G.V. (1982) Indian pottery from the Mississippi Valley: Coping with bad raw materials. Pp. 219–228 in: *Archaeological Ceramics* (J.S. Olin and A.D. Franklin, editors). Smithsonian Institution Press, Washington, D.C.
The American Ceramic Society (1992–1996) *Phase Equilibria Diagrams*. Vol. IX-XII. The American Ceramic Society, Westerville, Ohio, USA.
Tilley, C.E. (1938) Aluminous pyroxenes in metamorphosed limestones. *Geological Magazine*, **75**, 81–86.
Tite, M.S., Freestone, I.C., Meeks, N.D. and Bimson, M. (1982) The use of scanning electron microscopy in the technological examination of ancient ceramics. Pp. 109–120 in: *Archaelogical Ceramics* (J.S. Olin and A.D. Franklin, editors). Smithsonian Institution Press, Washington.
Tite, M.S., Freestone, I.C. and Wood, N. (2012) An investigation into the relationship between the raw materials used in the production of Chinese porcelain and stoneware bodies and the resulting microstructures. *Archaeometry*, **54**, 37–55.

Ullrich, B. and Ballmair, H. (2002) Fuldaer und Meissener Porzellan des späten 18. Jahrhunderts: ein technikhistorischer Vergleich, *Beiträge zur Naturkunde in Osthessen*, **38**, 5–13.

Ullrich, B., Lange, P. and Kerbe, F. (2010) Thüringer und Meissener Porzellan des späten 18. und frühen 19. Jahrhunderts – ein werkstoffhistorischer Vergleich. Pp. 281–289 in: *Porzellanland Thüringen, 250 Jahre Porzellan aus Thüringen*, Museumsverband Thüringen e.V.

Wachtman, J.B. (1969) *Mechanical and Thermal Properties of Ceramics*. National Bureau of Standards special publication, vol. 303. US Department of Commerce, NBS. 268 pp.

Walcha, O. (1981) *Meissen Porcelain*. G.P. Putnam's Sons, New York. 516 pp.

West, D.R.F. (1982) *Ternary Phase Diagrams*. 2nd edition. Chapman & Hall, London, New York.

Wijnen, M.H.J.M.N. (1981) The Early Neolithic I settlement at Sesklo: an early farming community in Thessaly, Greece. *Analecta Praehistorica Leidensia*, **XIV**, 1–146.

Williams, D. (1997) Ancient Greek pottery. Pp. 86–91 in: *Pottery in the Making. World Ceramic Traditions* (I. Freestone and D. Gaimster, editors), British Museum Press, London.

Wood, N. and Rastelli, S. (2014) Parallel developments in Chinese porcelain technology in the 13th–14th centuries AD. Pp. 225–234 in: *Craft and Science: International Perspectives on Archaeological Ceramics* (M. Martinón-Torres, editor), Bloomsbury Qatar Foundation, Doha, Qatar. http://dx.doi.org/10.5339/uclq.2014.cas.ch24.

Yin, M., Rehren, Th. and Zheng, J. (2011) The earliest high-fired glazed ceramics in China: The composition of the proto-porcelain from Zhejiang during the Shang and Zhou periods (c.1700–221 BC). *Journal of Archaeological Science*, **38**, 2352–2365.

Zanelli, C., Raimondo, M., Guarini, G. and Dondi, M. (2011)The vitreous phase of porcelain stoneware: composition, evolution during sintering and physical properties. *Journal of Non-Crystalline Solids*, **357**, 3251–3260.

EMU Notes in Mineralogy, Vol. 20 (2019), Chapter 7, 283–322

Mineral pigments: the colourful palette of nature

Ina REICHE

National Museum, Berlin and CNRS, Paris
e-mail: ina.reiche@chimieparistech.psl.eu

The use of minerals as pigments in art and on archaeological objects, from the use of ochre in prehistoric caves to the elaborate transformation and use in ancient and modern artist palettes, is reviewed in this chapter.

Starting from the purposes of the study of pigments, the chapter presents current trends in the study of coloured minerals in cultural heritage science. It emphasizes through the use of case studies the potential of these minerals in terms of information about former ways of life and especially the artistic techniques employed in ancient times.

This information is gained through knowledge of geological and physicochemical processes acting on minerals and on artefacts produced by human activities. Some new trends are presented as the state of the art of how to master most of the methods and techniques useful for investigating our common cultural heritage.

1. Introduction

Besides ancient glasses, ceramics and metals, mineral pigments are among the most studied materials in cultural heritage sciences. This is due to the fact that pigments play a crucial role in cultural heritage because of the wide colour palette they can provide. Beautiful colours are very decorative and persist over very long time scales so that they can even be preserved on objects dating back to the Stone Age. The available paint palette for artists evolved continuously from the use of coloured minerals or other naturally available materials such as charcoal to more elaborated paint palettes including synthetic pigments and colourants in modern and contemporaneous times. Today pigments and colourants, or mixtures of them, of nearly every perceptible hue are available.

Pigments have been used in paintings and drawings as well as to decorate various decorative objects, walls and outdoor architectural elements. From prehistoric times there are polychrome painted caves, painted stone or osseous tools. From later on, coloured glasses and glazed ceramics, illuminated manuscripts, drawings on parchment or paper as well as paintings on wooden supports such as the beautiful Fayoum portraits (Fig, 1; Röhrs *et al.*, 2017) are found. Today many different canvases are used for paintings with pigments in contemporaneous art.

Coloured decorations are often associated with strong symbolic meanings as humans make them on purpose. Therefore, the pigments and colourants used can be regarded as indicators of human thinking and technical choices. The choice of materials used for decorating art works is driven largely by the availability but also by the price of the raw

DOI: 10.1180/EMU-notes.20.7

Figure 1. Mummy portrait of a young woman, with an enlarged section and locations of the analysis (ANT Iinv. No. 31161,48). © Rathgen-Forschungslabor/Antikensammlung, Staatliche Museen zu Berlin.

materials. Some pigments, such as lapis lazuli and purple were highly valued in ancient times because natural blue and purple pigments were rare. Their use in art objects is thus an indicator of the social status of the commissioner and/or owner. All these reasons explain why the study of ancient pigments is of fundamental interest when studying ancient societies. A wealth of information can be gained from their study. Pigments hold in their chemical and mineralogical characteristics information that can be used to improve understanding of the artistic and archaeological purposes, socio-economical changes and societal status of objects because it is possible, for instance, to trace the materials chosen and the trading of particular pigments.

In the study of synthetic pigments, technological know-how and transfer can be obtained. Their date of production and introduction to the artistic market can be used as a *terminus ante quem*, meaning an indirect dating criterion for objects that contain a particular pigment and thus providing possible hints for confirming or denying the authenticity of objects. Alteration phenomena leading, for example, to fading or discolouration of sensitive pigments can be studied as well as strategies for conservation and restoration.

In the current chapter, the use of natural pigments in artistic and archaeological contexts from prehistory to contemporary times and their chemistry and mineralogy will be reviewed. Chemical and mineralogical analyses of pigments allow classification as a function of their properties, period of use, local knowledge, or geographical origin. The investigative methods range from classical to cutting-edge techniques: techniques based on sampling, techniques used in the field or on museum collections, techniques providing information at the micro and nanoscales as well as those allowing chemical imaging in the laboratory. Imaging devices used in the field are reviewed briefly. For more detailed information about analytical techniques the reader is referred to appropriate references such as *e.g.* Calligaro *et al.*, 2004; Dik *et al.*, 2008; Cotte *et al.*, 2009a, 2010; Bertrand *et al.*, 2012; Alfeld and Broekaert 2013; Reiche and Chalmin 2014; Alfeld and de Viguerie 2017.

Some case studies are discussed in order to illustrate the potential information gained from the scientific investigation of mineral pigments in the art-historical and archaeological contexts. As an example the composition and properties of Palaeolithic drawings in cave sites of southern France are presented. Furthermore, the study of art technology and the investigation of alteration phenomena of central Asian wall paintings in grottoes are quoted. Investigation and imaging of the chemical composition and layered structure of oil paintings are highlighted in a study of a Courbet painting with two hidden compositions.

2. Terminological definitions

2.1. Definition of minerals, pigments and colourants

Mineral pigments are referred to as naturally occurring coloured minerals. Mineralogically they are defined by their chemical, structural and physical properties. In the context of cultural heritage science the characteristic of outstanding importance is colour. That is the obvious reason why such minerals have been used for decorating various objects. They are generally crystalline and of inorganic origin although coloured biominerals such as corals, shells or ivory do exist. To name a few examples of typical minerals from heritage contexts, there are yellow orpiment, yellow goethite, red hematite, blue lapis lazuli, green malachite and black manganese oxides. In a wider sense synthetic coloured compounds can also be considered as 'mineral' pigments because they are all composed of inorganic constituents, although they are not minerals *sensu stricto*. Examples of synthetic inorganic pigments are Egyptian blue and Han purple or pigments composed of different constituents such as cobalt violet ($CoO \times$

Al_2O_3) introduced in 1775. There are also a few composite (hybrid) pigments such as Maya blue that are formed by a mineral (palygorskite) and an organic constituent (indigo) trapped in the clay crystal structure. In this chapter the focus is on the presentation of natural pigments. The reader is referred to the following references for more information on ancient and new synthetic pigments: Berke (2002, 2004, 2007), Hahn *et al.* (2006), Dejoie *et al.* (2010).

Colourants are defined here as coloured organic compounds that can be crystalline or amorphous, *e.g.* indigo or carmine.

2.2. Definition of paints

Polychrome surfaces or paintings can be created by dispersing pigments in a liquid binding medium on the surface forming so-called paint layers after drying. The layers vary in thickness, composition and properties as a function of the application technique, the binding medium and the type of support onto which they are applied. Liquid binding media, for example, may be drying oil, Arabic gum or egg white. Fillers may be added to paint; such fillers produce special drying or optical effects.

Generally, several superimposed paint layers create a 'stratigraphy' of polychrome surfaces and painting. Only particular mural paintings called frescos or prehistoric rock art figures are painted in one layer without binder as the lime mortar or the calcareous rock hold the pigments after drying. The nature and thickness of the 'stratigraphic' system can be very variable and range from a few tens of microns up to 1 mm, depending on the application technique and on the cultural and regional contexts.

2.3. Definition of colour

Because there are various ways of describing colours and coloured materials it is important to clarify the vocabulary used in this chapter. This is crucial because standardization of colour terminology did not exist in ancient times, meaning for instance that a mineral used to produce a coloured surface could show different hues and be named differently as a function of the context. Therefore, mineral colour alone is not a diagnostic feature, *i.e.* different minerals or mixtures may show almost the same visible colour, and the same mineral can yield very different colours depending on chemistry, structure, defects or particle size. In addition, the physical effects that produce a colour are numerous and depend on the texture and nature of the material to which the colour is applied or by which the pigments are applied. Thus, the same pigment or colourant can produce slightly different hues. Some secondary compounds can also nuance the hue of a colour.

Because of terminological complexity, the use of defined standards and internationally recognized systems as objective guidelines is crucial when approaching coloured materials of our cultural heritage. We can cite systems such as the NCS – *Natural Colour System*™®© of the Scandinavian Colour Institute AB or the Munsell colour system. The Munsell colour system was created by Albert H. Munsell (Munsell, 1912) and corresponds to a colour space based on three colour dimensions: hue, value (lightness) and chroma (colour purity). Hunter introduced another mathematical terminology in 1948, the Lab colour space (Sève, 2009). It describes all perceivable colours in three dimensions. L in Lab stands for lightness as well as the letters a and b for the colour extremes green–red and blue–yellow. Hunter Lab coordinates are the result of a square root transformation of colour data. In 1976 the CIEL*a*b* or CIELAB (Sève, 2009) system was created by a cube root transformation of the CIE XYZ colour data and represents an objective way of characterizing colours (Schelper *et al.*, 2016). The Lab colour space exceeds the complete subset of colours of the red-green-blue (RGB) model based on the human trichromacy and on the three cone cell types that respond to three bands of light: red, green and blue. It also exceeds the CMYK colour model, which is a subtractive colour system based on four colour dimensions often used in colour reproduction. In contrast to the later models the Lab model is device independent. Such a scientific approach enables all parameters which play a role in colour creation (such as the nature of colour origin, the presence of one or several colouring agents or other optical effects of the object surface) to be taken into account correctly. Here in this chapter, the term colour is used as a definition of a hue observed on objects.

2.4. Relevant archaeological and historical mineral pigments

Mineral pigments are well known in art-historical and archaeological contexts, as they are, with a few exceptions, very stable against alteration due to temperature, climatic changes, light and other environmental factors such as air pollution, so that they can be recognised even on very ancient objects.

A palette of a hundred different types of mineral, inorganic or hybrid pigments can be defined. They can be characterized by their colour, structure, grain size, geographical origin and region of use as well as by their type of use and their natural or synthetic origin (Eastaugh, 2004).

Minerals are divided into ten major chemical classes. Only a few coloured pigments such as lapis lazuli belong to the silicate class that represents >90% of minerals. Most pigments are oxides or mixed (hybrid) mineral compounds. This is linked to the fact that transition metal ions form coloured materials with oxides through optical absorption effects explained by crystal or ligand field splitting theory (Burns, 1993).

Selected common mineral pigments with their chemical composition or crystallographic phase may be identified unambiguously (Table 1). Indeed, some pigments show the same or a very similar chemical composition and can only be differentiated through identification of their mineralogical phases.

3. A general approach for analysing mineral pigments

Many archaeometric and art technological studies emphasize the importance of establishing analytical strategies appropriately adapted to the study of cultural heritage materials. The analysis of mineral pigments follows the same integrated reasoning. It is crucial to start the investigation with the question to be answered and to adapt the

Table 1. Selected mineral pigments of archaeological and art-historical relevance with the chemical formula of the colouring compound and, in the case of synthetic pigments, the date of production.

Colour and mineral classes with chemical formula and first date of use or invention in parentheses	White	Yellow	Orange	Red	Violet	Green	Blue	Brown/ black
Silicates						Egyptian Green Cu-glass	Lazurite/Lapis Lazuli $Na_{8-10}Al_6Si_6O_{24}S_{2-4fff}$ Cuprorivaite/ Egyptian blue $CaCuSi_4O_{10}$ Smalt Co-glass Chrysocolla $Cu_{2-x}Al_x(H_{2-x}Si_2O_5)(OH)_4)\cdot nH_2O$ $(x<1)$ Aerinite	
Sulfides		Orpiment As_2S_3 Cadmium Yellow (1825) cadmium sulfide CdS, C.I. Pigment Yellow 37 Greenokite $(Ca_4(Al,\text{-}Fe,Mg)_{10\text{-}}Si_{12}O_{35}(OH)_{12}CO_3\cdot\text{-}12H_2O)$	Cadmium sulfoselenide CdS + CdSe. Depending on the S/Se ratio, C.I. Pigment Orange 20	Realgar As_2S_3 Cinnabar HgS Cadmium red (Cadmium sulfoselenide) (1910) CdS + CdSe Depending on the S/Se ratio, C.I. Pigment Red 108		Zinc cadmium sulfide, CdS + ZnS, C.I. Pigment Yellow 35		

oxides and hydroxides	Anastase (1920) TiO_2 Rutile (1938) TiO_2 Antimony white (1920) senarmontite Sb_2O_3 Zinc white (1835) zincite ZnO	Goethite / yellow ochre α-FeOOH Lead-tin yellow Pb_2SnO_4 (type I)$PbSnO_3$ (type II) Massicot PbO		Haematite/ red ochre Fe_2O_3 Minium Pb_3O_4	Viridian $Cr_2O_3.nH_2O$ Paolo Veronese Green, Chrome green (1809) Atacamite / Paratacamite $Cu_2(OH)_3Cl$		Magnetite (Prehistory? /1920?) Fe_3O_4 Maghemite γ-Fe_2O_3 Manganese oxides and hydroxides (bixbyite, braunite, hausmannite, hollandite, polianite, pyrolusite, romane-chite...) MnO_2+Mn_2O_3 Jacobsite, $MnFe_2O_4$ Antimony black Sb_2O_3 Manganese and iron oxides $(Mn, Fe)_3O_4$ Cobalt oxide Co_2O_3
carbonates	Chalk $CaCO_3$ Lead white $2PbCO_3Pb(OH)_2$ Huntite $Mg_3Ca(CO_3)_4$				Malachite $CuCO_3(OH)_2$	Azurite $Cu(CO_3)_2Cu(OH)_2$	
nitrates		Cobalt yellow (1848) $K_3[Co(NO_2)_6]\cdot nH_2O$	Cobalt nitrate $Co(NO_3)_2\cdot 6H_2O$				
sulfates	Gypsum $CaSO_4\cdot 2H_2O$ Hemihydrate/ bassa-nite/ plaster of Paris $CaSO_4\cdot 0.5H_2O$ Anhydrite $CaSO_4$ Permanent white (1830) baryte $BaSO_4$				Posnjakite $Cu(SO_4)$		
Phosphates	apatite ivory or bone white $Ca_5(PO_4)_3OH$ zinc phosphate $Zn_3(PO_4)_2$				Bone turquoise or odontolite Mn^{5+} doped apa-tite $Ca_5(PO_4)_3OH$ Vivianite $Fe_3(PO_4)$ $8(H_2O)$		

arsenite					Scheele's Green copper hydrogen arsenite $CuHAsO_3$ (1775)		
organic compounds					Verdigris $Cu(CH_3COO)_2\cdot nCu(OH)_2$	Prussian blue (1704) $Fe_4[Fe(CN)_6]_3$ Indigo Indigotine $C_{16}H_{10}N_2O_2$	Carbon Graphite C
mixed / hybrid compounds	Lithopone (1853) $ZnO + BaSO_4$	Naples Yellow $Pb(SbO_3)_2$/ $Pb_3(SbO_4)_2$ Chrome yellow (1797) $2PbSO_4\cdot PbCrO_4$ Maya Yellow dehydro-indigo (indigoides) associated with palygorskite $(Mg,Al)_2$-$Si_4O_{10}(OH)\cdot 4(H_2O)$	crocoite chrome red (1809) $PbCrO_4\cdot Pb(OH)_2$	Cobalt violet (1775) $CoO\cdot Al_2O_3$	Cobalt green (1780) $CoO\cdot 5ZnO$ Emerald green (1814) $Cu(CH_3COO)_2\cdot 3Cu(AsO_2)_2$ Cadmium green (Cadmium yellow CdS + viridian $Cr_2O_3.nH_2O$	Manganese blue $BaSO_4\cdot Ba(MnO_4)_2$ Maya Blue indigo-palygorskite $(Mg,Al)_2$ $Si_4O_{10}(OH)\cdot 4(H_2O)$	Ivory or bone black C + $Ca_5(PO_4)_3OH$ (apatite)

analytical strategy to the type of object, the sampling options and the specificity and rarity of the object. In addition, the heterogeneity and state of preservation of cultural heritage materials need to be taken into account. Therefore, it is also of utmost importance to study these materials within their context so as to be able to select representative material and to obtain conclusive results from the selected analytical methods.

Recent technical developments allow more and more non-invasive analyses or chemical imaging *in situ*, thus keeping the object or sample intact and in its place during the analysis.

3.1. Laboratory investigation based on samples

3.1.1. Study of loose pigments

When a sample can be removed from a polychrome object, study of the pigment grains directly is possible. In this case the investigation should start with optical microscopy. Pigments are characterized on the one hand by their colour but on the other hand by: their particle size and size distribution (fine or coarse-grained pigments); particle shape (spherical, acicular, fibrous, bladed, lamellar, tabular, columnar, dendritic, pennate); the possible aggregation of several crystals (botryoidal habit, framboidal, cryptocrystalline); particle surface properties; fracture and cleavage properties; and degree of translucency. These properties are diagnostic and can be used for pigment identification through microscopic observation. Some of these characteristics can be observed using an optical microscope; other properties require the use of a polarized light microscope. If a crystal is pleochroic, change of colour is observed while rotating under plane-polarized light. The refractive index and relief are characteristic as well. Observation between crossed polars provides information about degree of anisotropy and extinction phenomena.

Optical micrographs of different pigments show characteristic mineralogical features of the crystal grains (Fig. 2).

Mineral pigments can also be studied by various techniques such as vibrational spectroscopies such as Fourier transform infrared spectroscopy (FT-IR) or Raman spectroscopy, X-ray diffraction, near infrared, visible and ultraviolet reflectance spectroscopy as well as X-ray fluorescence analysis (XRF).

FT-IR spectra of Prussian or Berlin blue and lead white enable pigment identification through characteristic absorption bands (Fig. 3). Lead white could be identified by FT-IR in the case of the pink area (Ba P3) of a double sided expressionist painting 'Barrikadenkampf (Revolution)' by Ludwig Meidner (Inv. No. NG 61/61, Neue Nationalgalerie, Staatliche Museen zu Berlin) but not vermilion, which was identified by micro-Raman spectroscopy (Fig. 4b). Figure 5 shows: (a) an XRF spectrum with its measurement point on a light blue area (b) of the Ludwig Meidner painting. Elements such as Fe, Ca, Pb, Si, Al, K, Ti, Zn, (Cu) may be detected indicating the presence of a mixture of Prussian blue, ultramarine, lead white and calcite (Röhrs *et al.*, 2014). Figure 6a shows an XRD pattern of modern Maya blue, indicating the presence of palygorskite, and Fig. 6b shows a Vis spectrum of an archaeological Maya blue sample

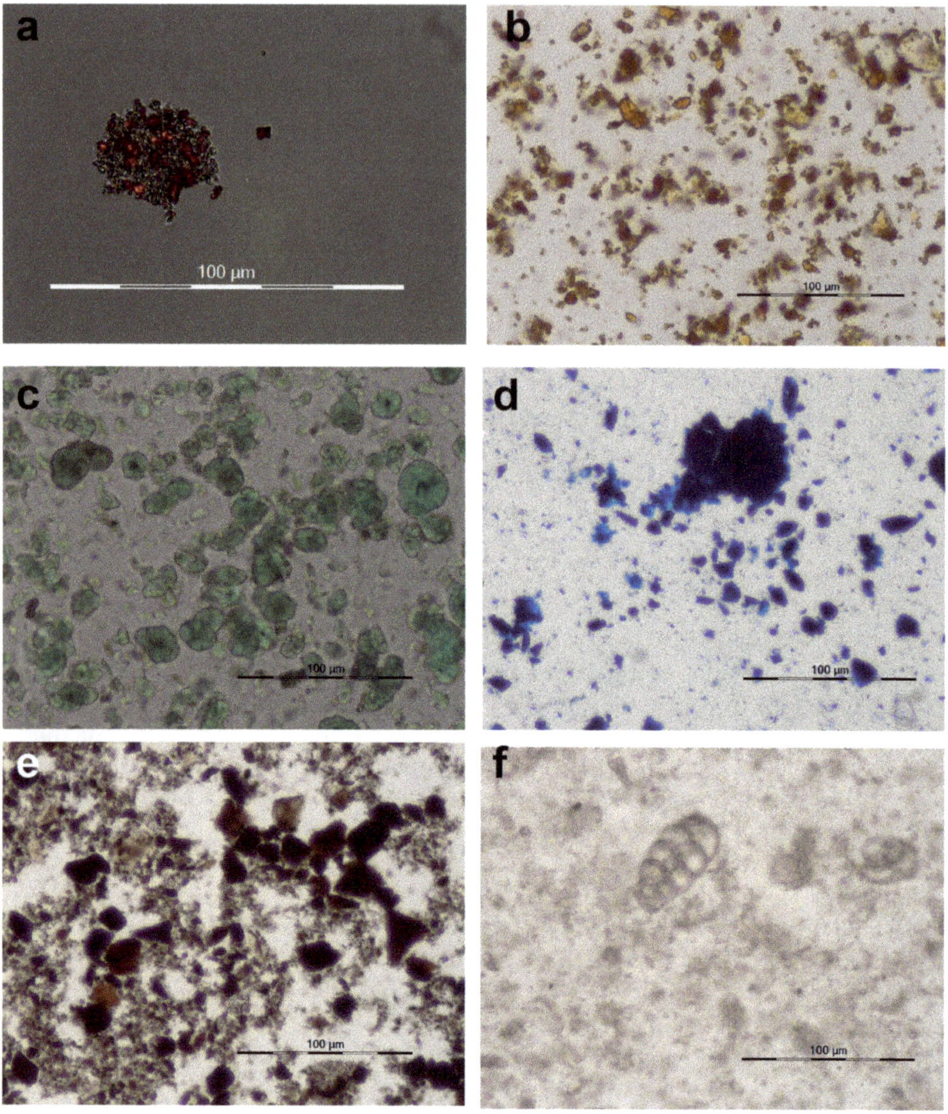

Figure 2. Optical micrographs of: (a) a sample from a pink area (Ba P3) containing white and red grains probably composed of lead white and vermilion, of a double sided expressionist painting 'Barrikadenkampf (Revolution)' by Ludwig Meidner, Inv. No. NG 61/61, Neue Nationalgalerie, Staatliche Museen zu Berlin-Preuβischer Kulturbesitz; (b) yellow (Terra di Siena); (c) green (Schweinfurter Grün); (d) blue (Prussian or Berlin blue); (e) black (bone and ivory black); and (f) white chalk (Rügener Kreide) pigments. All micrographs were taken with a × 50 objective. © Regine-Ricarda Pausewein, Rathgen-Forshcungslabor, Staatliche Museen zu Berlin-Preuβischer Kulturbesitz.

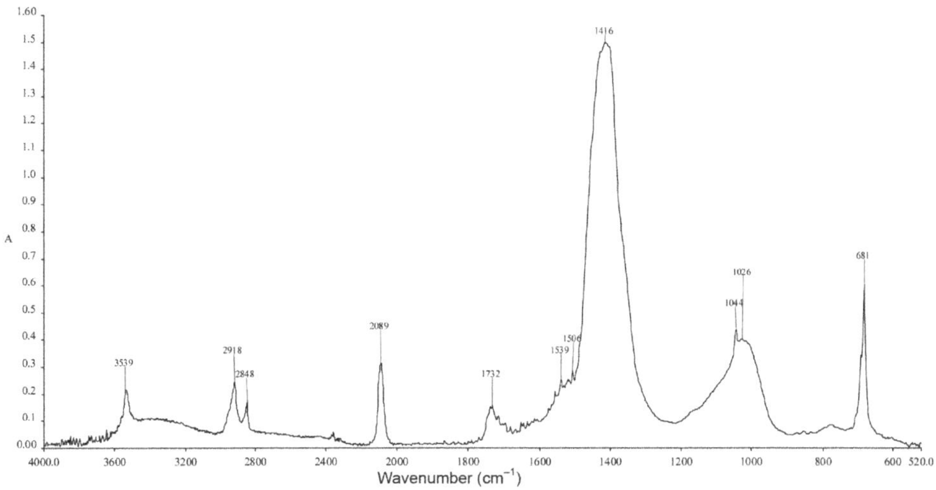

Figure 3. FTIR spectrum of lead white of a pink area (Ba P3) of a double-sided expressionist painting 'Barrikadenkampf (Revolution)' by Ludwig Meidner, Inv. No. NG 61/61, Neue Nationalgalerie, Staatliche Museen zu Berlin.

from a Ceramic head with painted blue headdress, Yucatan Mexico (Fig. 6c, Inv. No.: IV Ca 5119, Ethnologisches Museum, Staatliche Museen zu Berlin) in comparison to modern Maya blue and indigo references from Kremer (Kremer 36007 and 36003, respectively) (Morales-Merino *et al.*, 2016).

3.1.2. Study of pigments in paints

Mineral pigments are generally found in paint layers of paintings or of polychrome objects. The precious and rare nature of cultural heritage objects makes it necessary to study them in a non-invasive or micro-invasive way. Observations of the painting or objects under different lights (UV, Vis and IR imaging and false colour imaging, X-ray radiography, *etc.*) allow, in some cases, the pigment to be identified. Polychrome surfaces or paintings are composed of more or less complex layers of paint ranging from preparatory ground and priming layers to glazes and varnishes. To obtain a special hue, sometimes several paint layers are present and therefore several pigments are included in the layered structure of the paint. It is thus necessary to identify the pigment in each layer. The sample is thus prepared as a cross section allowing microscopic observation and analysis using micro-destructive or non-invasive techniques on each constituent layer (Fig. 8; Reiche *et al.*, 2015). The preparation of the cross section is key to obtaining a representative view of the paint stratigraphy as function of the kind of matter, its fragility, shape and size. Preparation also depends on the objective of the study: the art historical question to be answered.

After observation using light microscopy, scanning electron microscopy coupled with energy-dispersive X-ray analysis (SEM-EDX) is used to identify pigments from the characteristic chemical elements contained in each layer (Fig. 9). The particularly complicated Courbet painting *L'homme blessé* (Fig. 7) has two underlying

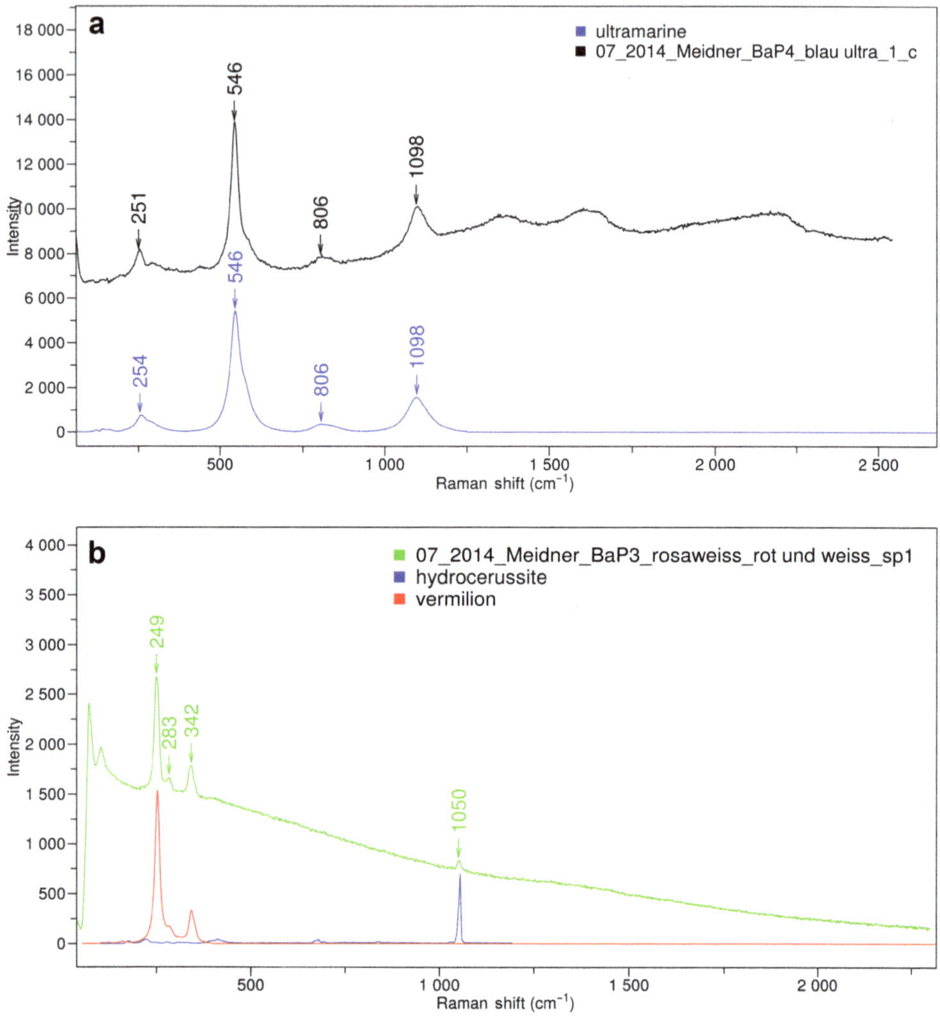

Figure 4. Micro-Raman spectra of (a) a blue area containing ultramarine (Ba P4) and (b) of a pink area (Ba P3) containing lead white and vermilion of a double-sided expressionist painting 'Barrikadenkampf (Revolution)' by Ludwig Meidner, Inv. No. NG 61/61, Neue Nationalgalerie, Staatliche Museen zu Berlin.

compositions: layer 1 of cross section 1089 contains lead white, iron oxide (orange) and very small amounts of barium sulfate (with Sr); layer 2 contains lead white, quartz, carbon-based black or brown and unidentified orange particles; layer 3 contains lead white, vermilion, Prussian blue, iron oxides (orange and red) with Al and Si and carbon-based black/brown; layer 4 contains lead white, iron oxides (orange and red) with As and Si, vermilion, cobalt blue (with Al) and organic red on an alum substrate (small Sn component); layer 5 contains lead white, iron oxides (orange) some with As and some

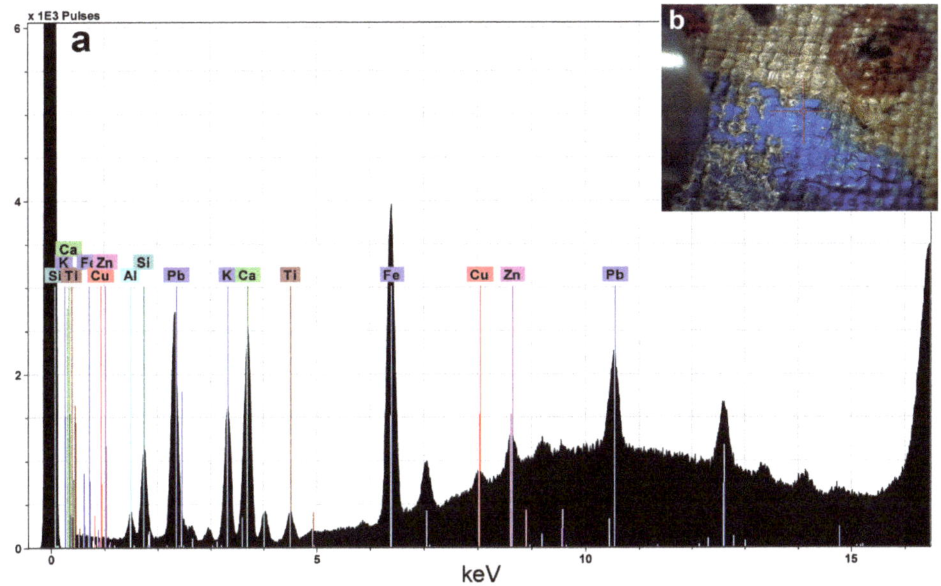

Figure 5. (a) Micro-XRF spectrum of a light blue area (b) containing Prussian Blue, ultramarine, lead white and calcite of a double-sided expressionist painting 'Barrikadenkampf (Revolution)' by Ludwig Meidner, Inv. No. NG 61/61, Neue Nationalgalerie, Staatliche Museen zu Berlin.

with Al, feldspar (K Al silicate), bone black (with Mg) and carbon-based black/brown; layer 6 contains bone black, umber, Mn in clay and umber: layer 7 contains lead white, barium sulfate (with Sr), bone black (with Mg), carbon-based black. The ground layer is missing in this cross section and layer 8 corresponds to the varnish of the actual composition (Reiche *et al.*, 2015).

The identity of pigments in the paint layers can only be inferred from EDX analysis. A precise identification of each pigment needs further phase analyses like FTIR in ATR mode (Fig. 10) or optical spectroscopy such as fibre optic reflectance spectroscopy (Röhrs and Stehr, 2014).

3.2. *In situ* pigment analyses using portable equipment

3.2.1. Digital microscopy

As for investigations in the laboratory, *in situ* analyses should start with microscopic observations. New digital devices are available that allow the study of the objects in the museum or at the archaeological site. Such investigations are of utmost importance for choosing the appropriate analysis zones for further *in situ* pigment analyses. It allows the definition of the different pigment hues present, the zones with overpaints, former reparation or restorations. In the case of the leather wall hanging 'The Island of Formosa' composed of ten different painted leather parts, twelve different hues can be distinguished (Fig. 11a). Patches sewn on the object and inlays glued to the surface can also be seen. Some areas show better protected colour and leather. Two areas are clearly

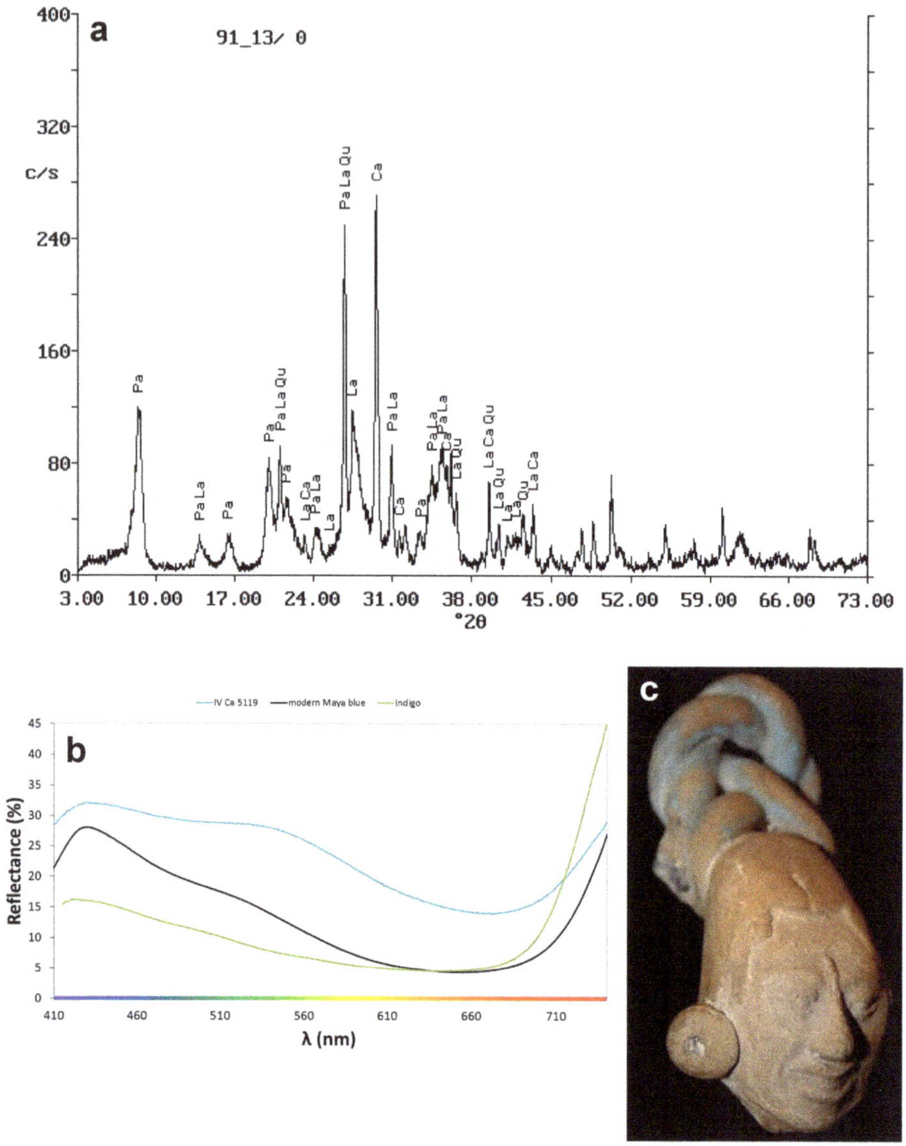

Figure 6. (a) XRD diagram of modern Maya blue (Kremer 36007) and (b) Vis spectrum of archaeological Maya blue of the object depicted in (c) ceramic head with painted headdress, Yucatan Mexico (Inv. No. IV Ca 5119, Ethnologisches Museum Berlin) with respect to a modern Maya blue and indigo reference.

lighter than the others. The microscopic analysis also confirmed that the blue corners are painted over the frame decorations, which might mean that they were added later. Some parts of the painting layer are missing. Further missing parts are holes, seam yarn and other details in the leather itself (Fig. 11b) (Aibéo *et al.*, 2016).

Figure. 7. (a) Courbet painting *L'homme blessé* (81.5 cm × 97.5 cm, Inv. No. RF 338, Musée d'Orsay, Paris) with a detail (b) of the photograph (and XRR superimposed) of the centre of the Courbet painting showing the two hidden compositions

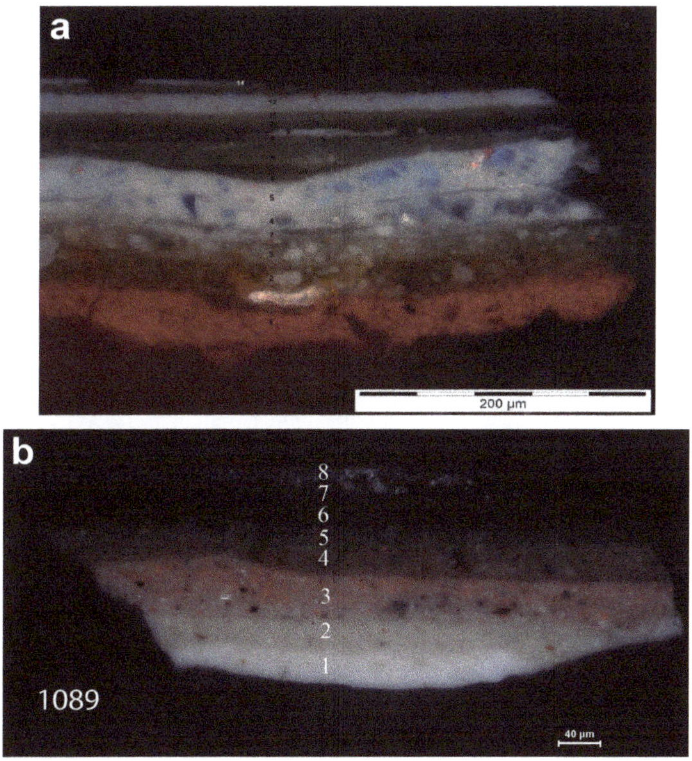

Figure 8. (a) Cross section of sample P3 showing up to 14 paint layers of the Caspar David Friedrich painting 'Mönch am Meer' (Inv. No. NG 9/85), Alte Nationalgalerie, Staatliche Museen zu Berlin and (b) Cross section no. 1089 showing up to ten paint layers of the Gustave Courbet painting 'L'homme blessé' (Inv. No. RF 338), Musée d'Orsay, Paris.

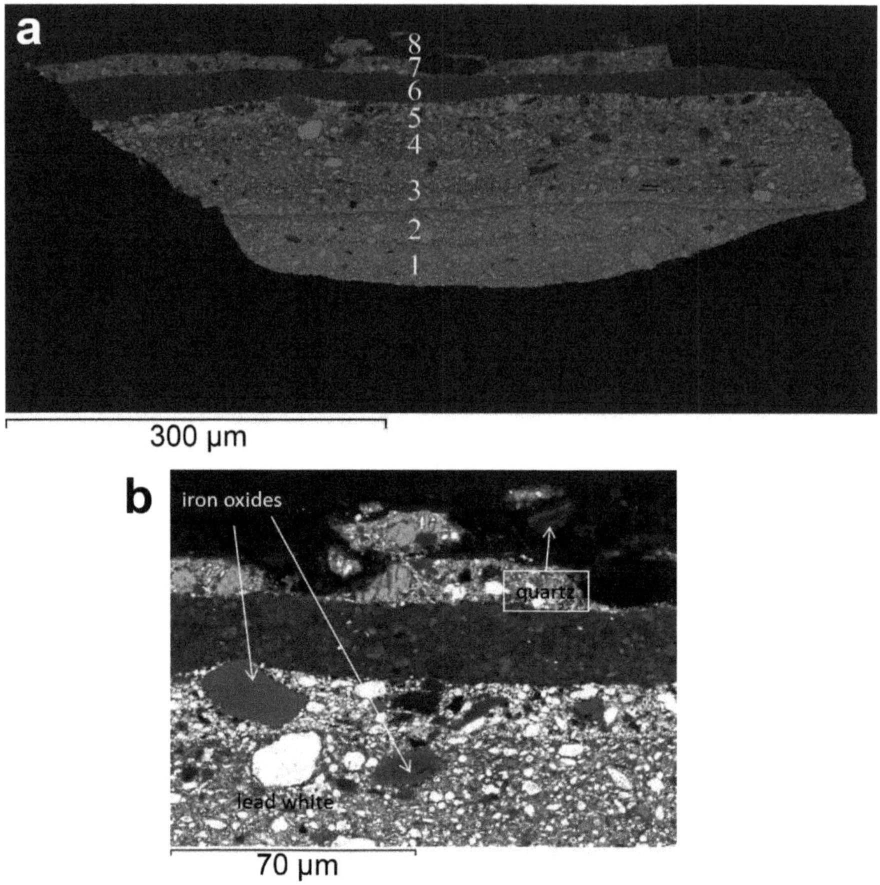

Figure 9. Electron micrograph (a) of the cross section no. 1089 of the Gustave Courbet painting *L'homme blessé* (Inv. No. RF 338), Musée d'Orsay, Paris and (b) of a detail of the layers 4 to 7 with the minerals present indicated.

3.2.2. *X-ray fluorescence analyses*

After definition of the analytical points, further *in situ* analyses are possible. First, chemical analyses using non-invasive portable XRF can be performed. As a matter of fact pXRF is almost ideal for studying paint layers, as they are generally flat, thin and free from absorbing coatings. There are several portable devices available ranging from a handheld XRF spectrometer with a spot size up to 1 cm down to a micro-XRF spectrometer with spot sizes of 100 μm. We also have a small device allowing *in situ* chemical imaging using pXRF on small zones up to 20 cm × 20 cm in size.

In situ XRF analysis of the pigments contained in the wall hanging 'The island of Formosa' was performed using the portable ArtTAX Pro spectrometer (Bruker AXS, Mo tube excitation). The elements from sodium (Na) to uranium (U) can be detected by the spectrometer. For better detection, a vacuum head was used; this minimizes the

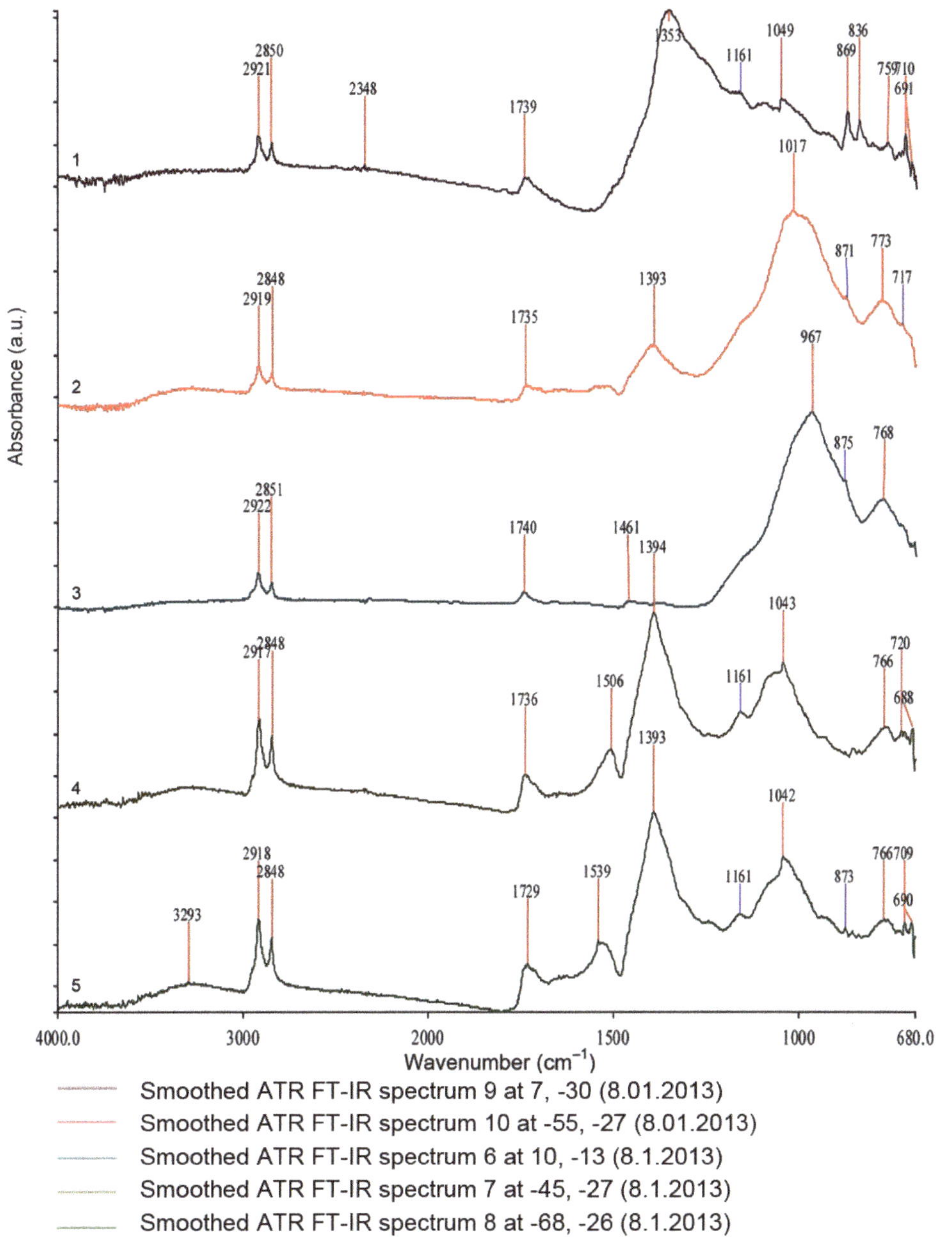

Figure 10. ATR FTIR point analyses for identification of the constituents of paint cross sections. Examples of several measured spectra on the cross section P3 of the Caspar David Friedrich painting 'Mönch am Meer', ANG showing the presence of smalt in layers 5 and 6. Characteristic absorption bands of the Co-glass smalt are at ~1080 cm^{-1}.

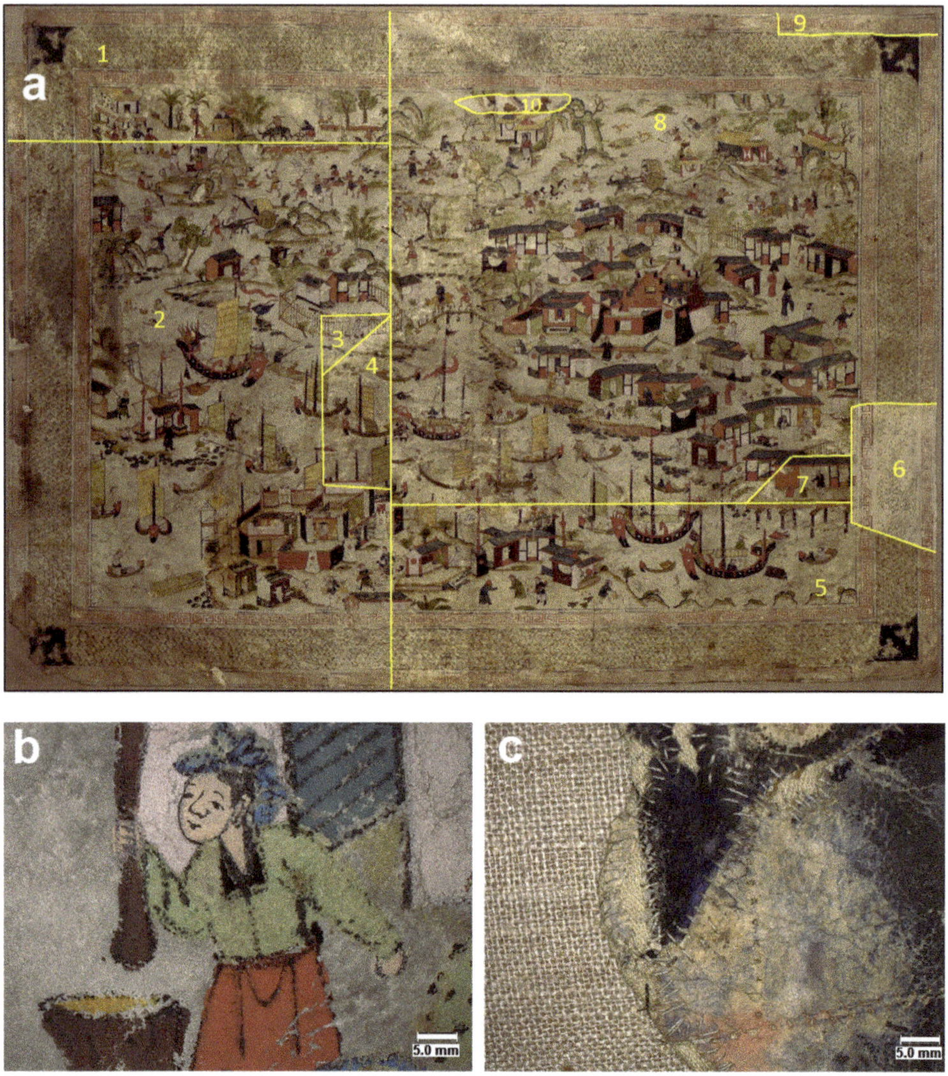

Figure 11. (a) Leather wall hanging 'The Island of Formosa', 17th c. kept in the Ethnologisches Museum in Berlin, Inv. No. I D 37597, digital micrographs of: (b) a detail (× 5 objective) showing different hues and (c) a repaired area.

amount of air through which the X-rays must pass and therefore increases its sensitivity to the lighter elements. A constant flow of He gas is used to minimize adsorption of the fluorescent X-rays by air. The area analysed has a diameter of ~0.1 mm (or 100 μm) thanks to focusing X-ray optics. This helps to study small details on the object. The detailed experimental conditions were 45 kV, 100 μA, 100 s life time, 0–50 keV spectral range and background suppression by 60 smoothing cycles. These conditions allowed us to infer a large range of pigments used for the decoration of the leather wall hanging including cinnabar and sometimes traces of minium, a mixture of minium and lead white, lead white alone, a mixture of cinnabar and lead white, goethite and a combination of azurite and lazurite. In the blue corners a special blue was found as it contains Co, Ni, As and Pb. It is probably the synthetic pigment cobalt blue (Fig. 12). Green, yellow and black pigments could not be identified using micro-XRF as they are mainly or partly composed of organic constituents (Aibéo *et al.*, 2016).

Even depth-resolved confocal microXRF analyses are possible *in situ* nowadays. They allow the investigation of the chemical composition of different paint layers without sampling. However, these analyses are restricted to the upper paint layers because of the absorption of the emitted X-rays by upper layers. The stability of the positioning of the depth-resolved analyses is still an issue (Müller *et al.*, 2017).

3.2.3. Micro-Raman analyses

For the direct identification of pigments, complementary *in situ* analyses are necessary using portable micro-Raman instruments. As for portable XRF devices, there is a wealth of modern Raman spectrometers that allow *in situ* pigment study. They are optimized to work in a horizontal set-up to work directly on the exhibited object. A Horiba XploRa micro-Raman system was used for the analyses of the leather wall

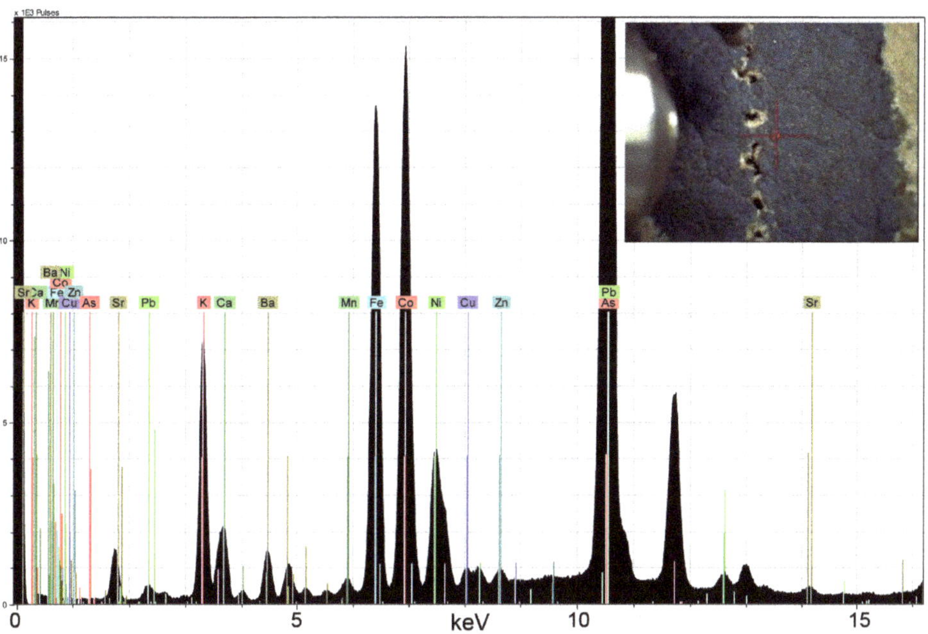

Figure 12. Micro XRF spectrum of 'The Island of Formosa' leather wall hanging kept in the Ethnologisches Museum in Berlin, Inv. No. (I D 37597), spot 43_16_139_blue in the blue corners, showing the presence of K, Fe, Co, Ni, As, Pb and traces of Ca, Ba, Cu and Zn.

hanging 'The island of Formosa'. This device is equipped with three lasers with wavelengths of 532, 638 and 785 nm. The maximum power of each laser is 25 mW (532 nm), 24 mW (638 nm) and 90 mW (785 nm). Moreover different filters permit control of the laser power (transmissions of 100%, 50%, 25%, 10% 1%, 0.1%) and the spectral resolution is controlled by the use of different gratings. A Si standard with a characteristic band at 520.5 cm^{-1} is used for wavelength-scale calibration. Data acquisition and spectra elaboration are performed with specific NGS *LabSpec* software. In order to be able to work *in situ* the set-up was modified with respect to the usual Raman microscopy arrangement. The laser path is deviated by 90° using a so-called macro accessory which contains a 45° sloped mirror for analyses. By means of these investigations indigo and carbon black (Fig. 13) could also be identified on the leather wall hanging and the presence of further pigments such as minium, cinnabar, lazurite, *etc.* could be confirmed (Aibéo *et al.*, 2016).

3.2.4. Complementary in situ analyses

These pigment analyses can be complemented by further *in situ* analyses using, for instance, portable Vis spectrometers and FTIR spectrometers working in reflection mode (Vetter and Schreiner, 2011). The field of technological development to miniaturize and to optimize spectroscopic methods to be portable so that they can be used *in situ* is very dynamic. Every day there are new options for *in situ* pigment analyses. Even new options for pigment identification using portable micro-XRD analyses are possible today (Gianoncelli *et al.*, 2008; Eveno *et al.*, 2010).

3.3. Investigations using imaging techniques

3.3.1. In the laboratory

As noted above the analysis of pigments in paintings is generally performed by means of optical (OM) and SEM-EDX on samples and cross-sections along with various non-invasive surface-sensitive analytical techniques (X-ray fluorescence, Raman spectro-

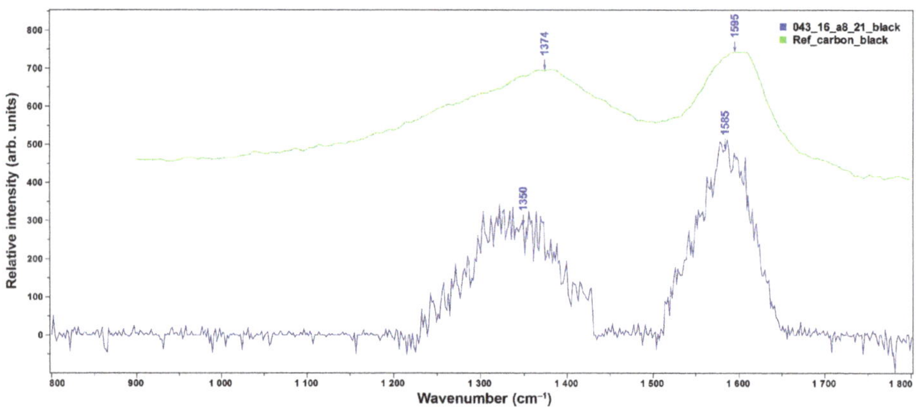

Figure 13. In situ micro-Raman spectrum of carbon black identified on the black area of 'The Island of Formosa' leather wall hanging kept in the Ethnologisches Museum in Berlin, Inv. No. I D 37597.

scopy or X-ray diffraction). This analysis is restricted to tiny zones. Therefore, the need arose to use imaging techniques to get a more representative view of the pigment distribution in paintings and polychrome objects. Classical painting imaging methods are UV imaging, infrared reflectography or X-ray radiography.

These imaging techniques can today be complemented by new chemical imaging techniques such as Raman and FTIR imaging, macro-XRF scanning (MA-XRF) and hyperspectral imaging techniques (Dik *et al.*, 2008; Janssens *et al.*, 2010; Alfeld *et al.*, 2011, 2013; Alfeld and de Viguerie, 2017). All these techniques are available as stationary laboratory equipment. Non-invasive two-dimensional scanning X-ray fluorescence (MA-XRF) spectroscopy provides the imaging of elements present in paintings down to trace concentrations. This is achieved by scanning the work area with an X-ray beam of a small diameter and recording spectra point by point. Numerous applications of this technique – even using mobile scanners as reviewed by Alfeld and de Viguerie (2017) have highlighted the potential of MA-XRF for scanning precious paintings from different periods revealing overpaints, ancient restorations, *etc*. Such set-ups have already demonstrated attractive potentials and capabilities on numerous art objects and especially paintings. The main advantage of this technique is that the chemical XRF image can be superimposed on the optical image. Therefore, MA-XRF imaging of paintings is today a well established method using different sources and also using mobile equipment. The MA-XRF imaging of the Courbet painting *L'homme blessé* shows the possibility of chemical imaging of paintings in the laboratory. The investigation was carried out with a prototype designed at the C2RMF. The detailed experimental description of the instrument is reported elsewhere (Ravaud *et al.*, 2016). Maps of 100 pixels × 100 pixels were acquired with a step size of 1.4 mm corresponding to a 140 mm × 140 mm area. Eleven areas could be scanned on the painting. They were selected according to the most interesting zones identified on the X-radiograph. The selected areas correspond to the zones with the most important changes in the three compositions of the painting. The data were processed with the open source program *PyMCA* (Solé *et al.*, 2007) which allows the generation of fitted peak area maps for various elements from elemental maps (in PNG format) derived using a program developed at the C2RMF called *DataImaging 3D* (Pichon *et al.*, 2015). Detailed descriptions of the chemical maps obtained of the paintings were presented by Reiche *et al.* (2016). As an example, the Hg L map of the case study presented in detail in Section 4.3 is shown indicating two hidden underlying painting compositions (Fig. 14). Besides the head of a male, visible today, two female heads are visible in the Hg L distribution map. The Hg L map shows deeper paint layers as the Hg L fluorescence line has a relatively high energy and thus a greater depth of information.

3.3.2. *In situ MA-XRF scanning analyses*

Chemical imaging studies are also possible today using portable XRF scanning devices such as the portable ELIO XRF device (XGlab®). The *in situ* XRF map of one area of the leather wall hanging 'The Island of Formosa' is shown as an example. The area analysed is at the heart of the wall hanging in the so-called area 2 on a selected boat. The area scanned is 2 cm × 4 cm (Fig. 15). The corresponding chemical maps of the Hg Lα

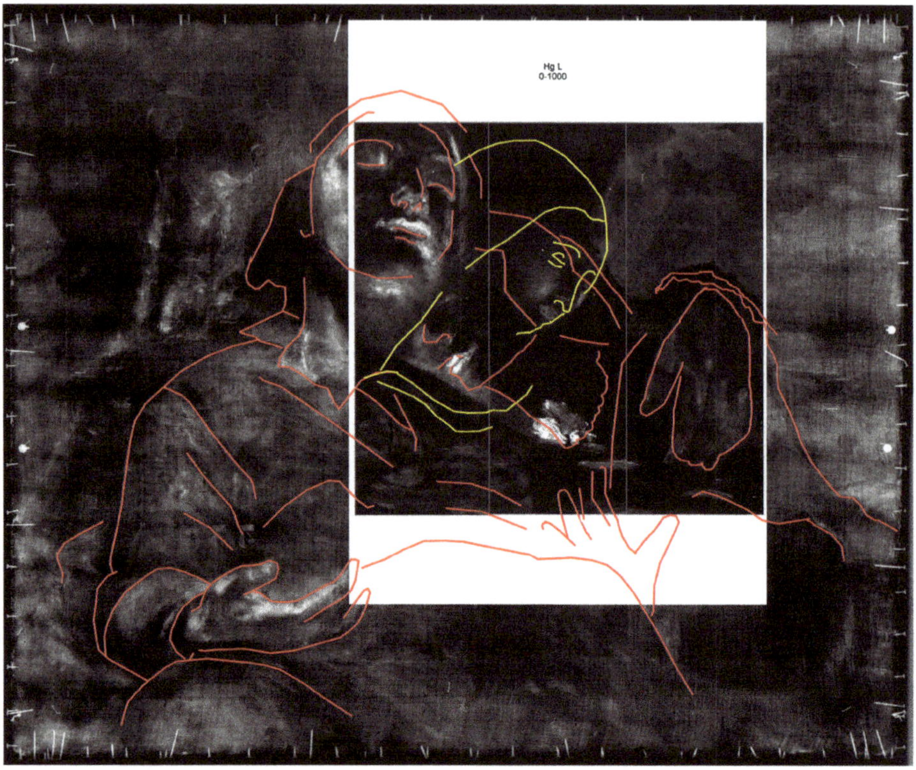

Figure 14. Macro-XRF Hg L map of eleven scanned zones superimposing the X-ray radiography of the Gustave Courbet painting 'L'homme blessé', Musée d'Orsay, Inv. No. RF 338 and the two hidden compositions indicated in yellow (first one) and red (second composition)

Figure 15. Photograph of the area of the XRF mapping in area 2 of the wall hanging 'The Island of Formosa'.

and the Pb Lα fluorescence lines (Fig. 16) show that the red area is rich in Hg, indicating vermilion and light yellow, red and pink zones contain significant amounts of lead indicating the presence of lead-containing pigments such as lead white and massicot. The dark yellow does not contain much Pb and is therefore likely to be composed of an organic yellow colourant.

3.3.3. On the micro- and nanoscale, combined imaging methods at large-scale facilities

For the very special purposes of pigment identification and analysis, imaging methods such as ion beam facilities or synchrotron sources are required. Such methods are necessary because of the intrinsic complexity and multi-scalar heterogeneity of painting materials, combined with the relative abundance of their individual components. In addition, the analysis of chemical elements at the trace and ultra-trace level, as well as determination of crystallographic phases at the micro- or even nanoscale is often necessary in order to answer questions related to provenance, production techniques, alterations and conservation of pigments. Several methods may need to be combined on a small scale in order to characterize individual constituents. This methodological challenge can only be met at large-scale facilities. In particular, synchrotron radiation sources offer polychromaticity, high flux, low divergence, small source size, stability, 'calculability' of the source, and polarization; these characteristics are well adapted to meet the aforementioned challenges. After preliminary laboratory investigations, such approaches using high-tech equipment can solve very specific problems in cultural heritage science.

It is not possible to cover exhaustively all significant on-going imaging developments at large-scale facilities. This chapter is, thus, restricted to the main methods used for high-definition imaging in cultural heritage.

Synchrotron-based XRF or accelerator-based PIXE imaging are, without doubt, the most commonly used methods at large-scale facilities. They offer non-invasive analysis of the chemical compositions of pigments from the macro- to the nanoscale, while being sensitive to major, minor and trace elements. Synchrotron XRF is very often combined with X-ray absorption spectroscopy and X-ray diffraction, because these methods can be performed simultaneously using the same X-ray source by

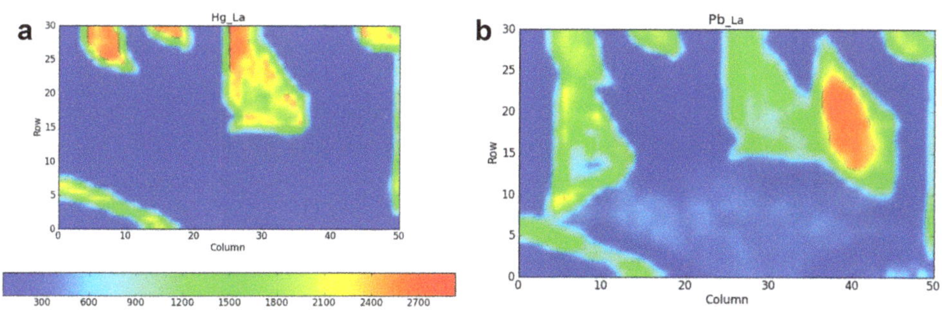

Figure 16. XRF-maps for (a) Hg Lα and (b) Pb Lα in area 2 of the wall hanging 'The Island of Formosa'. The intensity bar ranges from blue (low intensity) to red (high intensity) of the X-ray line.

adopting different imaging detectors. Further combinations are possible with other X-ray techniques such as small-angle X-ray scattering (SAXS) and computed tomography (CT) but molecular spectroscopic methods can also be performed simultaneously with X-ray imaging such as Raman and FTIR imaging. For more detailed information, the reader is referred to further literature (Dik *et al.*, 2008; Cotte *et al.*, 2009a,b, 2010; Janssens *et al.*, 2010; Bertrand *et al.*, 2012; Reiche and Chalmin, 2014; Albéric *et al.*, 2015; Gay *et al.*, 2015; Alfeld and de Viguerie 2017)[i6] which presents various possible combinations of large-scale facility imaging methods used in cultural heritage studies.

4. Examples highlighting the information obtained when studying mineral pigments in historical and archaeological contexts

The pigments and colourants selected and the complexity of the paint palette employed reflect differences in the choice of artistic technique and in the symbolic meaning or function of works of art. They can also be used as indicators of the social rank of the commissioner or owner or be specific to a social group.

Therefore, the technical study of pigments and colourants used in archaeological artefacts and in works of art allows us insights into the nature, genesis and origin of artefacts, into the production or transformation techniques used, and indirectly into the use and exchange of knowledge in past societies. Such techniques may also solve problems of authenticity.

Other important issues concerning the technical study of pigments and colourants in cultural heritage are the understanding of their alteration over time, the set-up of adequate conservation measures and the establishment of the best possible restoration strategies as well as in the detection of forgeries.

The following examples are intended to illustrate the analytical approach and the variety of questions that can be addressed while studying ancient mineral pigments.

4.1. Non-invasive *in situ* characterization of manganese oxides used for Palaeolithic cave art in the Rouffignac cave, Dordogne, south-western France. Differentiation of artists' contributions

Since its recognition as prehistoric art at the end of the 19th century, rock art preserved in limestone caves in southern France and in northern Spain have been subjected to extensive research that has been performed to study the style and technique of the paintings and drawings, to date the figures, and to determine their possible overlap and sequence of execution. Chemical analysis allows the identification of the constituent pigments as well as the 'paint pots' obtained by mixtures of different minerals and possibly also organic compounds that are today degraded. Paints or crayons used for cave art generally contain red hematite or red ochre (α-Fe_2O_3), yellow goethite (α-FeOOH) or yellow ochre; whereas black can contain either charcoal or various manganese oxides/hydroxides. In the very fortunate case of the use of charcoal or other C-based compounds the decoration of cave walls can be dated directly using the ^{14}C

method (Valladas *et al.*, 2001; Beck *et al.*, 2013; Quiles *et al.*, 2016). In the absence of black C-based pigments other indirect means have to be employed in order to gain more insights into the chronological order of prehistoric representations. Therefore, archaeologists seek the presence of C-based paints and pigments in caves. Further studies are needed of the artistic techniques to differentiate between various artists involved in painting caves to establish the chronological order for the representations. Iron oxides, various Mn oxides and hydroxides and C-based compounds can be studied using *in situ* techniques, which make sampling obsolete. On-site XRF analyses, XRD and Raman analyses after microscopic observations of the walls are possible today thanks to mobile spectrometers (Lahlil *et al.*, 2012; Beck *et al.*, 2014; Gay *et al.*, 2015, 2016). From the variety of paint and drawing composition and application techniques, it can be assumed that it is possible to distinguish several artists involved in the execution of such rock art. The XRF analysis of Fe-based red pigments is difficult because the rock substrate generally also contains Fe-bearing minerals. Concerning Mn oxides/hydroxides, their differentiation from the rock substrate is more straightforward because they are sufficiently different in composition. A differentiation or a classification into different pigment groups is possible by means of XRF analysis. However, there is a large variety of different mineralogical Mn oxides and hydroxides with similar chemical compositions so that *in situ* phase analysis is required to separate pigments used for different prehistoric representations. The challenge of pigment identification is enhanced by the fact that Mn oxides and hydroxides are generally poorly crystalline so that they show diffraction patterns with a small peak-to-noise ratio. However, several studies in the Rouffignac cave showed the feasibility of a combination of XRF/XRD and Raman analyses of the prehistoric figures in order to distinguish pigment groups and to identify different mineral phases as a function of the pigment group identified.

The Rouffignac cave situated in the Dordogne region in southern France is one of the key sites of Upper Palaeolithic art. It shows ~240 animal drawings and engravings on the walls attributed to the Magdalenian period (Barrière, 1982; Plassard, 1999; Plassard and Plassard, 2016). Because of the predominance of representations of mammoths, this animal is the distinctive element of the cave. It is also referred to as 'the hundred mammoths' cave'. Its largest drawing composition, the 'Grand Plafond', comprises 65 black animals corresponding to one third of the cave's rock art (Fig. 17a). On the 'Grand Plafond' (Great ceiling) the figures are arranged without any striking global organization, whereas in the 'Henri Breuil gallery' with 'the ten mammoths' frieze' and the 'rhino frieze' as well as the 'panel of the large Patriarche', although engraved, are all well aligned.

Four of the ten mammoths and one rhinoceros of the 'three rhinoceros' frieze' of the 'Henri Breuil gallery' and the two isolated 'faced mammoths' as well as 45 figures of the 'Great Ceiling' (Grand Plafond) have been studied systematically.

The methods used were necessarily non-invasive, allowing mineralogical and chemical study of the figures without sampling or touching the figures. The techniques used were portable micro-Raman spectroscopy and XRD as well as XRF analyses. Raman

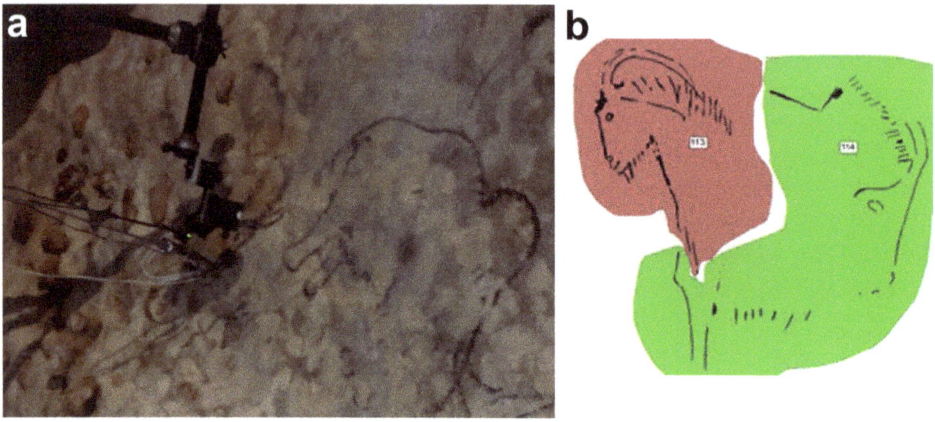

Figure 17. (a) Portable XRF image of a bison of the Great Ceiling at Rouffignac cave, Dordogne, France, (b) parts of the figures of ibex 113 and bison 114 reconsidered as a function of the chemical composition of the strokes.

spectroscopy, together with XRF, is one of the most widely used methods for the study of mineralogical phases in cultural heritage artefacts and pigments because of its non-invasive nature. It has already been used to study cave art several times (Hernanz *et al.*, 2006, 2008; Ospitali *et al.*, 2006). In the context of cave art it is of special value to be able to detect figures drawn with C-based pigments for dating purposes. Two very intense bands at 1390 and 1590 cm^{-1} (Tomasini *et al.*, 2012) allow graphite to be identified. In the case of Rouffignac, the Raman investigation determined mainly Mn oxide-based pigments (Lahlil *et al.*, 2012). Carbon-based materials have been identified in very localized areas, namely: (1) in the line, considered to be modern, outlining the head of the mammoth (# 66) located on the Great ceiling, (2) in small C particles on the mammoth (#180) of the two isolated 'faced mammoths', and (3) carotenoid-type organic C on the mammoth frieze and the rhino frieze. The carbon-based pigments cannot, however, be used for dating the figures because they are present in insufficient amounts and are likely to be from elsewhere and not related directly to the prehistoric pictures (Beck *et al.*, 2013). Therefore, the study should be complemented by the identification of the mineral phases in order to classify the figures according to different pigments or pigment mixtures. Using XRD and Raman spectroscopy, two main types of Mn oxides were identified among the different figures: pyrolusite (MnO_2) and romanechite ($(Ba,H_2O)_2Mn_5O_{10}$) containing ~20 wt.% of BaO. They were also identified as mixtures of pyrolusite or hausmannite (Mn_3O_4) and romanechite giving rise to an intermediate BaO group (Lahlil *et al.*, 2012; Beck *et al.*, 2014).

The figures of the 'ten mammoths frieze' studied by XRF show a homogeneous picture with respect to its composition. They contain a Mn oxide with ~10±3 wt.% of BaO. In addition, XRD results showed that quartz was present systematically either in the pigment or as a layer below the pigment (Beck *et al.*, 2014).

The work confirms, therefore, that 'ten mammoths' frieze' of the Henry Breuil Gallery is an homogeneous mineralogical group, which is in good agreement with the stylistic homogeneity of the drawings. Mineralogical results confirm the hypothesis of a unity of conception and realization. Despite the prevailing content of one type of Mn oxide, the black pigments used to draw the mammoth frieze seem to be a mixture of several mineralogical phases, an observation that is compatible with the use of untransformed raw materials from natural sources. Furthermore, romanechite is a natural Mn oxide, which cannot be obtained by transformation of other Mn compounds. This result confirms that the stylistic coherence of the 'ten mammoth frieze' is the result of one artist or group that worked over a relatively small period of time with the same raw material. In contrast, the figures in the 'rhinoceros frieze' are different from the 'ten mammoth frieze' and similar to some of the Great Ceiling. In the various animals (horse, mammoth, ibex and bison) of the Great Ceiling, all three types of Mn oxide-bearing pigments were identified in the different figures. These pigment groups cannot be matched to a type of animal representation or to stylistic groups. Research is currently under way to relate stylistic observations and superimposition studies to the chemical fingerprint of the studied figures. It seems obvious that several hands created the prehistoric artworks of the Great Ceiling in a relatively short time leading to a panel lacking an evident overall organization (Gay *et al.* submitted). The first results of some subsets show the value of combing different types of approaches, allowing the rereading of figures and subsets of figures of the Great Ceiling. As an example, two neighbouring drawings, the ibex 113 and the bison 114 of the panel are considered (Fig. 17b). The front legs of the ibex 113 show pigment compositions of ~7% BaO, corresponding to the intermediate BaO-containing MnO group. This composition is not in agreement with that obtained for the ibex muzzle, the back of its head and its horn (~3% BaO) corresponding to the very-low-BaO-containing MnO group. Interestingly, the chemical signature of the ibex legs matches perfectly that of the head of the neighbouring bison 114. The chemical characteristics, therefore, allow a new reading of these two figures and indicate that the strokes, initially interpreted as front legs of ibex 113, actually represent the front leg of bison 114 (Fig. 17b). As another example, different chemical signatures were found for several parts attributed to mammoth 122. Drawing strokes inside the mammoth and those on its legs illustrating the animal's fur differ from one of the contour strokes of the drawing (6% and 3%, respectively). It is probably a stylistic addition by the artist him/herself or, more likely, was added by another artist possibly at a later date.

Other examples discussed by Gay *et al.* (submitted) show the alternative use of different crayons to create more or less complex compositions of figures. This reinforces the idea that the 'Grand Plafond' was probably created in successive subsets of figures by small teams of artists without taking into account the general organization of the panel. The drawing of these subsets can be imagined the same way as the creation of figures of the mammoth or the rhinoceros friezes in the Henry Breuil Gallery drawn over a relatively short time period by a very small team. The superimposition of these different subsets of the great ceiling leads to its actual presentation.

4.2. Alteration mechanisms of pigments on central Asian wall paintings from the Northern Silk road, kept at the Asian Art Museum, Berlin

Land and maritime routes, linking East and West for thousands of years and known today under the terms 'Silk Roads' and 'Spice Routes' enabled not only the trade of merchandise such as silk, wool, gold, metals, ivory, spices and other precious items but also the exchange of cultures, ideas, knowledge and religious beliefs. Through these ancient routes, Buddhism could spread from India to China so that important Buddhist monasteries flourished over centuries in Central Asia, including the famous sites of Turfan and Kucha, situated on the northern edge of the Taklamakan desert or Khotan and Miran, south of the desert. Grottoes and temples, richly decorated with wall paintings and painted sculptures, served the Buddhist monks as lodging and centres of worship. The 'Silk Roads' inspired in the 19th and 20th century AD European archaeologists, such as Aurel Stein (British: 1862–1943), Paul Peillot (French: 1879–1945) or Albert Grünwedel (German: 1856–1935) and Albert von Le Coq (German: 1860–1930), who conducted expeditions around the Taklamakan desert (today: Xinjiang Province, China) (Grünwedel, 1912; Von Le Coq, 1922–26). The findings of the four German exploration voyages between 1902 and 1914 include amongst other artifacts mural paintings, painted sculptures and painted wooden structures which are today stored in the Asian Art Museum, Berlin (Ulrich, 1964; Riederer, 1977; Russell-Smith, 2005; Gabsch, 2012).

These mural paintings can be attributed artistically to four distinguishable styles: (1) Indo-Iranian style (4th to 6th c. AD), (2) Indo-Iranian style (7th to 8th c. AD although both styles could be contemporary), (3) Chinese style (8th to the 9th c. AD) and (4) the Uygur style (10th to 13th c. AD, Fig. 18). The region of interest named here as the Northern Silk Road represents the route between Kashgar and Turfan (Xinjiang, China).

Once the grottoes were carved into the cliff of the soft sandstone mountains, two layers of mud plaster (coarse underlayer and a fine upper layer) were applied to the walls, consisting of clay mixed with sand and straw or other fibres such as cotton, wool or hemp. A fine and smooth white preparation layer was then applied to the earthen plaster on which the paintings were depicted using inorganic pigments, organic binding media and organic colourants. The paintings were made using the *a secco* technique, in which an organic binder is required to fix the pigments onto the painting surface. This technique has been applied widely along the Silk Road; examples can be found in Bamiyan (Afghanistan) or in the Mogao grottoes at Dunhuang (China). In the case of the detached mural paintings, which are today kept at the Asian Art Museum in Berlin, the different conservation measures as well as the damage caused by inappropriate storage during World Wars affected the actual state of conservation of the mural paintings.

4.2.1. Identification of different mineral pigments

Pigments of 76 mural painting fragments from different Buddhist sites along the Northern Silk Road (from west to east: Tumshuk, Kizil, Kumtura, Shortshuk, Bezeklik, Turfan, Chotscho and Murtuk) have been studied by means of polarization light

Figure 18. Part of a mural painting from cave 8 (Chinese numeration, Inv. No. MIK III 8426), Asian Art Museum, Staatliche Museen zu Berlin, Stiftung Preuβischer Kulturbesitz.

microscopy (PLM), IR spectroscopy, emission spectrum analysis and XRD (Riederer, 1977) (Fig. 19). The analyses were performed on the bulk material. Only a few cross-sections, considered to be essetial nowadays for the investigation of paintings, were performed. More recent works by Egel and Simon (2011a,b, 2013), Gabsch (2012), Egel *et al.* (2015) and Schmidt *et al.* (2016) concentrate on fragments of three particular caves (Cave 8, Kizil; Cave 40, Sim-sim; and Temple α, Chotscho). Although based on a small series of tiny samples, the study performed by Riederer in the 1970s gives the most global overview to date of the pigments used along the Northern Silk Road through several centuries (6th–10th c. AD).

White pigments: Gypsum, anhydrite and more rarely lead white (basic lead carbonate) are the three pigments that were identified. Lead white was characterized in only five of 76 fragments analysed and was found predominantly in the Eastern branch of the Northern Silk Road (region of Turfan) rather than in the western region of Kucha. When found in the western part, *e.g.* on a fragment from Kumtura, then the mural paintings originated from a later period (8th–9th c., Fig. 19). The geographical tendency to find lead white at sites located in the east coupled with the fact that deposits of natural basic lead carbonate (hydrocerussite) are rare, suggest a synthetic origin for this pigment originating in China. But results reported by other authors do not support

Figure 19. Pigments found by Riederer (1977) as a function of the location of the Buddhist site and the time of origin of the paintings. Adapted from Egel *et al.* (2015).

this assumption. Indeed, lead white or lead-based white pigments were characterized from the west as well as from the east from the sites analysed by Riederer. At Buddhist sites to the west of those analysed by Riederer, lead white was found on mural painting from Afrasiab in Usbekistan (5th–8th c.; Kossolapov and Kalinina (2007)) and a white lead-based pigment was characterized in Bamiyan (Afghanistan) on mural paintings created no later than the 5th c. AD (Shoten, 2006). At Buddhist sites to the east, lead white was identified on mural paintings from the late Tang Dynasty (618–907 AD) in Cave 85 from the Mogao Grottoes site at Dunhuang (Schilling *et al.*, 2010) and several lead-based white pigments were identified in the Tiantishan Grottoes, situated between Dunhuang and Xi'an. Lead white, lead chloride, lead sulfate and the mineral leadhillite were found on samples assigned to the Tang period.

Yellow pigments: While yellow ochre was found frequently, the much brighter coloured orpiment was identified on only two fragments (Bezeklik and Kizil). Its use as a pigment is also reported for other Buddhist sites as in Penjikent (Tadjikistan), in Afrasiab (Usbekistan) and Dunhuang in China (Kossolapov and Kalinina, 2007; Schilling *et al.*, 2010. Important natural deposits of orpiment, already known in antiquity, were located in the Balkan basin, in Turkey and Persia, but also in China (Yunnan province). The orpiment used on the Northern Silk Road could thus originate from either eastern or western natural deposits. The rare identification of orpiment (As_2S_3) by Riederer could be due to its transformation into the white arsenic trioxide (As_2O_3) under certain circumstances, such as exposure to light or ozone. A transformation mechanism into a white compound with unknown structure, containing arsenic but no sulfur, also seems to be possible through the action of microorganisms (Schilling *et al.*, 2010).

Orange and red pigments: The orange pigment minium and red ochre were found throughout the geographical range studied. On the contrary, the red mineral cinnabar was only found on mural paintings from the east: between Shortshuk and Murtuk. Meanwhile, cinnabar was also found on wall paintings from Buddhist sites situated further to the west, *e.g.* in Sim-Sim (Schmidt *et al.*, 2016), in Afrasiab (Kossolapov and Kalinina, 2007) and Bamiyan (Shoten, 2006). The blackening of both pigments, minium and cinnabar, is known from the literature (Cotte *et al.*, 2006; Nöller and Helman-Wazny, 2013; Radepont *et al.*, 2015). If the darkening of minium is commonly observed in the mural paintings from the Asian Art Museum in Berlin, cinnabar was found intact, the quality of the pigment being decisive for the alteration reaction (Nöller and Helman-Wazny, 2013).

Blue pigments: Lapis lazuli, and more rarely azurite, were used as blue pigments. Lapis lazuli was found exclusively on earlier wall paintings ($6^{th}-7^{th}$ c.), whereas both pigments were characterized on the latest paintings ($8^{th}-10^{th}$ c.). The intensity of the blue colour can vary a lot from one fragment to the other (Fig. 20), which is probably due to the quality of the pigment.

Green pigments: Atacamite was the most widespread green pigment and was used to paint whole surfaces. Contours were accentuated with lines painted with chrysocolla. Malachite could be identified on only two fragments from Kumtura. More recent studies, performed on cross-sections, show that different green shades were also obtained mixing malachite with atacamite in a gypsum matrix or adding indigo and lapis lazuli to atacamite (Egel *et al.*, 2018).

Black pigments: Vine black, obtained from the carbonization of wood, and soot (lamp black) were the black pigments found, the latter probably being imported from China, where it was used as ink.

4.2.2. Alteration phenomena

If inorganic pigments are relatively stable in comparison to the organic binders and colourants, certain pigments still undergo alteration through oxidation or through other degradation mechanisms over long periods. The brown/red-violet colour found today on the mural paintings, the transformation of the blue pigment lapis lazuli into a greyish colour and the presence of oxalates are typical alteration phenomena observed.

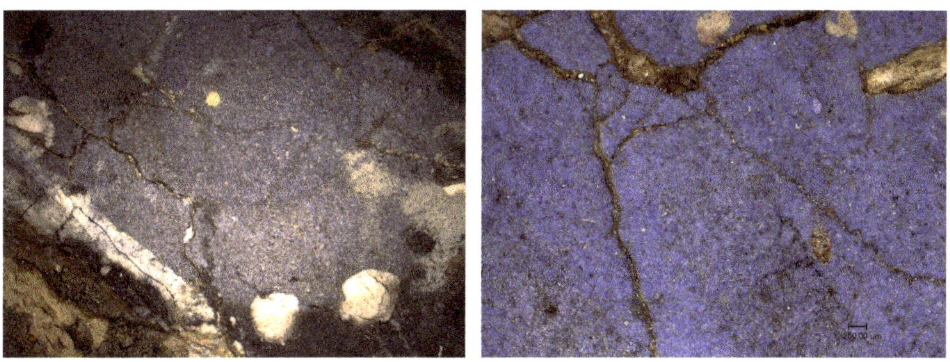

Figure 20. The quality of the lapis lazuli used can vary significantly resulting in variable colour intensity.

Lead oxide phases today show a brown-violet colour. Different formation processes of minium, plattnerite and massicot are observed (Riederer, 1977, pers. comm.). Plattnerite can be formed through oxidation of both massicot and minium. Riederer made the assumption that the brown-violet alteration product was originally the yellow pigment massicot.

For the blue colour made from lapis lazuli, a transformation into a greyish colour is observed, the underlying alteration process of which is not fully understood. Ultramarine is known to be sensitive to acids. Possible sources of acid could be: (1) oxalic acid (bio-colonization); (2) release of acetic acid in cases where the fragments consolidated with polyvinyl acetate (PVAc); or (3) colourant, which led to an acidic environment because of its degradation. Two alteration phenomena have been observed on carvings produced with lapis lazuli: first the formation of goethite, hematite and jarosite because of the oxidation of pyrite contained in the lapis lazuli and secondly the formation of a white mineral, enriched in aluminium and silicon, not further identified (Calligaro, 2014).

Oxalates were detected in numerous paint layers without restriction to any specific colour. The identification of metal oxalates has been reported many times in the literature over recent decades (Egel and Simon, 2011b; Egel *et al.*, 2018). Their formation could be the result of attack on sensitive pigments such as calcium carbonate, by oxalic acid. For the presence of oxalic acid, several sources are conceivable: it is a metabolism product of lichens and fungi, it can be produced by many plants and the photo-oxidation of organic binding media can lead to the formation of low molecular-weight dicarboxylic acids, which can again convert to the lowest unit of oxalate acid (Bordignon, 2008). The origin of oxalates in the paint layers from Central Asia is unclear. Even if mould growth on several pigments bound in animal glue has been proven (Stephan *et al.*, 2012), the formation of oxalates due to fungal infestation or due to the degradation of the binding medium on wall painting mock-ups has not.

In order to produce a rich palette of colours, painters have the the option to use large numbers of materials (inorganic pigments and organic colourants) as a mixture or as an overlay of several paint layers (Palitza, 2017). Organic colourants, prepared by extraction from a variety of biological sources (vegetable, animal), have been used to dye textiles or to prepare paints by precipitating the colourant on an inorganic substrate. Only indigo has been identified (Riederer, 1977) also mixed with ultramarine and atacamite, resulting in a deeper green (Egel, pers. comm.).

With respect to the question of colourants, violet-red tones are particularly interesting. A white surface layer made of gypsum and/or silicates on the top of a dark painting layer (hematite or minium partly oxidized) is often observed. An example is given in Fig. 21: the red-violet colour of the hair (1) is due to hematite (Fe_2O_3) and the decoration lines (2), showing today a brownish colour, are made of a white matrix (silicate) with few inclusions of orange particles (minium). The white mineral could have been used intentionally in the process as a ground for the dyestuffs (Palitza, 2017).

Figure 21. Red-violet area of the hair on the head of the figure depicted on the wall painting fragment III 4458b, Temple α, Khocho, Museum für Asiatische Kunst, Staatliche Museen zu Berlin).

4.3. Revealing the two underlying compositions of the Gustave Courbet painting '*L'Homme blessé*' using a combined non-invasive approach of MA-XRF imaging and CXRF

The study of paintings using non-invasive analytical imaging methods allows in-depth insights onto the chemical composition of paint layers, changes in composition, overpaints and former restorations. Under favourable circumstances, when paint layers of different chemical compositions are present, possible underlying paint layers and compositions can be identified. A well established method revealing the chemistry of paintings is two-dimensional (2D) macro-X-ray fluorescence imaging (MA-XRF). This method gives an averaged image of the chemical composition of all detectable paint layers. Only partial information on the paint layer stratigraphy can be gained by combining information from different non-invasive examinations and observations including X-ray radiography (XRR) and multispectral imaging methods (Fig. 7b). However, revealing distinct paint layers, their composition, thickness and succession is crucial to a better understanding of the genesis of the art work, the artist's process and in general the historical context of the painting, especially if several painting compositions are present on one canvas. To overcome this shortcoming, new approaches have been developed that allow non-invasive three-dimensional (3D) analyses of the chemical composition of the paint stratigraphy including depth-resolved analysis with confocal XRF set-ups (CXRF). The example presented here is the first comprehensive combined use of MA-XRF with CXRF applied to the investigation of a famous painting by Gustave Courbet *'L'Homme blessé'* kept in the Musée d'Orsay in Paris. According to former detailed studies, two other hidden compositions are present on the painting (Reiche *et al.*, 2016). By means of the examination by XRR and of a few cross-sections taken in the 1970s by microscopical methods, several hints of these previous compositions of the painting could be gained, including the painting palette used (see Section 3.1.2). The actual composition presents the painter itself; one hidden composition shows him with a woman, probably Virginie Binet, his wife at this time; and another hidden composition shows an unknown woman alone. Up to ten paint layers belonging to the three compositions are present on the canvas prepared with a calcium-based ground and two layers of lead-white mixed with iron oxides and quartz. Blue pigments are composed of Cobalt blue, Prussian blue and ultramarine. Red and orange hues contain iron oxides, vermilion, and an organic lake on alum substrate. Brown is obtained by iron oxides and umbers. Chalk, lead white, baryte and clay are found in white. Black is made of bone and carbon-based black. Green contains emerald or Scheele's green. As the different painting compositions were created around the 1840–50s, the artist's pigment palette changed for the final composition. At this time, artists began to use new synthetic pigments, thus allowing more detailed insights into the details of the hidden compositions by chemical imaging and depth profiling. The sequence of the three successive painting compositions needed to be clarified. In addition, it was not certain that the first composition was a completed version of the painting as the first composition is scarcely visible on the XRR (Fig. 7b). Therefore, additional insights into the hidden compositions were required.

Macro-XRF maps of relevant chemical elements were prepared in combination with depth profiles at strategic points crossing the three compositions. The chemical maps provided more details of the two hidden compositions. The Hg L map is very informative as the high X-ray energy of the Hg L fluorescence lines allows us to obtain information about the deepest layers of the painting. Mercury is the constituent element of the red pigment, vermilion (mercury sulfide, HgS). This pigment is found beside white pigments in the flesh colours and in the red wound of Courbet in the actual composition. Therefore, it reveals the hidden faces. In the composition of the man with the woman it shows the lips, the nose and also the eyelid of the woman's face. The hand embracing the young woman is also visible. It is surrounded by a cuff, which appears dark because it does not contain vermilion. The lips and the nose of the face of the woman alone, supposed to be the very first composition, is also visible in the Hg L map superimposed with the hair of the woman of the composition together with the man (inset in Fig. 14). Depth profiling using CXRF was performed on a point where both compositions overlap and show mercury-containing areas, such as parts of the faces indicating which of the two hidden compositions corresponds to the first and which to the second. Figure 22 shows the depth profile at location 11 of the painting (trunk of the background of the actual composition). Here, the deepest mercury-containing paint layer revealed probably corresponds to the first composition and thus to the woman alone because the XRR revealed her lips in this area. The Pb profile does not allow us to differentiate between the two compositions. Therefore,

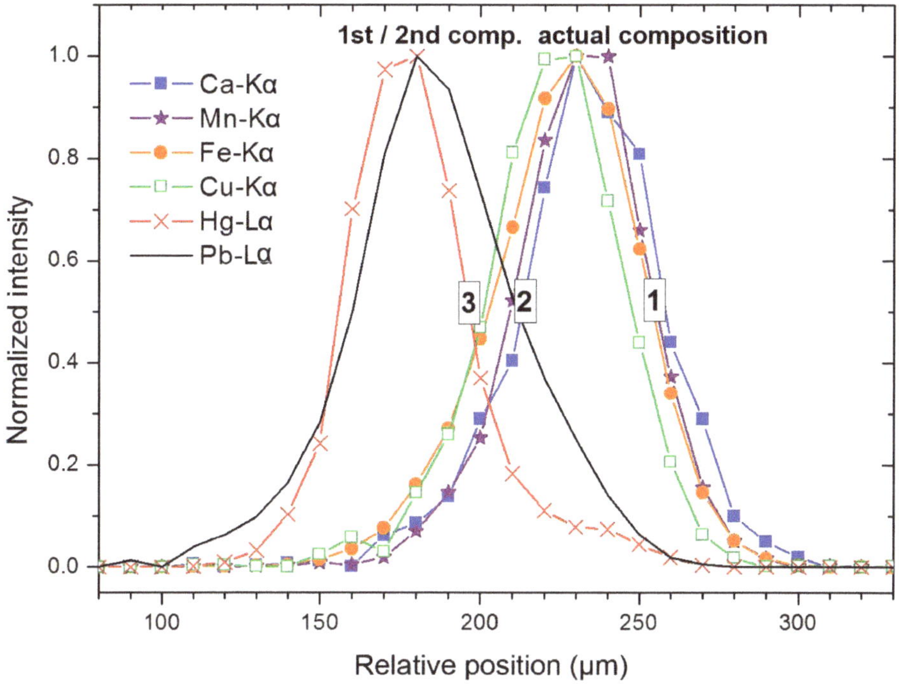

Figure 22. Depth profile at analysis location no. 11 on the Courbet painting '*L'homme blessé*' corresponding to the lips of the woman in one of the hidden compositions (graph modified from Reiche *et al.*, 2016).

it can be shown that the woman alone was painted first before Courbet painted himself with Virginie Binet in the second composition.

This study revealed the order of three successive compositions of this painting. The first composition was clearly unfinished as discussed in more detail by Reiche *et al.* (2016). It also highlights the importance of employing complementary imaging methods in order to obtain a complete three-dimensional vision of the chemistry and stratigraphy of paintings.

5. Conclusions and perspectives

This chapter reviews the current state of study of coloured minerals of our cultural heritage. This includes mineral pigments such as ochre or manganese oxides used in prehistoric contexts, pigments from historical periods to modern times. The range of pigments corresponding to minerals is emphasized. They are part of the increasingly diverse artistic pigment palette. Special terms used in this chapter are defined so that a common understanding of the chapter is possible.

The reasons for studying archaeological and historical pigments of a mineralogical origin and the information potential they bear were described here.

The methods that allow the study of the different aspects of the pigments are presented as well as current analytical developments from sample-taking techniques to cutting-edge, complementary, non-invasive imaging methods.

Three case studies highlight the information potential of studying mineral pigments in order to obtain new information not visible by optical means. Aspects such as the differentiation of several artists who created the artifacts, the evolution of the paint palette over time and by region along the Silk Road as well as alteration phenomena. Last, but not least, insights were provided into the genesis of a painting by revealing details and the sequence of two hidden painting compositions on a 19th century French painting by Gustave Courbet by new 3D imaging methods.

Some new trends are presented of how to master most of the methods and techniques useful for investigating our painted cultural heritage. These new trends focus in particular on the combination of molecular, structural and chemical investigations possible in imaging modes, such as macro-XRF scanning and hyperspectral imaging. They produce chemical maps with complementary information that can be superimposed on optical images and can be understood by non-specialists. This makes them much used and very useful techniques that can help to advance interdisciplinary research such as required for the in-depth study of painting and polychrome objects. Nevertheless, the basis for gaining new and precise insights into the history of our cultural heritage, into their meaning and the ways of life in ancient societies is in accord with the research questions addressed.

References

Aibéo, C.L., Sauvêtre, A. and Reiche, I. (2016) In situ scientific analyses of the leather wall hanging "Ilha Formosa". *Berliner Beitrage zur Archaometrie, Kunsttechnologie und Konservierungswissenschaft*, **24**, 55–68.

Albéric, M., Müller, K., Pichon, L., Moignard, B., Lemasson, Q., Pacheco, C., Fontan, E. and Reiche, I. (2015) Non-invasive quantitative micro-PIXE-RBS/EBS imaging reveals the lost polychromy and gilding of the Neo-Assyrian ivories from the Louvre collection. *Talanta*, **137**, 100–108.

Alfeld, M. and Broekaert, J.A.C. (2013) Mobile depth profiling and sub-surface imaging techniques for historical paintings ù A review. *Spectrochimica Acta Part B: Atomic Spectroscopy*, **88**, 211–230.

Alfeld, M. and de Viguerie, L. (2017) Recent developments in spectroscopic imaging techniques for historical paintings – A review. *Spectrochimica Acta Part B: Atomic Spectroscopy*, **136**, 81–105.

Alfeld, M., Janssens, K. Dik, J., de Nolf, W. and van der Snickt, G. (2011) Optimization of mobile scanning macro-XRF systems for the in situ investigation of historical paintings. *Journal of Analytical Atomic Spectrometry*, **26**, 899–909.

Alfeld, M., Pedroso, J.V., van Eikema Hommes, M., Van der Snickt, G., Tauber, G., Blaas, J., Haschke, M., Erler, K., Dik, J. and Janssens, K. (2013) A mobile instrument for in situ scanning macro-XRF investigation of historical paintings. *Journal of Analytical Atomic Spectrometry*, **28**, 760–767.

Barrière, C. (1982) *L'art pariétal de Rouffignac*. Picard, Fondation Singer-Polignac, Paris.

Beck, L., Genty, D., Lahlil, S. and Lebon, M. (2013) Non-destructive portable analytical techniques for carbon *in situ* screening before sampling for dating prehistoric rock paintings. *Radiocarbon*, **55**, 436–444.

Beck, L., Rousselière, H., Castaing, J., Duran, A., Lebon, M., Moignard, B. and Plassard, F. (2014) First use of a portable system coupling X-ray diffraction and X-ray fluorescence for in-situ analysis of prehistoric art. *Talanta*, **129**, 459–464.

Berke, H. (2002) Chemistry in ancient times: The development of blue and purple pigments. *Angewandte*

Chemie International Edition, **41**, 2483–2487.

Berke, H. (2004) Blau und Purpur. Die Erfindung von Farbpigmenten im Altertum. *Restauro*, **6**, 401–404.

Berke, H. (2007) The invention of blue and purple pigments in ancient times. *Chemical Society Reviews*, **36**, 15–30.

Bertrand, L., Cotte, M., Stampanoni, M., Thoury, M., Marone, F. and Schöder, S. (2012) Development and trends in synchrotron studies of ancient and historical materials. *Physics Reports*, **519**, 51–96.

Bordignon, F. (2008) The formation of metal oxalates in the painted layers of a medieval polychrome on stone, as revealed by micro-Raman spectroscopy. *Studies in Conservation*, **53**, 158–169.

Burns, R.G. (1993) *Mineralogical Applications of Crystal Field Theory*. Cambridge University Press, Cambridge, UK.

Calligaro, T. (2014) "Characterization of the lapis lazuli from the Egyptian treasure of Tôd and its alteration using external μ-PIXE and μ-IBIL. *Nuclear Instruments and Methods in Physics Research B*, **318**: 139–144.

Calligaro, T., Dran, J.-C., Salomon, J.and Walter, Ph. (2004) Review of accelerator gadgets for art and archaeology. *Nuclear Instruments and Methods in Physics Research Section B*, **226**, 29–37.

Cotte, M., Susini, J., Metrich, N., Moscato, A., Gratziu, C., Bertagnini, A. and Pagano, M. (2006) Blackening of Pompeian cinnabar paintings: X-ray microspectroscopy analysis. *Analytical Chemistry*, **78**, 7484–7492.

Cotte, M., Checroun, E., Mazel, V., Sole, V.A., Richardin, P., Taniguchi, Y., Walter, P. and Susini, J. (2009a). Combination of FT-IR and X-rays synchrotron-based micro-imaging techniques for the study of ancient paintings. a practical point of view. *ePreservation Science*, **6**, 1–9.

Cotte, M., P. Dumas, Taniguchi, Y., Checroun, E., Walter, P. and Susini, J. (2009b). Recent applications and current trends in cultural heritage science using synchrotron-based Fourier transform infrared micro-spectroscopy. *Comptes Rendus Physique*, **10**, 590–600.

Cotte, M., Susini, A., Dik, J. and Janssens, K. (2010) Synchrotron-based X-ray absorption spectroscopy for art conservation: Looking back and looking forward. *Accounts of Chemical Research*, **43**, 705–714.

Dejoie, C., Martinetto, P., Dooryhée, E., Strobel, P., Blanc, S., Bordat, P., Brown, R., Porcher, F., Sanchez del Rio, M. and Anne, M. (2010) Indigo@Silicalite: a New organic–inorganic hybrid pigment. *ACS Applied Materials & Interfaces*, **2**, 2308–2316.

Dik, J., Janssens, K., Van Der Snickt, G., van der Loeff, L., Rickers, K. and Cotte, M. (2008) Visualization of a lost painting by Vincent van Gogh using synchrotron radiation based X-ray fluorescence elemental mapping. *Analytical Chemistry*, **80**, 6436–6442.

Eastaugh, N. (editor) (2004) *Pigment Compendium, A Dictionary of Historical Pigments*. Elsevier Butterworth-Heinemann, Amsterdam.

Egel, E. and Simon, S. (2011a) *The graying phenomenon on mural paintings from Cave 8, Kizil, Xinjiang, China*. (International Conference on the Research and Conservation of the Kucha Caves, Kizil), Science Press, Beijing.

Egel, E. and Simon, S. (2011b). *Technological studies of copper pigments and degradation products in Cave 40, Simsim, Xinjiang, China*. (International Conference on the Research and Conservation of the Kucha Caves, Kizil). Science Press, Beijing.

Egel, E. and Simon, S. (2013) Investigation of the painting materials in Zhongshan Grottoes (Shaanxi, China). *Heritage Science*, **1**.

Egel, E., Mitschke, A., Gabsch, T. and Reiche, I. (2015) Untersuchung der Maltechnik und der Alterungsphänomene einer buddhistischen Wandmalerei aus der Tempelruine Alpha (10./11. Jahrhundert, Chotscho, Xinjiang, China) vom Museum für Asiatische Kunst, Staatliche Museen zu Berlin. *Berliner Beitrage zur Archaometrie, Kunsttechnologie und Konservierungswissenschaft*, **23**, 41–52.

Egel, E., Simon, S., Gabsch, T. and Reiche, I. (2018) A summary on material research on Buddhist mural paintings from the Northern Silk Road – Pigment degradation and laboratory simulation of their alteration mechanisms. Pp. 109–122 in: *Archaeology and Conservation along the Silk Road* (G. Krist and L. Zhang, editors). Böhlau Verlag, Vienna, Cologne, Weimar.

Eveno, M., Duran, A. and Castaing, J. (2010) A portable X-ray diffraction apparatus for in situ analyses of masters' paintings. *Applied Physics A*, **100**, 577–584.

Gabsch, T. e. (2012) *Auf Grünwedels Spuren, Restaurierung und Forschung an zentralasiatischen*

Wandmalereien. Koehler & Amelang GmbH, Leipzig, Germany.

Gay, M., Alfeld, M., Menu, M., Laval, E., Arias, P., Ontaón R. and Reiche, I. (2015) Palaeolithic paint palettes used at La Garma Cave (Cantabria, Spain) investigated by means of combined in situ and synchrotron X-ray analytical methods. *Journal of Analytical Atomic Spectrometry*, **30**, 767–776.

Gay, M., Müller, K., Plassard, F., Cleyet-Merle, J.-J., Arias, P., Ontaón, R. and Reiche, I. (2016) Efficient quantification procedures for data evaluation of portable X-ray fluorescence – Potential improvements for Palaeolithic cave art knowledge. *Journal of Archaeological Science: Reports*, **10**, 878–886.

Gay, M., Plassard, F., Müller, K. and Reiche, I. (submitted) Relative chronology of Palaeolithic drawings of the Great Ceiling, Rouffignac cave, by chemical, stylistic and superimposition studies. *Journal of Archaeological Science-Reports*.

Gianoncelli, A., Castaing, J., Ortega, L., Dooryhée, E., Salomon, J., Walter, P., Hodeau, J.L. and Bordet, P. (2008) A portable instrument for *in situ* determination of the chemical and phase compositions of cultural heritage objects. *X-ray Spectrometry*, **37**, 418–423.

Grünwedel, A. (1912) *Altbuddhistische Kultstätten in ChinesischTurkistan: Bericht über archäologische Arbeiten von 1906 bis 1907 bei Kuca, Qarašar und in der Oase Turfan*. Reimer, Berlin.

Hahn, O., Reiche, I. and Stege, H. (2006) Applications in arts and archaeology. Pp. 687–700 in: *Handbook of Practical X-Ray Fluorescence Analysis* (B. Beckhoff, B. Kanngieβer, N. Langhoff, R. Wedell and H. Wolff, editors). Springer-Verlag, Berlin, Heidelberg, New York.

Hernanz, A., Gavira-Vallejo, J.M. and Ruiz-López, J.F. (2006) Introduction to Raman microscopy of prehistoric rock paintings from the Sierra de las Cuerdas, Cuenca, Spain. *Journal of Raman Spectroscopy*, **37**, 1054–1062.

Hernanz, A., Gavira-Vallejo, J.M., Ruiz, J.F. and Edwards, H.G.M. (2008) "A comprehensive micro-Raman spectroscopic study of prehistoric rock paintings from the Sierra de las Cuerdas, Cuenca, Spain." *Journal of Raman Spectroscopy*, **39**, 972–984.

Janssens, K., Dik, J., Cotte, M. and Susini, J. (2010) Photon-based techniques for nondestructive subsurface analysis of painted cultural heritage artifacts. *Accounts of Chemical Research*, **43**, 814–825.

Kossolapov, A. and Kalinina, K. (2007) *The Scientific Study of Binding Media and Pigments of Mural Paintings from Central Asia*. Mural Paintings of the Silk Road – Cultural Exchanges between East and West. 29th Annual International Symposium on the Conservation and Restoration of Cultural Property, National Research Institute for Cultural Properties, Tokyo, Archetype, London.

Lahlil, S., Lebon, M., Beck,L., Rousselière, H., Vignaud,C., Reiche, I., Menu, M., Paillet, P. and Plassard, F. (2012) The first in situ micro-Raman spectroscopic analysis of prehistoric cave art of Rouffignac St-Cernin, France. *Journal of Raman Spectroscopy*, **43**, 1637–1643.

Morales Merino, C., Röhrs, S., Marten, S. and Reiche, I. (2016) Characterizing Mesoamerican blue pigments by non-invasive techniques. *Tagungsband der Jahrestagung für Archäometrie und Denkmalpflege*, **28**, September – 1. Oktober 2016, Göttingen, Germany.

Müller, K., Hartwieg, B., Stelzig, S. and Reiche, I. (2017) *In situ non-invasive study of paint layers on Early Netherlandish paintings: new insights into the paintings' history by portable CXRF*. ICOM-CC, Kopenhagen, Denmark, ICOM–CC.

Munsell, A.H. (1912) A pigment color system and notation. *The American Journal of Psychology*, **23**, 236–244.

Nöller, R. and Helman-Wazny, A. (2013) The materials of Turfan and Dunhuang manuscripts: Analysis of paper, pigments and dyes. *IDP Newsletter*, **41**, 6–7.

Ospitali, F., Smith, D.C. and Lorblanchet, M. (2006) Preliminary investigations by Raman microscopy of prehistoric pigments in the wall-painted cave at Roucadour, Quercy, France. *Journal of Raman Spectroscopy*, **37**, 1063–1071.

Palitza, U. (2017) *Mit Plinius auf der Seidenstraβe*. Studien zu Farbenfabrikation und Maltechnik der Antike, E.A. Semann, Leipzig, Germany, 144 pp.

Pichon, L., Calligaro, T., Lemasson, Q., Moignard, B. and Pacheco, C. (2015) Programs for visualization, handling and quantification of PIXE maps at the AGLAE facility. *Nuclear Instruments and Methods in Physics B*, **363**, 48–54.

Plassard, J. (1999) *Rouffignac, le sanctuaire des mammouths*. Le Seuil, Paris.

Plassard, F. and Plassard, J. (2016) Le grand plafond de Rouffignac. De nouveaux indices sur l'organisation des images. *PALEO* **Numéro spécial** (Hommage à Norbert Aujoulat): 143–151.

Quiles, A., Valladas, H., Bocherens, H., Delqué-Koli, E., Kaltnecker, E., van der Plicht, J., Delannoy, J.-J., Feruglio, V., Fritz, C., Monney, J., Philippe, M., Tosello, G., Clottes, J. and Geneste, J.-M. (2016) First high-precision chronological model for the decorated Upper Paleolithic cave of Chauvet-Pont d'Arc, Ardèche, France. *Proceedings of the National Academy of Sciences of the USA*, **113**, 4670–4675.

Radepont, M., Coquinot, Y., Janssens, K., Ezrati, J.-J., de Nolf, W. and Cotte, M. (2015) Thermodynamic and experimental study of the degradation of the red pigment mercury sulfide. *Journal of Analytical Atomic Spectrometry*, **30**, 599–612.

Ravaud, E., Pichon, L., Laval, E., Gonzalez, V., Gonzalez, V., Eveno, M. and Calligaro, T. (2016) Development of a versatile XRF scanner for the elemental imaging of paintworks. *Applied Physics*, A **122**, 17.

Reiche, I. and Chalmin, E. (2014) Synchrotron methods: Color in paints and minerals. Pp. 209–239 in: *Treatise on Geochemistry, Second Edition* vol. **12** (H.D. Holland and K.K. Turekian, editors). Elsevier, Oxford, UK.

Reiche, I., Müller, K., Mysak, E., Eveno, M. and Mottin, B. (2015) Towards a three-dimensional vision of the different compositions and the stratigraphy of the painting *L'Homme blessé* by Gustave Courbet: Coupling SEM-EDX and confocal micro-XRF. *Applied Physics A*, **121**, 903–913.

Reiche, I., Eveno, M., Müller, K., Calligaro, T., Pichon, L., Laval, E., Mysak, E. and Mottin, B. (2016) New insights into the painting stratigraphy of L'Homme blessé by Gustave Courbet combining scanning macro-XRF and confocal micro-XRF. *Applied Physics A*, **122**, 947.

Riederer, J. (1977) Technik und Farbstoffe der frühmittelalterlichen Wandmalereien Ostturkistans. *Beiträge zur Indienforschung*, **4**, 353–423.

Röhrs, S., Pausewein, R.-R., Schwerdtfeger, S., Simon, S. and Reiche, I. (2014) RF Report No. 7_013114: Doppelgemälde, *Apokalyptische Landschaft, Barrikadenkampf* von Ludwig Meidner (1884 – 1966), Inv. Nr. NG 61/61, datiert 1912/13.

Röhrs, S. and Stehr, U. (2014) Vis diffuse reflectance spectroscopy coupled to a zoom stereo microscope – a new tool for the analysis and documentation of miniature paintings. *Berliner Beitrage zur Archaometrie, Kunsttechnologie und Konservierungswissenschaft*, **22**, 47–56.

Röhrs, S., Schwerdtfeger, S., Schwarzmaier, A. and Reiche, I. (2017) Non-invasive micro-analytical study of five mummy portraits and the Severan Tondo of the Antikensammlung, Berlin. *Berliner Beiträge zur Archäometrie, Kunsttechnologie und Konservierungswissenschaft*, **25**, 137–148.

Russell-Smith, L. (2005) *Uygur Patronage in Dunhuang: Regional Art Centres on the Northern Silk Road in the Tenth Century*. Brill, Leiden and Bosten.

Schelper, T., Egel, E., Gassmann, J., Seyfried, F. and Reiche, I. (2016) Vom Original zum Malmodell: die Bemalung der neuen Replik der Nofretete-Büste der Gipsformerei, Staatliche Museen zu Berlin. *Berliner Beitrage zur Archaometrie, Kunsttechnologie und Konservierungswissenschaft*, **24**, 10–20.

Schilling, M.R., Mazurek, J., Carson, D., Bonin, S., Fan, Y. and Ma, Z. (2010) Analytical research in Cave 85. Pp. 438–448 in: *Conservation of Ancient Sites on the Silk Road. Proceedings of the Second International Conference of the Grotto Sites, Mogao Grottoes, Dunhuang, People's Republic of China, June 28th–July 3rd, 2004* (N. Agnew, editor). Getty Conservation Institute, Los Angeles, California, USA.

Schmidt, B.A., Ziemann, M.A., Pentzien, S., Gabsch, T., Koch, W. and Krüger, J. (2016) Technical analysis of a Central Asian wall painting detached from a Buddhist cave temple on the Northern Silk Road. *Studies in Conservation*, **61**, 113–122.

Sève, R. (2009) *Science de la couleur: Aspects physiques et perceptifs*. Chalagam, Marseille, France, pp. 139–146.

Shoten, A. (2006) *Study of Afghanistan's Displaced Cultural Properties, Materials and Techniques of the Bamiyan Mural Paintings*. Tokyo, Japan Center for International Cooperation in Conservation, National Research Institute for Cultural Properties, Tokyo.

Solé, V.A., Papillon, E., Cotte, M., Walter, P. and Susini, J. (2007) A multiplatform code for the analysis of energy-dispersive X-ray fluorescence spectra. *Spectrochimica Acta Part B: Atomic Spectroscopy*, **62**, 63–68.

Stephan, I., Gabsch, T., Dimke, T. and Egel, E. (2012) Schimmelpilzbewuchs auf Materialien in Anlehnung an zentralasiatische Wandmalereien aus dem 5./6. Jh. *Auf Grünwedels Spuren, Restaurierung und Forschung an zentralasiatischen Wandmalereien*, **12**, 164–171.

Tomasini, E.P., Halac, E.B., Reinoso, M., Di Liscia, E.J. and Maier, M.S. (2012) Micro-Raman spectroscopy of carbon-based black pigments. *Journal of Raman Spectroscopy*, **43**, 1671–1675.

Ulrich, F. (1964) Erfahrung bei der Wiederherstellung von Fresken und Lehmplastiken aus Ost-Turkistan. *Zeitschrift für Museumstechnik*, **10**, 73–78.

Valladas, H., Tisnerat-Laborde, N., Cachier, H. and Arnold, M. (2001) Radiocarbon AMS dates for Paleolithic cave paintings. *Radiocarbon*, **43**, 977–986.

Vetter, W. and Schreiner, M. (2011) Characterization of pigment-binding media systems – comparison of non-invasive in-situ reflection FTIR with transmission FTIR microscopy. *e-Preservation Science*, **8**, 2.

Von Le Coq, A.D. (1922–26) *Die buddhistische Spätantike in Mittelasien – Ergebnisse der Kgl. preußischen Turfan-Expeditionen*, Berlin.

EMU Notes in Mineralogy, Vol. 20 (2019), Chapter 8, 323–344

Gems and man: a brief history

George (Rip) RAPP

Department of Geosciences, University of Minnesota Duluth, Duluth, Minnesota 55812, USA
grapp@d.umn.edu

From shell beads in the Palaeolithic and stone beads in the Neolithic to beautiful artificial gems in modern times, the history of gems has roughly paralleled that of humans. In the beginning, myths and folklore about the healing properties of gemstones dominated the story. Today, the story is about scientific techniques making larger or more colourful gems and newly discovered mineral deposits revealing gemstone treasures. In the western world the written history of precious and semiprecious stones begins with the *On Stones* of the Greek philosopher and naturalist Theophrastus (*ca.* 315 BC) followed by the *Natural History* of the Roman historian Pliny (77 AD), which was the standard work on gems and minerals for more than a thousand years. The gemstones of the Old Testament and those of ancient East Asia tell their separate stories. Following a brief summary of these early works, this paper continues with individual descriptions of the major gems and semiprecious stones, focusing on their two most important attributes – colour and hardness – as well as where they are found. This is followed by a brief discussion of altered gems and a summary of modern interactions of gems and man. This paper concludes with some personal experiences of the author and a brief introduction to the geology of gem deposits.

1. Introduction

This article includes certain minerals in addition to those narrowly defined as gems because there exists a great deal of overlap (and confusion historically) among precious stones, semiprecious stones, and minerals that are considered valuable and sought after by collectors. It seems quaint today, but until the last few hundred years gems were sought as much for their healing or magical properties as for their value as ornaments (Fig. 1).

We begin with two early aspects of gems and man that are part of written history: (1) the first text on minerals and gems in the western world, the book *On Stones* written by the Greek Theophrastus *ca.* 315 BC, and (2) gems mentioned in the *Old Testament*. Both the ancient Egyptians (Lucas, 1962) and inhabitants of Mesopotamia (Moorey, 1985) used glazed minerals as beads and objects of adornment, but this short article cannot do more than mention this. We continue with a review of the main gems and minerals used by man, listing their hardness, colours, and the most significant localities where they have been found, beginning in antiquity. The final sections look at very recent times, including the development of synthetic gems, some personal experiences of this author, and a note on the geology of gem minerals.

DOI: 10.1180/EMU-notes.20.8

Figure 1. Serpentine dish with inlaid goldfish, pearls and other gemstones, 1st century BC or AD, with 9th century AD mounts from the Court of Charles the Bald. Not all gemstones were used for body ornaments.

The definition of a gem does not have the specificity of definition that is given to other earth materials, for example, a mineral. A gem (including precious and semiprecious stones) can be a mineral, or a rock (such as lapis lazuli), or a natural organic material (such as amber or pearl) that is used to make jewellery or other adornments. In his popular book on gemstones of North America, Sinkankas (1997) includes nearly everything that humans have carved, for example catlinite, an argillite used by Native Americans to make smoking pipes. There does not seem to be a limit to what polishers will polish into decorative objects, even graphic granite (unique intergrowths of quartz and feldspar). Most gem minerals are hard and durable, but even extremely hard substances such as rock crystal (quartz) have been painstakingly carved into elaborate objects such as jewellery and sealstones. The Babylonians carved seals into cylinder-shaped gems they used as ornaments (Whitlock, 1934). Some softer minerals have been used because of their colour, lustre, or other physical property that gives them aesthetic value. Rarity also lends value to a gemstone.

Many mineral names have changed over time. As a result, not all gems and minerals mentioned in early texts can, with confidence, be given modern names. This article uses modern names or else clarifies. Numerous references are readily available to sort out mineral and gem names (Bauer, 1968; Kunz, 1971; Roberts *et al.*, 1990; Blackburn and Dennen, 1997; de Fourestier, 1999; Bayliss, 2000). In addition, some minerals have many different names. For example, gemmologists use the name iolite, but geologists and mineralogists prefer the name cordierite. This same mineral has also been called dichroite and water sapphire. Early accounts sometimes ascribe the source(s) of gems and minerals, but often these were hearsay. This article discusses only source localities of which the author is reasonably confident.

Gemstones very early became associated with folklore and superstition. Nearly all early references to gems and minerals up through the Middle Ages (1066–1492) noted the magical or healing properties of the gem or mineral. Georgius Agricola (1494–1555) was perhaps the first major author to reject the attribution of magical or healing properties to gems and minerals (Agricola, 1546/1964). Colour, even among the most knowledgeable, was the dominant way to identify gems and minerals through the Middle Ages. Without knowledge of mineral formation or chemical composition, people developed myths as a way to explain the colours of gemstones.

Gemstone deposits certainly are not ubiquitous. Some, such as lapis lazuli, come from only one, or a very few, localities in the world. Although Afghanistan was the only known source of this mineral during antiquity, lapis artefacts have been found throughout the Old World (Herrmann, 1968), confirming that gemstones were a trade commodity from earliest times. The wealthy urban centres sought gemstones from around the world. For example, the Harappa of the Indus valley sent carnelian to the powerful kings in Mesopotamia following the Silk Road. Every conceivable commodity was traded along this famous route, including rock crystal (quartz). Thus gemstones have served archaeologists as valuable indicators of trade patterns.

2. Early works

2.1. Theophrastus

Theophrastus (*ca.* 371–287 BC), a Hellenistic Greek philosopher, composed treatises on many subjects. His treatise on stones (minerals, rocks and gemstones), variously known as Περί λίθων or *De Lapidibus* (Theophrastus, 1956), remained a definitive reference on the subject throughout the Middle Ages. Theophrastus described only sixteen minerals, but he laid the foundation for the sciences of mineralogy and gemmology, and the minerals he described led to the discovery of many of the chemical elements. He was an avid collector and stressed the importance of observation and of obtaining samples. He acquired information on the minerals of many regions of Asia Minor, Egypt, Nubia and the Persian Gulf. His treatise is composed of several sections that discussed primarily precious and semiprecious minerals. He focused on observable properties, such as size, colour, hardness, smoothness, reaction to fire, ability to attract magnetically and the location where the minerals were found. Theophrastus made the first known reference to the phenomenon of pyroelectricity; he noted that the mineral tourmaline becomes charged when heated. He described the precious stones emerald, amethyst, onyx, jasper. He also included what he considered to be a variety of sapphire that was blue with veins of gold; this was probably lapis lazuli. He mentioned only rarely the magical properties with which the ancients believed stones were endowed. He also devoted sections to organic materials such as amber, coral, pearl and ivory that were classified by the ancients in the same category as gems and minerals. He noted that pearls come from shellfish. He also considered the practical uses of various stones, such as which were necessary for the manufacture of glass, and the use of the touchstone for assaying gold and gold alloys. For good reason he is often called the father of mineralogy and gemmology.

2.2. Pliny

Gaius Plinius Secundus, better known as Pliny the Elder or just Pliny, was a Roman author, naturalist and natural philosopher. Pliny's last work was the *Naturalis Historia*, a thirty-seven volume work published in 77 AD, which served as an encyclopaedia of the world's knowledge at that time (Pliny, 1989). A number of different editions are available worldwide, including the Harvard Loeb Classical Library edition. Pliny compiled the work from a reported two thousand sources. He lumped minerals together by colour, mining location and – of questionable value – attributes passed on in legends. Pliny attempted to classify gems according to colour as the most observable attribute, which led him to the practice of lumping all stones of similar colour into the same category. For example, he classified most green stones as varieties of emerald. Regarding gemstones described earlier by Theophrastus, Pliny added very little scientific information beyond what he gleaned from his Greek predecessor. The reason for including this note on Pliny is that he became the most often quoted author on gems and minerals for nearly thirteen centuries.

2.3. Minerals of the Old Testament

The King James Version of the *Bible* has 1,704 references to gems and minerals under 124 Greek and Hebrew names (Wright and Chadbourne, 1970). Many of these gems and minerals cannot be identified from the biblical text, and others are obscure. One of the most detailed descriptions in the Old Testament is of a breastplate worn by the high priests of the temple; it contained twelve gemstones representing the twelve tribes of Israel. These gemstones were mounted in four rows. The first row was ruby, chrysolite and beryl; the second row was turquoise, sapphire and emerald; the third row was jacinth, agate and amethyst; the fourth row was topaz, onyx and jasper. In medieval times these same twelve gems became affiliated with the twelve signs of the zodiac. The modern custom of birthstones is derived from this tradition.

2.4. Ancient East Asia: China, Japan and India

In ancient China many precious and semiprecious stones were used in medicinal preparations. These included diamond, ruby, sapphire, emerald, jade, peridot (an olivine) and lizardite (a serpentine mineral). Minerals and rocks exploited for more than 3,000 years, at least back to the Late Shang at Anyang (Wang and Shen, 1986), included marble, turquoise, topaz, garnet, rock crystal, aquamarine, lapis lazuli, tourmaline, moonstone (a variety of feldspar), chrysoberyl, pearl and coral. Turquoise objects are known from the T'ang period (Laufer, 1974). In his multi-volume work on ancient China, Needham (1959), in sections 23 (geology) and 25 (mineralogy), discussed geodes, jade and precious stones. Rock crystal and agate were favourite semiprecious stones throughout early China. Favoured gemstones, like other customs, changed with time and place.

Japan is situated in a volcanic zone within the Pacific Ring of Fire. As a volcanic archipelago with a limited range of rock types, the range of gemstones is quite narrow. However, a number of unusual gemstones are found in Japan, including sugilite, a relatively rare pink to purple silicate mineral. One gemstone common to volcanic areas is obsidian. Indeed, there are >80 archaeological obsidian sources in Japan. In addition, transparent greenish grey topaz crystals are found on Honshu Island. It is worth noting that quartz crystals have been mined in Japan since the eighth century AD. In modern Japan there is a lively market in gems and semiprecious stones including topaz and pearl.

India was the source of many gems that were highly desired by Romans. Jade carving developed in India with the advent of the Mughal emperors, who brought knowledge of the craft from east central Asia and Iran. Diamonds in India were exploited for millennia, at least from the seventh century BC. However, according to Bauer (1968), most of the diamonds remained in India until the tenth century AD, when invaders plundered the area. India has also been the source of outstanding emeralds.

3. Gems and semiprecious stones

See Rapp (2008) for more complete descriptions and an abundant reference section. Much of the mineral data in the following sections are taken from that book. Hardness is based on the non-linear Mohs scale in which talc is the softest mineral with a hardness of one and diamond is the hardest at ten. Modern stainless steel has a hardness between four and five, and the hardness of glass is approximately 5.5.

3.1. Quartz family minerals

Quartz is the second most common mineral in the earth's crust, after feldspar. It is also the most widely distributed of all minerals in the near-surface environment. Quartz is truly ubiquitous, occurring in nearly all geological environments and in nearly all geographic areas. Since earliest antiquity, quartz in all of its many forms has been a common semiprecious gemstone often called rock crystal. It has a hardness of seven on the Mohs hardness scale. It was mined in Egypt as early as the Predynastic period (beginning ~6000 BC) and by the eighth century AD in Japan. The earliest lapidaries in India, *ca.* 1500 AD, dealt mainly with quartz gems.

3.1.1. Amethyst

Amethyst is quartz characterized by various shades of purple or violet. In Roman mythology, amethyst was believed to have been coloured purple by the god of wine and was thought to offer protection against drunkenness. It derives its name from a Greek word meaning not to intoxicate. Amethyst was mined in Egypt as early as the First Dynasty (2920–2770 BC) (Lucas, 1962). It is found in many areas. The large crystal clusters and geodes exploited today come mainly from Brazil. Catherine the Great of Russia was extremely fond of amethyst and amassed a large collection from mines in the Ural Mountains. Religious leaders still wear amethyst gems in rings.

3.1.2. Citrine

Citrine is pale yellow, yellow to yellow brown or light reddish brown quartz. These colours can be bleached with heating. It is transparent to translucent. The name originates from the French *citron* (lemon). It is sometimes erroneously called topaz.

Citrine has been used since antiquity as gems and seal stones. Much of the citrine available on the market today may be heat-treated amethyst.

3.1.3. Chalcedony

Chalcedony (Fig. 2) is a cryptocrystalline variety of quartz that can be white, greyish, blue, brown or black. It has a waxy lustre and is transparent to translucent. The name is derived from Chalcedon, an ancient city located in what is now part of Istanbul, Turkey. Ancient sources of the mineral included India, Yemen, Iraq, Egypt's Eastern and Western Deserts, and the Sinai. It was used there as a gemstone as early as Predynastic times, which began about 6000 BC.

3.1.4. Chrysoprase

Chrysoprase is a green cryptocrystalline quartz, a variety of chalcedony. Usually jade green, its colour can vary from olive green to a pale sea-green. The green colour is imparted by small amounts of nickel. Chrysoprase has a hardness of 6–7, which adds to its value as a gemstone. It has been identified in ancient Egyptian artefacts, some of which have been dated to the Predynastic period (Lucas, 1962). The first significant deposit of chrysoprase was found in Szklary, Poland, in 1425 AD.

3.1.5. Carnelian

Carnelian is a translucent, red to reddish brown cryptocrystalline quartz. It has been used since the third millennium BC in the wide region from Mesopotamia to India. Carnelian was used for beads by the early second millennium BC; since then it has been used as gems, amulets, seal stones and as inlay for jewellery.

3.1.6. Agate

Agate is a varicoloured microcrystalline quartz, often with irregular or concentric coloured bands. It is one of the most widely distributed materials at the earth's surface.

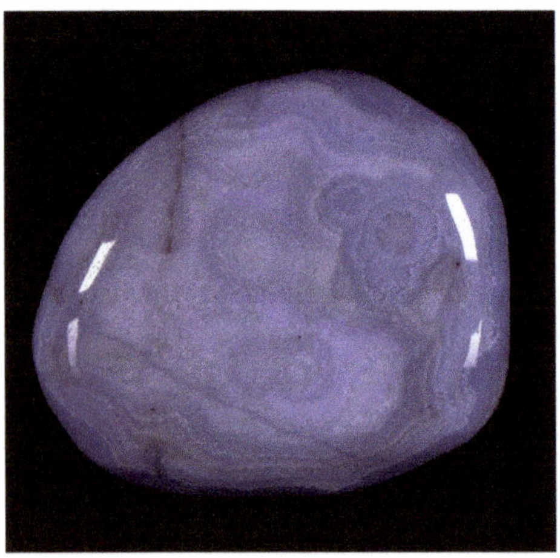

Figure 2. Cut and polished chalcedony gemstone. Like all high-silica minerals chalcedony takes a high polish.

The earliest evidence for working with agates comes from Mesopotamia, where agate ornaments and tools have been recovered. Theophrastus stated that the name was taken from one of the principal sources for the stone, the Achates River in Sicily. Wright and Chadbourne (1970) pointed out that early civilizations used semiprecious stones such as agate amulets blessed by a priest to ward off dangers from everything from thunderstorms to poisonous spiders. During the Classical period of the Eastern Mediterranean, agate was a popular stone for carved signet rings and for objects of art. The Greeks and Romans colourized agates by soaking in sugar solution followed by acid treatment. The Roman emperor Nero was a collector of agate cups. Sources of agate mentioned by Pliny include Crete, India, Phrygia, Egypt, Cyprus, Persia and various places in Greece. The area of Idar-Oberstein in Germany may be the oldest gem-cutting centre in the world. It supplied cut and polished agate to the Greeks and Romans and remained an important centre for agate carving and polishing through the Middle Ages

3.2. Non-quartz silicate minerals

3.2.1. Beryl

Beryl has a hardness of 7.5–8. Common beryl is green, pale green or pale greenish blue and is often opaque. It may occur in gem forms as emerald (green), aquamarine (blue) or morganite (red). Beryl is highly weather-resistant, so it is commonly found as one of the weathering products of pegmatites. The colour of beryl can be altered by irradiation or heating.

Emeralds were Cleopatra‘s favourite gemstone. The earliest emeralds were mined from the legendary Cleopatra’s Mines in the Sikait-Zabara region of Egypt. Until the Europeans discovered emeralds in the New World (Colombia and Brazil), Egypt was the only source. In India’s most famous temple, dedicated to the goddess Minakshi, her idol is made of emerald, probably carved out of a single crystal.

Aquamarine is a beautiful sea-blue variety of beryl (Fig. 3). The colour is caused by iron ions within the crystals. Aquamarine can appear either blue or colourless when viewed from different angles. Gem-quality aquamarine is found as flawless hexagonal crystals, which may have striations along the length of the crystal. The best of the gem-

Figure 3. Aquamarine (blue beryl) crystal. Sri Lanka and Burma were the major sources of these gems in antiquity. (Photo of aquamarine sample from Pakistan, collection of the Museum of Mineralogy, Università di Padova.)

quality aquamarine is found in Brazil, where it occurs in pegmatites and associated alluvial deposits of gravel. Other localities include Russia's Ural Mountains, Afghanistan, Pakistan, India and Madagascar. Some aquamarines are heated to high temperatures to enhance the colour or heat-treated to change a blue green or teal gemstone to a pure blue. This results in a permanent colour change.

3.2.2. Topaz

Topaz has a hardness of 8. It is transparent to translucent and occurs in a wide range of colours from colourless to white, grey, bluish, greenish, yellowish, orange, purple, and pinkish to reddish. The name topaz is derived from the Sanskrit word for fire. Note that the topaz of the biblical Old Testament may have been chrysolite (an olivine). The Romans believed topaz had power to improve eyesight, and the ancient Egyptians wore it as an amulet to protect them from injury. The classic localities of Mursinka in the Ural Mountains and Schneckenstein in Germany were discovered only in the seventeenth century.

3.2.3. Garnet

Garnets are a group of silicates with Mohs hardness between 6.5 and 7.5. The gem varieties are grossular (also named rosolite, hessonite), almandine, pyrope, andradite, spessartine and uvarovite. Colour is the most obvious feature of gem garnets. The garnet group shows one of the widest varieties of colour in the mineral world, ranging from colourless, through white, grey, yellow, yellowish green, green, yellowish brown, brown, pink, red and black. Only blue is extremely rare. Principally used as a gemstone, the name grossular originates from the Latin *grossularium* (gooseberry). During the Middle Ages, deep blood-red almandine garnet was preferred as a gemstone for decorating religious artifacts. Gem-quality garnets commonly develop as a result of regional metamorphism.

3.2.4. Zircon

Zircon has a hardness of 7.5 and is quite durable. It also has been known as cyrtolite, and its ancient name was jacinth or hyacinth. Zircon is widespread as an accessory mineral in silicic igneous rocks. Because of its durability, gem varieties are often found in gravel deposits. The high refractive index and the optical dispersion of zircon result in a brilliancy approaching that of diamond. The colourless variety looks more like diamond than any other natural stone because of this brilliance and light dispersion. The variety hyacinth includes the clear and the transparent red, orange and yellow varieties. Gem-quality zircon occurs in Australia, Cambodia, Myanmar, Sri Lanka, Thailand and many other countries.

3.2.5. Tourmaline

Tourmaline (Fig. 4) is a chemically complex silicate that occurs in several varieties as a result of chemical substitution. Chromium ions colour tourmaline green, which sometimes causes it to be mistaken for emerald. It has a hardness of 7 to 7.5. Practically

Figure 4. Tourmaline crystal.

all gem-quality tourmalines occur in rocks formed in the last stages of granite pegmatite crystallization. The use of tourmaline as a gemstone is documented in ancient Greece and Rome after Alexander the Great's conquests in the East provided a source. The Egyptians wore it as an amulet to protect them from injury. The Romans believed it had power to improve eyesight.

3.2.6. Jade

Jade is the gem term used for two different minerals: jadeite and nephrite. When worked and polished, jadeite and nephrite can usually be distinguished by their appearance. The lustre of nephrite is oily, while the lustre of jadeite is vitreous. The apple-green or emerald-green varieties are usually jadeite. Jades have been worked into implements and ornaments since Neolithic times in China.

Nephrite is an actinolite amphibole with a hardness of 5 to 6. It is light green to greenish black in colour. A number of ancient Egyptian artefacts have tentatively been identified as nephrite. For a mineralogical study of Chinese nephrite jade, see Wen and Jing (1996). Laufer's (1974 reprint) extensive treatise on Chinese jade was first published in 1912, so it is quite dated. The ancient Chinese acquired their nephrite from Burma (Myanmar) and Turkestan. It was used in China from the Neolithic in the eighth millennium BP through the Han Dynasty. Nephrite has been employed for carvings, decorative and ceremonial objects, personal ornaments, and gems. In North America the Canadian province of British Columbia is the major source of nephrite.

Jadeite is a pyroxene mineral with a hardness of 6–7. It is usually light to dark green. The Aztecs, Mayas, and other precontact peoples of Mexico and Central America carved green jadeite for ornaments and amulets. Only the Montagua Valley of Guatemala has been established definitively as a source for pre-Columbian jadeite in the New World. There is considerable speculation and research directed at the identification of other Mesoamerican sources, but none has yet been confirmed.

3.2.7. Opal

Opal has a hardness varying from 5.5 to 6.5. In its chemically purest form, opal is milky white to colourless. However, through admixture of pigmenting material it can be almost any colour, including yellow, brown, orange, green and blue, usually in pale shades (Fig. 5). The principal gem variety, termed precious opal, shows a rich internal play of colours in reflected light and has a pearly opalescence if it is viewed from different angles. The name opal may have derived from Sanskrit *upala*, meaning precious stone. The use of precious opal appeared in ancient Greece and Rome after Alexander the Great's conquests in the East. For the Romans, opal ranked second only to emerald among noble gems. From the fifth century BC on, many of the opals from ancient Europe came from Slovakia near Dubnik. Pliny described the opals from this area, >1,000 km from Rome. Unfortunately, these opals became known as Hungarian (possibly because a second site of production was located in northern Hungary). Production from Slovakia continued through the Middle Ages. The ancient cultures of Mexico had access to many deposits of precious opal including those of Queretaro. Mexican opal was taken back to Europe by explorers as early as 1520, where it was used for engraved gems. Fire opal gets its 'fire' from the diffraction of light where units of silica are arranged in a geometric pattern of the proper size to diffract light entering the material. Fire opal first came from Europe centuries ago, with Hungary providing the best material.

3.2.8. Opalite

Opalite should not be confused with opal. Opalite has been used as a name for two different gem stones: neither is the mineral opal. One type of opalite is a rock composed of dolomite and opalized fluorite and may contain quartz and chalcedony. It is prized for its dramatic colours, particularly dark purple and lavender. It forms as mineralized nodules. The other type is a modern synthetic opalized glass that simulates the mineral opal, it became widespread from the 1960s.

3.2.9. Lapis lazuli

Lapis lazuli has a hardness of 5–5.5. It is known for its blue colour, which can vary from deep blue, indigo blue, azure blue, and violet blue to greenish blue. Its name is derived from the Latin word for stone and a Persian word for blue colour. In the fifth century AD, the mineral was introduced into Europe with the name *ultramarinum*, which became ultramarine. A word of caution: the word *lapis* comes from the Latin for stone, so it has been used as a part of many gemstone names (*e.g. lapis indicus* is

Figure 5. Cut and polished opal. Gem quality opal occurs in many parts of the world.

nephrite). The only proven ancient source of lapis lazuli in the Old World is Badakshan, Afghanistan. Badakshan is mentioned in Chinese writings as early as the sixth century AD. Lapis from Badakshan is almost always associated with pyrite. Lapis lazuli found its way to the Indo-Pakistani region by Neolithic times. The mines of Badakshan were described by Marco Polo, who (reputedly) visited them in the thirteenth century (Bauer, 1968). Importation of lapis began as early as the Third Dynasty at Ur for cylinder seals and beads. It found its way into Egypt by Predynastic times. The Romans believed lapis to be an aphrodisiac.

3.3. Non-silicate minerals

3.3.1. Diamond

Diamond, with a hardness of 10, is the hardest natural mineral. It can be colourless, but it is also found in white to blue white, grey, various shades of yellow, brown, orange, pink, red, lavender blue, green and black. Diamond's name originates from the Latin word *adamas* (invincible). From that name we get *adamantine* lustre. Pliny stated that there were as many as six different stones called *adamas*. However, it is unlikely that all of these were mineralogically diamond. Craftsmen in India used diamonds to drill holes in hard gems like beryl to make beads. Europeans were not aware of India's diamonds until Marco Polo visited the area in the thirteenth century AD. Because diamond forms in a rare igneous rock called kimberlite, its occurrences are limited. The United States Geological Survey estimates that the average diamond in an engagement ring is the product of the removal and processing of 200 to 400 million times its volume of kimberlite. Diamond is highly resistant to weathering, so it is often found in alluvial deposits. Africa currently produces most of the world's diamonds, largely from the countries of South Africa, Angola and Botswana. The diamonds from Sierra Leone in Africa have been labelled 'blood diamonds' because they have funded one of the most savage rebel movements in modern history. Russia is the largest producer in Asia; some diamonds still come from India. Only a small percentage of modern diamond production comes from the rest of the world.

3.3.2. Corundum

Corundum, with a hardness of 9, is one of the hardest minerals known. It is found in two gem varieties: ruby and sapphire. It occurs in crystalline limestone and dolomite, schists and gneisses, and granitic igneous rocks, although gem-quality material is relatively rare in igneous rocks.

Ruby is red in colour and takes its name from the Latin *rubeus* (red). According to medieval Arab scholars, this stone was mined in Sri Lanka. Pliny referred to red stones by the general term *carbunculus*. He listed sources as India, North Africa, Caria, Ethiopia, Miletus, Orchomenos, Chios, Corinth, Troezen (in the northeastern Peloponnese, Greece), Marseilles and Lisbon. Although ruby was mentioned in some translations of the Old Testament, there is no evidence that this mineral was known to the ancient Hebrews. Nevertheless it was widely available as a gemstone. A discontinuous band of marble formed along a tectonic belt running 1,800 miles along the southern slope of the Himalayas from Tajikistan through Afghanistan, Pakistan, and Nepal to China contains many important ruby deposits.

Sapphire is found in shades of blue and sometimes yellow. Sapphire is dichroic – in other words, the colour changes with the angle from which it is viewed. Very minute, regularly arranged inclusions of the mineral rutile are responsible for the asterism of star sapphire. Sapphire is an ancient term of uncertain origin, possibly related to Hebraic *sappir* and Sanskrit *sanipruja*. Sapphire was used as a gemstone beginning in the early first millennium BC. Pliny's *chrysolithus* from India may have been yellow sapphire. According to Pliny, *hyacinthus* (blue corundum) came from Ethiopia, although India may have been its original source. *Nilos* may have also have been blue corundum. Pliny listed its source as India, Galatia and Egypt. Today sapphires are found in Africa, India, Sri Lanka, Burma, Australia, Nigeria, Madagascar, Kenya, Thailand, Cambodia, Vietnam, Brazil and the USA. Sapphire is particularly abundant in Sri Lanka. It has been mined in Australia for >100 years. Gem-quality sapphires in the United States are found in the state of Montana. Because of its hardness and durability, many gem sapphires have been found in gravels and placer deposits.

3.3.3. Fluorite

Fluorite, sometimes called fluorspar, has a hardness of 4 and a perfect octahedral cleavage. Fluorite can be colourless, but it is also found in shades of purple, blue, green, yellow, white, pink, crimson to brownish red, brown and bluish black (Fig. 6). Its name comes from the Latin *fluor* (to flow). It is widely distributed geologically and has long been used as a gemstone. Predynastic Egyptian beads of fluorspar are known (Lucas, 1962). Beads of fluorspar have been found in the ancient ruins of Tiahuanaco, near Lake Titicaca, on the Bolivian high plateau (Bauer, 1968). One of the most famous fluorite localities is in Derbyshire, England. This gemstone is called Derbyshire Blue John. It is a purple blue fluorite extracted from several mines or caves, including the famous Blue John Cavern. In the nineteenth century, this fluorite was highly valued as an ornamental gemstone.

3.3.4. Chrysoberyl

Chrysoberyl, with a hardness of 8.5, has long been a favourite gemstone because of its beauty and hardness. It is found in two varieties, Alexandrite and cat's eye. It has an extensive colour range and can be found in various shades of yellowish green, yellow, grey, brown, blue green and emerald green. Its name is taken from the Greek word

Figure 6. Green fluorite crystals on a quartz-amethyst matrix. The word fluorescence comes from the brilliant blue luminescence found in fluorite specimens from England.

χρυσος (golden) and from the mineral beryl, in allusion to its colour and beryllium content. This may be the mineral Pliny called *chrysoprasus*; however, he listed no source where it was found. The variety Alexandrite is unique because the colour depends on the light source. It is green or blue green in daylight and under fluorescent light, but shades of red when seen in incandescent light. Alexandrite was discovered in the 1830s in Russia's Ural Mountains and named after the future Czar of Russia, Alexander II.

3.3.5. Turquoise

Turquoise has hardness between 5 and 6. The colour varies from bright blue, pale blue, sky-blue, bluish green, apple-green, to greenish grey. Turquoise is one of the oldest known talismanic and curative gems, prized by many cultures worldwide. Its name originates from the French word for Turkish. The famous Persian turquoise mines were located near Nishapu in the province of Khorasan. Turquoise was used in ancient Egypt prior to the First Dynasty with mines operated in the Sinai Peninsula by the fourth millennium BC. Pliny mentioned green stones that may have been turquoise, but there seems to have been some confusion between lapis lazuli and turquoise. Turquoise was mentioned by the Arab mineralogist ibn-Mansur in 1300 AD (Bauer, 1968). Turquoise has been used in China for more than 3,000 years, at least back to the late Shang at Anyang. There are a significant number of turquoise deposits in the southwestern United States that have been worked since Pre-Columbian times. Perhaps the most exploited is located near Chaco Canyon, New Mexico, where over 56,000 pieces of turquoise were recovered from two burials at Pueblo Bonito. Much additional turquoise has been recovered from other sites in Chaco Canyon dating from the period 900–1150 AD (Mathien, 2001). Turquoise was a major commodity in Mexico from the late first millennium BC through modern times. Cortez reported that Aztec chieftains wore turquoise stones in their lower lips. There were numerous turquoise mines operating in northern Mexico. In pre-Columbian Mesoamerica worked turquoise artefacts are abundant throughout its prehistory beginning in the early Preclassic (*ca.* 2000–1000 BC). Its possession signified the status of nobility. It was prized for gems and jewellery.

3.3.6. Malachite

Malachite, a copper hydroxycarbonate, has hardness between 3.5 and 4. It is bright green to dark or blackish green in colour. Its name is derived from the Greek μολοχιτις. It often occurs with distinctive bands in varying shades of green. It is quite soft for a semiprecious stone, which makes it easy to carve into beads, gems and other decorative objects. Malachite was used in the Old World at least as early as the fifth millennium BC. It was used by the ancient Egyptians, Greeks and Romans to make jewellery and amulets. Pliny mentioned a number of minerals that could have been malachite. The most important deposits exploited in early modern times were in the Ural Mountains of Russia. Currently the Democratic Republic of the Congo is the most important malachite producer. The richest deposits are found in Katanga Province.

3.4. Organic gem materials

3.4.1. Amber

Amber is not a mineral, but rather fossilized resin exuded by the conifer *Pinites succinifer* (see Angelini, 2010). The name is taken from Arabic '*anbar*'. Its varieties are called succinite and retinite. Amber has a hardness of only 2.5–3.0. Artefact amber is quite susceptible to weathering and to poor conservation techniques. The colour ranges from golden yellow through orange, red, brown, green, violet and black. It can be transparent to opaque. Amber from various sources, when exposed to ultraviolet light, will fluoresce with different colours including red, green, blue and yellow. Most modern amber from the Dominican Republic is golden brown or iridescent blue depending on whether it is viewed under or against direct sunlight.

Ornaments found in Sixth Dynasty (3200 BC) Egyptian tombs have been labelled amber. Although there are several early references to amber artefacts in Egypt dating as early as 3500 BC, the analyses of Lucas (1962) showed that these were resins derived from coniferous trees in the Levant other than *Pinites succinifer*. In Mesopotamia, amber has been found in sites dated as early as the mid-third millennium BC. By the Bronze Age amber was traded throughout Europe, including the British Isles. The best European amber is found around the shores of the Baltic Sea; however, it has also been found in Sicily and Romania. The Phoenicians traded bronze for amber with the Ligurians (Liguria is a coastal region in northwest Italy) between the thirteenth and sixth centuries BC. The Etruscans traded in amber as early as 1500 BC. Homer wrote about amber beads in the *Odyssey*. Most amber found in the Aegean area and Italy came from the Baltic region. The earliest true amber found in Greek sites was from Mycenae and Pylos at the end of the Middle Helladic period (*ca.* 1900–1600 BC). Amber artefacts have been found in Crete, other Aegean islands, Cyprus and Syria dating to the end of the Late Helladic (*ca.* 1500–1100 BC). In the Western Hemisphere in recent times Chiapas State in Mexico and the Dominican Republic have been major suppliers of amber. The deposits in Mexico have been exploited since pre-Columbian times (Sinkankas, 1997).

3.4.2. Pearl

Pearl is composed of the calcium carbonate mineral aragonite produced within the soft tissue of a living shelled mollusc. The highest quality pearls come from the genus *Pinctada*. Pearls are characterized by their translucence and lustre with an intricate play of surface colours. The iridescence results from the interference and/or diffraction effects of light through many layers. The colour of pearls varies with the type of mollusc species as well as the water environment it grew in. Pearls can be found in white, yellowish white, pinkish white, bluish white, reddish and blackish grey. The name originates from Latin *perma* (a type of shellfish). Although pearls have long been a valued gem, they generally were not used in Egypt until Ptolemaic times. The earliest use of pearl as a gem appears to be in Europe and dates from the Palaeolithic. Pearls were highly prized on the Arabian Peninsula where there was a highly developed pearl diving and trading industry. Freshwater pearls were used by Native Americans. For example, they have been found in the ancient burial mounds of the Hopewell people in Ohio. Conversely, aboriginal peoples of Australia regard pearls as a nuisance because they occasionally cracked their teeth on them while dining on north coast oysters (Ward, 1985). Pearl-oysters and pearl-mussels are found on the Red Sea coast of Egypt, in the Persian Gulf and off the coast of Sri Lanka. Freshwater pearls have also been found in rivers in the British Isles and continental Europe. In modern times, saltwater pearls (mostly cultured) are harvested from the northwestern Pacific Ocean, particularly the waters around Japan. The technique for growing cultured pearls was developed by the Japanese in the early twentieth century.

3.4.3. Shell

Shell ornaments are found throughout the world. The origin of the name is from the Teutonic root *skal* (to peel off or separate), the Icelandic *skel* (a tile) and Middle English *schelle*. Shell disk beads have been found at ancient Egyptian sites, which were occupied until the Twenty-second Dynasty (943 BC). Shell ornaments were also favoured throughout precontact North America, and evidence of their use extends into the continental interior, suggesting heavy trade. Shell was used for beads and other personal ornaments.

Mother of pearl, also called nacre, is an iridescent material from the shell lining of many molluscs including pearl-oysters and abalone. Nacre was used as an inlay in jewellery. Mother of pearl comes in many natural colours but frequently is bleached and dyed for gemstone use.

Abalone is a gastropod shellfish. Gemstones made of its shell are sometimes referred to as mother of pearl because they often display an iridescence superior to that found in any other molluscs. Abalone shell is extremely tough and durable due to its unique structure of calcium carbonate clusters held together by a tough, sticky protein. These shell gems have been found in archaeological sites around the world. One site in South Africa is dated at 75,000 BP. There also is a rare gem called abalone pearl, found only in nature. These are not spherical but have odd shapes and are multi-coloured in intriguing patterns.

Ammolite is a gemstone made of the fossilized shells of ammonites. These gemstones, which resemble opal, are composed primarily of aragonite.

3.4.4. Jet

Jet, an organic material related to coal, was just one of a range of black lithic materials used in ancient times for jewellery. Jet is a lustrous black substance that is easily worked and will take a high polish (Fig. 7). Its hardness can vary from 2.5 to 4.0. According to Pliny (*De natura fossilium*), the word jet derives from *gagates* and then from Gagas, a river of Lycia on the southern coast of Turkey. It has sometimes been called black amber, and the term 'jet black' comes from this material. Jet occurs as sporadic lenticular masses and is derived from pieces of drifted wood that were buried in isolation and did not pass through a peat phase. Such material was not coalified, but rather underwent decomposition and retained a cellular structure. Good jet has been a

Figure. 7. Left: a polished jet stone. Right: A Victorian cameo carved from jet.

minor gemstone for amulets and pendants since Roman Britain. According to Pliny, jet was procured from Pamphylia and Lycia.

3.5. Other semiprecious stones

3.5.1. Obsidian

Obsidian is not a mineral; it is a glassy volcanic rock that forms as the result of rapid quenching of dacitic and rhyolitic magmas. It has been used as a semiprecious stone for millennia. Obsidian has the hardness of high-silica glass: 5–5.5. It is normally black or brown and has a vitreous lustre. It sometimes contains inclusions that create a silver or gold sheen. Some obsidian contains needle-like inclusions, gas bubbles or cristobalite crystals that look similar to snowflakes, giving rise to the name 'snowflake obsidian'. According to Pliny the name *obsianus* came from Obsius who discovered the material in Ethiopia. This may well be hearsay. In addition to its use as a tool because it has a sharp cutting edge, obsidian was used to manufacture ceremonial objects, beads and gems. The ancient Egyptians used obsidian as early as Predynastic times (5500–3100 BC). It was not available locally and so was imported, probably from Abyssinia (Lucas, 1962). A gold and obsidian necklace, dated to 1810–1700 BC, was excavated by Howard Carter and is now displayed at the New York Metropolitan Museum of Art. Obsidian deposits are plentiful in the USA and Mexico. In North America, rounded nodules of obsidian called 'Apache tears' were highly valued by the indigenous people. Apache tears look opaque in reflected light, but translucent when held up to light. Obsidian gemstone beads can be made with ease by modern technology. Shirataki, Hokkaido, Japan is one of the largest obsidian sources in northeast Asia.

3.5.2. Jasper

Jasper is impure red chert, 75% to 95% silica in the form of cryptocrystalline or microcrystalline quartz of roughly equidimensional crystals. It is normally red and often opaque. Its name is derived from Latin *iaspis*, corresponding to the Persian *iashm* and *jashp*. Pliny called the many varieties of agate or jasper in use at that time *iaspachates*. Because of its large silica content, gem jasper takes a high polish. It is widely distributed geographically, which partly accounts for its popularity. For thousands of years black jasper was in use to determine the gold content of gold-silver alloys. Rubbing the alloys on the jasper produced a streak, the colour of which indicated the gold content. Jasper was plentiful in the Egyptian Eastern Desert (Lucas, 1962) and exploited in Egypt since Predynastic times. Jasper has been used since prehistoric times in North America and Japan. Like most precious and semiprecious stones it has its own legends, myths and purported medicinal properties. One legend says that jasper would drive away evil spirits and protect against snake and spider bites. In the fourth century AD it was used in attempts to bring rain.

4. Altered, fake and counterfeit gems

Rocks and minerals that have been altered by humans are found throughout the archaeological record from very early periods (Nassau, 1984). For example, the ancient Romans valued darker chalcedony over lighter varieties, and they learned that burning light-coloured chalcedony would cause it to darken. To prevent fracturing, this had to be done carefully, without heating or cooling the stone too quickly (Nassau, 1994). Methods for heat-treating gemstones became a closely guarded trade secret. In dealing with the colour of minerals in antiquity we do not know when early artisans discovered that the colour of many minerals could be changed by heating. Normally it is not possible to distinguish a naturally coloured gem mineral from a heat-treated one with the naked eye, but laboratory analysis can reveal microscopic fractures that result from exposure to high temperatures.

Examples of altered colour can also be found in obsidian and tourmaline. Obsidian occurs naturally in black, grey, brown and as the snowflake variety. It does not generally occur in reds, blues or greens, but such colours can be found in rock shops. Colourless tourmaline can be turned pink by radiation although the colour will fade with time, particularly if exposed to sunlight. Alexandrite may be natural, *i.e.* a rare variety of the mineral chrysoberyl, or synthetic. The synthetic one is made of corundum laced with vanadium to produce the colour change, it is also known as alexandrite. Pale to medium blue topaz is enhanced by irradiation to produce a more intense blue colour.

The first German editions of Bauer's monumental two-volume work on precious stones in 1895 and 1896 included a discussion of artificial and counterfeit gems (Bauer, 1968). Modern technology, driven by large industrial research laboratories, has provided a host of methods to duplicate natural gems and semiprecious stones. All manner of gem crystals from beads to large pendants can be produced on a massive scale. For example, Peruvian blue opal is the national stone of Peru. It is found only in the Peruvian Andes, but in recent years dyed stones have been mass marketed by some dealers as Peruvian Blue Opal.

The belief in the healing power of gems and crystals is not limited to the pre-scientific era. Today one can buy books with titles such as: *Healing Crystals and Gemstones; Gemstones and Crystals to Reduce Stress, Attract Money, and Increase Energy; Gemstone Healing; Lucky Gemstone and Crystal Talismans, Charms, and Amulets for*

Health, Wealth, and Happiness. As the market for mineral crystals has developed, people have found ways to simulate many types of crystals. Perhaps one of the easiest is fluorite. Perfect fluorite crystals grow as octahedrons, and the cleavage is also octahedral. With practice one can take a specimen of fluorite and cleave it into perfect octahedrons, but these are not crystals with naturally occurring faces, they are only cleavage fragments. Natural fluorite crystals are always attached to some matrix, so the best way to spot cleavage fragments is to look for matrix. If no amount of attached matrix is present, it is a cleavage octahedron.

5. Recent times

Being able to identify fake gems and semiprecious stones is theoretically easy as each has a unique set of physical and chemical properties. The synthetics are designed to have the look and colour of a natural gemstone, but most do not have either the chemical or physical characteristics of the natural material. It remains possible to identify the fakes with the proper scientific equipment, particularly optical, but it is difficult if the only equipment is the unaided eye with a 10-power hand lens, a hardness set of some sort, and a way to accurately measure the density. Highly polished and highly reflecting materials such as gemstones are especially difficult to identify without analytical tools. As every gemmologist and mineralogist knows, by far the most difficult job is to try to identify a gem or mineral from a photograph. College, university and museum mineralogists are also regularly asked if a specimen is an example of some local variety name used by collectors (*e.g.* Maw-Sit-Sit from Burma). Local variety names must run into the hundreds of thousands, and few professionals keep those names in their head. Large computer databases may help to alleviate this problem.

Artificial gemstones are made to imitate natural gemstones. For example, several synthetic minerals look similar to diamonds. A cubic zirconium oxide known as cubic zirconia is manufactured to simulate diamond (this is not artificial zircon, which is a zirconium silicate, though the similarity of names may cause confusion) . Another example is synthetic moissanite. Moissanite is silicon carbide and occurs as a rare natural mineral. Moissanite has a higher refractive index than diamond, so it has more sparkle than the diamond, and so it is sold as a diamond look-alike. Other synthetic gems currently made to simulate diamond are made of high refractive index glass, and even an organic acrylic glass of high reflectivity.

Artificial gemstones may have a more vivid colour than their natural counterparts. Synthetic corundum, including those simulating ruby and sapphire, cost only a fraction of the natural gemstones and can be manufactured to have chemical and physical characteristics that are identical to the naturally occurring gemstones. Artificial amethyst can be produced if some iron is included with the silica and the resulting crystal irradiated with gamma rays (O'Donoghue, 1987).

In >60 years as a (sometimes) mineralogist I have visited more than three dozen mineral dealers (rock shops) in perhaps a dozen U.S. states and a few foreign countries (*e.g.* Egypt). The Geological Society of America, at its annual meetings, has a large exhibit hall where two or more mineral dealers exhibit for sale a broad range of minerals including gems, semiprecious stones and related minerals of interest to collectors. I cannot recall an exhibit where there was not at least one altered mineral, usually altered colour. Perhaps it is fitting that the book *Encyclopedia of Fraud* is published by Obsidian Publishing (Wells, 2002). Although obsidian is a volcanic rock (not a mineral), it is found in most private mineral collections. Readers interested in gems and man might want to read *Fifty Minerals that Changed the Course of History* (Chaline, 2012).

6. A note on the geology and geography of gem mineral deposits

Precious and semi-precious minerals form in many rock types: pegmatites (a coarse-grained intrusive igneous rock), metamorphic rocks (*e.g.* garnets are commonly found as crystals in gneiss and mica schist), hydrothermal veins (*e.g.* emerald deposits of Colombia), and both primary (*e.g.* agate) and secondary (*e.g.* beaches and placers) sedimentary deposits.

Gemstones are rare in part because they tend to be scattered sparsely throughout relatively large rock masses or to occur as small aggregates filling veins and small cavities. My own limited experience with precious and semiprecious stones has been with pegmatites in Minas Gerais State in Brazil and the Black Hills in the U.S. state of South Dakota (Roberts and Rapp, 1965), and with surface deposits in the states of South Dakota and Minnesota as well as small amethyst workings in northern Minnesota.

Brazil is especially rich in precious and semi-precious mineral deposits, with more than 100 gemstone deposits with significant output of emerald, ruby, sapphire, tourmaline, topaz, amethyst, citrine, agate, opal and diamond. Many of the gem minerals come from pegmatites, which are the final product of the last fluid of a crystallizing magma. The final magmatic liquid is rich in chemical elements that did not find a place in the more common minerals. This liquid contains most of the minor elements such as lithium and beryllium, so pegmatites generate gem minerals including beryl, tourmaline, topaz and sometimes spessartite garnet. Gem pegmatites have a somewhat different overall composition from non-gem pegmatites, including lower silica contents. Most gemstones in pegmatite are found in relatively small vugs or cavities. However, pegmatites contain the largest crystals of any geological deposit, including many spectacular gem minerals. The largest gem crystal I had seen (until I visited the giant crystal exhibit at the Earth and Man Museum in Sofia, Bulgaria) was the famous imperial topaz of Oro Preto, Brazil, found in the heart of the Minas Gerais pegmatites.

Not all countries have pegmatites, but surface and near-surface sediments that host mineral deposits are ubiquitous. Tourmaline, beryl and other precious and semiprecious minerals are often concentrated in stream deposits. Sapphires have been recovered from gravels being worked for gold. Namibia's Atlantic shore beach deposits and sediments carry appreciable quantities of diamond. Fluvial reworking of glacial sediments in British Columbia, Canada, has concentrated sapphires and garnets.

A significant diamond deposit in China's Hunan Province was formed in an alluvial regime. Agates are frequently found in glacial gravels and beach deposits of lakes.

Afghanistan is famous as the source of the world's finest lapis lazuli, which has been mined there continuously for >6000 years. But Afghanistan also has other important gem minerals, notably emerald, ruby, tourmaline, aquamarine, amethyst and turquoise. Most of these gemstones come from northeast Afghanistan. While the mining of lapis lazuli goes back millennia, the mining of emerald, ruby and tourmaline began in only the early 1970s. Afghanistan has become known for its blue tourmaline and sea-foam blue green tourmaline, but exploitation of gem deposits has been hampered by decades of war.

7. Conclusion

Humans have valued gems and precious stones since the Palaeolithic. Favoured gems have changed over time as fashions, availability and new discoveries vary the picture. Beauty, hardness and availability determine which gemstones will dominate the market at any one time and place. The increasing skills of gemsmiths have also broadened and enriched the story of gems and man. Today, artificial gemstones are popular. In the next decade – who knows? – they may be called ersatz or fakes and devalued. The only thing sure is that humans seemingly will always fancy ornaments as part of the good life.

References

Agricola, G. (1546/1964) *De Natura Fossilium.* (Textbook of Mineralogy). Translated by M. Bandy and J. Bandy, Dover Publications, New York. 240 pp.

Angelini, I. (2010) Amber and resins. Pp. 367–384 in: *Scientific Methods and Cultural Heritage* (G. Artioli, editor). Oxford University Press.

Bauer, M. (1968) *Precious Stones* (in two volumes). Dover Publications, New York.

Bayliss, P. (2000) *Glossary of Obsolete Mineral Names.* The Mineralogical Record, Tucson, Arizona, USA, 235 pp.

Blackburn, W.H. and Dennen, W.H. (editors) (1997) *Encyclopedia of Mineral Names.* The Canadian Mineralogist Special Publication **1**. Mineralogical Association of Canada, Ottawa, Ontario, 358 pp.

Chaline, E. (2012) *Fifty Minerals that Changed the Course of History.* Firefly Books Ltd., Buffalo, New York, USA, 223 pp.

de Fourestier, J. (1999) *Glossary of Mineral Synonyms.* The Canadian Mineralogist Special Publication **2**, Mineralogical Association of Canada, Ottawa, Ontario, 434 pp.

Herrmann, G. (1968) Lapis lazuli: the early phases of its trade. *Iraq*, **30**, 21–57.

Kunz, G.F. (1971) *The Curious Lore of Precious Stones.* Dover, New York, 406 pp.

Laufer, B. (1974) *Jade: A Study in Chinese Archaeology and Religion.* Reprint of 1912 edition. Dover Publications, New York, 370 pp.

Lucas, A. (1962) *Ancient Egyptian Materials and Industries.* revised by J.R. Harris. Histories and Mysteries of Man Ltd, London, 523 pp.

Mathien, F.J. (2001) The organization of turquoise production and consumption by the prehistoric Chacoans. *American Antiquity*, **66**, 103–118.

Moorey, P.R.S. (1985) *Materials and Manufacture in Ancient Mesopotamia: the Evidence of Archaeology and Art. Metals and Metalwork, Glazed Materials and Glass.* BAR International Series 237. British

Archaeological Reports, Oxford, UK, 302 pp.

Nassau, K. (1984) The early history of gemstone treatments. *Gems and Gemology*, Spring, 22–23.

Nassau, K. (1994) *Gemstone Enhancement.* Butterworth, London, 252 pp.

Needham, J. (1959*) Science and Civilization in China*, volume III Mathematics and the Sciences of heaven and earth. Cambridge University Press, Cambridge, 874 pp.

O'Donoghue, M. (1987) *Quartz.* Butterworths, London, 110 pp.

Pliny (1989) *Natural History* (10 volumes). W.H.S. Jones (translator). Harvard University Press, Cambridge, MA.

Rapp, G.R. (2008) *Archaeomineralogy.* 2nd edition. Springer, Heidelberg, Germany, 348 pp.

Roberts, W.L. and Rapp, G.R. (1965) *Mineralogy of the Black Hills.* Bulletin 18, South Dakota School of Mines and Technology, Rapid City, South Dakota, USA, 268 pp.

Roberts, W.L., Campbell, T.J. and Rapp, G.R. (1990) *Encyclopedia of Minerals.* Van Nostrand Reinhold, New York, 979 pp.

Sinkankas, J. (1997) *Gemstones of North America* Volume III. Geoscience Press, Tucson, Arizona, USA, 675 pp.

Theophrastus (1956, 315 BC) *Theophrastus On Stones: Introduction, Greek text, English translation, and Commentary.* Translated by John F. Richards and Earle Radcliffe Caley. Ohio State University, Columbus, Ohio, USA, 238 pp.

Wang, G. and Shen, K. (1986) The recognition of main copper-bearing minerals in ancient China. *Earth Science* (Journal of Wuhan College Geology), **11**, 111–115.

Ward, F. (1985) The pearl. *National Geographic.* August 1985, 193–223.

Wells, J.T. (2002) *Encyclopedia of Fraud.* Obsidian Publishing, Austin, Texas, USA, 876 pp. plus Appendices.

Wen, G. and Jing, Z. (1996) Mineralogical studies of Chinese jade. *Acta Geologica Taiwanica*, **32**, 55–83.

Whitlock, H.P. (1934) *Jade and the Antique Use of Gems.* American Museum of Natural History, New York. 23 pp.

Wright, R.V. and Chadbourne, R.L. (1970) *Gems and Minerals of the Bible.* Keats Publishing Inc, Connecticut, USA, 148 pp.

EMU Notes in Mineralogy, Vol. 20 (2019), Chapter 9, 345–366

Gemmology in the service of archaeometry

M. Pia RICCARDI[1], Loredana PROSPERI[2], Serena C. TARANTINO[3] and Michele ZEMA[4]

[1]*Dipartimento di Scienze della Terra e dell'Ambiente, Università degli Studi di Pavia, via Ferrata 1, Pavia Italy*
mariapia.riccardi@unipv.it
[2] *Istituto Gemmologico Italiano, Piazza San Sepolcro 1, Milan, Italy*
l.prosperi@igi.it
[3] *Dipartimento di Scienze della Terra e dell'Ambiente, Università degli Studi di Pavia, via Ferrata 1, Pavia Italy*
serenachiara.tarantino@unipv.it
[4] *IUCr, International Union of Crystallography, 5 Abbey Square, Chester CH1 2HU, UK*
mz@iucr.org

Archaeometric studies of ancient artifacts containing gems or gem-quality geological materials have an intrinsic complexity. The scientific questions to be answered are related not only to the type of material used (*e.g.* a mineral or poly-mineral geological material), but also to their age and provenance.

The answers can derive only from multidisciplinary study which combines experimental observations with information and clues from different disciplines.

This paper presents three case studies in which mineralogical knowledge and the non-invasive approach typical of gemmological analyses solve the problem of gem identification. The answer aboout the origin of gems and/or minerals is more complex because little is known about the *chaîne opératoire* that precedes the use of gems in artifacts. Little is known currently about the geological complexity of ancient ores, some are now exhausted. Moreover, the criteria used in choosing raw material are not known.

A multidisciplinary approach can lead to identification of the sources of supply of the material, understanding the choices made for the realization of the artifacts, defining the links with the geological, geographical and cultural realities that complete their context of origin. Correct archaeometric investigation must follow the 'four C's' rule and keep in mind the 'complexity' of the artifact, answer 'congruent' questions to the study, aggregate different 'competences' and be open to 'collaborations'.

1. Introduction

The use of geological materials as ornaments has evolved over time with a trend parallel to the increase in social complexity (Jessup, 1950; Arrhenius, 1988; Collareta, 2003, 2011; Mottana, 2006; Aimone, 2011; Hilgner, 2017; Bertelli, 2017a; Caselli, 2017). Gemmological materials used in the past can be restricted to a few dozens of mineral species with their many varieties, to some rocks and to a few products of the plant and animal kingdoms (Rapp, 2019). However, these must be juxtaposed with

DOI: 10.1180/EMU-notes.20.9

several artificial and synthetic materials, including glass (De Michele and Manzini, 1993; Henig, 2008; Riccardi *et al.*, 2017), as well as with composite gems and with all those many gems that have undergone treatments to improve their appearance, transparency or colour (Nassau, 1994; UNI 10245: 1993; CIBJO, Confédération International de la Bijouterie, Joaillerie, Orfèvrerie des Diamants, Perles et Pierres, 2012).

Given the knowledge of the analytical methods applied to the study of geological materials for gemmological use (*e.g.* Artioli, 2010; Domínguez-Bella, 2012; Carter, 2016; Garrison, 2016), the current chapter is focused on an interpretative approach to these materials and derived objects with the aim of answering historical or archaeological questions. Such an approach goes beyond the simple recognition of the natural (mineral) or artificial phase used in the making of the object itself. When such historical objects are analysed it must be remembered that they represented a 'status symbol', and therefore the materials used are often rare in order to exalt this status.

As is the case for all geological materials used in ancient production cycles, including gems and gem materials, it is necessary to deal with the intrinsic complexity of geological contexts (Abdalla and Mohamed, 1999; Krippner *et al.*, 2014; Stutenbecker *et al.*, 2017; Aurisicchio *et al.*, 2018; see also: *The Canadian Mineralogist*, 2017, vol. **55**), both in terms of the methodological approach and in the interpretation of data. The most common questions are related to the origin of the materials and, if the material underwent some manipulation with respect to the natural, what was the operational chain to obtain it or to mold it into an object.

The main question that drives archaeometric studies based on gems is identification of their origin. In fact, this information helps to outline routes and contacts for the exchange of goods and thus define trade patterns. This knowledge, together with historical and/or archaeological elements, provides clues to the economic systems and dynamics of societies in different historical periods. What is the path of raw material from its extraction to the use of the gem in an artisan workshop? Were those who extracted the rough gem also those who worked it? If not, where were the workshops? Were these workshops also hubs for the exchange of gems?

The link between object to be made, commissioning and the gem itself is very tight and changes over time. Consequently, the study of the provenance cannot be separated from the knowledge of the local and historical contexts and of the society within which the object and the material is used. Furthermore, it is often revealed that many gems have been used multiple times (De Michele and Manzini, 1993; Superchi, 1999; Calligaro *et al.*, 2007; Riccardi *et al.*, 2017). In the practice of the goldsmith's art, the gold of old jewels is recast and the gems are recovered and reused in a spiral that we could define as 'autarchic'. This is also demonstrated by the extreme variability of the dimensions and shapes the new bezels must fit.

A further complication is provided by the 'provenance of raw materials' concept itself. The believe behind the studies of the source of any raw materials has been explained in a work concerning a gem material, such as turquoise (Weigand *et al.*,

1977): "it is possible to recognize the source of raw materials if the chemical or mineralogical differences between different natural sources are greater than they are within each source". This postulate highlights the importance of geological and mineralogical-petrographic knowledge when carrying out provenance studies. Nevertheless, much work remains to be done. Geo-mineralogical studies of mineral formation contexts must be conducted at an observation scale much more detailed than normally done for geological studies, being aware that the material extracted and harvested in the past may not be available today.

An example that can clarify the complexity of an archaeometric study of the origin of raw materials is that of garnets, which were used intensively in antique jewellery and in particular in Merovingian artifacts. Garnets are silicates comprising multicomponent solid solutions with the general formula $A_3^{2+}B_2^{3+}Si_3O_{12}$, where A may be Ca^{2+}, Mg^{2+}, Fe^{2+} or Mn^{2+} and B may be Al^{3+}, Fe^{3+} or Cr^{3+}. Garnets are grouped in two series: those in which the A cation is Ca^{2+} (uvarovite, grossular, andradite) and those in which A is not Ca^{2+} and B is Al^{3+} (pyrope, almandine, spessartine). In nature, garnets are found commonly in metamorphic and igneous rocks and as debris in many sediments. Garnets have a broad chemical and geochemical footprint, depending on various geo-petrographic factors: the physical parameters of geological processes (pressure, temperature and time), the type of the parent rock, the composition of fluids, the microstructure of host rock and local chemical-physical equilibrium (microdomains). This compositional variability may be an aspect that complicates the search for sources of supply of raw minerals. In the archaeometric literature, there are many studies dealing with the characterization of the garnets of historical and archaeological objects, adopting non-invasive or micro-invasive analytical techniques (Arrhenius, 1971, 1985, 1988; Greiff, 1999; Hilgner, 2017), such as PIXE (Particle-Induced X-ray Emission) and Raman spectroscopies (Farges, 1998; Calligaro *et al.*, 2002; Dran *et al.*, 2004; Perin *et al.*, 2007; Mathis *et al.*, 2008), electronprobe microanalysis (Velde and Courtois, 1983; Rosch *et al.*, 1997; Quast and Shussler, 2000), X-ray fluorescence (Bimson *et al.*, 1982), X-ray diffraction (Schussler *et al.*, 2001) and laser ablation inductively coupled plasma mass spectrometry (LA-ICP-MS) (Carter and Dussubieux, 2016; Carter, 2016). An important archaeometric study carried out at the laboratories of the Louvre Museum on >1000 specimens of garnets belonging to Merovingian goldsmithery has enabled the construction of a comprehensive compositional database for garnets of gemmological use. The database has been used to identify five compositional groups (I to V) well defined in terms of their major-element contents, some minor elements and crystalline inclusions (Farges, 1998; Calligaro *et al.*, 2002, 2003, 2007; Perin *et al.*, 2007; Horvath and Bendo, 2011). Comparison with data obtained from the literature concerning the composition of garnets (Greiff, 1999; Quast and Schüssler, 2000) and the reading of historical sources made it possible to identify the areas of origin: almandines of group I and group II come from India, group III from Sri Lanka and group V from the Bohemian Massif. Some uncertainties remain for group IV, which could be of European origin (Calligaro *et al.*, 2007; Carter, 2016). If the definition of the geochemical footprint is completed with the quantification of the minor and trace elements, and if the geological sampling is expanded, the chemical compositional scenario becomes more complicated (Carter and Dussubieux, 2016): in the geographical area that includes south India and Sri Lanka, almandine garnets, indistinguishable in terms of minor elements, are subdivided into homogeneous groups according to their specific geographical areas of origin. However, the comparison on chemical and geochemical bases of these garnets with those used for the realization of almandine garnet beads of archaeological sites in the south Asia area (Thailand and South India) is not always reflected (Carter, 2016). Some geological sources are still partly unknown. The surprising fact is that all the data confirm that the archaeological garnets constitute very homogeneous compositional groups (Farges, 1998; Calligaro *et al.*, 2007; Horvath and Bendo 2011; Carter 2016; Bugoi *et al.*, 2016). This fact can be the expression of selection upstream of the processing and trade of raw material. A multidisciplinary study of this aspect could outline the criteria for the choice of geological garnets (colour, transparency, shape, dimensions, workability ...) and whether in the past, first- or second-quality materials were traded and for what uses. This would lead to an understanding of the conscious choices made in the past along the production chain of 'luxury goods', entering the minds of operators and artisans who managed part of the operational chain.

2. Some historical sources

A thorough examination of historical sources is of paramount importance in tackling the longstanding issues concerning the origin of gems. Archaeological studies are no less important, in particular those concerning the sites of departure, transit and arrival of luxury goods such as gems.

In analysing historical sources, the ambiguity of language used by different scholars may be an obstacle; for some, a given term defines a very precise mineral, while for others the same term represents a series of geological materials, similar in colour or ornamentation. Some terms are not easily interpretable.

In the Europe of the early Middle Ages, the first and only reliable written evidence on the eastern origin of certain gems is the *Christian Topography* by Cosma Indicopleuste (Wolska-Conus, 1968), a writer of the first half of the 6th century AD. A navigator, connoisseur of the Red Sea and the Persian Gulf, Cosma was stationed in the port and commercial centre of Adulis (60 km southeast of Massaua in Eritrea), the centre of commercial traffic not only with the Indies, but also with the southeastern and perhaps Kenyan–Somali coasts of Africa. Cosma was the first to record that the 'alabandinon' came from the east coast of India, and the 'hyacinth' from Taprobane, Sri Lanka. Important and extensive analytical researches on early medieval garnets of the Rhenish area led S. Greiff to conclude that the deposits of red garnet were indeed in India and Sri Lanka (Greiff 1999).

The testimony of Cosma confirms and completes what was written centuries earlier by Pliny the Elder, but to obtain more precise indications, it is necessary to read the Arab writers of the 8th–9th centuries, obviously unknown to Latin contemporaries and rediscovered starting from the 18th century. The famous doctor Yuhanna Ibn Masawayh

(777–857), known in Europe as Mesue, who lived in today's Iraq, published a book on gems and their characteristics, printed in Arabic in Cairo only in 1977 and translated into French in 1998 by G. Troupeau (1998). The deposits of India and Sri Lanka are described there, also recalling that on that island they were located at the foot of Adam's Peak (Adam Peak or Sri Pada, 2243 m), where the rich gem district of Ratnapura occurs.

More exhaustive is the treatise on gems of Abu al-Rayhan Mohamed ibn Ahmad al-Biruni (973–1048), translated by Fritz Krenkow in 1936 from the existing manuscripts, and since then reprinted several times in Muslim countries (Al-Beruni, 2007). Philosopher, scientist and traveller, Al-Biruni had direct knowledge of the Indian and Sinhalese deposits and also introduced the use of density for better characterization of gems. His essay confirms the origin of the garnet and sapphire sold in those countries around the year 1000.

The Egyptian monopoly of the emerald trade was witnessed widely by the aforementioned Muslim writers and by Al-Masudi (896–956), historian, geographer and traveller, who lived in today's Iraq. In his monumental work, entitled *The Meadows of Gold and Mines of Gems*, known since the 19th century (Macoudi, 1861), important references are also found to the Egyptian emerald mines, still active at that time.

Thanks to the exceptional discovery of the 'Geniza' annexed to the synagogue in Cairo (a deposit of documents discarded but not destroyed, because of the Holy Name), S.D. Goitein and collaborators have published letters and documents from the centuries immediately after the year 1000. In these documents, are highlighted the dense network of commercial relations among the Egyptian Jewish community, often in societies with Muslim and Christian operators, with other communities scattered throughout the Mediterranean (Goitein, 1973, 2002; Goitein and Friedman, 2011).

3. The study protocol

The identification of the types of gems present in a historical artifact must follow a non-invasive approach. Very often the object cannot be transported from the place where it is stored (museum) to the laboratory, and this further limits the investigation protocol (the 'don't move, don't even touch' protocol). Traditional gemmological analysis is a method of study that meets the needs described above. It is also a quick, economic and non-destructive method to identify the gems found commonly in historical artifacts with sufficient security. It represents the premise of more complex investigations. Gemmological investigation is aimed at detecting certain simple physical parameters of minerals, rocks and other specimens.

If the gems are 'loose', *i.e.* not mounted in jewellery or other objects, it is possible to make optimal observations (*e.g.* measurement of density and refractive index). In most cases, however, the gems are mounted and cannot be removed from their settings other than for restoration needs. In case of mounted gems, the data collected are necessarily incomplete and this leads to a considerable level of uncertainty in the diagnosis. Fortunately, the ancient and medieval jewellery artifacts, especially the more voluminous ones, have only a few mineral species embedded, such as red garnet, quartz in all its varieties, corundum, and beryl, easily identifiable even from partial detection of their physical properties.

In the two case studies illustrated below, gemmological analyses were carried out using common gemmological instruments: stereoscopic microscope with fibre optic illumination; refractometer; spectroscope; calcite dichroscope; polariscope; Chelsea filter; and short (256 nm) and long (360 nm) UV lamps. For a brief description of these instruments refer to Manutchehr-Danai (2009). The nomenclature adopted is that stipulated in the norms UNI10173 and UNI10245 with minor discrepancy concerning the aspect of cutting forms, not fully considered in the Standards. The evaluation of colour, for reasons of brevity, does not always correspond to the denominations contained in the reference standard (UNI9810). Measurement of density and/or refractive index is fundamental for the recognition of the gemmological species. One example is monocrystalline quartz in its most common varieties (colourless, citrine, amethyst). To facilitate this measurement, in many cases a minimum intervention in terms of the data collection (density and/or refractive index) of the settings was authorized.

The analytical protocol, therefore, started with the measurement of dimensional parameters by using digital callipers: the size, shape and type of cut of gems are important indications to evaluate (or estimate, in case the stone is still mounted) the carat weight. The use of the optical microscope or, where not possible, of a triplet lens has allowed determination of the conditions of the collet and to identify inclusions and/or defects in the gem, as well as the condition of its surface. The observation of the gem and its inclusions is a fundamental step: the association of inclusions represents an 'identity card' of the gem and sometimes proves decisive in defining its origin (Figs 1 and 2).

When possible, the measurements of refractive index and absorption spectrum have been associated with these observations. The refractive index is a discriminating measurement used to recognize a large number of gems by simply determining whether a gem is monorefringent or birefringent. Absorption spectra have been observed through a manual spectroscope (Fig. 3). They are diagnostic in the identification of the mineralogical species as well as of the gems embedded in an ancient jewel or artifact. Use of a spectroscope will distinguish a ruby from an almandine immediately (despite both being red) (Fig. 3a, c), an emerald from an olivine (green), or a sapphire from a synthetic blue spinel (Fig. 3b, d).

Note that, when studying gems of a complex artifact consisting of several materials, such as ancient goldsmith objects, the correct interpretation of the gems within the artifact also depends on the knowledge of other materials, metals above all, and on the set of historical and archival documents referring to the same object.

4. Case study: the Iron Crown

The Iron Crown (Fig. 4), so called for a dark grey metal strip on the inner side and considered to be one of the nails of Christ's cross, is kept at the altar of the Teodolinda chapel, in the Cathedral of Monza, Italy. Even today, it remains an enigmatic object in its historical interpretation.

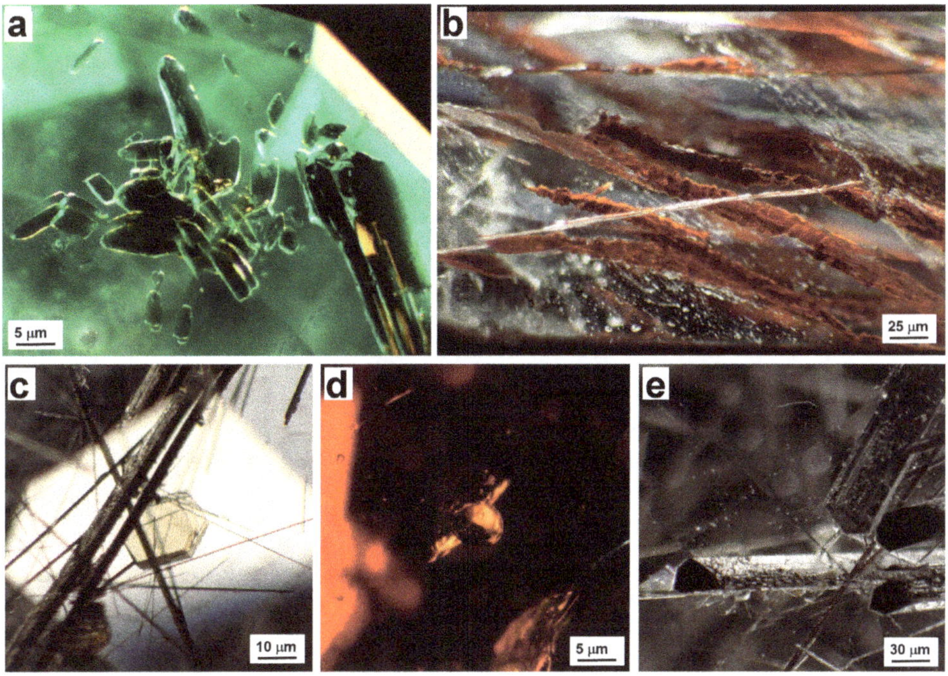

Figure 1. Optical microscopy images, dark field, of some inclusions in gems: (a) biotite in emerald; (b) hematite in quartz; (c) mica and rutile in quartz; (d) zircon with tension halons in garnet; and (e) tourmaline in quartz.

An interdisciplinary study was conducted over the years 1994 and 1995 (Lusuardi Siena, 1998; Bertelli, 2017b). The results of scientific investigations have raised new hypotheses and doubts about both the history of the object itself and its historical

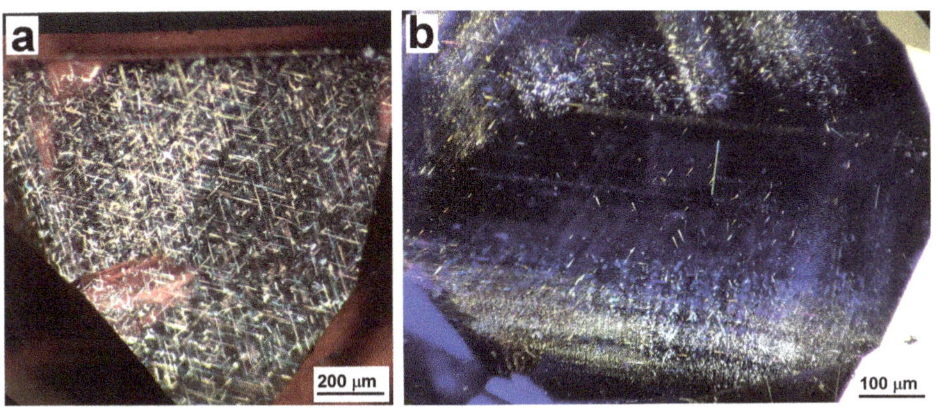

Figure 2. Optical microscopy images, dark field, of some inclusions in gems: (a) rutile in ruby and (b) rutile in sapphire.

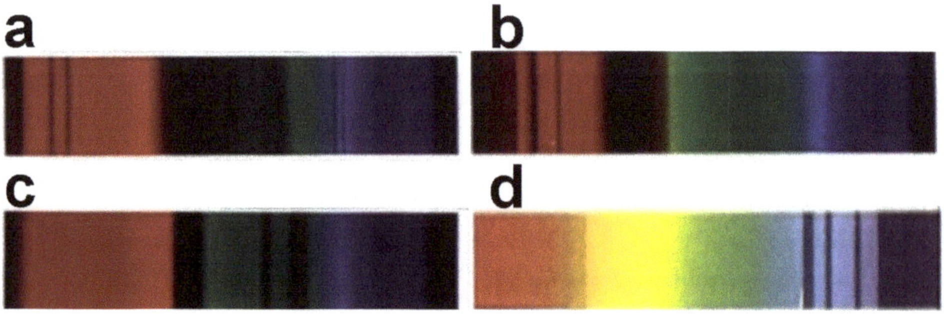

Figure 3. Examples of absorption spectra of some gems: (a) ruby; (b) emerald; (c) almandine garnet; and (d) olivine.

interpretation. The Crown shows signs of restorations to the structure, replacement of glazed plates and likely substitutions or perhaps only temporary removals of the gems. The questions about its interpretation which remain unresolved are summarized in the work of Bertelli (2017a,b), which provides an update on the historical studies of the object.

The Iron Crown is composed of six rectangular plates, P1–P6, attached to one another with their edges coffered on the outside to hold decorative enamels, rosettes and gems (Lusuardi Siena, 1998; Maspero, 2003, 2016). Plate P1 shows only two gems (Fig. 5), one placed in the centre with four rosettes and four enamels around it, the other is located between two rosettes in a narrow lateral strip. On each of the other five plates (Fig. 5 for P2 and P3; Fig. 6 for P4–P6) there are four gems: again, one occupies the

Figure 4. The Iron Crown (diameter 15 cm and height 5.5 cm) (photo by G. Monistier).

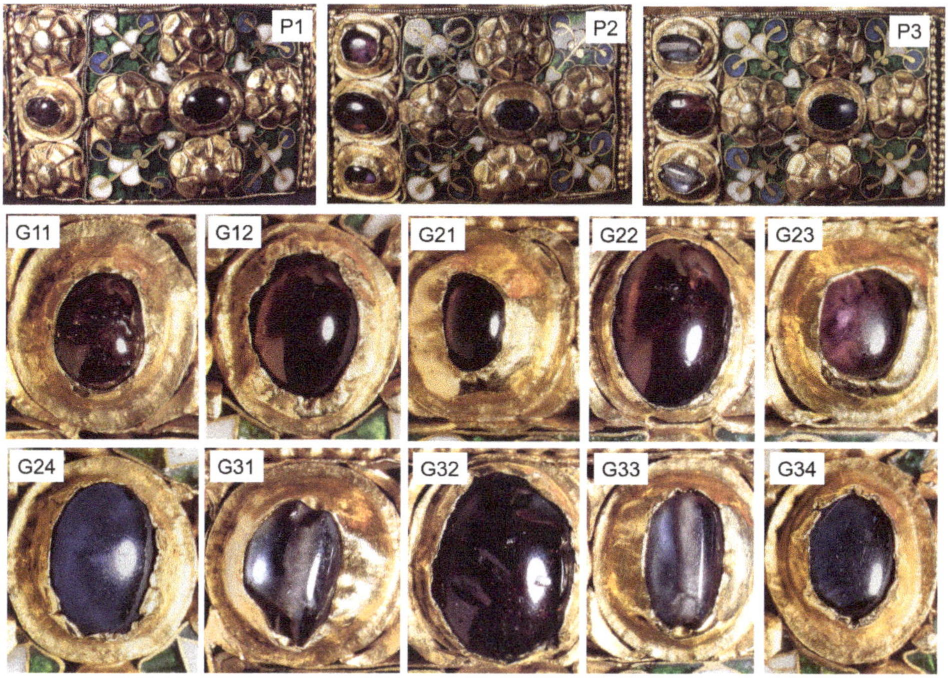

Figure 5. Plates P1, P2 and P3 of the Iron Crown and related gems. The dimensions of each element are given in the text and in Table 1 (photos by G. Monistier).

central position amongst four rosettes and enamels, while the other three are arranged vertically in the narrow lateral sector. We note that the Crown is currently assembled with plates P1, P2 and P3 (Fig. 5) with the jewelled border on the left, and P4, P5 and P6 (Fig. 6) with the jewelled border on the right. P1 and P6 are facing each other on their jewelled border sides (Fig. 4). Historians have hypothesized that the plates were originally assembled with the border all on the same side, and that the Crown had at least one more plate (Marimonti, 1841; De Michele *et al.*, 1986). All the gems are fixed with semi-occluding truncated cone settings with a wide collar, very similar to those of the middle border on the reliquary of Saint John the Baptist (Monza) and the reliquary-bag of Saint Stephen (Vienna), both attributed to the 9th century.

There are 22 gems set on the crown (Figs 5 and 6): corundum (7), garnet (7), quartz (4) and artificial glasses (4).

Corundum: The corundum gems in the crown are quite light-coloured and almost all are perforated longitudinally (Table 1). The hole shows helicoidal marks left by the drill, and in the case of gem G31 (Fig. 5) there is some encrusted material which, if analysed, might identify the abrasive used. This thus suggests that it is again a case of recycled material, with physical characteristics that may allow determination of its place of origin in Sri Lanka (Hughes, 1990; Gubelin and Koivula, 1992). Stone G42 (Fig. 6) has, in fact, kept the typical pebble shape of products from the gemmiferous gravels of this island.

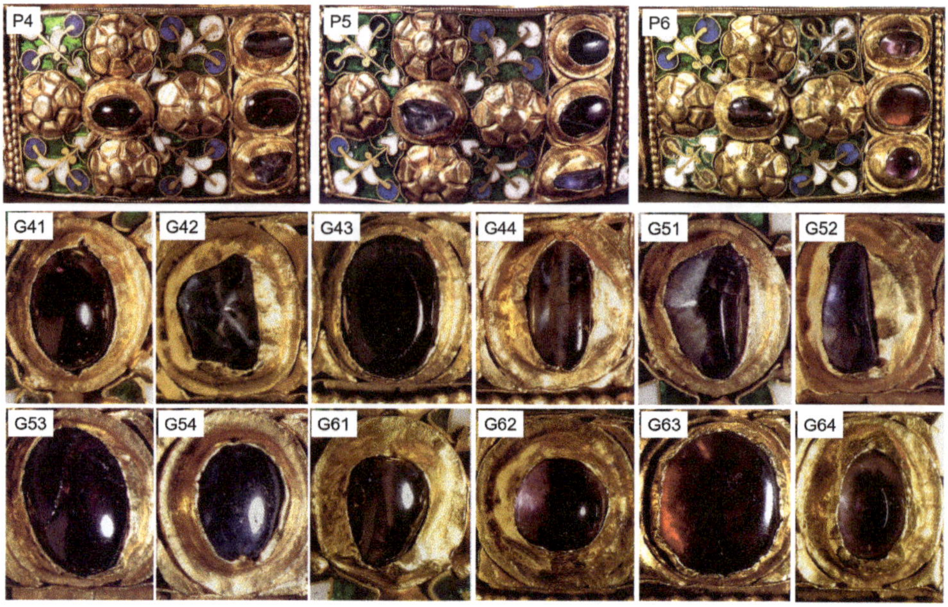

Figure 6. Plates P4, P5 and P6 of the Iron Crown and related gems. The dimensions of each element are given in the text and in Table 1 (photos by G. Monistier).

Garnets: The garnets of the Iron Crown are pyralspite (rhodolite), very close to almandine, as borne out by the refractive index and absorption spectra (Gunther, 1988; Stockton and Manson, 1985; Rouse, 1986). Inclusions indicate that the stones are from Sri Lanka (Gubelin and Koivula, 1992). The presence of surface grooves might be interpreted as a way to remove showy and heavy solid inclusions and give vivacity to the gem, but they are generally indicative of a re-use of the material.

Quartz: Quartz appears in an amethyst variety of a light violet colour (Table 1). One of the four specimens has an oval shape with mixed cabochon and table cut (Figs 5 and 6).

Glasses: Three specimens are coloured blue (from cobalt, as shown by the absorption spectra), and one is orange with a mass of aligned bubbles.

The gems of the Iron Crown have undoubtedly suffered losses and substitutions, as signs of damage to the gold lamina of the mountings testifies. Moreover, Fontanini, in the third edition of his *Dissertatio* (Fontatini, 1719?) writes that in 1299 "*unus lapis pretiosus*" was missing, having been substituted with a triangular stone which scarcely matched the quality of the others. Moreover, Bellani (1819) recorded that in the inventory drawn up in 1353, the Crown was found to have seven gems missing, yet at the same time the treasure included batches of loose gem-stones. Because the shapes of cuts underwent no changes until the 1500s, recognizing a substitution means relying on a reconstruction of the original chromatic pattern assigned to the gems and enamels. However, if the substitution respects the original colour of the gem to be replaced, even this criterion becomes invalid.

Table 1. Gemmological characteristics of the gems of the Iron Crown.

Gem	Dimensions (mm)	Colour	Clarity	Morphological and textural features	Measures	Notes	Classification
G11	10 × 7.8	Red	TP	oval; cabochon	AS; RI	fractures	garnet
G12	13.5 × 9	Red	TP	oval; cabochon	AS; RI		garnet
G21	9.8 × 7	Purple	TP	oval; cabochon	RI	inclusions	amethyst
G22	16.8 × 11	Red	TP	oval; cabochon	AS; RI	inclusions	garnet
G23	11 × 9	light purple	TP	irregular olav; cabochon	RI		amethyst
G24	12 × 8.2	Blue	TP	oval; cabochon	AR; RI		glass
G31	13.5 × 9	pale blue	TP	oval; cabochon	RI	crosswise hole	corundum
G32	17 × 12	Red	TP	oval; cabochon	AS; RI; incl.	superficial grooves, inclusions	garnet
G33	14 × 7	pale blue	TP	oval; cabochon	AS; RI	crosswise hole	corundum
G34	14 × 10	Blue	TL	oval; cabochon	AS		glass
G41	11.8 × 9.5	Red	TP	oval; cabochon	AS; RI; incl.		garnet
G42	12 × 8/8.5	pale blue	TL	irregular olav; cabochon	H > 8	crosswise hole	corundum
G43	16 × 12	Red	TP	oval; cabochon with table	AS; RI	inclusions	garnet
G44	14 × 7.5	pale blue	TP	oval; cabochon	AS, RI	crosswise hole	corundum
G51	14 × 9	pale blue	TP	irregular olav; cabochon	AS; RI; FL		corundum
G52	14.5 × 5.5	pale blue	TP	irregular oval	RI; FL	fracture; crosswise hole	corundum
G53	15.5 × 11	Red	TP	oval; cabochon	AS; RI	fracture, inclusions	garnet
G54	12 × 9.5	Blue	OP	oval; cabochon	AS		glass
G61	12.5 × 5/8	pale blue-violet	TP	irregular oval	AS; RI; incl.; FL	crosswise hole	corundum
G62	9.5 × 8.5	light purple	TP	oval; cabochon	RI		amethyst
G63	16.5 × 11.5	Orange	TP	oval; low cabochon	FL	Inclusions	glass
G64	12 × 7.5	light purple	TP	oval; cabochon with table	RI		amethyst

TP: transparent; TL: translucent; OP: opaque; AS: absorption spectrum; RI: refractive index; H: hardness; FL: fluorescence.

Certainly, identification of the materials constituting the Iron Crown has not solved all the interpretative problems that even today keep scholars busy dealing with medieval goldsmithery. The small number of gems, many of which are reused from older artifacts, the association of a few varieties of minerals, such as garnet, quartz (amethyst) and corundum, as well as the use of glass, show clearly that it is not the gems which are the 'star' materials but probably the enamels, which are very abundant in all the elements of the Crown and the metals.

For these materials and for metal alloys, data collected using portable XRF are available (Milazzo and Ciccardi, 1998). These first data, as often happens in archaeometric studies, raise new doubts: the composition of metals showed that, despite the name, in the Crown there is no iron and that the inner lamina is made of silver. The enamels in the slabs of the most ancient series (Lusuardi Siena *et al.*, 1998; Milazzo and Cicardi, 1998) seem to have been made with a potassium flux, while the restoration plates show the use of a sodium flux. At the same time, the glazes of the oldest slabs have tin as opacifier, while those of restoration contain Ca antimonates; the cobalt-blue coloured restoration enamels also contain Zn, perhaps associated with Pb. This in turn would indicate north European mines as a source of supply (Gratuze *et al.*, 1992), known only from the 10th–11th centuries (De Launay, 1913). Could the constant presence of Pb, detected by surface measurements, be justified by the intrinsic enamel composition or by the superficial processing of glazes?

5. Case study: King Tutankhamun's pectoral and Libyan Desert Glass

Libyan Desert Glass (LDG) (Fig. 7) is a natural glass consisting mainly of silica (SiO_2 ~98 wt.%) and traces of Al, Fe, Mg and other metals (Fudali, 1981; Koeberl, 1986, 1997; Barrat *et al.*, 1997; Greshake *et al.*, 2010, 2018). The H_2O content is ~0.1 wt.%. It was P.A. Clayton, in 1932, who first saw and reported the presence of abundant fragments of glass, some large, in the western part of the Egyptian desert, in the area between the Siwa oasis to the north and the Gif Kebir plateau in the south (Clayton, 1933, 1998; Clayton and Spencer, 1933, 1934). LDG can be opaque, translucent or transparent; the colour is mainly greenish yellow and variable from dark green to blackish.

The raw material can be found on the surface, in the inter-dune corridors of the Great Sand Sea, as fragments of various sizes, sometimes even >30 kg, with the surfaces worked by the wind. Flaking is absent and the fracture is concoidal.

Prehistoric man used LDG in artifacts (Oakley, 1952; Roe *et al.*, 1982; Negro and Damiano Appia, 1992). Despite its local use in pre-dynastic times, this glass does not occur in historical Egyptian objects (Lucas, 1962).

LDG is not an important gemmological material, but because of its likely genesis, it has taken on a very strong symbolic meaning. Apart from the first gemmological indication by L.J. Spencer in 1933 in the journal *The Gemmologist*, and an extended quotation in a gemmological text (Spencer, 1936), the glass of the Libyan Desert has not been mentioned further. The reasons for this are its scarce use in historical goldware

Figure. 7. Chromatic and textural varieties of Libyan Desert Glass (photos by V. De Michele).

artifacts, and, in part, its widely agreed lack of aesthetic properties. To obtain gems with intense but bright colours (brown, yellow-brown, brown-green), macroscopic samples showing dark a colour, almost black, must be chosen, while rough samples of a beautiful green colour are weakly yellow or even colourless after the cut (Fig. 8). The commercial value of the LDG does not reside in aesthetic properties, therefore, but in its symbolic meaning, as discussed below.

The discovery of the use of this material for the funeral of the great Egyptian king Tutankhamun was the result of a happy intuition, almost a doubt, during the examination of King Tutankhamun's pectoral (Fig. 9) held at the Cairo Museum (De Michele, 1998). The scarab-shaped central motif of one of Tutankhamun's pectoral

Figure 8. Gems obtained from some Libyan Desert Glass (photos by V. De Michele).

Figure 9. King Tutankhamun's pectoral (photos by J. Bodsworth/Wikipedia).

(14th century B.C.) was carved from a material strikingly similar to LDG in appearance. The pectoral with the large translucent green-yellowish scarab (18 mm × 28 mm) was found by Howard Carter in a chest (n° 267) in the so-called 'treasury' room beyond Tutankhamun's burial chamber (Carter, 1972). The true nature of the scarab material was not established with certainty. In book texts, this material was classified as chalcedony, a 'fibrous' variety of quartz (Desroches-Noblecourt, 1963; El Mallakh and Brackman, 1978; Aldred, 1979; Saleh and Sourouzian, 1987; Andrew, 1990). The material is cloudy, translucent, and shows a vitreous lustre. The lines showing the insect anatomical shape are fine, sharp grooves, indicating that the raw material is harder than the 5th degree on Mohs' scale. These features are compatible with chalcedony, but on the other hand, the colour is not compatible with any known variety of chalcedony.

The colour is a peculiar feature: fairly intense green-yellow, known by those who have had the opportunity to examine the glass pieces in the LDG area. By using optical fibres and a 10 × loupe, and by observing the textural details in the scarab, flow texture is observed, accompanied by small, round, white inclusions, very similar to spherules of cristobalite occurring in LDG (Fig. 10). The scarab is optically isotropic. The refractive index of the scarab's curved surface was measured using a refractometer (liquid n = 1.79) with the spot method. The spot method is a rapid method to read the refraction index on facets or cabochon-like curved surfaces. Precision is low, though the index values can be useful for identification. The refractive index of LDG was measured firstly by Spencer (1933; 1936), who obtained, by sodium light, a value of 1.4624; Weeks *et al.* (1984) obtained values ranging from 1.4590 to 1.4650 using the Becke line test. The refractive index measurements were particularly difficult because of the large size of the pectoral, therefore they were repeated several times and gave average values of ~1.48. In comparison, the refractive index of a fairly intense green-yellow LDG cabochon was measured using the same method and instrument and a similar value of 1.47–1.48 was obtained. Subsequent measurements conducted on other LDG products were carried out with the same refractometer and registered values ranging between 1.462 and 1.470. By considering the values reported in the literature and those measured on the artifact and on materials collected *in situ*, it can be concluded tthat the refractive index of LDG can vary from 1.459 to 1.470, according to the bulk chemical composition as well as the textural variations of the material itself (Table 2).

The data collected automatically exclude chalcedony, which would have registered a value of 1.53–1.54 (Table 2). Similarly, green Pakistan alabastrite, which is a concretional calcite, can be excluded. Other minerals characterized by a similar green-yellow colour show refractive index values very different from LDG (Table 2). Moreover, artificial glass would give a minimum value for the refractive index of 1.50–1.52, therefore an unlikely green obsidian, never found in Nature, can be also excluded.

Autoptic observation, inclusion and optical properties therefore point to LDG as being the raw material used to carve Tutankhamun's scarab. This material obviously went through the Nile Valley from the remote area >700 km away, to be used as a rarity

Figure 10. Textural details of some samples of Libyan Desert Glass: (a) green-yellowish variety with inclusions of cristobalite (1); and (b) transparent green-yellowish varieties with a fluid texture showing an alignment of bubbles (2) and dark brown bands (3).

Table 2. Gemmological materials with a yellow/yellowish tinge similar to that of Libyan Desert Glass.

Mineral	Refractive index	Hardness
Singly refracting materials		
LDG	1.459–1.470	6
Fluorite	1.434	4
Opal	1.44–1.45	5–6
Obsidian	1.49–1.50	5.5
Glass (artificial)	1.50–1.70	5
Hydrogrossular var Transvaal 'Jade'	1.72–1.73	7
Double refracting materials (spot methods)		
Talc var. Steatite	1.54	1–2.5
Chalcedony	1.53–1.54	7
Serpentine	1.56–1.57	4
Muscovite	1.58	3
Pyrophyllite var. Agalmatoite	1.58–1.59	1
Nephrite and other Actinolites	1.6–1.61	6
Prehnite	1.63	6
Apatite	1.64–1.65	5
Alabastrite	1.64–1.65	5
Jadeite	1.66	6.5
Vesuviana var. Californite	1.72	6.5

* Eppler, 1984; Gunther, 1988; Author's personal observations

by the New Kingdom pharaohs. This finding from Tutankhamun's tomb is evidence of the appreciation the young pharaoh reserved for the latest technological discoveries and resources of the mineral world.

The extraterrestrial nature of the iron used to make King Tutankhamun's dagger, suggested in a recent paper (Comelli *et al.*, 2016), and perhaps also the other 19 iron objects (Carter, 1996), offers an imaginative and romantic suggestion about the choice of materials for the king's grave goods, a choice linked with strong symbolism, perhaps to indicate the long journey of the king after his death. There were also unique funerary objects for kings of the Bronze Age.

6. Concluding remarks

The archaeometric study of historical artifacts, and in particular of the goldsmith's objects, has the merit of highlighting complex problems related to the characterization of objects and materials, to which doubts and questions are added. This is because, although it is sometimes forgotten, the methods of science, as well as literary acquisitions, open new horizons for investigation, rather than immediate certainties. However, this concept is included in the same definition of 'History', which is revived when there is convergence of attentions from a plurality of points of view.

Thus, while on the one hand, the materials offer interesting insights for the understanding of the object, on the other hand, they ask for further in-depth study, not only for the understanding of their intrinsic characteristics, closely related to the choices made for their use, but also in terms of their geological, geographical and cultural realities that complete their context of origin.

For a correct investigative method, we suggest here the 'four C's' rule (Fig. 11):

- *Complexity*: never forget that the object of study is complex, has a long history, is almost always poly-material, and each material preserves the memory of human and technical processes that include many variables. The reconstruction work itself is complex: investigators must take into account changes in time scale and changes in landscapes and territories. Materials are expressions of technical knowledge, even in their choice and express the historical memory of the communities. An object is always the communication of the strong link between territory, community and culture.
- *Congruity*: ask the right question with respect to the objectives of the investigation, and the right question must be calibrated in a highly interdisciplinary context.
- *Competence*: everyone contributes to the survey with their own specificities and the baggage of their personal and professional experiences.
- *Collaboration*: being open to other points of view and receptive to other disciplines.

Figure 11. The four C's Rule.

In this perspective, the definition of 'archaeometry' (Mannoni, 1990) is still current and appropriate: "The more you get rid of the blinders of the history of idealistic art, and study using the scientific method any type of product, even those considered trivial, the more you realize that almost always the choices of the materials were optimal in relation to the resources of a given territory, and that already in the prehistory, very rare materials, which even today can be considered the best to produce the objects for which they were intended, were sought and exchanged over very long distances".

Acknowledgements

The authors are very grateful to Vincenzo De Michele who has made available his studies conducted on the Iron Crown and on Tutankhamun's pectoral. Without his contribution this work would not have been written. They are also grateful for the discussions and the review. The authors thank the Italian Gemological Institute for making available the iconographic and technical material of the diagnostic campaign conducted on the Iron Crown, during the studies on the Treasury of the Monza Cathedral. They also thank G. Monistier for the detailed images of the individual gems on the artifact and shown in Figs 4, 5 and 6. They are also grateful to G. Dal Negro and R. Serra for having granted the use of some of their images related to the King Tutankhamun's pectoral and Libyan Desert Glass. Special thanks are due to the Museum of Natural History of Milan for having allowed the study of some fragments of the Libyan Desert Glass used as a comparison standard for the evaluation of the refractive index of raw and processed materials.

References

Aimone, M. (2011) *Nuovi dati sull'oreficeria cloisonné in Italia tra V e VI secolo. Ricerche stilistiche, indagini tecniche, questioni cronologiche. Archeologia Medievale*, **XXXVIII**, 369–418.

Abdalla, H.M. and Mohamed F.H. (1999) Mineralogical and geochemical investigation of emerald and beryl mineralization, Pan-African Belt of Egypt: genetic and exploration aspects. *Journal of African Earth Sciences*, **28**, 581–598.

Al-Beruni (2007) *Al-Beruni The book most comprehensive in knowledge on Precious Stones.* Translated from the Arabic and with foreword by H.M. Said. Adam Publisher, New Delhi.

Aldred, C. (1979) *I gioielli dei faraoni.* Tecniche Nuove, Milan, Italy.

Andrews, C. (1990) *Ancient Egyptian Jewelry.* British Museum, London.

Arrhenius, B. (1971) *Granatschmuck und Gemmen aus nordischen Funden des fruhen Mittelalters. Mit einem rontgenkristallografischen.* Beitrag von Diego Carlstrom und Zeichnungen von Bengt Handel, *Acta Universitatis Stokholmiensis*, series B, Theses and paper published in offset, volume 3 of Studies in North European Archaeology.

Arrhenius, B. (1985) *Merovingian Garnet Jewellery.* Almkvist and Wiksell International, Stockholm.

Arrhenius, B (1988) Why the king needed his own goldsmith. *Laborativ Arkeologi*, **10–11**, 109–111.

Artioli, G. (2010) *Scientific Methods and Cultural Heritage: An Introduction to the Application of Materials Science to Archaeometry and Conservation Science.* Oxford University Press.

Aurisicchio, C., Conte, A.M., Medeghini, L., Ottolini, L. and De Vito, C. (2018) Major and trace element geochemistry of emerald from several deposits: Implications for genetic models and classification schemes. *Ore Geology Reviews*, **94**, 351–366.

Barrat, J.A., Jahn, B.M., Amosse, J., Rocchia, R., Keller, F., Poupeau, G.R. and Diemer, E. (1997) Geochemistry and origin of Libyan Desert glasses. *Geochimica et Cosmochimica Acta*, **61**, 1953–1959.

Bellani, A. (1819) *La Corona Ferrea del Regno d'Italia considerata I° come monumento d'arte, II° come monumento storico, III° come monumento sacro.* Memoria apologetica, Tipografia Sartori, Milan, Italy.

Bertelli, C. (2017a) *Una croce veneziana in Lombardia.* Pp. 23–40 in: *La Croce di Chiaravalle: approfondimenti storico scientifici in occasione del restauro* (G. Benati and D. Di Martino, editors). Booktime Ed., Milan, Italy.

Bertelli, C. (2017b) *La Corona Ferrea.* Skira Ed., Ginevrea-Milano, Italy.

Bimson, M., La Neice, S. and Leese, M. (1982) The characterisation of mounted garnets. *Archaeometry*, **24**, 51–58.

Bugoi, R., Oanta-Marghitu, R. and Calligaro, T. (2016) IBA investigations of loose garnets from Pietroasa, Apahida and Cluj-Someseni treasures (5th century AD). *Nuclear Instruments and Methods in Physics Research B*, **371**, 401–406.

Calligaro, T., Colinart, S., Poirot, J.P. and Sudres, C. (2002) Combined external-beam PIXE and μ-Raman characterization of garnets used in Merovingian jewellery. *Nuclear Instruments and Methods in Physics Research B*, **189**, 320– 327.

Calligaro, T., Dran, J.C. and Poirot J.P. (2003) Del studio analytico de las gemas: Estudio por acelerador de las gemas del Museo Nacional de la Edad Media, Cluny. Pp. 277–286 in: *El tesoro visigodo de Guarrazar* (A. Perea, editor). CSIC editions, Madrid.

Calligaro, T., Perin, P., Vallet, F. and Poirot, J.P. (2007) Contribution à l'étude des grenats mérovingiens (Basilique de Saint-Denis et autres collections du musée d'Archéologie nationale, diverses collections publiques et objets de fouilles récentes). Nouvelles analyses gemmologiques et géochimiques effectuées au Centre de Recherche et de Restauration des Musées de France. *Antiquités Nationales*, **38**, 111–144.

Carter, H. (1972) *Tutankhamen.* Garzanti Ed., Milan, Italy.

Carter, H. (1996) *Tutankhamen's Treasure.* Sphere Books, London.

Carter, A.K. (2016) Determining the Provenience of garnet beads using LA-ICP-MS. Pp. 235–266 in: *Recent Advances in Laser Ablation ICP-MS for Archaeology* (L. Dussubieux, M. Golitko and B. Gratuze, editors). *Natural Science in Archaeology*, Series Edition. Springer-Verlag, Berlin Heidelberg.

Carter, A.K. and Dussubieux, L. (2016) Geologic provenience analysis of agate and carnelian beads using laser ablation-inductively coupled plasma-mass spectrometry (LA-ICP-MS): a case study from Iron Age Cambodia and Thailand. *Journal of Archaeological Sciences*, **6**, 321–331.

Caselli, L. (2017) Le ragioni dello stile: note sparse. Pp. 41–86 in: *La Croce di Chiaravalle: approfondimenti storico scientifici in occasione del restauro* (G. Benati and D. Di Martino, editors). Booktime Ed., Milan, Italy.

Collareta, M. (2003) Oreficeria e tecniche orafe. Pp. 549–560 in: *Arti e storia nel Medioevo. II. Del costruire: tecniche, artisti, artigiani, committenti* (G. Sergi and E. Castelnuovo, editors). Giulio Enaudi Ed., Turin, Italy.

Collareta, M. (2011) Storia dell'arte medievale, Intervento introduttivo. *Reti Medievali*, **12**(2), 29–41.

CIBJO (2012) The Gemstone Book. Gemstones, Organic Substances and Artificial Products – Terminology and Classification. Colored stone commission, **1**, 2–61.

Clayton, P.A. (1933) Letter dated May 9th to the editor A.R.H. of The Monthly Record news and account of his discovery. *The Geographical Journal, London*, **82**, 374–376.

Clayton, P.A. and Spencer, L.J. (1933) Communication on the Libyan Desert Glass given the 9th November 1933 at the occasion of the Anniversary meeting of the Mineralogical Society. *Nature*, p. 278.

Clayton, P.A. and Spencer, L.J. (1934) Silica glass from the Libyan Desert. *Mineralogical Magazine*, **23**, 501–508.

Clayton, P.H. (1998) *Desert Explorer. A bibliography of Colonel. P.A. Clayton.* Saltash, Zerzura Press. Pp. 193, **ISBN-13:** 978-0953135004

Comelli, D., D'Orazio, M., Folco, L., El-Halwagy, M., Frizzi, T., Alberti, R., Caporosso, V., Elnaggar, A., Hassan, H., Nevin, A., Porcelli, F., Rashed, G. and Valentini, G. (2016) The meteoritic origin of Tutankhamun's iron dagger blade. *Metoritics & Planetary Science*, 1–9.

De Launay (1913) *Gìtes minéraux et métallifères.* Librarie Polytechnique Ch. Béranger, Editeur, II and III vol., Paris et Liège.

De Michele, V. (1998) The "Libyan Desert Glass" scarab in Tutankhamen's pectoral. *Sahara*, **10**, 107–110.

De Michele, V. and Manzini, G. (1993) *Munera Imperalia. Analisi delle gemme del Duomo di Monza.* Istituto Gemmologico Italiano, Milan, Italy.

De Michele, V., Manzini, G. and Pavan, A. (1986) Le gemme della Corona Ferrea. *La Gemmologia, Istituto Gemmologico Italiano, Milano*, **XI**, 20–31.

Desroches-Noblecourt, C. (1963) *Vita e morte di un faraone. Tutankhamen.* Silvana Editore, Milan, Italy.

Dominguez-Bella, S. (2012) Archaeomineralogy of prehistoric artifacts and gemstones. Pp. 5–28 in: *Archaeometry and Cultural Heritage: The contribution of Mineralogy*. Seminarios de la Sociedad Espanola de Mineralogia, **9**.

Dran, J.C., Salomon, J., Calligaro, T. and Walter, P. (2004) Ion beam analysis of art works: 14 years of use in the Louvre. *Nuclear Instruments and Methods in Physics Research B*, **219/220**, 7–15.

El Mallakh, E. and Brackman, A.C. (1978) The gold of Tutankhamen. *Newsweek Book*, New York.

Eppler, W.F. (1984*) Praktische Gemmologie.* Ruhle-Diebener, Stuttgart, Germany.

Farges, F (1998) Mineralogy of the Louvres Merovingian garnet cloisonne′ jewelry; origins of the gems of the first kings of France. *American Mineralogist*, **83**, 323–330.

Fontanini, J. (1719?) *Dissertatio de Corona Ferrea Langobardorum.* Sumptibus Petri van der Aa, LugduniBatavorum, Editio tertia.

Fudali, R.F. (1981) The major element chemistry of Libyan Desert Glass and the mineralogy of its precursor. *Meteoritics*, **16**, 247–259.

Garrison, E. (2016) Petrography for archaeological geology. Pp. 145–178 in: *Technique in Archaeological Geology*. Natural Science in Archaeology, Series Edition, Springer-Verlag, Berlin, Heidelberg.

Goitein, S.D. (1973) *Letters of Medieval Jewish Traders.* Princeton University, New Jersey, USA.

Goitein, S.D. (2002) *Una società mediterranea.* A. Cura di Jacob Lassner, Milano, Bompiani XXIII, 639 pp.

Goitein, S.D. and Frienman, A. (2011) *India Traders of the Middle Ages. Documents from the Cairo Geniza.* India Book, Leiden, The Netherlands.

Gratuze, B., Soulier, I., Barrandon, J.N. and Foy, D. (1992) De l'origine du cobalt dans le verres. *Revue d'Archéométrie*, **16**, 97–108.

Greiff, S. (1999) Naturwissenschaftliche Untersuchungen zur Frage der Rohsteinquellen für frühmittelalterlichen Almandingranatschmuck rheinfränkischer Provenienz. *Jahrbuch des Römisch-Germanischen Zentral Museums Mainz*, **45**, 599–645.

Greshake, A., Koeberl, C., Fritz, J. and Reimold, W.U. (2010) Brownish inclusions and dark streaks in Libyan Desert Glass: Evidence for high-temperature melting of the target rock. *Meteoritics & Planetary Science*, **45**, 973–989.

Greshake, A., Wirth, R., Fritz, J., Jakubowski, T. and Bottger, U. (2018) Mullite in Libyan Desert Glass: Evidence for high-temperature/low-pressure formation. *Meteoritics & Planetary Science*, **53**, 467–481.

Gubelin, E.J. and Koivula, J.I (1992) *Photoatlas of Inclusion in Gemstones.* ABC ED., Zurich.

Gunther, B. (1988) *Bestimmungstabellen fur Edelsteine, Synthetische, Imitationen.* Lenzen E., Kirschweiler, Germany.

Henig, M. (2008) The re-use and copying of ancient intaglios set in Medieval personal seals mainly found in England: an aspect of the Renaissance of the 12th century. Pp. 25–34 in: *Good Impressions: Image and Authority in Medieval Seals* (N. Adams, J. Cherry and J. Robinson, editors). British Museum, London, Research Publication, **168**.

Hilgner, A. (2017) Oreficeria cloisonné nell'alto medioevo: inquadramento cronologico sulla base degli aspetti tecnologici e dell'origine delle materie prime. Pp. 9–29 in: *Small finds e cronologia (V-IX sec.*) (J. Pinar Gil, editor). Esempi, medoti e risultati.

Horváth, E. and Bendo, Z. (2011) Provenance study on a collection of loose garnets from a Gepidic Period grave in Northeast Hungary. *Archeometriai Mühely*, **1**, 17–32.

Hughes, R.W. (1990) *Corundum.* Butterworth-Heinemann, London.

Jessup, R.A. (1950) *Saxon Jewellery.* Faber (M0005997SH), London, Wilson Collection 124, 200349501.

Koeberl, C. (1986) Geochemistry of tektites and impact glasses. *Annual Review of Earth and Planetary Sciences*, **14**, 323–350.

Koeberl, C. (1997) Libyan Desert Glass: geochemical composition ad origin. Pp. 121–158 in: *Proceedings of the "Silica 96" Meeting on Libyan Desert Glass and related desert events (Bologna, 1996)* (V. De Michele, editor). *Sahara*, Segrate, Milan, Italy.

Krippner, A., Meinhold, G., Morton, A.C. and von Eynatten, H. (2014) Evaluation of garnet discrimination diagrams using geochemical data of garnet derived from various host rock. *Sedimentary Geology*, **306**, 36–52.

Lucas, A. (1962) *Ancient Egyptian Materials and Industries.* Histories and Mysteries of Man, reprint 1989, London.

Lusuardi Siena, M.S. (1998) L'identità materiale e storica della Corona. Pp. 173–249 in: *La Corona Ferrea nell'Europa degli Imperi, II.* Alla scoperta del Prezioso Oggetto, tomo II, Scienza e Tecnica, Giorgio Mondadori Ed, Milano, Italy.

Maçoudi (1861) Les prairies d'or. Texte et traduction par C. Barbier de Meynard and Pavé de Courteille, I-V, Paris.

Mannoni, T. (1990) Introduzione all'Archeometria. Pp. 27–35 in: *Scienze in Archeologia. II Ciclo di Lezioni sulla Ricerca applicata in Archeologia (Certosa di Pontignano 1988)* (T. Mannoni and A. Molinari, editors). Quaderni del Dipartimento di Archeologia e Storia delle Arti, Sezione Archeologica, All'Insegna del Giglio Ed, Florence, Italy.

Manutchehr-Danai, M. (editor) (2009) *Dictionary of Gems and Gemology.* Springer, Berlin, Heidelberg.

Marimonti, G. (1841) *Memorie storiche della città di Monza compilate sull'opera del Canonico Anton-Francesco Frisi.* Tipografia L. Corbetta, Monza, Italy.

Maspero, V. (2003*) La Corona Ferrea. La storia del pi antico e celebre simbolo del potere in Europa.* Vittone Ed., Italy.

Maspero, V. (2016) *Memorie di una millenaria. La Corona Ferrea racconta storie di potenti, folli e santi.* Libraccio Ed.

Mathis, F., Vrielynck, O., Laclavetine, K., Chene, G. and Strivay, D. (2008) Study of the provenance of Belgian Merovingian garnets by PIXE at IPNAS cyclotron. *Nuclear Instruments and Methods in Physics Research, B*, **266**, 2348–2352

Milazzo, M. and Cicardi, C. (1998) X-ray fluorescence characterization of the Corona Ferrea. *Archaeometry*, **40**, 351–360.

Mottana, A. (2006) Italian gemology during the Renaissance: A step toward modern mineralogy. Pp. 1–21 in: *The Origins of Geology in Italy* (G.B. Vai and W.G.E. Caldwell, editors). Geological Society of America, Special Paper **411**.

Nassau, K. (1994) *Gemstone Enhancement: History, Science and State of the Art* (Second Edition). Oxford and London (Butterworth Heinemann), XIV, 252.

Negro, G. and Damiani-Appia, M. (1992) Il "Silica Park": un centro di lavorazione del LDSG nel Great Sand Sea. *Sahara*, **5**, 105–108.

Oakley, K.P. (1952) Dating the Libyan desert glass. *Nature*, **170**, 4324, 447–449.

Perin, P., Calligaro, T., Vallet, F., Poirot, J.P. and Bagault, D. (2007) Provenancing Merovingian garnets by PIXE and µ-Raman spectrometry. Pp. 69–76 in: *Post-Roman Towns, Trade and Settlement in Europe and Byzantium, Vol* **1**, *The Heirs of the Roman West* (J. Henning, editor). Walter de Gruyter, Berlin.

Quast, V. and Schussler, U. (2000) Mineralogische Untersuchungen zur Herkunft der Granate merowingerzeitlicher Cloisonne'arbeiten. *Germania*, **78**, 75–96.

Rapp, G. (2019) Gems and man: A brief history. Pp. 323–344 in: *The Contribution of Mineralogy to Cultural Heritage* (G. Artioli and R. Oberti, editors). EMU Notes in Mineralogy, **20**. European Mineralogical Union and Mineralogical Society of Great Britain & Ireland, London.

Riccardi, M.P., Cucini, C., Lazzarini, L., Baroni, S. and Frizzi, T. (2017) Oro, argento, diaspro e vetro. Il racconto dei materiali della Croce di Chiaravalle. Pp. 201–236 in: *La Croce di Chiaravalle: approfondimenti storico scientifici in occasione del restauro* (G. Benati and D. Di Martino, editors). Booktime Ed., Milan, Italy.

Roe, D.A., Olsen, J.R., Underwood, J. and Giegengack, R.F. (1982) A handaxe of Libyan desert glass. *Antiquity*, **LVI**, 88–92.

Rosch, C., Hock, R., Schussler, U., Yule, P. and Hannibal, A. (1997) Electron microprobe analysis and X-ray diffraction methods in archaeometry: investigations on ancient beads from the Sultanate of Oman and from Sri Lanka. *European Journal of Mineralogy*, **9**, 763–783.

Rouse, J.D. (1986) *Garnet.* Butterworth, London.

Saleh, M. and Sourouzian, H. (1987) Official Catalogue. In: *The Egyptian Museum*. Von Zabern, Cairo.

Schussler, U., Rosch, C. and Hock, R. (2001) Beads from ancient Sri Lanka – first results of a systematic material analysis. Pp. 227–242 in: *Ancient Ruhuna. Sri Lankan-German archaeological project in the southern province*, **1** (H.J. Weisshaar, H. Roth and W. Wijeyapala, editors). Von Zabern, Mainz, Germany.

Spencer, L.J. (1933) Two new gem stones. *The Gemmologist*, **3**, 110–113.

Spencer, L.J. (1936) *A Key to Precious Stones.* Blackie, London.

Stockton, C.M. and Manson, D.V. (1985) A proposed new classification for gem-quality garnets. *Gems and Gemology*, **21**, 205–218.

Stutenbecker, L., Berger, A. and Schlunegger, F. (2017) The potential of detrital garnet as a provenance proxy in the Central Swiss Alps. *Sedimentary Geology*, **351**, 11–20.

Superchi, M. (1999) Le gemme dell'Evangeliario di Ariberto. Pp. 149–157 in: *Evangeliario di Ariberto* (A. Tomei, editor). Silvana Editore, Milan, Italy.

Troupeau, G. (1998) Le premier traité arabe de minéralogie: le livre de Yuhanna Ibn Masawayh sur les pierres précieuses. *Annales Islamologiques*, **32**, 219–238.

UNI Ente Nazionale Italiano di Unificazione (2013) *Materiali gemmologici, Nomenclatura.* UNI 10245, Milan, Italy.

Velde, B. and Courtois, L. (1983) Yellow garnets in roman amphorae: a possible tracer of ancient commerce. *Journal of Archaeological Science*, **10**, 531–539.

Weeks, R.A., Underwood, J.R., Jr. and Giegengack, R. (1984) Libyan Desert glass: A review. *Journal of Non Crystalline Solids*, **67**, 593–619.

Weigand, P.C., Harbottle, G. and Sayre, E.V. (1977) Turquoise source area in Mesoamerica and the southwestern USA. Pp. 15–34 in: *Exchange Systems in Prehistory*. Accademic Press, New York.

Wolska-Conus, W. (1968) *Cosmas Indicopleustès.* Topographie Chrétienne, Paris.

EMU Notes in Mineralogy, Vol. 20 (2019), Chapter 10, 367–392

Ancient Mediterranean polychrome stones

LORENZO LAZZARINI

Laboratorio di Analisi dei Materiali Antichi - Sistema dei laboratori Università IUAV di Venezia, San Polo 2468 – I-30125 Venezia, Italy
e-mail: lorenzo@iuav.it

The Romans, like the Egyptians and much more than the Greeks, used polychrome stones for decorative purposes in architectural elements, floor and wall facings and statuary. Throughout their Mediterranean provinces they systematically searched for and exploited a very large number of beautiful lithotypes, many of which they distributed to all corners of their empire. The most important of these stones were often re-used later in medieval-to-modern times; some of them are still offered on the market. They include granitoid rocks (granites, granodiorites/tonalities, gabbros, quartz-monzonites), a few lavas, many metamorphites (impure marbles, metabreccias and metandesites) and several sedimentary rocks (limestones, lumachellas, conglomerates, calcareous alabasters/travertines). The 40 most important and widespread of these lithotypes are considered here as regards their origin, the history of their use and their minero-petrographic characteristics, which can contribute to better knowledge of single species, to determination of the original quarries and to archaeometric solutions of several provenance problems.

1. Introduction

Anyone travelling along the coasts of the Mediterranean Sea or in the interior of many countries belonging in antiquity to the Roman Empire (which included non-Mediterranean areas such as *Britannia*, the Atlantic coasts of *Iberia* and *Gallia*, and central Europe), will often come across beautiful coloured stones in monuments of the Roman or later periods. The use of such stones started at the end of the Republic when the Romans conquered Greece and Carthage, and inherited the kingdom of Pergamum in Asia Minor; they thus became acquainted with the famous marble monuments of the most important towns of Macedonia, Attica and Peloponnesus, and they first discovered the decoration of the Hellenistic palaces with coloured stones. It was, however, under Augustus (27 BC–14 AD) that the stones were imported on a large scale, so that the appearance of Rome changed considerably: we know from Suetonius that Augustus was very proud to have received (from Julius Caesar) a city made of bricks and tufa, and to have left it built of marble. This goal became possible especially after the conquest of Alexandria and Egypt (30 BC) and the consequent access to the very numerous Pharaonic and Ptolemaic stone-quarries of the Eastern Desert and of the Aswan area (Gnoli, 1988).The expansion of the empire, often called the 'Romanisation' of the

DOI: 10.1180/EMU-notes.20.10

Mediterranean during the *pax romana*, and later on, saw the start of the parallel phenomenon of 'marbleing' of Roman towns, first of all in public buildings such as the forum and basilica, the theatre and amphitheatre, the *macella* and temples, *etc.*, which were the subjects of the evergetism of emperors and of rich benefactors. Beautiful coloured stones were imported, often from distant and impenetrable parts of the ancient world, to serve as an expression of imperial *munificentia* and *propaganda* (Pensabene, 2014). This phenomenon was soon accompanied by the use of precious and exotic stones by wealthy private individuals, who employed marble as a status symbol of their success and power: floors and walls faced with thin slabs of coloured stones, called *opera sectilia*, were far more expensive than mosaics, and proudly exhibited to friends and visitors. The demand for beautiful stones from the most exotic places in the empire increased considerably amongst the ruling classes, and became a substantial source of income for the emperors who owned the most important quarries, such as those of the red porphyry and the granites of Aswan and *Mons Claudianus* (Gebel Fatira) in Egypt, of the breccias of Chios and Larisa, as well as of the *cipollino verde*, in Greece, and of *africano* in Asia Minor. Starting from the 1st century AD, efficient exploitation-production of blocks, columns, pillars, and other semi-finished artefacts in the largest quarries was followed by a sophisticated transport and distribution organisation allowing big columns to cover very long distances, *e.g.* from Aswan in Egypt to *Italica*, near Seville, in Spain. The most important quarries (*officinae, metalla*) like those of red porphyry, of the granite of *Mons Claudianus*, and of *marmor carystium*, *chium*, *luculleum*, *numidicum* and *phrygium*, were imperial property (*patrimonium caesaris*). They were well organised into sectors (*bracchia*, *loci*), and these into cuts (*caesurae*), so as to be able to identify the provenance of a block with precision. Other quarries were owned by municipalities, or by private individuals. In each case there was a person with overall responsibility (*procurator montium*), usually a freed slave (*libertus*) with experience in the rational exploitation of stones, and a technician (*machinarius*) for the most difficult task, the moving of blocks and large monoliths in the quarry and for shipment. Slaves (*damnati ad metalla*) or paid workers (*lapicidae*) did the hard work of cutting and the initial shaping of artefacts in the quarry. In most cases transport took place on carriages drawn by oxen or donkeys as far as the nearest harbour, where ships (*naves onerariae*), often purpose-built to carry single monoliths, transported their cargoes of marble to the various Mediterranean destinations. When not covered by a specific order, the marble was stockpiled in great store-yards (*stationes marmorum*) like those of Ostia (Italy), Alexandria (Egypt), Nicomedia and Miletus (Izmit and Balat, Turkey, respectively,), which operated as distribution centres for the empire (Lazzarini, 2002c, 2007; Pensabene, 2014).

As a consequence of this efficient organisation, we know that the majority of the most prestigious and expensive coloured stones (those listed in Diocletian's Edict on maximum prices of 301, Gnoli, 1988: the green and red porphyries, the *cipollino verde*, the *africano*, the *giallo and rosso antico*, *etc.*) are almost ubiquitous in Roman towns, at least as small slabs (Lazzarini, 2007, 2009): when not present, we often find them imitated by fresco paintings or replaced by similar local stones.

From the very beginning the use of polychrome stones was limited mostly to small, irregular slabs inserted into mosaic floors called *scutulata pavimenta*. Later on, larger slabs of geometrical shapes were joined together for facing floors and walls, and architectural elements such as capitals and columns were introduced in public and private buildings. The use of polychrome marbles was then extended, especially from the 2nd century AD on, to statuary, tubs and furniture.

Towards the end of the Roman empire, with the implosion of its organisation including the 'logistics' of the marble market, and under the pressure of barbarian invasions, many quarries were abandoned, sometimes also because of the excessive quantity of stone accumulated at several *stationes marmorum*, and because of the reuse of marble pieces formerly installed in ruined buildings, a practice that was already underway in the 3rd century. Such reuse became almost the rule in the western Middle Ages, and lasted for centuries, often until modern times: Roman monuments were spoliated and new buildings, such as Christian churches, erected and/or decorated using ancient columns, capitals and other elements, often re-cut and reworked. The same phenomenon occurred in the East, in some instances, especially in Early Byzantine times, with the addition of new coloured lithotypes. In both cases the reuse of ancient marbles led to their further spread and distribution, so that we now find typical Roman polychrome stones almost everywhere in the Mediterranean area, in archaeological sites as well as in churches, mosques, castles and palaces: hence the importance of the determination of their original quarries. Such a determination may give immediate information on the provenance and economic value of an artefact (a statue or a part of a monument); it may help in reconstructing ancient commercial traffic and trade routes (the identification of marbles forming part of sunken cargoes enables ancient routes to be reconstructed quite precisely); the location of the quarry from which a damaged marble came makes it possible to find sound material for the purpose of replacement, restoration, copies, *etc*. Dated artefacts and monuments, in turn, may help in fixing the interval of use of a certain quarry and related stone. The precise identification of the provenance of white and coloured marbles used in antiquity remains an issue of fundamental interest for archaeologists, architects and art historians, and continues to engage scientists of various disciplines, but especially of Earth sciences.

The identification of coloured stones is, in general, much easier than that of white marbles (Lazzarini, 2004) because for most of them it may be determined through what we may call a 'visual connoisseurship', which can be acquired after systematic visits to ancient quarries (where one can observe the colour and texture variation of a specific stone), studies of ancient stones in monuments, museums and archaeological areas, as well as in historical lithological collections (Mielsch, 1985; Price, 2007). Once a possible area of origin of an unknown stone is determined, local inquiries with regional geologists and people working in marble factories may prove extremely useful, as can observation of the stone used in modern buildings as, very often, ancient materials have been re-exploited in modern times. This integrated provenancing method (Lazzarini, 2002a) has been applied successfully in the identification of several hitherto unknown ancient quarries (*e.g.* the granite from Mysia, and various breccias). When this method fails, then a much more time-consuming and not always successful methodology is applied, starting from a scientific approach with laboratory analyses. Ultimately, the establishment of a reference database is of fundamental importance. This is now, in large part, available thanks to the numerous specialist studies performed over the last 40 years, although its development is not comparable with the much more detailed database established for white, pure, marbles (Antonelli and Lazzarini, 2016). A database is also a pre-requisite for solution of the most difficult identification problems deriving from macroscopically very similar coloured lithotypes.

2. The archaeometric problems of polychrome stones

As cited above, the positive identification of several coloured stones that look alike macroscopically may be reached only with an archaeometric approach. It is worth mentioning here the problems related to some of the lithotypes considered here: first of all the magmatites, then the impure marbles, some breccias and finally the calcite alabasters:

- the pinkish variety of the Aswan granite may (to some extent) be confused with the *granito sardo* (Sardinian granite) and the Egyptian granite from Fawakhir/Wadi el-Sid (Gnoli, 1988; Brown and Harrell, 1995; Klemm and Klemm, 1991, 2008): a distinction may be based on the K/Rb and Ba/Zr ratios (Galetti *et al.*, 1992);
- the grey tonalite of *Mons Claudianus* (*granito del foro*) is macroscopically identical to the granite from Nicotera (province of Cosenza, Italy) used in Roman times in central and southern Italy for columns and pillars, probably as a substitute for the more famous (and expensive) Egyptian counterpart. The macroscopic distinction of these two lithotypes is very difficult, and identification should be based on the petrographic analysis of a thin section: the two-mica granite from Nicotera is quite distinct from the classical one-mica (biotite) tonalite of Gebel Fatira (Antonelli *et al.*, 2010);
- the *granito elbano* (Elba granite) is very similar to the Mysian granite: a first differentiation may be based on the presence of large (centimetric) white plagioclase megacrysts in the former, and on that of small (millimetric) hornblende prismatic crystals in the latter: both may be detected by an expert naked eye (or with a magnifying lens in the case of hornblende). A more reliable identification, however, may be based on the presence of traces of muscovite and tourmaline in the Elba granite, both of which are missing from the Mysian granite, and on the larger Rb, Zr and smaller Ba and Sr contents of the former with respect to the latter (De Vecchi *et al.*, 2000);
- the quartz-monzonite from ancient Troas (*marmor troadense*) is very similar to a quartz-monzonite from Corsica. Although used only in the Renaissance (by the Medici family), and thus posing no problems of

confusion with ancient Roman usage, the latter can be distinguished by its coarser crystals (megacrysts of 2–4 cm) of grey-violet K-feldspar;

- the exact determination of the original quarry of sarcophagi made of *lapis sarcophagus* is of considerable importance because they could have been made at Assos (now Behramkale, province of Ayvacık, Turkey), in the nearby island of Lesbos, or in the further away town of Pergamum (now Bergama). Petrography and geochemistry combined may solve provenancing problems using an existing reference database (Lazzarini and Visonà, 2009);
- it is very difficult to separate the uniformly red variety of *rosso antico* quarried in the Mani peninsula (Greece) from the equivalent variety of *cipollino rosso* exploited near ancient Iasos in Turkey. Differentiation criteria are grain size (a little bit larger for the latter) and Fe/Ni ratio (Gorgoni *et al.*, 2002). It is worth mentioning the existence on the island of Rhodes (Greece) of a red limestone which is macroscopically very similar to both marbles [*rosso antico* and *cipollino rosso*?], and which is also used for statuary and architectural elements from Hellenistic to Roman times (Herrmann, 1988): differentiation is possible, again by the microscopic examination of thin sections;
- the separation of the *cipollino verde euboico* of the district of Karystos from that of Styra, and of these *cipollini* from that of Mani (*cipollino verde tenario*) may be achieved from the analysis of their stable C and O isotope ratios (Lazzarini, 2007);
- it is often difficult (especially in small artefacts) to distinguish between some varieties of the *breccia di settebasi* (quarried in the Island of Skyros, Greece) from the *breccia medicea* of the Apuan Alps (Tuscany, Italy), a stone which is sometimes used as a substitute for it: the presence of chloritoid, easily determined in thin section, is indicative of an Italian origin for the artefact in question (Lazzarini, 2007);
- the difference between the very common *verde antico* (*marmor thessalicum*) and *verde di Tinos* (Tinos green), which was much more rarely used in Roman times, may be based on the brecciated fabric and presence of white marble clasts in the former, lacking in the latter (Lazzarini, 2007; Melfos, 2008);
- the distinction of the *giallo antico brecciato* from the *breccia dorata* (originating in the Montagnola Senese, Italy: Bruno and Lazzarini, 1999) may be achieved through a petrographic study: the former is made up of clasts of micrite/microsparite and frequently contains plagioclase among the accessory minerals; the latter are composed of clasts of slightly recrystallized calcite belonging to a low-metamorphic-grade marble, without plagioclase;

- the identification of the many calcite alabasters used by the Romans, mostly at a regional level, and their differentiation from the common Egyptian alabaster remains unresolved: the best contribution towards the solution of this problem at present is obtained from the determination of the ^{87}Sr/^{86} Sr ratio (Barbieri *et al.*, 2002; Lazzarini and Çolak, 2002; Lazzarini *et al.*, 2006, 2012).

3. The coloured stone catalogue

The present chapter is intended to make a general contribution towards the establishment of a basic database of the most important coloured stones introduced by the Romans for building and/or decorating public and private buildings through an overview of the provenance, periods and typologies of use together with an essential minero-petrographic characterization. Those considered here are the really 'international' lithotypes that travelled in antiquity from their home country, often to reach most (if not all) of the provinces of the Roman Empire. Given the considerable number of stone species that were distributed widely by the Romans, a choice has been made of the 40 most commonly used lithotypes. They have been grouped according to their petrographic nature into the three great classes of magmatites (granitoids and lavas), metamorphites (excluding the white and grey, pure crystalline marbles) and sedimentary rocks (breccias, limestones, lumachellas, alabasters/travertines). Note that several other 'international' stones were used by the Romans, the origin of many of which, especially of breccias and alabasters, remains unknown.

For each of the stone species considered here are given:

- a photograph of the holotype showing the typical colour and texture of the rock (in a few cases, when it features considerable macroscopic variation, more than one specimen is shown): this photo is sufficient in most cases to allow the identification of coloured lithotypes used in Mediterranean antiquity in Roman or later monuments.
- the Latin name of the stone (when known from ancient sources)
- a photomicrograph of a thin section of the holotype. This photomicrograph taken between crossed polars at different magnifications (according to the grain-size of the rock) may be of considerable help in confirming the macroscopic identification of stones of uncertain provenance
- two definitions, one with known ancient and modern synonyms of the stone (some of these are quite old Italian names, traditionally used since the Early Renaissance and still employed internationally by archaeologists and art historians), the location of its quarries (also located on a map, see Figs 1 and 2), the period and typology of use (the latter referred only to antiquity), the distribution in the ancient world (only through bibliographical references), an indication of the frequency of ancient use (according to the experience and direct

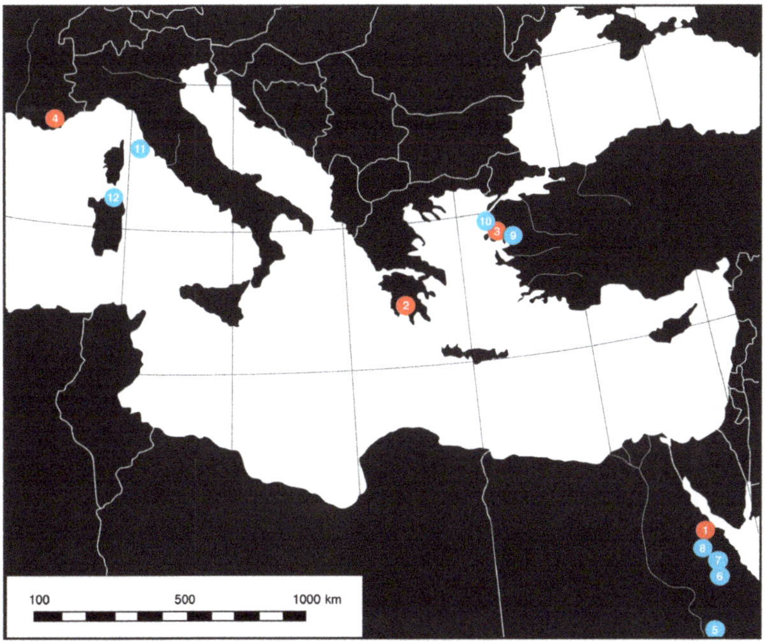

Figure 1. Geographic location of the quarries of magmatic extrusive (red) and intrusive (blue) rocks reported in the catalogue.

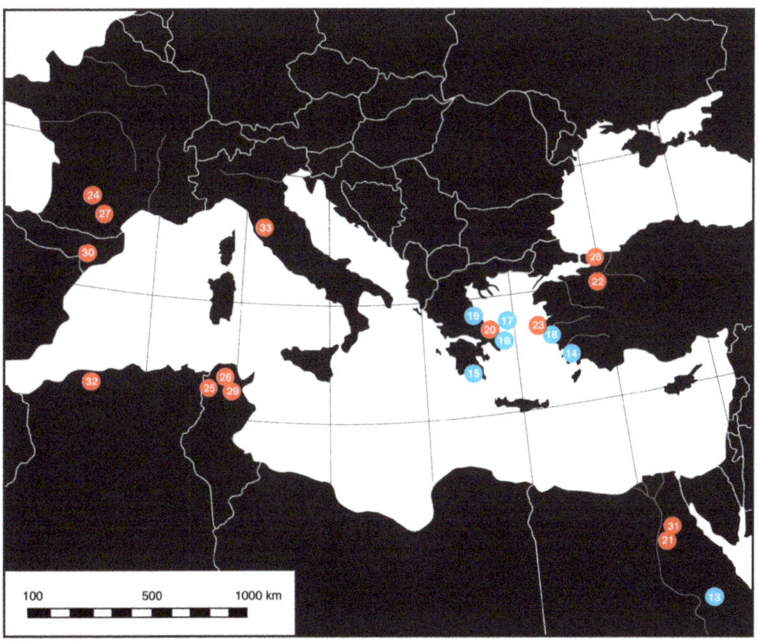

Figure 2. As for Fig.1, but for sedimentary (red) and metamorphic (blue) rocks.

knowledge of the present writer); and another with the geological age (when known), petrographic classification, type of macro/microtexture and grain size (fine = up to 2 mm; medium = 2–5 mm; coarse = >5 mm), and mineralogical composition, with a semi-quantitative evaluation of the main (primary, and/or characterizing) minerals and an indication of the accessory minerals. Both definitions also record the relative important bibliographic references.

Note: where scale bars are provided, each black or white rectangle is 10 mm in the long dimension. 'Field of view' applies to the horizontal dimension of the image.

Acknowledgments and dedication

The author is grateful to the editors of this volume for their useful comments, and to Wolters Kluwer Italia for authorizing the reproduction of the majority of the macrophotographs taken from chapter 7 of the book *Pietre e marmi antichi*, CEDAM, Castenaso (BO), 2004. This paper is dedicated to the memory of Carlo Gorgoni and Bruno Turi, dear friends and colleagues of the Universities of Modena and Rome "La Sapienza", respectively, with whom the author has shared several studies on some of the stones considered in here. This manuscript was accepted for publication in February 2016.

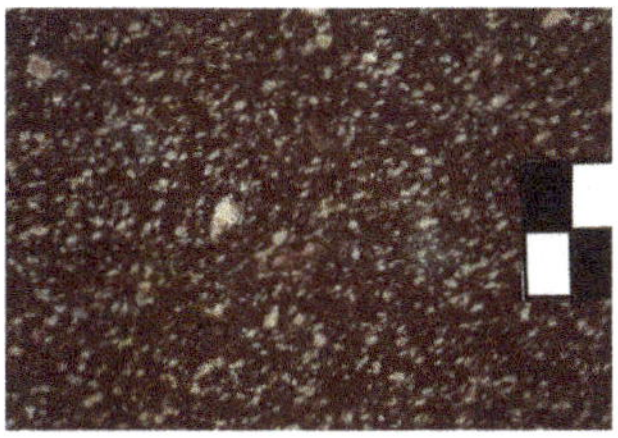

Stone 1 *LAPIS PORPHYRITES*
LITHOS ROMAIOS; PORFIDO ROSSO ANTICO
Mons porphyrites, *Mons Igneus*, Gebel Abu Dokhan, Eastern Desert, Egypt (1 in Fig. 1)
Late Ptolemaic period-mid V c. AD, then reused. Lazzarini (2009).
Slabs for *opera sectilia* >> columns > tubs and vases > statuary>sarcophagi
Maxfield and Peacock (2001); Del Bufalo (2012).

Proterozoic (~600 Ma)
Weak-to-medium grade metamorphic andesite/dacite (metandesite/metadacite)
Porphyritic, with small (mm) white/pink phenocrysts in a purplish aphanitic groundmass
Plagioclase >>> hornblende/oxyhornblende > biotite > accessories (magnetite, hematite, apatite, piemontite); the main minerals often (severely altered) into secondary minerals
Brown and Harrell (1995); Klemm and Klemm (2008); Makovicky *et al.* (2016a,b).
Field of view = 2.2 mm

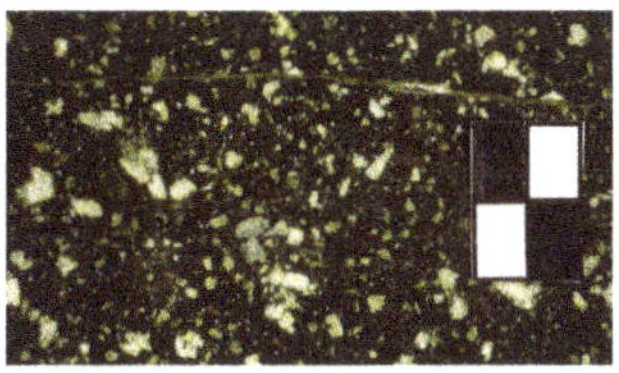

Stone 2 *LAPIS HIERACITES*
PORFIDO VERDE EGIZIANO
Eastern quarries of the Gebel Abu Dokhan, Eastern Desert, Egypt (1 in Fig. 1)
Roman Imperial period (I-II c.), then reused
Slabs for *opera sectlia* >>> small columns > vases
Gnoli (1988); Maxfield and Peacock (2001).

Proterozoic (~600 Ma)
Medium-grade metamorphic / metasomatic andesite/dacite
Porphyritic, with small (mm) white phenocrysts in a green-greenish aphanitic groundmass
plagioclase (moderately altered into clay and epidote) >> hornblende and biotite, (severely altered to chlorite, epidote, iron oxides) > accessories and secondary minerals as above
Brown and Harrell (1995); Klemm and Klemm (2008).
Field of view = 1.3 mm

Stone 3 *LAPIS LACEDAEMONIUS*
KROKEATIS LITHOS, VERDE DI LACONIA, SERPENTINO, PORFIDO VERDE ANTICO, PORFIDO VITELLI
Stefanià, Krokea, Peloponnesus, Greece (2 in Fig. 1)
Minoan-Mycenaean times; Late Republican-Late Imperial periods, then reused; Lazzarini, 2007
Seals and vases (Min.-Myc.); Slabs of *opera sectilia* >> small columns and capitals >> statuary > *cuticulae* (palettes)
Gnoli (1988); Lazzarini (2007).

Mid-Triassic
Metasomatized basaltic andesite/trachyandesite
glomero-porphyric, with green aphanitic groundmass, often with amygdales of bluish/red/brown chalcedony
plagioclase >>> chloritized pyroxene, chlorites (delessite, chamosite) > epidotes > accessories (titanite, magnetite and other iron oxides, pyrite, calcite)
Pe-Piper and Kotopouli (1981); Lazzarini (2007).
Field of view = 1.7 mm

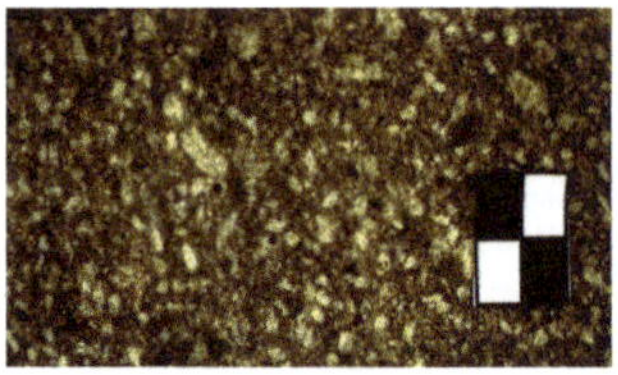

Stone 4 *LAPIS SARCOPHAGUS*
Greek Archaic period-IV c. AD
Ancient Assos (Troas), now Behramkale, Turkey (3 in Fig. 1); Lazzarini and Visonà (2009).
Sarcophagi (also used locally as building material)
Ward-Perkins (1992); Lazzarini (1994).

Late Miocene
Andesite/trachyandesite
Glomeroporphyric, medium-grained, porous, with microcrystalline/aphanitic brown groundmass
Plagioclase >> pyroxene (clino and ortho) > biotite >> accessories (Fe-Ti oxides, apatite)
Lazzarini (1994); Lazzarini and Visonà (2009)
Field of view = 4.5 mm

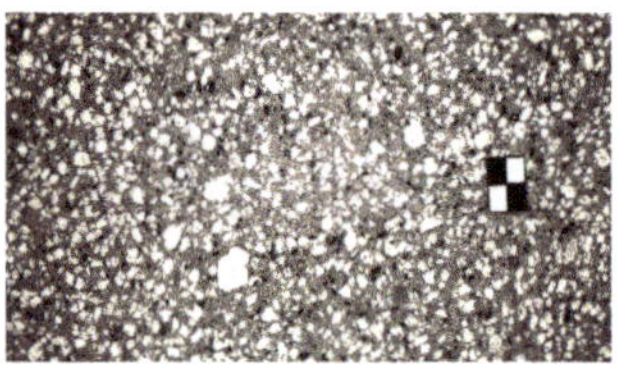

Stone 5
GRANITO BIGIO, GRANITO A MORVIGLIONE, ESTERELLITE
Saint Raphael, Boulouris, Dramont, province of Fréjus, France (4 in Fig. 1); Mazeran (2004).
II- V c. AD, then reused
Slabs for *opera sectilia* >> columns
Gnoli (1988); Mazeran (2004).

Oligocene
Dacite
Porphyric/glomeroporphyric with grey microcrystalline groundmass
Plagioclase (often zoned) >> quartz, biotite, hornblende >>> accessories (apatite, titanite, Fe-oxides/hydroxides)
Brentchaloff and Mazeran (1999); Rehault *et al.* (2012).
Field of view = 2.2 mm

Stone 6 *LITHOS PYRRHOPOECILOS, LAPIS THEBAICUS, LAPIS SYENITES*
SIENITE; GRANITO ROSSO EGIZIANO
Immediate SE outskirts of the town of Aswan, Egypt (5 in Fig. 1)
III Dynasty- V c. AD, then reused, Lazzarini (2009).
Still quarried
Columns and Pillars > statuary > obelisks > tubs > slabs for *opera sectilia* > sarcophagi
Gnoli (1988); Kelany *et al.* (2007); Klemm and Klemm (2008)..

Proterozoic (~565 Ma)
Normal-to-slightly alkaline granite-to monzogranite
granular, mostly inequigranular, coarse-grained; a fine-grained equigranular variety is rare, although used since pharaonic times
K-feldspar(perthite) > plagioclase > quartz > biotite >> hornblende > accessories (Fe-ores, titanite, apatite, allanite, zircon)
Klemm and Klemm (2008).
Field of view = 3.5 mm

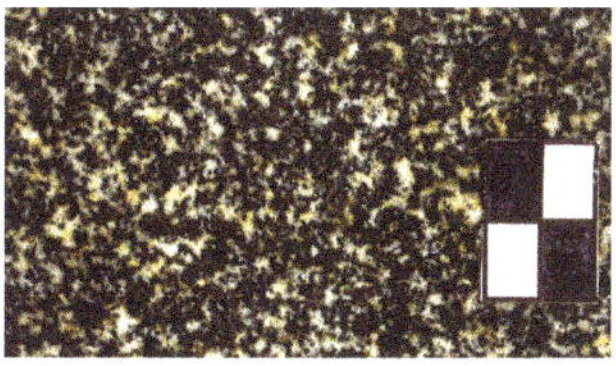

Stone 7
DIORITE EGIZIANA, GRANITO NERO
Gebel Nagug and immediate SE outskirts of the town of Aswan, Egypt (5 in Fig. 1)
III Dynasty-III c. AD, then reused, Lazzarini (2009). Still quarried
Statuary > columns and pillars > tubs > slabs > mortars and millstones
Klemm and Klemm (2008).

Proterozoic (~580 Ma)
granodiorite, passing to tonalite in the darker varieties
inequigranular, medium-to coarse-grained
Hornblende > plagioclase > biotite > K-feldspar (perthite) >> accessories (apatite, allanite, titanite, chlorite, calcite)
Kelany *et al.* (2007); Klemm and Klemm (2008).
Field of view = 7.0 mm

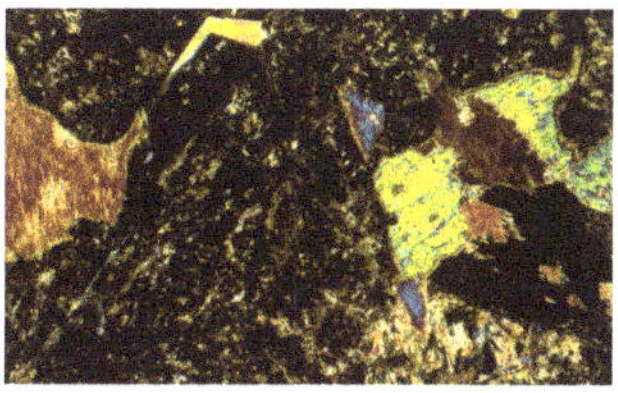

Stone 8 *LAPIS OPHYTES*
GRANITO VERDE DELLA SEDIA, OFITE
Wadi Umm Vikala, Eastern Desert, Egypt (6 in Fig. 1)
Pre-Early Dynastic periods; Imperial times (end of Ic. BC-III c. AD), then reused; Lazzarini (2009).
Funerary vases; slabs for *opera sectilia* >> tubs and vases > small columns > *trapezophoroi* > *cuticulae*
Gnoli (1988); Ashton *et al.* (2000)

Precambrian
Metagabbro
Equigranular, fine-grained (*sedia di S.Lorenzo* variety) to inequigranular medium-grained often with pegmatitic zones (*sedia di S. Pietro* variety)
Plagioclase (pervasively altered to saussurite) = pyroxene (± altered augite) > hornblende (mostly chloritized) >> magnetite >>> accessories (uralite, secondary quartz)
Brown and Harrell (1995), Klemm and Klemm (2008).
Field of view = 1.7 mm

Stone 9
GABBRO EUFOTIDE, GRANITO VERDE PLASMATO
Wadi Maghrabiya, Eastern Desert, Egypt (6 in Fig. 1)
Pre-Early Dynastic periods; Imperial times (end of I c. BC- end of I c. AD)
Funerary Vases; small slabs for *opera sectilia* > small objects (furniture)
Gnoli (1988); Harrell *et al.* (1999).

Precambrian
Metagabbro
Inequigranular, medium- to coarse-grained, with pegmatitic zones
Plagioclase (labradorite), altered significantly to saussurite = pyroxene (augite), much altered to uralite and chlorite
Harrell *et al.* (1999).
Field of view = 2.35 mm

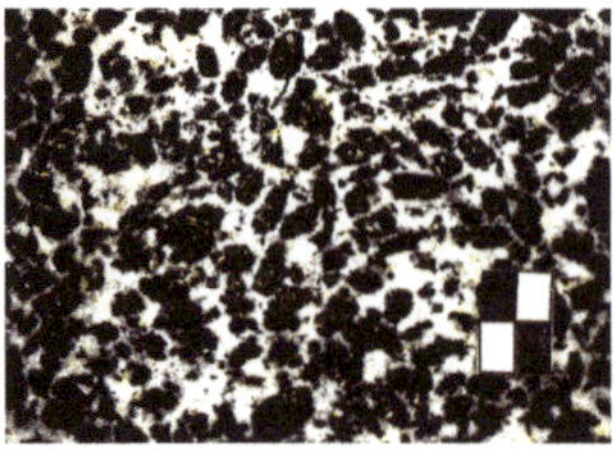

Stone 10 *MARMOR TIBERIANUM*
GRANITO BIANCO E NERO
Wadi Barud, Eastern Desert, Egypt (7 in Fig. 1)
Pre-Early Dynastic periods; Imperial times (end of I c. BC- end of I c. AD), then reused
Funerary vases; slabs for *opera sectilia* > columns
Gnoli (1988); Harrell and Lazzarini (2002).

Proterozoic (~680 Ma)
Quartz diorite: medium-to-mainly coarse-grained (Santa Prassede variety); fine-grained ("del Cairo" variety)
Inequigranular
Plagioclase (oligoclase-andesine), somewhat altered = hornblende (often chloritized) >> quartz >>> biotite >>> accessories (apatite, magnetite, titanite and zircon)
Brown and Harrell (1995); Harrell and Lazzarini (2002).
Field of view = 7.0 mm

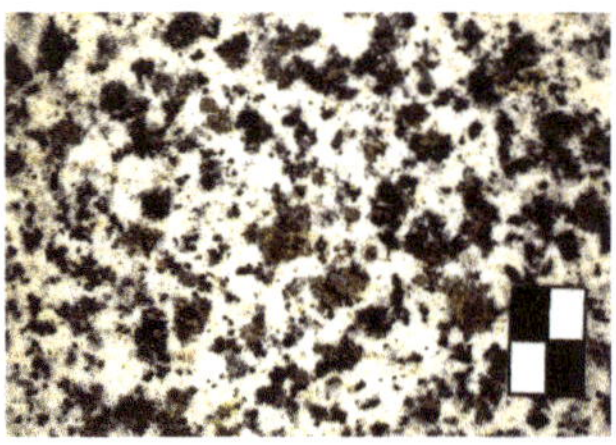

Stone 11 *MARMOR CLAUDIANUM*
GRANITO DEL FORO
Mons Claudianus, Gebel Fatira (tonalite), Wadi Umm Huyut (tonalite-gneiss), Eastern Desert, Egypt (7 in Fig. 1)
I–IV (?) c. AD, then reused; Lazzarini (2009)
Columns >> tubs > slabs for *opera sectilia*
Gnoli (1988); Brown and Harrell (1995); Peacock and Maxfield (1977); Harrell *et al.* (1999); Maxfield and Peacock (2001); Klemm and Klemm (2008).

Proterozoic (~680 Ma)
Tonalite/tonalite-gneiss
Inequigranular, with clamps of intergrown hornblende and biotite; fine-to-medium grain size
Plagioclase (oligoclase) >> quartz >> hornblende > biotite>> K-feldspar (microcline) >> accessories (apatite, magnetite, titanite and zircon)
Brown and Harrell (1995); Harrell *et al.* (1999); Klemm and Klemm (2008).
Field of view = 3.5 mm

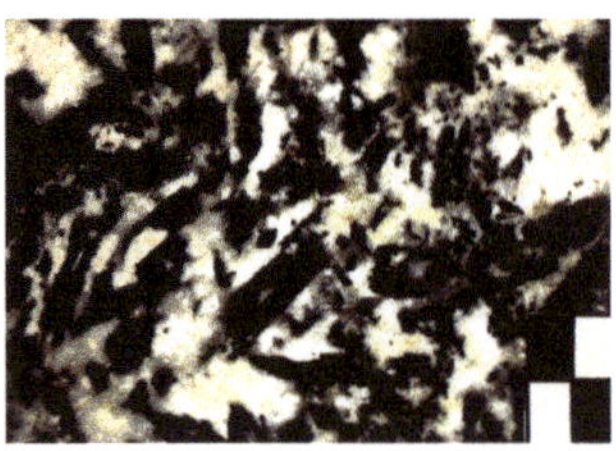

Stone 12
GRANITO DELLA COLONNA
Wadi Umm Shegilat, Eastern Desert, Egypt (8 in Fig. 1)
Pre-Early Dynastic periods; I–II c. AD, then reused
Funerary vases; Columns >> *trapezoforoi* > slabs for *opera sectilia*
Gnoli (1988); Brown and Harrell (1995).

Proterozoic
Pegmatitic diorite-to-gabbro
Inequigranular, very coarse grain size
Plagioclase, strongly altered to clay and sericite > hornblende, altered severely to chlorite (clinochlore) and iron oxides >>> magnetite
Lazzarini (1987); Brown and Harrell (1995).
Field of view = 3.5 mm

Stone 13
GRANITO VERDE FIORITO DI BIGIO
Wadi Umm Balad, Eastern Desert, Egypt (8 in Fig. 1)
Pre-Early Dynastic; I–II c. AD, then reused
Funerary vases; slabs for *opera sectilia* > columns > tubs and vases
Gnoli (1988); Brown and Harrell (1995).

Proterozoic
Quartz diorite
Equigranular, homogeneous fine grain size
Plagioclase (oligoclase-andesine), moderately altered into clay, sericite, epidote and calcite >>> biotite, often altered severely to chlorite and iron oxides >> quartz > accessories (apatite, magnetite, titanite and zircon)
Brown and Harrell (1995).
Field of view = 3.5 mm

Stone 14
MARMO MISIO, GRANITO MISIO
Kozak Dâg, province of Bergama, Turkey (9 in Fig. 1)
II–VI (?) c. AD, then reused; Lazzarini (2009).
Columns >> slabs for *opera sectilia* > sarcophagi
Lazzarini (1992); Lazzarini (1998).

Early–Middle Miocene
Amphibolic granite/granodiorite
Equigranular, with fine and homogeneous grain size
Plagioclase > K-feldspar > quartz > biotite>> hornblende >> accessories (magnetite, zircon, allanite, titanite)
De Vecchi *et al.* (2000).
Field of view = 3.5 mm

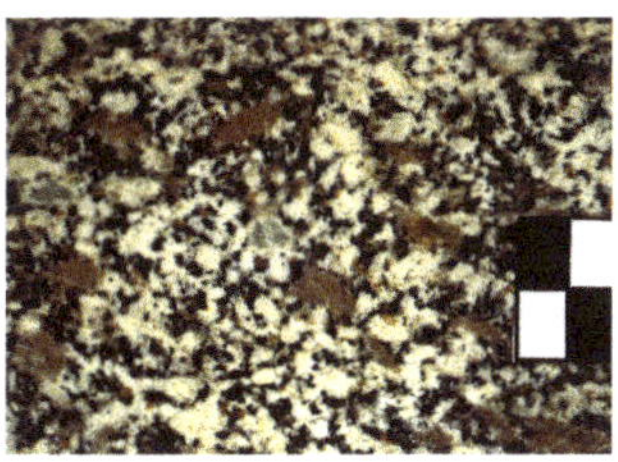

Stone 15 *MARMOR TROADENSE*
GRANITO VIOLETTO
Çigri Dâg, province of Ezine, Turkey (10 in Fig. 1)
II–VI c. AD, then reused; Lazzarini (2009).
Columns >>> pillars > slabs for *opera sectilia*
Lazzarini (1987); Ponti (1995).

Miocene (~21 Ma)
Quartz-monzonite
Inequigranular, medium grain size, often with K-feldspar megacrysts (2–4 cm)
K-feldspar = plagioclase > hornblende = quartz >> biotite >> accessories (apatite, magnetite, titanite, epidote, chlorite)
Lazzarini (1987); Birkle and Satir (1994).
Field of view = 1.7 mm

Stone 16
GRANITELLO, GRANITO ELBANO
Secchetto, Cavoli, ecc. Southern area of Monte Capanne, Elba Island (Italy) (11 in Fig. 1)
Late Augustan Age–Late Middle Ages, often reused; Lazzarini (2009). Still quarried
Columns >>> slabs for *opera sectilia* > tubs
Gnoli (1988); Tedeschi Grisanti (1992).

Late Miocene
Granodiorite
Equigranular, fine grain size, often with single large white, euhedral, plagioclase crystals (porphyroblasts); plagioclase (oligoclasic/andesinic) > quartz > K-feldspar, biotite >> accessories (apatite, titanite, zircon, chlorite, tourmaline, muscovite)
Marinelli (1965); De Vecchi *et al.* (2000).
Field of view = 3.5 mm

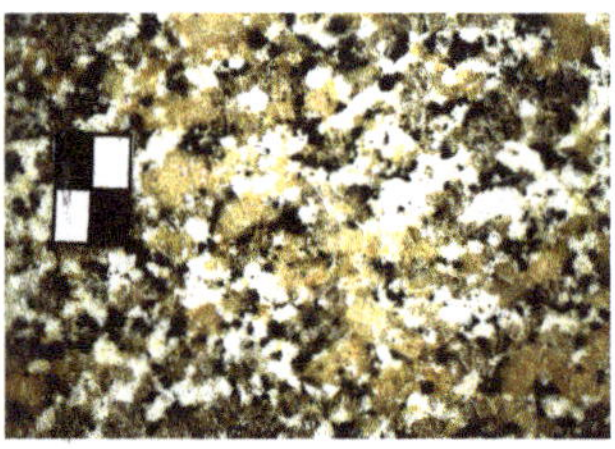

Stone 17
GRANITO SARDO
Capo Testa, Marmorata, Bocche di Bonifacio (San Bainzo), N Sardinia, Italy (12 in Fig. 1)
II–IV c. AD, then reused; Lazzarini, 2009. Still quarried
Columns > pillars >> slabs for *opera sectilia* > tubs
Gnoli (1988); Wilson (1988); Poggi and Lazzarini (2005).

Late Carboniferous–Permian
Monzogranite
Inequigranular, medium- to coarse-grain size
K-feldspar > plagioclase > quartz> biotite >> accessories (apatite, titanite, zircon, chlorite, Fe oxides)
Maccioni *et al.* (1968).
Field of view = 3.5 mm

Stone 18 *LAPIS HEKATONTALITHON, LAPIS HEXAKONTALITHON*
CENTOPIETRE, BRECCIA VERDE EGIZIANA
Mons Basanites, Wadi Hammamat, Eastern Desert, Egypt (13 in Fig. 2)
Early Dynastic; New Kingdom; Late Augustan Age-IV c. AD (?), then reused
Sarcophagi, vases; columns > slabs for *opera sectilia*
Gnoli (1988); Harrell *et al.* (2002); Klemm and Klemm (2008).

Late Precambrian
Polymictic metaconglomerate
Clastic, with pebble-to-cobble size (3–30 cm across), often inbriccated and isoriented
Volcanic/metavolcanic rocks (tuffs, lavas, mostly sialic) >> sedimentary rocks (greywackes, siltstones, rare chert and limestones) = plutonites (felsic granitoids, rare tonalities and diorites) >> metamorphites (serpentinites) and quartz
Willis *et al.* (1988); Hassan and Hashad (1990); Harrell *et al.* (2002).
Field of view = 1.7 mm

Stone 19 *BEKHEN, LAPIS BASANITES*
BASANITE
Mons Basanites, Wadi Hammamat, Eastern Desert, Egypt (13 in Fig. 2)
Pre-Dynastic–I c. AD
Small objects (palettes, scarabs, *etc.*); statuary (portraits)
Lucas and Rowe (1938); Andrew (1939); Gnoli (1988).

Late Precambrian
Greywacke/metagreywacke-to-siltstone
Arenaceous, laminated; siltitic
Metagraywacke = quartz in a calcite-sericite-chlorite-epidote matrix, with tourmaline and zircon as accessories; Siltstone = quartz, in a intergrown sericite-chlorite-calcite matrix, with epidote and zircon as accessories
Willis *et al.* (1988); Hassan and Hashad (1990); Klemm and Klemm (2008).
Field of view =0.9 mm

Stone 20 *MARMOR CARIUM, MARMOR IASSENSE*
CIPOLLINO ROSSO, AFRICANONE, veined variety (right), brecciated variety (left)
Kiykislaçik (ancient Iasos), province of Mula, Turkey (14 in Fig. 2)
II c. BC; III c.–Early Byzantine period, then reused; Lazzarini (2009).
Slabs for *opera sectilia* >> columns > table tops
Gnoli (1988); Andreoli *et al.* (2002).

Cretaceous (Campanian-Maastrictian)
Hematite-marble/metabreccia, with white/grey veins/clast in a matrix coloured red by hematite
granoblastic, heteroblastic, clastic, fine-to-medium grain size
calcite >>>> hematite >> quartz, plagioclase > muscovite, chlorite
Gorgoni *et al.* (2002).
Field of view =1.7 mm

Stone 21 *MARMOR TAENARIUM RUBRUM*
ROSSO ANTICO
Prophytis Elias, Paganea, Làghia, Kokkinoghia, Mianes, *etc.*, Mani peninsula, Peloponnesus, Greece (15 in Fig. 2)
Middle Minoan–Mycenaean period; Late II c. BC–Late Roman Empire, then reused; end of XIX c.–~1960; Lazzarini (2007).
Slabs for *opera sectilia* > cornices >> statuary > columns > table tops > vases and tubs
Gnoli (1988); Lazzarini (2007).

Senonian–Priabonian
Impure marble coloured red by hematite
granoblastic-heteroblastic with fine grain size
calcite >>> hematite > quartz, plagioclase (albitic) > muscovite, chlorite > accessories (Fe hydroxides, apatite, epidote, piemontite)
Calogero *et al.* (2000); Lazzarini (2007).
Field of view = 0.9 mm

Stone 22 *MARMOR CARYSTIUM, MARMOR STYRIUM*
CIPOLLINO VERDE EUBOICO, CIPOLLINO BIGIO EUBOICO
Styra-Karystos, southern-western Eubea, Greece (16 in Fig. 2)
Late II c. BC–Middle Byzantine period, then reused; end of XIX c. to date; Lazzarini (2007).
Columns > slabs for *opera sectilia* >> tubs > statuary
Gnoli (1988); Lazzarini (2007).

Late Jurassic–Early Cretaceous
Cipollino, impure chlorite/graphite marble
Homeo/heteroblastic often passing to lepidoblastic, with fine grain size
Calcite >>> quartz > plagioclase (albitic), muscovite/phengite, chlorite > accessories (Fe oxides/hydroxides, ilmenite, rare pyrite, epidote, apatite, titanite)
Lazzarini (2007).
Field of view = 1.7 mm

Stone 23
BRECCIA DI SETTEBASI, SEMESANTO (variety with mm clasts)
Aghios Panteleimon, Valaxa, Treis Boukes, Koprissies (also for semesanto), Skyros Island (Greece) (17 in Fig. 2)
Late I c. BC–IV c. AD, then reused; end of XIX c. to date; Lazzarini (2009).
Slabs for opera sectilia >> columns > tubs > statuary
Gnoli (1988); Lazzarini (2007).

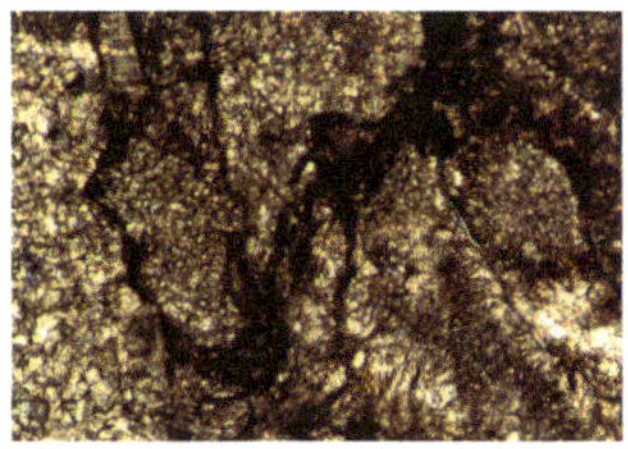

Middle Trias–Jurassic
Carbonatic metabreccia
Clastic with angular mm-dm clasts formed by slightly recrystallized, fine-grained calcite/dolomite
Calcite >>> dolomite >> accessories (hematite, Fe hydroxides, quartz, muscovite, chlorite, albitic plagioclase)
Lazzarini (2007).
Field of view = 2.2 mm

Stone 24 *MARMOR LUCULLAEUM*
AFRICANO
Sigacik (ancient Teos), province of Izmir, Turkey (18 in Fig. 2)
Early I c. BC–Late II c. AD, then reused; Lazzarini (2009).
Slabs for *opera sectilia* >> columns >> tubs > statuary
Gnoli (1988); Pensabene and Lazzarini (1998).

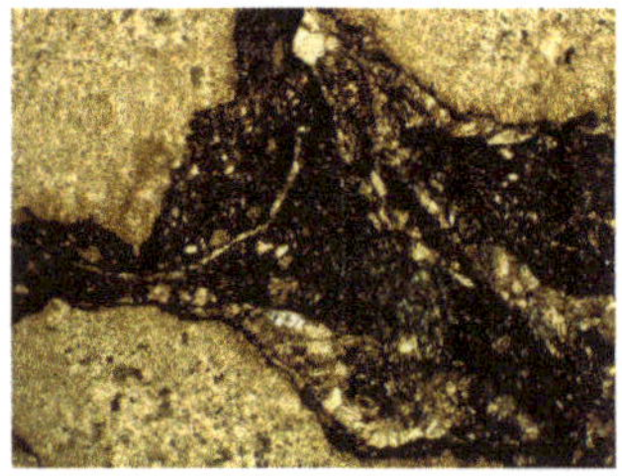

Cretaceous
Tectonic, carbonatic breccia/metabreccia
Clastic, with cm–dm clasts, angular to subrounded, with very fine grain size and rare macrofossils (Rudists, Crinoids)
mudstones >> dolostones; micrite >>> accessories (quartz, chlorite, muscovite, hematite, Fe hydroxides)
Lazzarini and Sangati (2004).
Field of view =1.7 mm

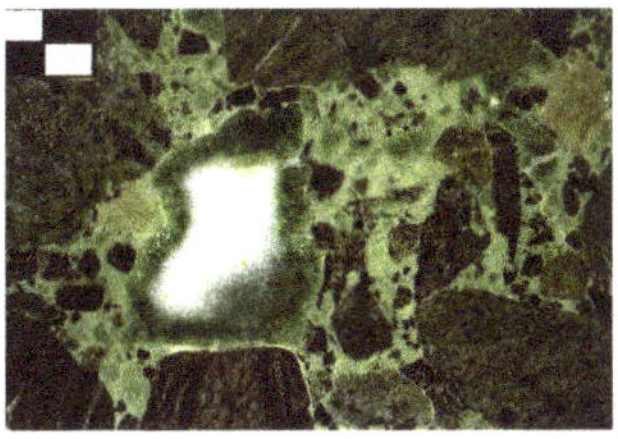

Stone 25 *MARMOR THESSALICUM, LAPIS ATRACIUS*
VERDE ANTICO
Mount Mopsion, Chasabali, province of Larisa, Greece (19 in Fig. 2)
beginning of the II c. AD–Middle (?) Byzantine period; end of XIX c. 1980; Lazzarini (2007).
Columns > Slabs for *opera sectilia* >> tubs and vases > statuary
Gnoli (1988); Lazzarini (2007).

Late Triassic
Ophycalcite breccia
Clastic, with mm–dm angular to subrounded clasts of dark green antigorite/antigoritic serpentinite and white marble in a mixed antigorite-calcite matrix
Antigorite > calcite >> magnetite >> accessories (Fe oxides/hydroxides, chromite, tremolite, asbestos, chlorite, talc, epidote, millerite)
Lazzarini (2007); Melfos (2008).
Field of view = 1.7 mm

Stone 26 *MARMOR CHALCIDICUM*
FIOR DI PESCO
Eretria, Island of Eubea, Greece (20 in Fig. 2)
III–I c. BC (local); I–IV (?) c. AD, then reused especially in the Baroque period; 1950 to date
Slabs for *opera sectilia* >> columns > tubs > *trapezophoroi*, sculptures
Lazzarini (2007); Russell and Farchard (2012).

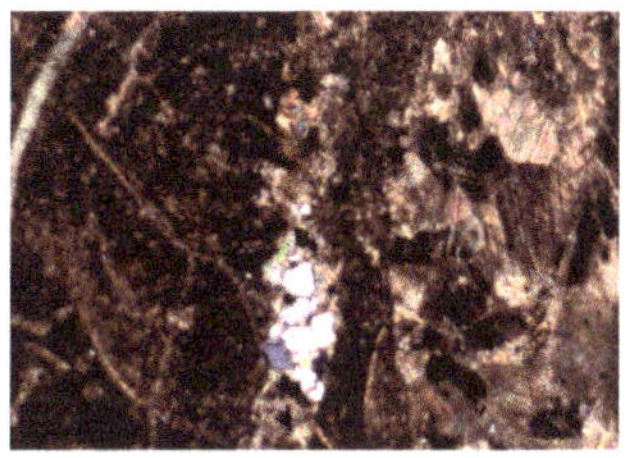

Trias
Cataclastic limestone, slightly metamorphosed
Cataclastic, with many recrystallized areas and cavities of *stromatactis* type
Mudstones (with rare fossils: ammonoids, corals, filaments)>blastic calcite areas;micrite >> accessories (hematite, sericite, chlorite, quartz, plagioclase, K-feldspar, zircon)
Lazzarini (2007).
Field of view = 4.5 mm

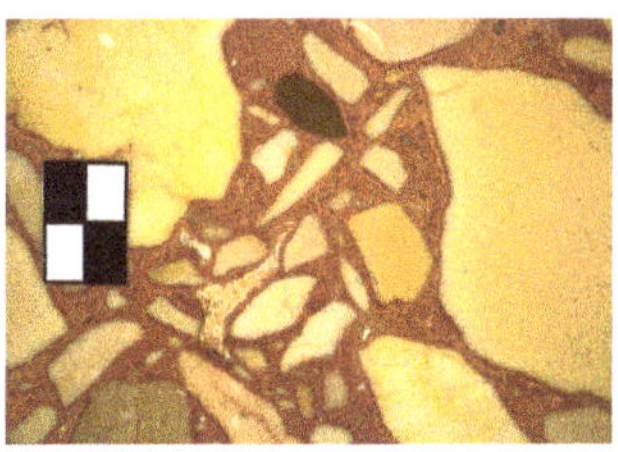

Stone 27 *LAPIS KNEKITES ?*
BRECCIA ROSSA E GIALLA
Wadi Imu and Wadi Abu Gelbana, province of Sohag, East Bank, Egypt (21 in Fig. 2)
Pre-Dynastic-Early Dynastic; I–II c. AD ?, then reused.
Funerary vases and other small objects; sarcophagi and tubs, slabs for *opera sectilia*
Klemm and Klemm (2008); Lazzarini (2002b)

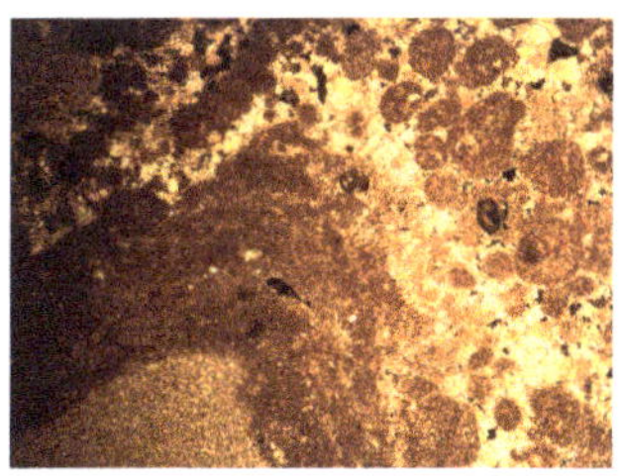

Late Miocene ?
Polygenic carbonatic breccia
Clastic, mm–cm subangular to subrounded pale-yellow/yellow and grey clasts
Mudstone to packstone clasts, composed of microsparite in a micritic-clayey cement with peloids containing rare reworked microfossils and detrital quartz, coloured by Fe hydroxides/oxides
Klemm and Klemm (2008).
Field of view = 2.35 mm

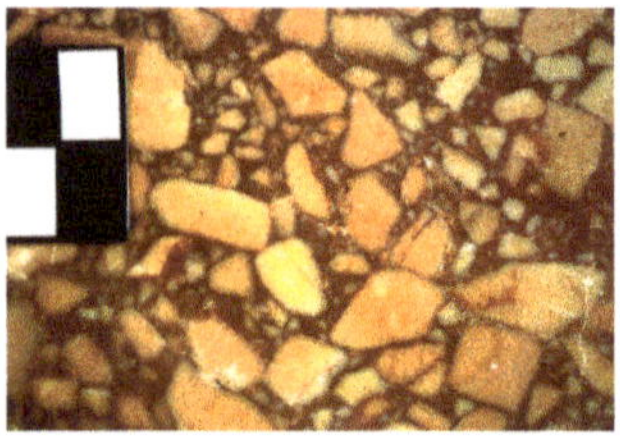

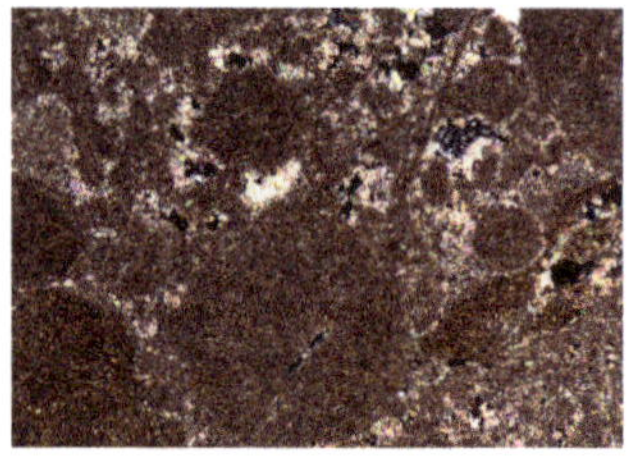

Stone 28 *MARMOR SAGARIUM*
BRECCIA CORALLINA, BRECCIA NUVOLATA, BROCCATELLONE (variety)
Vezirhan, province of Bilecik; Balikliova, Toprak Alinmis, Karga and Azmak Tepe, Karaburun Peninsula, Turkey (22, 23 in Fig. 2)
Late I c. BC–V (?) c. AD, then reused; newly quarried from 1980 to date; Lazzarini (2009).
Slabs for *opera sectilia* >> columns > tubs > statuary
Gnoli (1988); Lazzarini (2002c); Bruno *et al.* (2012).

Cretaceous
Monogenic carbonatic breccia
Clastic, with angular to subangular cm–dm clasts
mudstone sometimes with peloids; micrite >>> accessories (quartz, muscovite, hematite, Fe hydroxides)
Lazzarini (2002c).
Field of view = 2.2 mm

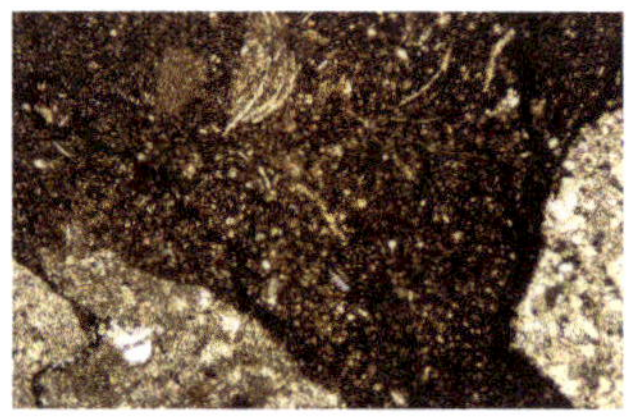

Stone 29 *MARMOR CHIUM*
PORTASANTA
Latomi, Chios town, Island of Chios, Greece (23 in Fig. 2)
IV c. BC–XI c. AD (with interruptions in the Late Antiquity–Early Byzantine periods), then reused; Lazzarini (2007).
Slabs for *opera sectilia* >> columns > tubs > bases > statuary
Lazzarini (2007).

Early–Middle Trias
Tecton ic carbonatic breccia, mono/digenic
Clastic, with cm–dm angular/subangular clasts, with small amounts of matrix and veins of secondary calcite
mudstones (sometimes with filaments) >> grainstone (often with peloids and rare bioclasts) >> dolostone; micrite >> dolomite >> accessories (quartz, plagioclase, muscovite/illite, chlorite, hematite)
Lazzarini (2007).
Field of view = 1.7 mm

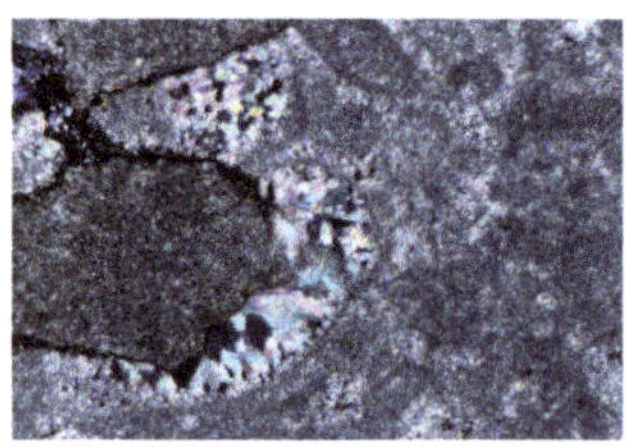

Stone 30
BRECCIA DI ALEPPO
Kariés, province of Chios (town), Island of Chios, Greece (23 in Fig. 2)
Late I c. BC–Late I c. AD, then reused
Slabs for opera sectilia > columns > *trapezophoroi*, stelae, statuary
Lazzarini (2007).

Early–Middle Trias
Multicoloured polygenic breccia
Clastic, with cm–dm grey, red, yellow angular to subrounded clasts, sometimes fossiliferous (ammonoids, corals, *etc.*), in a grey/red cement
mudstones >> bufflestones; micrite >>> accessories (quartz, sericite, Fe oxides/hydroxides)
Lazzarini (2007).
Field of view = 2.2 mm

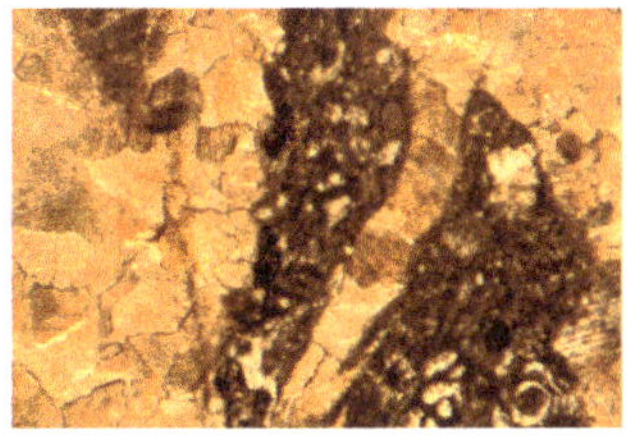

Stone 31 *MARMOR CELTICUM*
MARMO DI AQUITANIA, BIANCO E NERO ANTICO, GRAND ANTIQUE
Aubert, Cap de la Bouiche, Pyrenées, France (24 in Fig. 2)
III AD–Protobyzantine period, then reused; 1844–1940; Lazzarini (2009).
Slabs for *opera sectilia* >> columns
Lazzarini (2005).

Early Cretaceous
Tectonic, carbonatic breccia
Clastic, formed by black mm–dm angular clasts of a black carbonaceous limestone in a white sparitic cement mudstone/wackestone with abundant microforams, rare bivalves and brachiopods >> dolostone; micrite >> dolomite >> accessories (carbonaceous matter, Fe hydroxides)
Lazzarini (2005).
Field of view = 0.05 mm

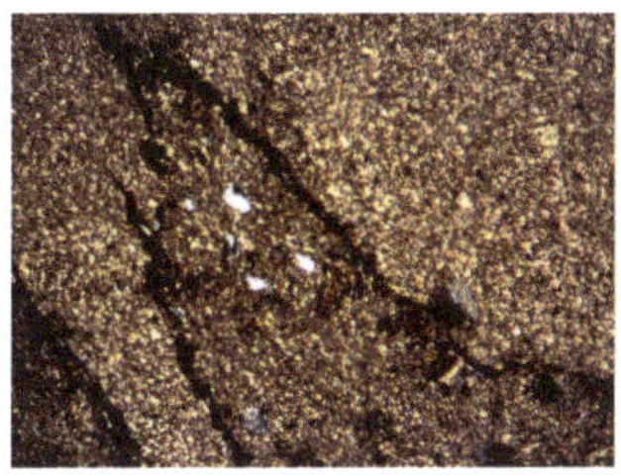

Stone 32 *MARMOR NUMIDICUM*
GIALLO ANTICO
Djebel Chemtou, Chemtou (ancient *Simitthus*), Tunisia (25 in Fig. 2)
II c. BC–IV c. AD, then reused XX c.; Lazzarini (2009).
Slabs for *opera sectilia* >>> columns > statuary > tubs and vases
Gnoli (1988); Rakob (1993).

Jurassic
Limestone/carbonatic breccia
More or less clastic with yellow, pink angular to subrounded clasts in a yellow/ brown/red cement
Mudstone/sparstone with micrite >> sparite >>> accessories (Fe oxides/hydroxides, plagioclase, quartz, illite/muscovite)
Zagrami *et al.* (2000).
Field of view = 1.7 mm

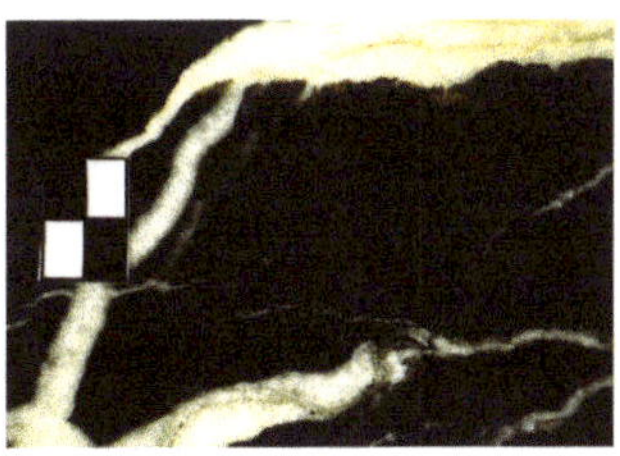

Stone 33 *LAPIS NIGER*
BIGIO MORATO (variety of NERO ANTICO)
Djebel Aziz, province of Tunis, Tunisia (26 in Fig. 2)
I c. BC–V (?) c. AD, then reused; 1980 to date
Slabs for *opera sectilia* > statuary
Gnoli (1988); Lazzarini *et al.* (2007).

Early Trias
Carbonaceous limestone
Oolithic/ooid crystalline, grain supported, with abundant veins of sparitic calcite
grainstone, locally passing to packstone with ooids/ooliths in a micritic/orthosparitic cement, with rare, much reworked microfossils and abundant carbonaceous matter
Agus *et al.* (2007).
Field of view = 1.7 mm

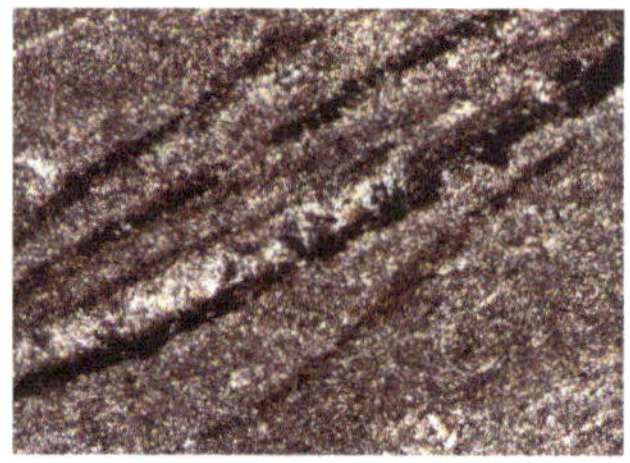

Stone 34
CIPOLLINO MANDOLATO, GRIOTTE, MARBRE CAMPAN
Campan (Haute-Adour), Pont de la Taule (Couflens, Seix), Pyrenees, France (27 in Fig. 2)
Late I–V c. AD; XIX c. to date; Lazzarini (2009).
Slabs for *opera sectilia* >> small columns
Antonelli and Lazzarini (2000); Antonelli (2002).

Late Devonian (Famennian)
Nodular limestone, with rarely preserved macrofossils (*Goniatites sp.*)
Microsparitic nodules in a clayey-micritic matrix coloured green by chlorite, or red by hematite
Sparite >>> K-mica/illite, chlorite >> accessories (quartz, plagioclase, titanite, pyrite, Fe oxides)
Antonelli and Lazzarini (2000); Antonelli (2002).
Field of view = 1.1 mm

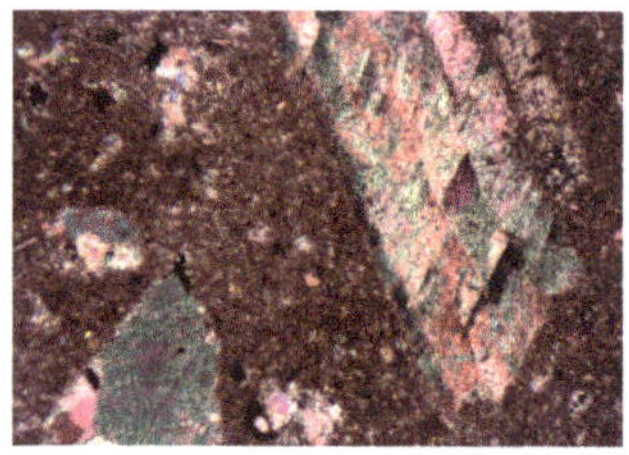

Stone 35 *MARMOR TRIPONTICUM*
OCCHIO DI PAVONE
Kutluca, province of Izmit, Turkey (28 in Fig. 2)
III–VII c. AD, then reused; 1950–90; Lazzarini (2009).
Slabs for *opera sectilia* > columns > tubs and vases > sarcophagi
Gnoli (1988); Lazzarini (2002c).

Cretaceous
Fossiliferous (Rudists) limestone (lumachella)
Bioclastic
Micrite >> sparite >>> (hematite, Fe hydroxides, quartz)
Lazzarini (2002c).
Field of view = 2.2 mm

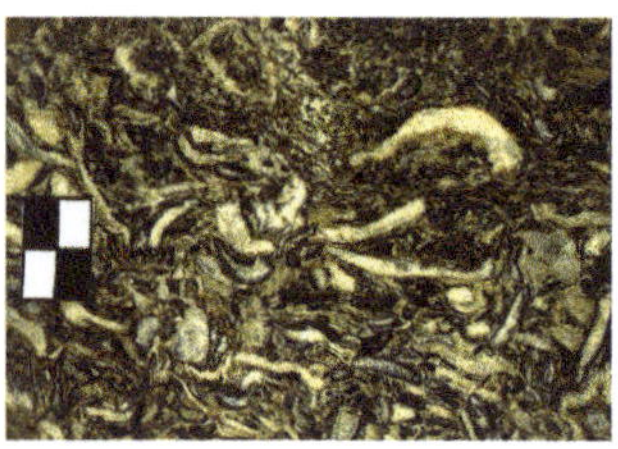

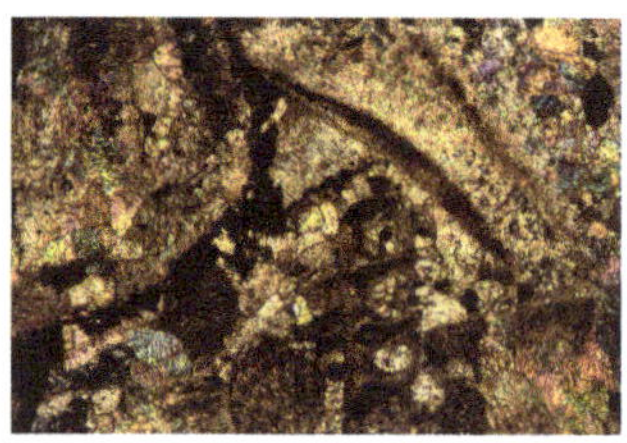

Stone 36
LUMACHELLA ORIENTALE, LUMACHELLA D'EGITTO, PIETRA PIDOCCHIOSA
Djebel Oust, province of Tunis (29 in Fig. 2)
Late I c. BC–III c. AD, then reused
Slabs for *opera sectilia* and tabletops >> statuary > tubs
Gnoli (1988); Lazzarini and Mariottini (2012).

Late Jurassic / Early Cretaceous
Fossiliferous (Rudists, Bivalves, Foraminifera) limestone
Bioclastic
Micrite > microsparite >>> quartz>> accessories (glauconite, Fe hydroxides/oxides)
Lazzarini and Mariottini (2012).
Field of view = 1.7 mm

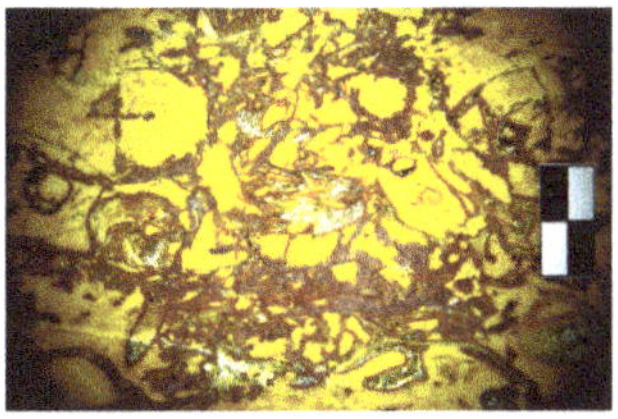

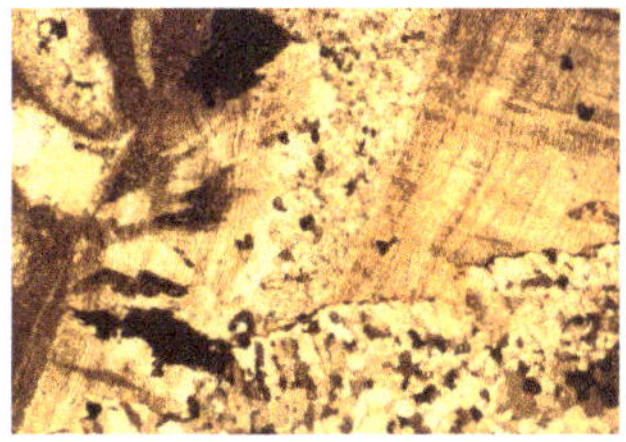

Stone 37
BROCCATELLO DI SPAGNA, JASPI DE LA CINTA
La pedrera de la Cinta, Els Valencians, province of Tortosa, Spain (30 in Fig. 2)
Middle I–V c. AD, then reused; XVI–XX c.; Lazzarini (2009)
Slabs for *opera sectilia* >> columns >> inscribed stelae
Gnoli (1988); Roda (1997); Falcone and Lazzarini (1998); Muñoz i Sebastià and Rovira i Gómez (1997).

Cretaceous (Aptian)
Fossiliferous (Rudists, Echinids, Algae, *etc.*) limestone (lumachella) (Rudstone)
Bioclastic
Micrite >> sparite >>> accessories (hematite, goethite, Fe hydroxides, quartz)
Falcone and Lazzarini (1998).
Field of view = 3.5 mm

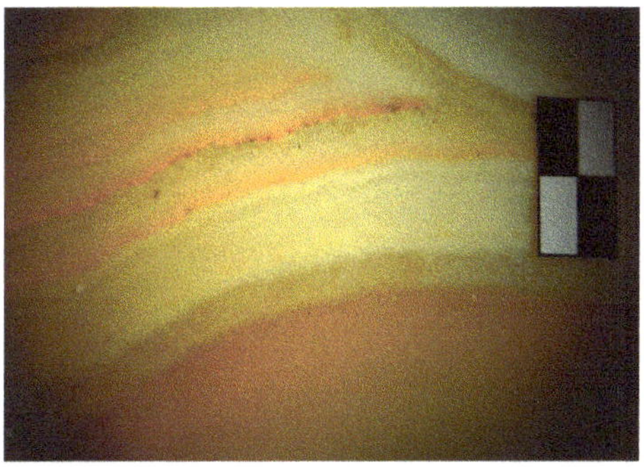

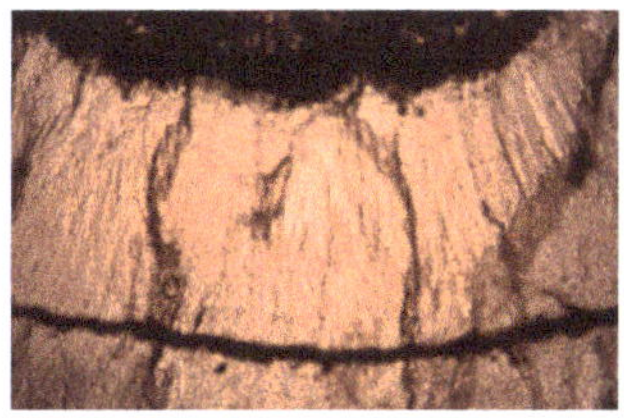

Stone 38 *LAPIS ALABASTRITES*
ALABASTRO MELLEO, ALABASTRO COTOGNINO, ALABASTRO EGIZIANO
Hatnub, Wadi Gerrawi, Wadi Sannur, Zawiet Sultan, *etc.*, Middle Egypt (31 in Fig. 2)
Late Neolithic to date; Lazzarini (2009).
Small objects, vases (*alàbastra*) >> slabs for *opera sectilia*> columns > statuary
Gnoli (1988); Klemm and Klemm (1991); Klemm and Klemm (2008); Shaw (2010).

Quaternary
Calcite alabaster/travertine
Concretionary with thick radial-fibrous/dendritic levels of sparite alternated with thin micritic ones
Sparite >> micrite >> micritic aragonite
Barbieri *et al.* (2002); Klemm and Klemm (2008); Pentecost (2010).
Field of view = 2.2 mm

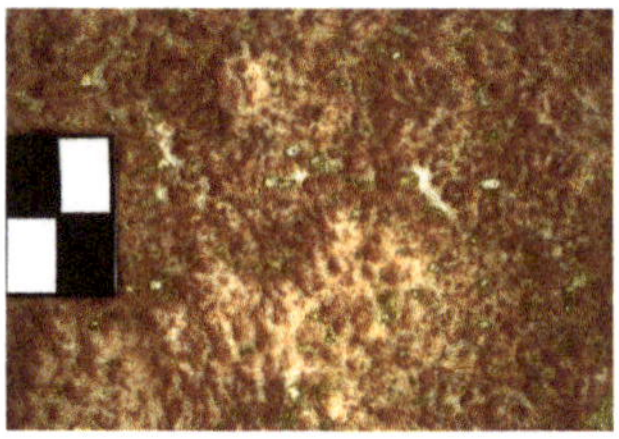

Stone 39 *ALABASTRO A PECORELLA*
Bou Hanifa, province of Oran, Algeria (32 in Fig. 2)
I–IV c. AD; Lazzarini (2009)
Slabs for *opera sectilia* >> columns > *trapezophoroi*, statuary
Gnoli (1988); Herrmann *et al.* (2012).

Quaternary
Calcite alabaster/travertine
Concretionary with dendritic feather crystals/laminated and mosaic areas
Calcite >>> hematite
Lazzarini *et al.* (2006); Pentecost (2010); Herrmann *et al.* (2012).
Field of view = 2.35 mm

Stone 40
ALABASTRO DI JANO DI MONTAIONE, ALABASTRO DI PALOMBARA, ALABASTRO TARTARUGATO, ALABASTRO CINERINO
Iano di Montaione, province of Florence, Italy (33 in Fig. 2)
II–V c. AD; XVI–XX c.
Slabs for *opera sectilia* >> columns > table tops
De Michele and Zezza (1979); Gnoli (1988); Lazzarini *et al.* (2006).

Quaternary
Calcitic alabaster/travertine
Concretionary with laminated areas containing brown/bluish alabaster clasts
Sparite >> micrite >> micritic aragonite
Lazzarini *et al.* (2006).
Field of view = 2.35 mm

References

Agus, M., Cara, S., Lazzarini, L. and Mola, M. (2007) A laboratory characterization of black limestones (Neri Antichi) from Zeugitania (Tunisia). *Marmora*, **2** (2006), 71–82.

Andreoli, A., Berti, F., Lazzarini, L. and Pierobon Benoit, R. (2002) New contributions on Marmor Iassense. Pp. 13–18 in: *ASMOSIA VI. Interdisciplinary studies on ancient stone.* Proceedings of the Sixth International Conference, Venice, June 15–18, 2000, (L. Lazzarini, editor), Bottega d'Erasmo, Padova, Italy.

Andrew, G. (1939) The greywackes of the Eastern Desert of Egypt. *Bulletin de l'Institut d'Egypte*, **21**, 153–190.

Antonelli, F. (2002) I marmi della Gallia e dell'Iberia importati a Roma. Pp. 267–275 in: *I marmi colorati della Roma imperiale* (M. De Nuccio and L. Ungaro, editors), Marsilio, Veggiano (PD).

Antonelli, F. and Lazzarini, L. (2000) Le “Marbre Campan” (Cipollino Mandolato: histoire, diffusion, archéométrie. *Revue d'Archéométrie*, **24**, 111–128.

Antonelli, F., Lazzarini, L. (2016) An updated petrographic and isotopic reference database for white marbles used in antiquity. *Rendiconti Lincei Scienze Fisiche e Naturali*, **26**, 399–413.

Antonelli, F., Lazzarini, L. and Cancelliere, S. (2010) “Granito del foro” and “Granito di Nicotera”: petrographic features and archaeometric problems owing to similar appearance. *Archaeometry*, **52**, 919–936.

Ashton, B.G., Harrell, J.A. and Shaw, I. (2000) Stone. Pp. 5–77 in: *Ancient Egyptian Materials and Technology* (P.T. Nicholson and I. Shaw, editors). Cambridge University Press, Cambridge, UK.

Barbieri, M., Lilyquist, C. and Testa G. (2002) Provenancing Egyptian and Minoan calcite-alabaster artifacts through $^{87}Sr/^{86}Sr$ isotopic ratio and petrography. Pp. 403–414 in: *ASMOSIA VI. Interdisciplinary studies on ancient stone*. Proceedings of the Sixth International Conference, Venice, June 15-18-2000, (L. Lazzarini, editor), Bottega d'Erasmo, Padova.

Birkle, P. and Satir, M. (1994) Geological Aspects of the use of Kestambol quartz-monzonite intrusion (Troas/ Turkey) as a constructing material in archaeological sites around the Mediterranean sea. *Studia Troica*, **4**, 143–155.

Brentchaloff, D. and Mazeran, R. (1999) Exploitation antique de l'estérellite: mise au point petro-archéologique. *Memoirs de l'Institut de préhistoire et d'archéologie Alpes Mediterranée*, **XLI**, 17–26.

Brown, V.M. and Harrell, J.A. (1995) Topographical and petrological survey of ancient Roman quarries in the Eastern Desert of Egypt. Pp. 221–234 in: *The Study of Marble and other Stones used in Antiquity. ASMOSIA III*, Transactions of the 3rd International Symposium (Y. Maniatis, N. Herz and Y. Basiakos, editors), Archetype, London.

Bruno, M. and Lazzarini, L. (1999) Discovery of the Sienese provenance of *Breccia Dorata*, and *Breccia Gialla Fibrosa*, and the origin of *Breccia Rossa Appenninica*. Pp. 77–82 in: *Archéomatériaux, Marbres et Autres Roches* (M. Schroerer, editor). Actes de la IV Conférence Internationale ASMOSIA, Bordeaux-Talence 9–13 October 1995, La Nef-Chastrusse, Bordeaux, France.

Bruno, M., Elçi, H., Yavuz, A.B. and Attanasio, D. (2012) Unknown ancient marble quarries of western Asia Minor. Pp. 562–572 in: *Interdisciplinary Studies on Ancient Stone*, Proceedings of the IX ASMOSIA Conference (A. Gutierrez Garcia, M.P. Lapuente and I. Roda, editors), Tarragona 2009, Industries Grafiques Gabriel Gilbert, Tarragona, Spain.

Calogero, S., Bertelle, M., Antonelli, F. and Lazzarini, L. (2000) A Mössbauer study of some coloured marbles (*cipollino mandolato, rosso antico* and *fior di pesco*): implications on the nature of their colour. *Journal of Cultural Heritage*, **1**, 429–444.

Del Bufalo, D. (2012) *Porphyry. Red Imperial Porphyry, Power and Religion*. Allemandi & C., Turin, Italy.

De Michele, V. and Zezza, U. (1979) Le pietre ornamentali di Roma Antica della Collezione Borromeo nel Museo Civico di Storia Naturale di Milano. *Atti della Società Italiana di Scienze Naturali*, **120(1-2)**, Milano, 67–110.

De Vecchi, G., Lazzarini, L., Lnel, T., Mignucci, A. and Visonà, D. (2000) The genesis and characterization of "Marmor Mysium" from Kozak (Turkey), a granite used in antiquity. *Journal of Cultural Heritage*, **1**, 145–153.

Falcone R., Lazzarini L. (1998) Note storico-scientifiche sul broccatello di Spagna. Pp. 87–97 in "Marmi Antichi II, cave e tecnica di lavorazione, provenienze e distribuzione" (P. Pensabene, editor), *Studi Miscellanei*, **31**.

Galetti, G., Lazzarini, L. and Maggetti, M. (1992) A first characterization of the most important granites used in antiquity. Pp. 167–177 in: *Ancient Stones: Quarrying, Trade and Provenance. Acta Archaeologica Lovaniensia*, **4** (M. Waelkens, N. Herz and L. Moens, editors), Leuven University Press, Leuven, Belgium.

Gnoli, R. (1988) *Marmora Romana*, 2a Edizione, L'Elefante, Roma.

Gorgoni, C., Lazzarini, L. and Pallante, P. (2002) New archaeometric data on Rosso Antico and other red marbles used in antiquity. Pp. 199–206 in: *ASMOSIA VI. Interdisciplinary Studies on Ancient Stone* Proceedings of the Sixth International Conference, Venice, June 15–18, 2000 (L. Lazzarini, editor), Bottega d'Erasmo, Padova, Italy.

Harrell, J.A., Brown, V.M. and Lazzarini, L. (1999) Two newly discovered Roman quarries in the Eastern Desert of Egypt. Pp. 285–292 in: *Archéomatériaux, Marbres et Autres Roches* (M. Schroerer, editor). Actes de la IV Conférence Internationale. ASMOSIA, Bordeaux-Talence 9-13 oct. 1995, La Nef-Chastrusse, Bordeaux, France.

Harrell, J.A. and Lazzarini L. (2002) A new variety of granito bianco e nero from Wadi Barud, Egypt. Pp. 47–51 in: *ASMOSIA 5, Interdisciplinary studies on ancient stone*, Proceedings of the Fifth International Conference (J. J. Herrmann, Jr., N. Herz and R. Newman, editors), Boston 1998, Archetype, London.

Harrell, J.A., Brown, V.M. and Lazzarini, L. (2002) Breccia Verde Antica: sources, petrology and ancient uses.

Pp. 207–218 in: *ASMOSIA VI. Interdisciplinary studies on ancient stone*, Proceedings of the Sixth International Conference, Venice, June 15–18, 2000 (L. Lazzarini, editor), Bottega d'Erasmo, Padova, Italy.

Hassan, M.A. and Hashad A.H. (1990) Precambrian of Egypt. Pp. 201–245 in: *The Geology of Egypt* (R. Said, editor), Balkema, Rotterdam, The Netherlands.

Herrmann, A. (1988) Rhodian red limestone sculptures. Pp. 244–246 in *Kanon: Festschrift Ernst Berger zum 60. Geburtstag gewidmet* (M. Schmidt, editor). Vereinigung der Freunde antiker Kunst, Basel.

Herrmann, J.J., van den Hoek, A. and Tykot, R.H. (2012) Alabastro a pecorelle, Ain Tekbalet, and Bou Hanifia, Algeria: a preliminary report. Pp. 463–470 in: *Interdisciplinary Studies on Ancient Stone*, Proceedings of the IX ASMOSIA Conference (A. Gutierrez Garcia, M.P. Lapuente, and I. Roda, editors). Tarragona 2009, Industries Grafiques Gabriel Gilbert, Tarragona, spain.

Kelany, A., Negem, M., Tohami, A. and Heldal, T. (2007) Granite quarry survey in the Aswan region, Egypt: shedding new light on ancient quarrying. Pp. 87–98 in: *Quarryscapes: Ancient Stone Quarry Landscapes in the Eastern Mediterranean*. Geological Survey of Norway, Special Publication, **12**.

Klemm, D. and Klemm, R. (1991) Calcit-alabaster oder travertin? Bemerkungen zu Sinn und Unsinn petrographischer Bezeichnungen in der Egyptologie. *Göttinger Miszellen*, **122**, 57–70.

Klemm, R. and Klemm, D. (2008) *Stones and Quarries in Ancient Egypt*. The British Museum Press, London.

Lazzarini, L. (1987) I graniti dei monumenti italiani e i loro problemi di deterioramento. Pp. 157–172 in: "Materiali lapidei, II" (A. Burreca, M. Laurenzi Tabasso and G. Palandri, editors), *Bollettino D'Arte*, Suppl.al N°4.

Lazzarini, L. (1992) Des pierres pour l'éternité: les granits utilisés dans l'antiquité classique, in "Les marbres dans l'antiquité", *Les Dossiers d'Archéologie*, **173**, 58–67.

Lazzarini, L. (1994) *Lapis sarcophagus*: an historical and scientific note. Pp. 103–116 in: "Conservazione del Patrimonio Culturale", Contributi del Centro Linceo Interdisciplinare "Beniamino Segre", Accademia Nazionale dei Lincei, Roma.

Lazzarini, L. (1998) Sul marmo misio, uno dei graniti più usati anticamente, in "Marmi Antichi II, cave e tecnica di lavorazione, provenienze e distribuzione" (P. Pensabene, editor). *Studi Miscellanei*, **31**, 111–117.

Lazzarini, L. (2002a) La determinazione della provenienza delle pietre decorative usate dai romani. Pp. 223–225 in: *I marmi colorati della Roma imperiale* (M. De Nuccio and L. Ungaro, editors). Marsilio, Veggiano (PD).

Lazzarini, L. (2002b) Di alcune pietre inedite e rare della Basilica di San Marco a Venezia. Pp. 107–115 in: *Hadriatica. Scritti in onore di Wladimiro Dorigo* (E. Concina, G. Trovabene and M. Agazzi, editors). Il Poligrafo, Padova, Italy.

Lazzarini, L. (2002c) The origin and characterization of *breccia nuvolata*, *marmor Sagarium*, and *marmor Triponticum*. Pp. 58–67 in: *ASMOSIA 5, Interdisciplinary studies on ancient stone*. Proceedings of the Fifth International Conference (J.J. Herrmann, Jr., N. Herz and R. Newman, editors) Boston 1998. Archetype, London.

Lazzarini, L. (2004) Archaeometric aspects of white and coloured marbles used in antiquity: the state of the art. *Periodico di Mineralogia*, **73**, 113–125.

Lazzarini, L. (2005) Un marmo francese a Venezia. Pp. 61–71 in: *Der unbestechliche Blick, Lo sguardo incorruttibile. Festschrift zu Ehren von Wolfgang Wolters* (M. Gaier and U. Nicolai, editors), Porta Alba Verlag, Trier.

Lazzarini, L. (2007) *Poikiloi Lithoi, Versiculores Maculae*, I marmi colorati della Grecia Antica, Storia,uso, diffusione, cave, geologia, caratterizzazione scientifica, archeometria, deterioramento, Fabrizio Serra Editore, Pisa-Roma.

Lazzarini, L. (2009) The distribution and re-use of the coloured marbles in the Roman Empire. Pp. 459–484 in: Proceedings of ASMOSIA VII (Y. Maniatis, editor). *Bulletin de Correspondance Hellénique*, Supplement 51, De Boccard, Paris-Athènes.

Lazzarini, L. and Çolak, M. (2002) Quarries and characterization of a hitherto unknown alabaster and marble from Thyatira (Akhisar, Turkey). Pp. 35–40 in: *ASMOSIA VI. Interdisciplinary studies on ancient stone*, Proceedings of the Sixth International Conference, Venice, June 15–18, 2000, (L. Lazzarini, editor), Bottega d'Erasmo, Padova, Italy.

Lazzarini, L. and Mariottini, M. (2012) A first study of some lumachelle (fossiliferous stones) used in Roman antiquity. Pp. 445–451 in: *Interdisciplinary Studies on Ancient Stone*. Proceedings of the IX ASMOSIA Conference (A. Gutierrez Garcia, M.P. Lapuente, and I. Roda, editors), Tarragona 2009, Industries Grafiques Gabriel Gilbert, Tarragona, Spain.

Lazzarini, L. and Sangati, C. (2004) I più importanti marmi e pietre colorati usati dagli antichi. Pp. 73–100 in: *Pietre e marmi antichi. Natura, caratterizzazione, origine, storia d'uso, diffusione, collezionismo* (L. Lazzarini, editor), CEDAM, Castenaso (BO).

Lazzarini, L. and Visonà, D. (2009) Lapis sarcophagus and the provenance of its Mediterranean sarcophagi. Pp. 369–388 in: *LEUKOS LITHOS, Marbres et autres roches de la Méditerranée: etudes interdisciplinaires*, Actes du VIIIe Coll. Internationale de l'ASMOSIA (Ph. Jockey, editor), Aix-en-Provence 12–18 June, 2006, Maisonneuve & Larose, Paris.

Lazzarini, L., Villa, I.M. and Visonà, D. (2006) Caratterizzazione e identificazione di alabastri usati in antico. Pp. 273–281 in: *Archeometria del Costruito. L'edificato storico: materiali, strutture e rischio sismico* (G.M. Crisci and C. Gattuso, editors),. Edipuglia, Bari, Italy.

Lazzarini, L., Agus, M. and Cara, S. (2007) The ancient quarries of the Neri Antichi (black limestones) from Zeugitania (Tunisia). *Marmora*, **2** (2006), 59–70.

Lazzarini, L., Visonà, D., Giamello, M. and Villa, I. (2012) Archaeometric characterisation of one Tunisian and two Italian calcareous alabasters used in antiquity. Pp. 436–444 in: *Interdisciplinary Studies on Ancient Stone*. Proceedings of the IX ASMOSIA Conference (A. Gutierrez Garcia, M.P. Lapuente and I. Roda, editors), Tarragona 2009, Industries Grafiques Gabriel Gilbert, Tarragona.

Lucas, A. and Rowe, A. (1938) The ancient Egyptian bekhen-stone. *Annales du Service des Antiquités de l'Egypte*, **38**, 127–156.

Maccioni, L., Oppi, G. and Rossetti, V. (1968) Osservazioni geo-petrografiche sui graniti di Capo Testa. *Resoconti Associazione Mineraria Sarda*, LXXIII, **8**, 5–49.

Makovicky, E., Frei, R., Karup-Moller, S. and Bailey, J. (2016a) Imperial porphyry from Gebel Abu Dokhan, the Red Sea Mountains, Egypt, Part I, Mineralogy, Petrology, Geochemistry, use and occurrence, *Neues Jahrbuch für Mineralogie, Abh.*, **193**, 1–27.

Makovicky, E., Frei, R., Bailey, J. and Karup-Moller, S. (2016b) Imperial porphyry from Gebel Abu Dokhan, the Red Sea Mountains, Egypt, Part II, Geochemistry. *Neues Jahrbuch für Mineralogie, Abhandlungen*, **193(1)**, 29–44.

Marinelli, G. (1965) Le intrusioni terziarie dell'isola d'Elba. *Atti della Società toscana di Scienze Naturali*, serie A, **66**, 50–253.

Maxfield, V. and Peacock, D. (2001) *The Roman imperial quarries. Survey and excavation at Mons Porphyrites, 1994–1998, Vol.* **1***: Topography and Quarries*. Egypt Exploration Society, London.

Mazeran, R. (2004) Les granitoides employés en decoration dans les monuments romaines et paléochrétiens du Sud-Est de la France. Pp. 219–230 in: *Carrières et Constructions en France et dans les pays limitrophes. IV* (J. Lorenz and J.-P. Gély, editors). Editions du Comité des Travaux Historiques et Scientifiques, Toulouse.

Melfos, V. (2008) Green Thessalian Stone: the Byzantine quarries and the use of a unique architecture material from Larisa area, Greece. Petrographic and geochemical characterization. *Oxford Journal of Archaeology*, **27**, 387–405.

Mielsch, H. (1985) *Buntmarmore aus Rom im Antikenmuseum Berlin*. Staatliche Museen Preussischer Kulturbesitz, Berlin.

Muñoz i Sebastià, J.H. and Rovira i Gómez, S.J. (1997) La industria del Jaspi de Tortosa a l'etad moderna (segles XVI-XVII). *Nous Colloquis I, Tortosa*, 33–55.

Peacock, D.P.S. and Maxfield, V. (1997) *Mons Claudianus*, Survey and excavations 1987-1993, Vol. I Topography and quarries. Institut Francais d'Archéologie Orientale, Fouilles de L'IFAO 37, Cairo.

Pensabene, P. (2014) *I marmi nella Roma antica*. Carocci editore, Roma.

Pensabene, P. and Lazzarini, L. (1998) Il problema del Bigio Antico e del Bigio Morato: contributo allo studio delle cave di Teos e di Chios. Pp. 142–173 in: *Marmi Antichi II, cave e tecnica di lavorazione, provenienze e distribuzione* (P. Pensabene, editor), *Studi Miscellanei*, 31.

Pentecost, A. (2010) *Travertine*. Springer Verlag, Berlin-Heidelberg.

Pe-Piper, G. and Kotopouli, C.N. (1981) Very low grade metamorphism of Triassic volcanic, West Hellenic nappes, southern Peloponnese, Greece. *Geological Society of America Bulletin*, **92**, 1762–1806.

Poggi, D. and Lazzarini, L. (2005) Il granito sardo: cave e cavatura, usi, diffusione e aspetti archeometrici, *Marmora*, **1**, 49–68.

Ponti, G. (1995) *Marmor troadense.* Granite quarries in the Troad. *Studia Troica*, **5**, 291–320.

Price, M. (2007) *Decorative Stone. The Complete Sourcebook.* Thames and Hudson, London.

Rakob, F. (1993) *Simitthus I.* Die Steinbrche und die antike Stadt, Mainz am Rhein, Germany.

Rehault, J.P., Hontlaas, C., Guennoc, P., Bellon, H., Ruffet, G., Cotton, J., Sosson, M. and Maury, R.C. (2012) Offshore Oligo-Miocene volcanic fields within Corsica-Liguria Basin: magmatic diversity and slab evolution in the Western Mediterranean sea. *Journal of Geodynamics*, **58**, 73–95.

Roda, I. (1997) Los marmoles romanos de Hispania. *Histria Antiqua*, **3**, 47–56.

Russell, B. and Fachard, S. (2012) New work on quarrying in the territory of Eretria, Euboea. Pp. 612–618 in: *Interdisciplinary Studies on Ancient Stone.* Proceedings of the IX ASMOSIA Conference (A. Gutierrez Garcia, M.P. Lapuente and I Roda, editors), Tarragona 2009, Industries Grafiques Gabriel Gilbert, Tarragona.

Shaw, I. (2010) Hatnub: quarrying travertine in ancient Egypt. Egypt Exploration Society, Hainault, Essex.

Tedeschi Grisanti, G. (1992) Il granito dell'Elba a Pisa: uso e riuso nell'XI e XII secolo. Pp. 43–51 in: *Niveo de Marmore. L'uso artistico del marmo di Carrara dall'XI al XV secolo* (E. Castelnuovo, editor). Edizioni Colombo, Genova, Italy.

Ward-Perkins, J. B. (1992) "Africano" marble and "Lapis Sarcophagus". Pp. 55–60 in: *Marble in Antiquity. Collected papers of J.B. Ward-Perkins*, Archaeological Monographs of the British School at Rome 6 (H. Dodge and B. Ward-Perkins, editors), Whitstable Litho, Kent.

Willis, K.M., Stern, R.J. and Clauer, N. (1988) Age and geochemistry of Late Precambrian sediments of the Hammamat Series from the Northeastern desert of Egypt. *Precambrian Research*, **42**, 173–187.

Wilson, R.J.A. (1988) Ancient granite quarries on the Bocche di Bonifacio. Pp. 103–112 in: *Classical Marble: Geochemistry, Technology, Trade* (N. Herz and M. Waelkens, editors). Nato ASI Series 153, Kluwers, Dordrecht-Boston, London.

Zagrami, M.F., Gaied, M.E. and Chaabani, F. (2000) Petrographie des roches marbrieres de Tunisie. Pp. 73–76 in: *Les pierres marbrieres de Tunisie, Annales des mines et de geologie*, **38**. La Charguia, Tunis.

EMU Notes in Mineralogy, Vol. 20 (2019), Chapter 11, 393–409

Obsidian and volcanic glass shards: Characterization and provenancing

DONATELLA BARCA, GINO M. CRISCI and DOMENICO MIRIELLO

Department of Biology, Ecology and Earth Sciences (DiBEST), Ponte P. Bucci 12b, University of Calabria, 87036 Arcavacata di Rende (CS), Italy
e-mail: d.barca@unical.it, crisci@unical.it, miriello@unical.it

Obsidian is a volcanic product that forms under particular geological conditions, and hence occurs in limited areas of the Earth. In ancient times, obsidian was used successfully by various peoples to produce artistic artefacts, but also to make tools and weapons used in everyday life. For this reason, obsidian was transported from geological sources to other locations. The study of methods used to identify the provenance of obsidian artefacts has become crucial for understanding commercial relations between distant ancient populations. Other volcanic products, generally associated with obsidian, are volcanic glass shards. Glass shards were used in Mexico as aggregates to produce plasters, and recent studies have shown that they were also transported along commercial routes. This chapter presents an introductory overview of the sources of obsidian in the Tyrrhenian area, showing how minor, trace and rare earth elements can be used to solve provenance problems. A case study regarding the provenance of glass shards inside archaeological plasters taken from Teopacazco (Teotihuacan, Mexico) is also presented.

1. Introduction

Obsidian is a volcanic glass, generally aphyric and chemically homogeneous, formed from a rapid solidification of rhyolitic magma. Because of its properties, such as fracture predictability and exceptional cutting-edge quality, obsidian was an important raw material extracted and processed to produce blades, razors, knives and weapons before the discovery of metallurgy. The importance of obsidian in human history is also testified by the diffusion of obsidian tools in most archaeological contexts. Each source of obsidian has a distinct chemical fingerprint, and obsidian artefacts are chemically stable on an archaeological time-scale; hence, the chemical characterization of obsidian artefacts is central to studies focused on long-distance trade and the development of market systems.

Although the major-element composition of obsidian from different origins may present some variability that could help to identify their geological provenance, trace-element concentrations, which vary strongly from source to source, allow a sound discrimination of sources and identification of obsidian raw material used for artefacts.

To characterize geochemically obsidian for archaeometric purposes, non-destructive analytical techniques are preferred, such as X-ray fluorescence (XRF) applied in a non-

DOI: 10.1180/EMU-notes.20.11

destructive way, proton-induced X-ray emission (PIXE), neutron activation analysis (NAA) and scanning electron microscopy (SEM). Among the various techniques applied by researchers for provenance studies, Laser Ablation-Inductively Coupled-Plasma Mass Spectrometry (LA-ICP-MS), a micro-destructive analytical methodology, has also proved to be a powerful tool for *in situ* determination of trace-element patterns in archaeological obsidian fragments.

Recent studies have also shown that other techniques, such as portable-XRF (pXRF) and Raman micro-spectroscopy, have have also been deployed for obsidian analyses. In particular, pXRF analysis combines non-destructivity with rapidity, and can be done also within museums (Craig *et al.*, 2007; Millhauser *et al.*, 2011; Forster and Grave, 2012; Tykot *et al.*, 2013). Raman micro-spectroscopy can also be considered to be a promising tool to determine the provenance of the obsidian (Bellot-Gurlet *et al.*, 2004; Carter *et al.*, 2009).

The aim of this chapter is to give a brief overview of the geochemical characteristics of obsidian sources in the Tyrrhenian area, focusing on the geochemical differences that can discriminate among source areas and establish their provenance for archaeometric purposes. In addition, the study will focus on the provenance of volcanic glass shards of rhyolitic composition (Barca *et al.*, 2013) which are used as an aggregate in the plasters of Teotihuacan (the so called "City of Gods"), the main city in Central Mexico during the Classic Period (200–600 AD). The comparison between the data from the analyses of the shards in the floor plasters and the data from all Mexican obsidian outcrops has allowed us to identify the magmatic systems exploited by Mesoamerican peoples.

Hence, this chapter will show that the study of provenance is a powerful tool for testing models regarding prehistoric trades, people's interactions and access to resources.

2. The origin and geochemical classification of obsidian

Obsidian is the name given to a glassy rock formed from a polymerized silicate melt that was cooled too rapidly for crystallization to occur (Best, 2003). Obsidian is usually dark in colour; however, because of the presence of impurities and the variable iron and magnesium contents, it can occur in different shades of colour varying from grey-green to black. Rare obsidian samples show a mixture of deep purple and black.

Obsidian composition is extremely felsic, usually with 70% or more of SiO_2. The oxide percentages of other major element (Al_2O_3, Na_2O, K_2O and CaO) and the trace- and rare earth -element (*REE*) contents vary significantly among the various geological sources; for this reason, the chemical characterization of obsidian can be used to discriminate among them and is particularly productive in the provenance research field (Cann and Renfrew, 1964; Crisci *et al.*, 1994; Tykot, 1997; Acquafredda *et al.*, 1999).

The most common analytical methods employed to characterize obsidian are partially destructive and, thus, cannot always be used for archaeometric research. Even non-destructive methods such as XRF applied in a non-destructive way, PIXE, NAA and SEM have various limitations related to the shape of the surface, the size or the weight of the object studied, and it is sometimes impossible to determine the concentrations of all the elements which are useful for identifying the various obsidian sources for archaeometric purposes. The LA-ICP-MS technique combines micro-destructivity with the capacity to analyse a great number of trace elements and *REE* with high sensitivity in a very short time. The only limitation is that the sample size must be compatible with the ablation cell (Barca *et al.*, 2007, 2012).

Although natural, massive, high-silica glass (or obsidian) in hand specimen often appears to have zero crystallinity, few natural glasses do not contain any crystals at all (Best, 2003). Indeed, high magnification almost always reveals that obsidian contains crystallites. The nature of these microphenocrysts can also be used to discriminate obsidians from different sources (Acquafredda *et al.*, 1999).

3. Distribution and geochemical characteristics of obsidian sources in the Mediterranean area (the peri-Tyrrhenian)

In the Western Mediterranean, only four islands are known to have obsidian sources likely to have been exploited in Neolithic times: Lipari, Palmarola, Pantelleria and Sardinia (Fig. 1) (Barca *et al.*, 2007, 2008; De Francesco *et al.*, 2008; Tykot, 1997, 2002).

3.1. Lipari

With an area of 37 km^2, Lipari is the largest of the Aeolian Islands (Aeolian Archipelago, Southern Tyrrhenian Sea). The subaerial volcanic activity on Lipari started 223 ka ago with the emission of mafic products (Gillot and Villari, 1980), and ended in 1230±40 AD with the emplacement of the Rocche Rosse rhyolitic lava flow (Tanguy *et al.*, 2003). Rhyolites are characteristic of volcanic events younger than 42 ka. The early rhyolitic subaphyric products belong to the Valle Muria Synthem (42–20 ka; Tranne *et al.*, 2002). The renewal of the activity is dated at 10 ka, with pyroclastic surges and obsidian lava flows of the Vallone Fiume Bianco Synthem. During historical times, an unusual Strombolian-type explosive activity formed the Monte Pilato pumice cone, followed, in 1230 AD, by the Rocche Rosse eruption.

The early rhyolitic products cover large areas in the southern part of the island, whereas those erupted during the youngest phase of activity dated at 10 ka cover almost the whole northeastern part of Lipari. Considering the age of the rhyolitic products, the obsidians erupted in historical times (ForgiaVecchia and Rocche Rosse) were not available for exploitation in prehistoric time. For this reason, only older obsidian may have provided raw material for Neolithic artefacts (Buchner, 1949; Pichler, 1980; Bigazzi and Bonadonna, 1973; Cortese *et al.*, 1986; Gillot and Cornette, 1986; Lefèvre and Gillot, 1994). In addition, analyses conducted by fission-track on 66 archaeological obsidian samples indicate that Vallone Gabellotto (Fiume Bianco) was the main outcrop exploited in the past (Arias-Radi *et al.*, 1972; Bigazzi and Radi, 1981; Arias *et*

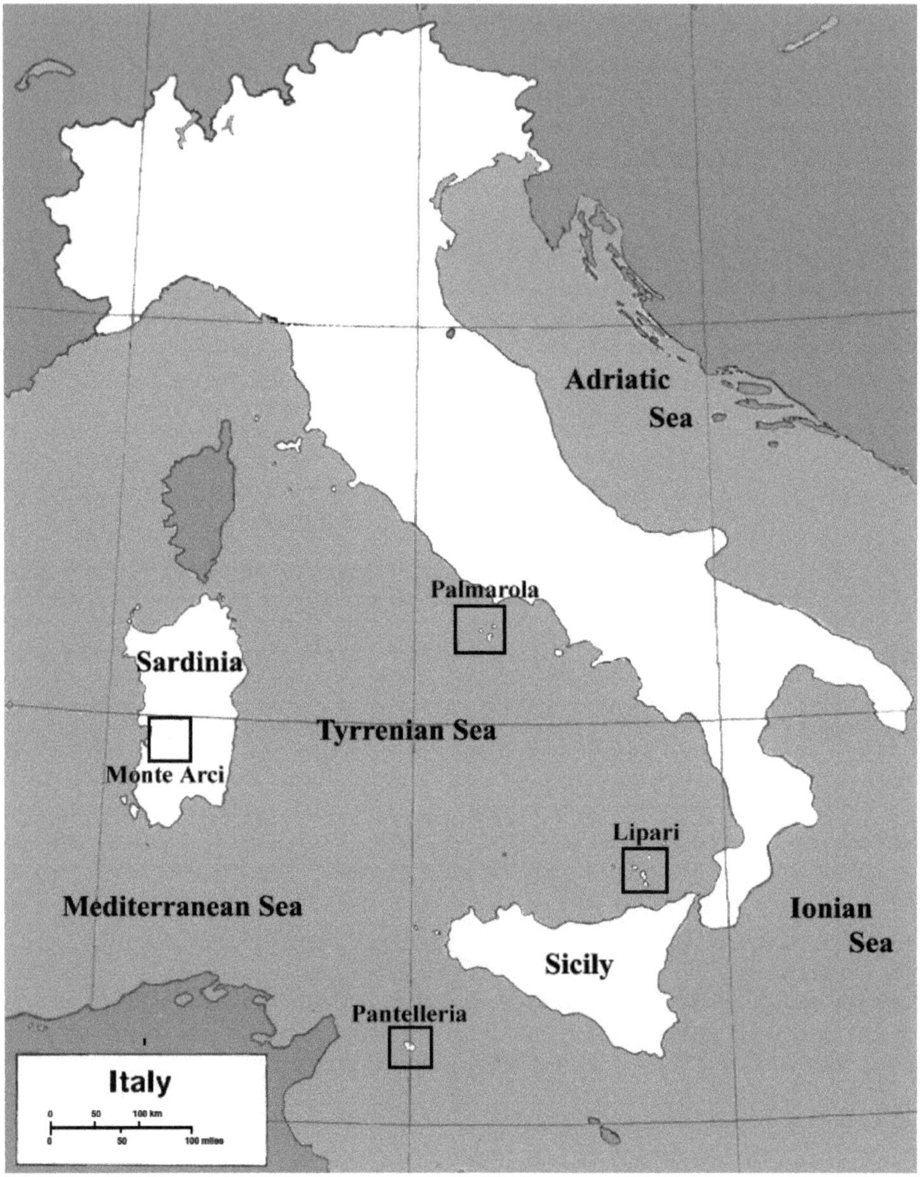

Figure 1. Locations of obsidian outcrops (squares) in the peri-Tyrrhenian area.

al., 1986; Tykot *et al.*, 2013). In addition, a recent study published by Forni *et al.* (2013) on the geological evolution of the Lipari volcanic complex, examines the various eruptions that produced obsidian, and highlighted the importance of the outcrops at Gabellotto and Canneto Dentro.

The Lipari obsidian has a composition varying from alkali-rhyolite to rhyolite with calcalkaline affinity; these geochemical characteristics led scientists to correlate Aeolian volcanism with the subduction of the Ionian Plate below the Calabrian Arc (Beccaluva *et al.*, 1982; Ellam *et al.*, 1989). Nevertheless, recent works have suggested that the geochemical evolution of the magmas could reflect the transition from collisional volcanism to post-collisional/rifting volcanism (Crisci *et al.*, 1991; De Astis *et al.*, 2003).

Lipari obsidian is of excellent quality, black in colour, very shiny and sometimes perlitic. It is generally subaphyric and contains clinopyroxene micro-phenocrysts (10–50 m), almost always in synneusis with magnetite and olivine micro-phenocrysts (Acquafredda *et al.*, 1999).

3.2. Sardinia: Monte Arci

In Sardinia, obsidian crops out only at Mt Arci, a volcanic complex covering ~30 km^2 and located in the hinterland of the gulf of Oristano. The volcanic activity developed during two distinct cycles in the Pliocene and Pleistocene, and therefore it belongs to the latest volcanism in Sardinia. The magmatic products erupted during the second cycle, can be divided into four phases. The lavas erupted in the first phase were very rich in silica, and consisted mainly of rhyolites, either massive or perlitic-obsidianaceous. Dacites and andesites, trachytes and trachyrhyolites followed in chronological succession, until the last stages of volcanic activity, which were characterized by quiet eruptions of basaltic lava flows (Piras, 2002; Bigazzi *et al.*, 2005).

The obsidian outcrops were described for the first time in the 19th century by De La Marmora (1839–40). Subsequently, in the 1980s, several independent studies contributed to the characterization of the multiple Monte Arci obsidian outcrops. Unfortunately, the results of these studies are available only in brief conference papers (Francaviglia, 1986; Mackey and Warren 1983) or in an unpublished dissertation (Herold, 1986).

More recently, considering the geochemical features the obsidian of Monte Arci, Tykot (2002) subdivided them into four groups, SA, SC, SB1 and SB2. The obsidian sampled near Conca Cannas and Uras are clustered within the SA group; those sampled near Pau, Perdas Urias and Sonnixeddu belong to the SC group; those sampled at Santa Maria Zuarbara and Marrubiu are in the SB1 and SB2 groups, respectively (Tykot, 2002; Barca *et al.*, 2007; De Francesco *et al.*, 2008). Due to their geochemical similarity, SB1 and SB2 have been grouped under the name SB in the present study.

Monte Arci obsidian populations are characterized by large biotite micro-phenocrysts (50–200 μm), abundant crystals of feldspar (plagioclase and alkali feldspar) 50 μm in size, orthopyroxene, magnetite, monazite and ilmenite (Acquafredda *et al.*, 1999).

3.3. Palmarola

Palmarola is a small volcanic island (<3 km^2 in area) located in the western part of the Pontine Islands Archipelago in the Gaeta Gulf. Its volcanic activity can be related to the stretching of the Tyrrhenian Sea (De Rita *et al.*, 1989; Acquafredda *et al.*, 1999). Obsidian crops out in the southern part of Mt Tramontana, as a domal crust crossing the island from Il Porto to Scoglio Spermatore and on the eastern coast. In addition, abundant secondary obsidian deposits have been identified on the southeastern edge of the island at Punta Vardella (Buchner, 1949; Herold, 1986).

Geochronological data, obtained by fission-track analyses, indicate an age of 1.7±0.3 Ma for obsidian from Monte Tramontana (Bigazzi *et al.*, 1971; Bigazzi and Radi, 1981). In a recent work Tykot *et al.* (2005) provided a detailed geochemical study of 80 samples, and were able to distinguish three source localities: Punta Vardella, the northern end of Punta Vardella and Monte Tramontana. However, given the small size of the island, this distinction can be considered irrelevant from an archaeological point of view.

Palmarola obsidian contains micro-phenocrysts of clinopyroxene (5–20 μm) and biotite. It is generally black in colour, glassy, poorly shiny and semi-opaque (Acquafredda *et al.*, 1999). However, a small amount of highly transparent obsidian was found at Punta Vardella by Tykot *et al.* (2005).

3.4 Pantelleria

Pantelleria, a small island of ~83 km^2, is located in the NW–SE part of the Sicily Channel Rift Zone (SCRZ), ~90 km east of Cape Bon, Tunisia. Pantelleria is famous for its peralkaline rocks, and especially for its greenish obsidian enriched in sodium and iron, known as Pantellerite (Civetta *et al.*, 1998; Acquafredda *et al.*, 1999). Pantelleria has a bimodal distribution of magmatic products. Mafic lavas, exposed in the NW corner of the island, include transitional basalt and hawaiite (from ~46 to 49 wt.% SiO_2). Felsic lavas and tuffs, which include metaluminous trachyte, peralkaline trachyte and pantellerite (from ~62 to 72 wt.% SiO_2), prevail in the SE sector (White *et al.*, 2009). K–Ar determinations of mafic lavas done on different basaltic units give ages of 118 ± 9, 83 ± 5 and ~29 ka BP (Civetta *et al.*, 1984). Ages determined on felsic volcanic rocks range from 324 ka BP to 4 ka BP (Civetta *et al.*, 1984, 1988, 1998; Mahood and Hildreth, 1986). The volcanic history of the island is characterized by large explosive eruptions, some of which produced caldera collapses, alternating with periods dominated by less energetic eruptions (Civetta *et al.*, 1998). The oldest caldera, named La Vecchia, is dated at 114 ka BP (Mahood and Hildreth, 1986); the youngest caldera, named the Monastero caldera by Cornette *et al.* (1983) and the Cinque Denti caldera by Mahood and Hildreth (1983), is related to the eruption of the Green Tuff (50 ka BP; Orsi and Sheridan, 1984). The more recent (post-50 ka) history of the island has been subdivided by Civetta *et al.* (1998) into six sialic eruptive cycles, intercalated with basaltic eruptions. The Green Tuff is considered the first of these six cycles. All the others are dated at around 35–29, 22, 20–15, 14–12 and 10–4 ka BP, respectively (Civetta *et al.*, 1998).

Francaviglia (1988) distinguished five groups of Pantelleria obsidian, which are characterized by different chemical compositions: three of the groups are from "vertically differentiated" mines exposed at Balata dei Turchi; the other two are in Gelkhamar and Lago di Venere. Chemical analyses of artefacts from Pantelleria, Malta and Sicily showed that the upper Balata dei Turchi mine (*i.e.* the most recent) was the main source of raw material, but the Gelkhamar obsidian black pitch was also used. Finally, the lower (*i.e.* at sea level) obsidian from Balata dei Turchi was not used for tool-making, at least during the Bronze Age (Francaviglia, 1988; Francaviglia and Piperno, 1987; Tykot, 1995, 1996; Tykot *et al.*, 2013).

Obsidian from Pantelleria is grey-greenish in colour, mildly shiny and transparent and easily distinguishable from other European obsidian by its peculiar colour. Samples are generally poor in micro-phenocrysts, which comprise relatively large crystals of anorthoclase (50–150 μm) and apatite; clinopyroxene, magnetite and olivine are also present (Acquafredda *et al.*, 1999).

4. Provenance studies of archaeological obsidian in Italian Neolithic settlements

As stated peviously, obsidians can be distinguished on the basis of their chemical compositions. A large spectrum of techniques is generally used in order to determine such compositions and a great number of elements is determined ranging from major to minor, trace and rare earth.

Figure 2 shows the results obtained by Acquafredda *et al.* (1999), De Francesco *et al.* (2008) and Tykot (2002) on obsidian samples from different localities in the Tyrrhenian Sea; they are plotted on an alkalies *vs.* silica diagram (Le Bas *et al.*, 1992). Lipari, Palmarola and Monte Arci (the SA, SB and SC groups, respectively) show similar silica and alkali contents; Pantelleria is characterized by higher alkali and lower silica contents than the other sources.

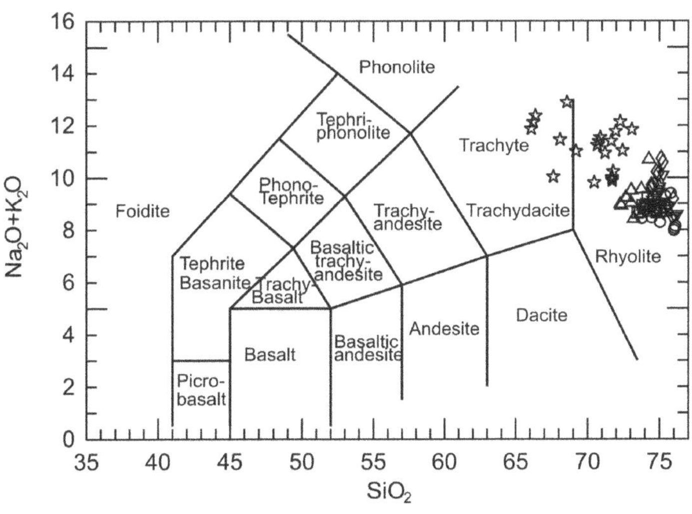

Figure 2. Total alkalies-silica diagram (Le Bas *et al.*, 1992) of obsidian cropping out around the Tyrrhenian Sea. Data from Acquafredda *et al.* (1999), De Francesco *et al.* (2008) and Tykot (2002) are also plotted. Inverted triangles: Lipari; diamonds: Palmarola; stars: Pantelleria; circles: SA Mt Arci; squares: SB Mt Arci; triangles: SC t Arci.

In contrast, trace- and rare-earth-element compositions are extremely discriminatory and allow a precise separation of the sources. The binary diagram Zr/Y *vs*. La (Fig. 3) can be used to distinguish the six sources accurately. The diagram was constructed from data collected by Barca *et al.* (2007) and Tykot (2002), using different analytical techniques. Obsidians from Lipari and Palmarola show compositional homogeneity within their respective groups; obsidians from Pantelleria and from the other three groups belonging to the Monte Arci complex in Sardinia show a major dispersion of the data, but they are easily differentiated from one another, with significant variations among trace-element concentrations. A good discrimination of all geological sources (Lipari, Palmarola and Monte Arci SA, SB and SC groups) can also be obtained using the Aitchison approach to the geostatistical analysis of compositional data (Aitchison, 1982, 1983, 1986), by transforming the data in centred-log-ratio (clr). An example is shown in Fig. 4, where a three-dimensional scatterplot of the clr concentrations of Ce, Ba and Sr is plotted.

5. The obsidian sources in Mesoamerica

The Mesoamerican region is characterized by numerous obsidian-rich volcanoes. In detail, the area is dominated by the presence of two important volcanic provinces: the volcanic succession of the Sierra Madre Occidental (SMO) and the Trans-Mexican Volcanic Belt (TMVB) (Ferrari *et al.*, 1999, 2005). The SMO is an ~1000 m thick succession, dominated by silicic ash flows, emplaced during the Eocene, Oligocene and Early Miocene.

The TMVB is a volcanic arc built on the southern edge of the North American plate. It runs east-west across Mexico (Ferrari *et al.*, 1999) and was formed in response to subduction of the Cocos plate along the Acapulco trench beginning in the Middle Miocene (Ferrari *et al.*, 1999). The TMVB volcanism is characterized by a wide range

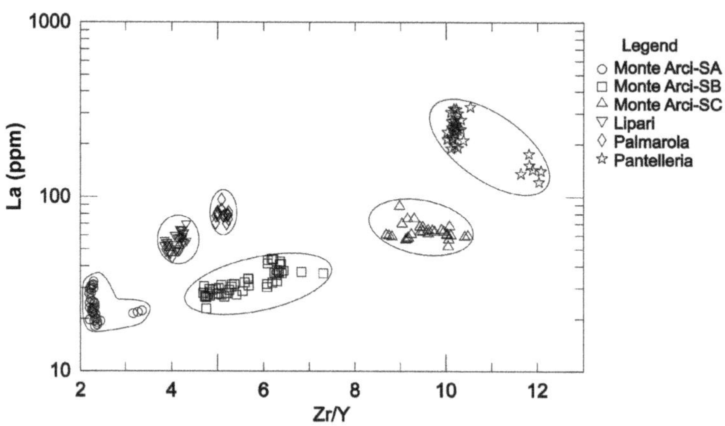

Figure 3. Zr/Y *vs*. La binary diagram for obsidian cropping out around the Tyrrhenian Sea. The diagram was drawn using data provided by Barca *et al.* (2007) and Tykot (2002).

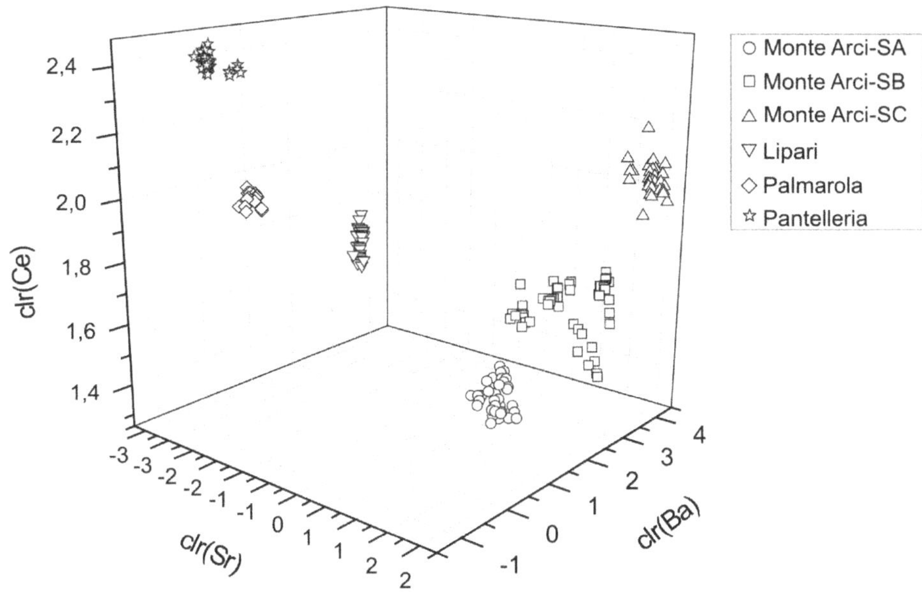

Figure 4. Three-dimensional scatterplot of the clr of the concentrations of Ce, Ba and Sr. The diagram was drawn using data provided by Barca *et al.* (2007) and Tykot (2002).

of chemical compositions. The emplacement of large volumes of rhyolites and minor ignimbrites occurred during the second magmatic episode between 8 and 5 Ma (Ferrari *et al.*, 1999).

The complexity of the area is far greater in eastern Mexico; in Central Veracruz, products of alkaline volcanism (called the Eastern Alkaline Province) alternate with arc-related lavas (Ferrari *et al.*, 2005). The extinct late-Pliocene peralkaline volcanic centre of Sierra de Pachuca crops out at the northern edge of the eastern TMVB (Lighthart Ponomanenko, 2004).

The rhyolitic components are interesting from an archaeological point of view, because they are often associated with obsidian lavas, exploited extensively by the Pre-Hispanic inhabitants of Mesoamerica.

Considering the rarity of obsidian in the world, the abundance of source zones in a relatively restricted area is of great interest from both geological and archaeological perspectives. In particular, the identification of the major obsidian sources in Central Mexico and their geochemical characterization has provided a database to be used for archaeological specimens (Cobean *et al.*, 1991; Cobean, 2002; Carballo *et al.*, 2007; Gazzola *et al.*, 2010; Barca *et al.*, 2013).

Using Instrumental Neutron Activation Analysis (INAA), Cobean *et al.* (1991) provided high-precision trace-element analyses of 208 geological samples collected from 25 Mesoamerican obsidian sources. Based on their geographical distribution, the Mexican sources can be divided into two groups: (1) sources cropping out on Mexico's

Gulf Coast (Pico de Orizaba Veracruz, Altotonga Veracruz, Guadalupe Victoria Puebla, Derrumbadas Puebla, Zaragoza Puebla and Oyameles Puebla); and (2) sources cropping out in Central Mexico (Sierra de Pachuca Hidalgo, Tulancingo and Rancho Tenango Hidalgo, Paredon, Zacualtipan, Otumba State of Mexico, Zinapécuaro and Ucaréo Michoacan, Teuchitlan, Tequila and Magdalena Jalisco, Penjamo Guanajuato, El Paraiso and Fuentezuelas Querétaro; Fig. 5).

The abundance of outcrops in the area, as well as visual attractiveness and physical properties, made obsidian the raw material preferred by Mesoamerican inhabitants, who used it to make cutting tools such as knives, arrowheads and scrapers.

Nowadays, the different trace-element compositions observed in the different sources has allowed the reconstruction of trade routes and cultural exchange networks along which the materials and artefacts were transported (Cobean *et al.*, 1991; Cobean, 2002; Carballo *et al.*, 2007; Barca *et al.*, 2013). In particular, an interesting study on obsidian procurement in Central Mexico has been conducted by Carballo *et al.* (2007). Firstly, these authors determined the chemical composition of the majority of Mexican obsidian sources using LA-ICP-MS. Secondly, they characterized the obsidian finds from four archaeological sites, and assigned a provenance to each sample. Among the

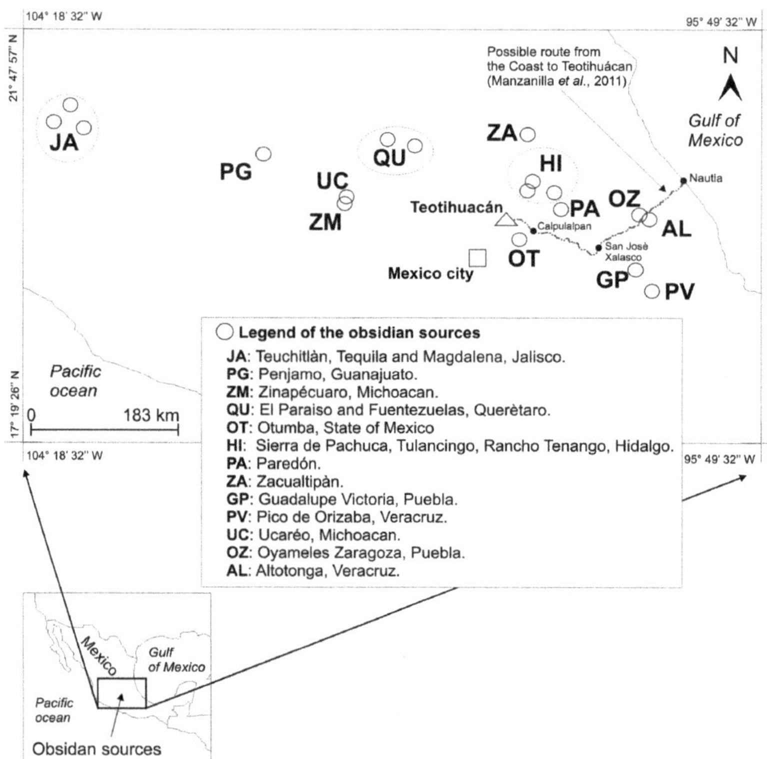

Figure 5. Map of the obsidian outcrops in Mexico (modified after Barca *et al.*, 2013).

archaeological sites studied, three were rural villages of a few hectares, known as Amomoloc (900–600 BC), Tetel (700–400 BC) and Las Mesitas (500–350 BC), which are located in the modern state of Tlaxcala. The fourth is Teotihuacan, the most important city during the Classic Period (AD 350–550), which is located 45 km NE of Mexico City.

The study demonstrates a variable exploitation of the sources among the villages. It was extremely flexible at Amomoloc, including five or six quarries in both Central and Oriental source regions. The inhabitants of Tetel also collected obsidian from five different sources, but their procurement was focused on the source of Paredon. In the site of Las Mesitas, the exploitation of Central sources was conducted similarly, and Paredon was the preferred source. Finally, in Teotihuacan, 95% of obsidian was acquired from sources in the Central area and the favourite sources were Otumba, Pachuca and Tulancingo (Carballo *et al.*, 2007). A further study conducted by Gazzola *et al.* (2010) on the use of obsidian in the pre-Hispanic city of Teotihuacan, confirmed that Teotihuacan controlled the two largest obsidian sources in Central Mexico: Otumba, which was located 10 km from the city and provided grey obsidian and Sierra de Pachuca (also known as Sierra de las Navajas), which was located 50 km from the city and was famous for green obsidian (Fig. 6).

6. Obsidian as a raw material in Mesoamerica – a particular case: The use of volcanic ash as an aggregate in plasters

Research carried out on Teotihuacan have shown that obsidian raw materials and handmade products were valuable goods. Their circulation represented the main aspect of the economic and social development of the city. In addition, obsidian was a central

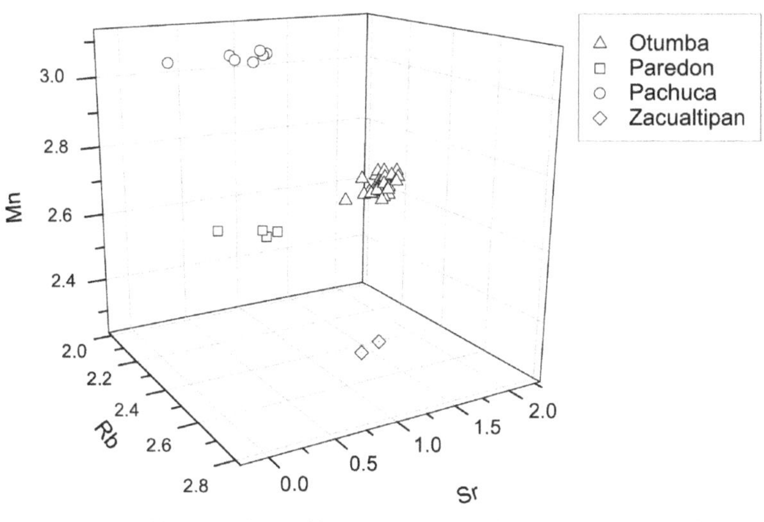

Figure 6. 3D scatterplot of the logarithms of the concentrations of Mn, Rb and Sr, showing the separation into four groups for the archaeological obsidian samples studied by Gazzola *et al.* (2010). Modified.

component in political rituals and it was essential for the arming of the city's military forces (Carballo, 2011).

Recent studies have shown also that people from Teotihuacan appreciated other types of volcanic material. Indeed, the study carried out in the courtyard of Teopancazco, a neighbourhood centre of Teotihuacan, has shown that volcanic glass shards of rhyolitic composition were used to produce plasters (Barba *et al.*, 2009; Murakami, 2010; Barca *et al.*, 2013).

Aiming to identify the provenance of glass shards used as an aggregate in plasters, Barca *et al.* (2013) compared their composition with those of obsidian from various sources. This comparison was justified by the authors by assuming that the same magma that produced obsidian lava flows could also generate pyroclastic deposits, rich in glass shards, during an explosive eruption. In order to obtain accurate analyses on micrometric glass shards within the plasters, two analytical techniques were used: SEM-EDS (scanning electron microscopy coupled with energy-dispersive X-ray spectrometry) and LA-ICP-MS. As stated above, Carballo *et al.* (2007), Gazzola *et al.* (2010), Pastrana (1998, 2010) and Spence (1981) identified the sources of Otumba and Sierra de Pachuca as those preferred for Teotihuacanos. Based on these previous studies, Barca *et al.* (2013) had hypothesized that the source material for glass shards was probably related either to the magmatic system of Otumba or Sierra de Pachuca. Nevertheless, comparison of the results obtained from the glass shards with the database of the geological obsidian sources, showed that the glass materials were collected from the magmatic system of Altotonga (Veracruz) although it is located 180 km. away from Teotihuacan along the route to the Gulf Coast (Fig. 5). In fact, the plot of Ba *vs.* Mn obtained by combining the analyses of the shards and bibliographic data from all Mexican obsidian outcrops (Fig. 7) shows clearly that the shards in the plaster samples are consistent with Altotonga (Veracruz). The authors propose two possible reasons to explain this choice. The first is technological: people from Teotihuacan appreciated the good quality of the Altotonga glass shards for plaster production. The second is, perhaps, symbolic: the inhabitants of Teopancazco wanted to build their compound with materials coming from the route they had followed from the Gulf Coast to Teotihuacan, tracing their identity to that distant land (Barca *et al.*, 2013).

7. Reconstruction of an ancient route using obsidian

The acquisition of raw materials implies an accurate knowledge of the territory, and the movement needed to obtain them must be part of a network of routes also used for the acquisition of other resources required by the group. In the particular case of Teotihuacan, the results obtained by Gazzola *et al.* (2010) and by Carballo *et al.* (2007) support the idea that most of the obsidian artefacts found in that site were made of obsidian derived from the Otumba and Sierra de Pachuca sources. However, the results also indicate that some obsidian was obtained from other sources, such as Paredón (Puebla). Considering the presence of obsidian finds from other sources located on the way to Veracruz, Gazzola *et al.* (2010) suggested an early relationship with other

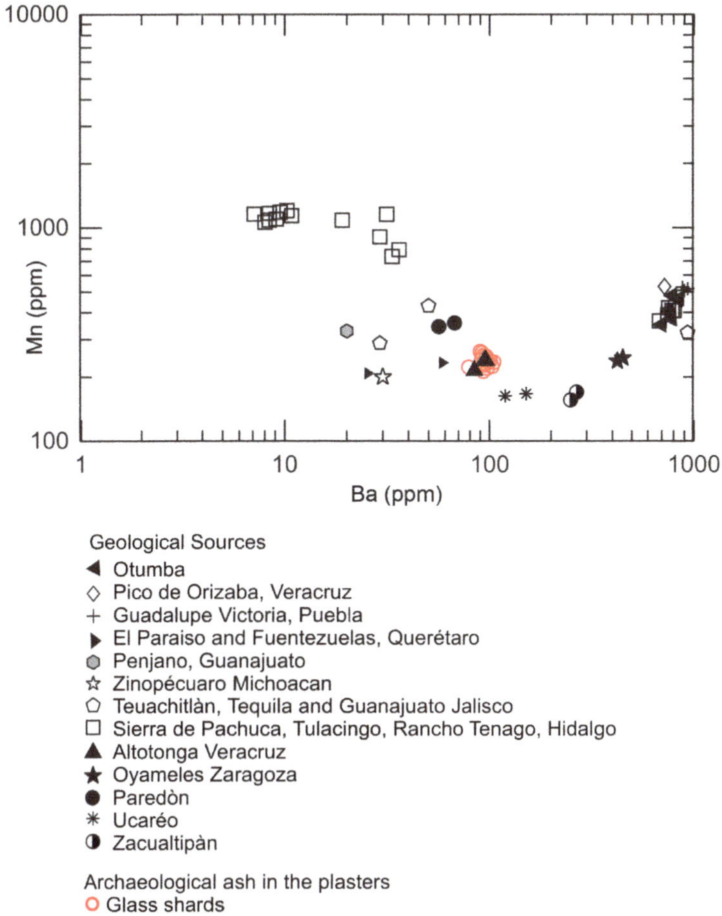

Figure. 7. Binary diagram Ba *vs.* Mn used to discriminate between the Mexican geological sources of obsidian. The diagram compares the data published by Carballo *et al.* (2007), Cobean *et al.* (1991), Cobean (2002) and Barca *et al.* (2013). The diagram shows that Altotonga (Veracruz) is the source of the glass shards within the plasters collected in Teotihuacan (modified after Barca *et al.*, 2013).

communities inhabiting the Central Plateau, and an exchange of resources among these regions. In addition, they hypothesize the control exerted by Teotihuacanos on other sources, such as Zacualtipan and Paredón.

The evidence from the Teopancazco excavations indicates that there was an important link between Teopancazco and the Gulf Coast of Mexico (Manzanilla, 2011). Many archaeological materials recovered at the site, such as ornaments and garments, were produced with feathers, shells and other materials brought from the Gulf Coast. In addition, isotopic studies of human skeletons recovered at Teopancazco indicate the presence of people from the Gulf area, and from sites located along the corridor running from Teotihuacan to the coast (Manzanilla, 2012).

In general, these researches have revealed the close relationship between communities from Teopancazco and the Gulf Coast. In particular, the results obtained by Barca *et al.* (2013) suggested that people travelling from the coast to Teopancazco took glass shards from an Altotonga outcrop with them. The reason for using this material is still unknown. It is suggested that either they were aware of its properties or they used the material for symbolic reasons.

Acknowledgements

The authors are grateful to L. Barba, J. Blancas, R. De Luca, L. Manzanilla, A. Ortiz, A. Pastrana and A. Pecci whose constructive comments and suggestions contributed to the quality of the manuscript.

References

Acquafredda, P., Andriani, T., Lorenzoni, S. and Zanettin, E. (1999) Chemical characterization of obsidians from different Mediterranean sources by non-destructive SEM-EDS analytical method. *Journal of Archaeological Science*, **26**, 315–325.

Arias, C., Bigazzi, G., Bonadonna, F.P., Cipolloni, M., Hadler, J.C., Lattes, G.M.C. and Radi, G. (1986) Fission track dating in archeology. A useful application. Pp. 151–159 in: *Scientific Methodologies Applied to Works of Arts* (PL Parrini, editor). Montedison Progetto Cultura. Plenum Press, New York and London.

Arias-Radi, G., Bigazzi, G. and Bonadonna, F.P. (1972) Le tracce di fissione. Un metodo per lo studio delle vie di commercio dell'ossidiana. *Origini*, **6**, 155–170.

Aitchison, J. (1982) The statistical analysis of compositional data (with discussion). *Journal of the Royal Statistical Society, Series B (Statistical Methodology)*, **44**, 139–177.

Aitchison, J. (1983) The principal component analysis of compositional data. *Biometrika*, **70**, 57–65.

Aitchison, J. (1986) The statistical analysis of compositional data. In: *Monographs on Statistics and Applied Probability*. Chapman & Hall Ltd, London Reprinted (2003) with Additional Material by The Blackburn Press, Caldwell, New Jersey, USA.

Barba, L., Blancas, J., Ortiz, A., Manzanilla, L., Barca, D., Crisci, G.M., Miriello, D. and Pecci, A. (2009) Provenance of the limestone used in Teotihuacan (Mexico). A methodological approach. *Archaeometry*, **51**, 525–545.

Barca, D., De Francesco, A.M. and Crisci, G.M. (2007) Application of Laser Ablation–ICP-MS for characterization of obsidian fragments from peri-Tyrrhenian area. *Journal of Cultural Heritage*, **8**, 141–150.

Barca, D., De Francesco, A.M., Crisci, G.M. and Tozzi, C. (2008) Provenance of obsidian artefacts from site of Colle Cera, Italy, by LA-ICP-MS method. *Periodico di Mineralogia*, **77**, 41–52.

Barca, D., Lucarini, G. and Fedele, F. (2012) Provenance of obsidian artefacts from the Wādī Ath-Thayyilah 3 Neolithic site (eastern Yemen plateau) by LA-ICP-MS. *Archaeometry* **54**, 603–622.

Barca, D., Miriello, D., Pecci, A., Barba, L., Ortiz, A., Manzanilla, L.R. and Crisci, G.M. (2013) Provenance of glass shards in archaeological lime plasters by LA-ICP-MS: implications for the ancient routes from the Gulf of Mexico to Teotihuacan in Central Mexico. *Journal of Archaeological Science*, **40**, 3999–4008.

Beccaluva, L., Rossi, P.L. and Serri, G. (1982) Neogene to recent volcanism of the southern Tyrrhenian–Sicilian area: implications for the geodynamic evolution of the Calabrian Arc. *Earth Evolution Science*, **3**, 222–238.

Bellot-Gurlet, L., Le Bordonnec, F.X., Poupeau, G. and Durbenet, S. (2004) Raman micro-spectroscopy of western Mediterranean obsidian glass: one step towards provenance studies? *Journal of Raman Spectroscopy*, **35**, 671–677.

Best, M.G. (2003) *Igneous and Metamorphic Petrology*. Blackwell Science Ltd., Oxford, UK.

Bigazzi, G. and Bonadonna, F.P. (1973) Fission track dating of the obsidian of Lipari Island (Italy). *Nature*, **242**, 322–323.

Bigazzi, G. and Radi, G. (1981) Datazioni con le tracce di fissione per l'identificazione della provenienza dei manufatti di ossidiana. *Rivista di Scienze Preistoriche*, **36**, 223-250.

Bigazzi, G., Bonadonna, F.P., Belluomini, G. and Malpieri, L. (1971) Studi sulle ossidiane italiane IV. Datazione con il metodo delle tracce i fissione. *Bollettino della Società Geologica Italiana*, **90**, 469–480.

Bigazzi G., Oddone M. and Radi G. (2005) The Italian obsidian sources. *Archeometriai Muhely*, **1**, 1–13.

Buchner, G. (1949) Ricerche sui giacimenti e sulle industrie di ossidiana in Italia. *Rivista di Scienze Preistoriche*, **4**, 162–186.

Cann, J.R. and Renfrew, C. (1964) The characterization of obsidian and its application to the Mediterranean region. *Proceedings of the Prehistoric Society*, **30**, 111–133.

Carballo, D. (2011) Obsidian and the Teotihuacan State. Weaponry and Ritual Production at the Moon Pyramid. *University of Pittsburgh Memoirs in Latin American Archaeology*, **21**, 198.

Carballo, D.M., Carballo, J. and Neff, H. (2007) Formative and classic period obsidian procurement in central Mexico: a compositional study using laser ablation-inductively coupled plasma-mass spectrometry. *Latin American Antiquity*, **18**, 27–43.

Carter, E.A., Hargreaves, M.D., Kononenko, N., Graham, I., Edwards, H.G.M., Swarbrick, B. and Torrence, R. (2009) Raman spectroscopy applied to understanding Prehistoric Obsidian Trade in the Pacific Region. *Vibrational Spectroscopy*, **50**,116–124.

Civetta, L., Cornette, Y., Crisci, G., Gillot, P.-Y., Orsi, G. and Requejos, C.S. (1984) Geology, geochronology and chemical evolution of the island of Pantelleria. *Geological Magazine*, **121**, 541–668.

Civetta, L., Cornette, Y., Gillot, P.-Y. and Orsi, G. (1988) The eruptive history of Pantelleria (Sicily Channel) in the last 50 ka. *Bulletin of Volcanology*, **50**, 47–57.

Civetta, L., D'Antonio, M., Orsi, G. and Tilton, G. (1998) The geochemistry of volcanic rocks from Pantelleria Island, Sicily Channel: Petrogenesis and characteristics of the mantle source region. *Journal of Petrology*, **39**, 1453–1491.

Cobean, R.H. (2002) *A World of Obsidian: The Mining and Trade of a Volcanic Glass in Ancient Mexico*. University of Pittsburgh and Instituto Nacional de Antropología e Historia, Mexico City.

Cobean, R.H., Vogt, J.R., Glascock, M.D. and Stocker, T.L. (1991) High precision trace element characterization of major Mesoamerican obsidian sources and further analyses of artefacts from San Lorenzo Tenochtitlan, Mexico. *Latin American Antiquity*, **2**, 69–91.

Cornette, Y., Crisci, G.M., Gillot ,P.Y. and Orsi, G. (1983) Recent volcanic history of Pantelleria: a new interpretation. *Journal of Volcanology and Geothermal Research*, **17**, 361–373.

Cortese, M., Frazzetta, G. and La Volpe, L. (1986) Volcanic history of Lipari (Aeolian Islands, Italy) during the last 10,000 years. *Journal of Volcanology and Geothermal Research*, **27**, 117–133.

Craig, N., Speakman, R.J., Popelka-Filcoff, R.S., Glascock, M.D., Robertson, J.D., Shackley, M.S. and Aldenderfer, M.S. (2007) Comparison of XRF and PXRF for analysis of archaeological obsidian from southern Peru. *Journal of Archaeological Science*, **34**, 2012–2014.

Crisci, G.M., De Rosa, R., Esperanca, S., Mazzuoli, R. and Sonnino, M. (1991) Temporal evolution of a three component system: the Island of Lipari (Aeolian arc, southern Italy). *Bulletin of Volcano*logy, **53**, 207–221.

Crisci, G.M., Ricq-De Bouard, M., Lanzafame, U. and De Francesco, A.M. (1994) Nouvelle méthode d'analyse et provenance de l'ensemble des obsidiennes néolithiques du Midi de la France. *Gallia Préhistoire*, **36**, 299–327.

De Astis, G., Ventura, G. and Vilardo, G. (2003) Geodynamic significance of the Aeolian volcanism (southern Tyrrhenian Sea, Italy) in light of structural, seismological and geochemical data. *Tectonics*, **22(4)**, 1040. doi:10.1029/2003TC001506.

De Francesco, A.M., Crisci, G.M. and Bocci, M. (2008) Non-destructive analytic method using XRF for determination of provenance of archaeological obsidians from the Mediterranean area: a comparison with traditional XRF methods. *Archaeometry*, **50**, 337–350.

De La Marmora, A. (1839-40) *Voyage en Sardaigne*. Turin J. Bocca.

De Rita, D., Funiciello, R., Pantosti, A. and Velonà, M. (1989) Caratteristiche geologico-strutturali delle isole Pontine nord-occidentali. Pp. 43–64 in: *Atti Convegno* "Incontro con la Geologia Circeo, Sabaudia 1984".

Ellam, R.M., Hawkesworth, C.J., Menzies, M.A. and Rogers, N.W. (1989) The volcanism of southern Italy: role of subduction and the relationship between potassic and sodic alkaline magmatism. *Journal of Geophysics Research*, **94**, 4589–4601.

Ferrari, L., López-Martínez, M., Aguirre-Díaz, G. and Carrasco-Núñez, G. (1999) Space–time patterns of Cenozoic arc volcanism in central Mexico: from the Sierra Madre Occidental to the Mexican Volcanic Belt. *Geology*, **27**, 303–307.

Ferrari, L., Tagami, T., Eguchi, M., Orozco-Esquivela, M.T., Petrone, C.M., Jacobo-Albarrán, J. and López-Martínez, M. (2005) Geology, geochronology and tectonic setting of late Cenozoic volcanism along the southwestern Gulf of Mexico: The Eastern Alkaline Province revisited. *Journal of Volcanology and Geothermal Research*, **146**, 284–306.

Forni, F., Lucchi, F., Peccerillo, A., Tranne, C.A., Rossi, P.L. and Frezzotti, M.L. (2013) Stratigraphy and geological evolution of the Lipari volcanic complex (central Aeolian archipelago). Pp. 213–279 in: *The Aeolian Islands Volcanoes* (F. Lucchi, A. Peccerillo, J. Keller, C.A. Tranne and P.L. Rossi, editors). Memoir **37**. Geological Society, London.

Forster, N. and Grave, P. (2012) Non-destructive PXRF analysis of museum-curated obsidian from the Near East. *Journal of Archaeological Science*, **39**, 728–736.

Francaviglia, V. (1986) Characterization of Mediterranean obsidian sources by classical petrochemical methods. *Preistoria Alpina-Museo Tridentino di Scienze Naturali*, **20**, 311–332.

Francaviglia, V. (1988) Ancient obsidian sources on Pantelleria Italy. *Journal of Archaeological Science*, **15**, 109–122.

Francaviglia, V. and Piperno, M. (1987) La repartition e la provenance de l'obsidienne archeologique de la Grotta dell'Uzzo et de Monte Cofano (Sicile). *Revue d'Archéométrie*,**11**, 31–39.

Gazzola, J., Sánchez Del Río, M., Solís, C. and Calligaro, T. (2010) Particle-induced X-ray emission (PIXE) analysis of obsidian from Teotihuacan. *Archaeometry*, **53**, 343–354.

Gillot, P.Y. and Cornette, Y. (1986) The Cassignol technique for potassium-argon dating, precision and accuracy: example from late Pleistocene to recent volcanics from southern Italy. *Chemical Geology*, **59**, 205–222.

Gillot, P.Y. and Villari, L. (1980) K/Ar geochronological data on the Aeolian arc volcanism. A Preliminary Report. IIV, IIV Open File Report 3/80.

Herold, G. (1986) *Mineralogische, Chemische und Physikalische Untersuchungen an den Obsidianen Sardiniens und Palmarolas, Grundlagenzur Rekonstruktion Prähistorischer Handelswegeim Mittelmeerraum*. Ph.D. thesis, Universität (TH) Fridericiana, Karlsruhe.

Le Bas, M.J., Le Maitre, R.W. and Woolley, A.R. (1992) The construction of the total alkali-silica chemical classification of volcanic rocks. *Mineralogy and Petrology*, **46**, 1–22.

Lefèvre, J.C. and Gillot, P.Y. (1994) Datation potassium-argon de roches volcaniques du Pléistocène Supérieur et de l'Holocène: exemple de l'Italie du Sud; application à l'archéologie. *Bulletin de la Societe Prehistorique Francaise*, **91**, 94–99.

Lighthart Ponomanenko, A., (2004) The Pachuca obsidian source, Hidalgo, Mexico: a geoarchaeological perspective. *Geoarchaeology: an International Journal*, **19**, 71–91.

Mackey M. and Warren, S.E. (1983) The identification of obsidian sources in the Monte Arci region of Sardinia. Pp. 420–431 in: *Proceedings of the 22nd Symposium on Archaeometry* (A. Aspinall and S.E. Warren, editors). 30 March–3 April, 1982. University of Bradford, Bradford, UK.

Mahood, G.A. and Hildreth, W. (1983) Nested calderas and trapdoor uplift at Pantelleria, Strait of Sicily. *Geology*, **11**, 722–726.

Mahood, G.A. and Hildreth, W. (1986) Geology of the peralkaline volcano at Pantelleria, Strait of Sicily. *Bulletin of Volcanology*, **48**, 143–172.

Manzanilla, L.R. (2011) Sistemas de control de mano de obra y del intercambio de bienes suntuarios en el corredor teotihuacano hacia la Costa del Golfo en el Clásico. *Anales de Antropología*, **45**, 9–32.

Manzanilla, L.R. (2012) *Estudios arqueométricos del centro de barrio de Teopancazco en Teotihuacan*. Coordinación de la Investigación Científica-Coordinación de Humanidades, Universidad Nacional Autónoma de México, México.

Millhauser, J.K., Rodríguez-Alegría E. and Glascock, M.D. (2011) Testing the accuracy of portable X-ray fluorescence to study Aztec and Colonial obsidian supply at Xaltocan, Mexico. *Journal of Archaeological Science*, **38**, 3141–3152.

Murakami, T. (2010) *Power relations and urban landscape formation: a study of construction labor and resources at Teotihuacan*, Ph.D. thesis, Arizona State University, USA.

Orsi, G. and Sheridan, M.F. (1984) The Green Tuff of Pantelleria: rheoignimbrite or rheomorphic fall? *Bulletin Volcanologique*, **47**, 611–626.

Pastrana, A. (1998) La *explotación azteca de la obsidiana en la Sierra de las Navajas*. INAH, México, DF, Colecion Cientifica, Serie Arqueologia.

Pastrana, A. (2010) L'obsidienne de Teotihuacan. Pp. 99–107 in: *Teotihuacan. Cité des Dieux*, Musée Quai Branly, Paris.

Pichler, H. (1980) The island of Lipari. *Rendiconti della Società Italiana di Mineralogia e Petrologia*, **36**, 415–440.

Piras, G. (2002) I geositi ed i paesaggi geologici del Monte Arci come valenze per il turismo ambientale e culturale. In: *l'Ossidiana del Monte Arci nel Mediterraneo: recupero dei valori di un territorio, Atti del Convegno* Pau 29/11–1/12 2002.

Spence, M.W. (1981) Obsidian production and the state in Teotihuacan. *American Antiquity*, **32**, 507–514.

Tanguy, J.C., Le Goff, M., Principe, C., Arrighi, S., Chillemi, V., Paliotti, A., La Delfa, S. and Patanè, G. (2003) Archeomagnetic dating of Mediterranean volcanics of the last 2100 years: validity and limits. *Earth and Planetary Science Lett*ers, **211**, 111–124.

Tranne, C.A., Lucchi, F., Calanchi, N., Lanzafame, G. and Rossi, P.L. (2002) Geological map of the island of Lipari (Aeolian Islands). University of Bologna and INGV, L.A.C., Firenze.

Tykot, R. (1995) *Prehistoric trade in the Western Mediterranean: the sources and distribution of Sardinian Obsidian.* Ph.D. thesis Harvard University, Ann Arbor, Michigan, USA. (University microfilms).

Tykot, R. (1996) Obsidian procurement and distribution in the central and western Mediterranean. *Journal of Mediterranean Archaeology*, **9**, 39–82.

Tykot, R. (1997) Characterization of the Monte Arci (Sardinia) obsidian sources *Journal of Archaeological Science*, **24**, 467–479.

Tykot, R. (2002) Chemical fingerprinting and source tracing of obsidian: the central Mediterranean trade in black gold. *Accounts of Chemical Research*, **35**, 618–627.

Tykot, R., Setzer, T., Glascock, M.D. and Speakman, R.J. (2005) Identification and characterization of the obsidian sources on the Island of Palmarola Italy. *Geoarchaeological and Bioarchaeological Studies*, **3**, 107–111.

Tykot, R., Freund, K.P. and Vianello, A. (2013) Source analysis of prehistoric obsidian artifacts in Sicily (Italy) using pXRF. *ACS Symposium Series*, **1147**, 196–210.

White, J.C., Parker, D.F. and Ren, M. (2009) The origin of trachyte and pantellerite from Pantelleria, Italy: Insights from major element, trace element, and thermodynamic modelling. *Journal of Volcanology and Geothermal Research*, **179**, 33–55.

EMU Notes in Mineralogy, Vol. 20 (2019), Chapter 12, 411–444

Synchrotron Radiation InfraRed microspectroscopy and imaging in the characterization of archaeological materials and cultural heritage artefacts

Augusto MARCELLI[1] and Gianfelice CINQUE[2]

[1] *INFN – LNF, Via E. Fermi 40, I-00044 Frascati, Italy*
e-mail: augusto.marcelli@lnf.infn.it
[2] *Diamond Light Source, Harwell Science and Innovation Campus, Chilton, Didcot, Oxfordshire OX11 0DE, UK*
e-mail: gianfelice.cinque@diamond.ac.uk

Fourier-transform infrared (FTIR) spectroscopy is a widespread and highly sensitive analytical method for the identification and characterization of a wide range of materials *via* their infrared (IR) absorption bands. Until now, the potential of IR micro-spectroscopy and imaging for the characterization of works of art or other objects of cultural heritage significance has been only partially exploited; in particular the use of the synchrotron radiation (SR) IR microprobe to study, at the micron scale, materials of interest for archaeological and cultural heritage studies has become popular only in the past decade.

One of the main requirements imposed on the studies of ancient and/or valuable materials is that the techniques applied must be non-destructive. In this scenario, SR-based FTIR methods are perfectly suitable. Moreover, IR spectroscopy and imaging are emerging techniques that combine the assets of IR in terms of molecular specificity with the unique properties of synchrotron light. SR-FTIR micro-spectroscopy offers great advantages over conventional methods because it provides a broader spectrum (down to THz) and higher spectral quality (signal/noise ratio) at the highest spatial resolution (diffraction limited). This is due to the high brilliance and collimation of SR-IR, while still being non-damaging to the investigated system. The unique SR-IR parameters are essential for the compositional analysis of the tiny, sub-millimetric samples characteristic of ancient materials, which are heterogeneous by nature, and with complex molecular distributions at extremely variable concentrations.

SR-FTIR spectroscopy and imaging can be applied successfully to the characterization of organic and inorganic materials *via* so-called IR fingerprinting, as well as for their compositional quantification. The range of materials investigated is very broad and encompasses painting materials, stones, glasses, ceramics, coatings on metals, paper and wooden materials, canvas or other textiles, organic colourants, resins, varnishes, cosmetics, and binding media such as glues, waxes, oils, *etc*. SR-IR-based methods can also be used to understand the historical technologies and to identify the raw materials used to produce archaeological artefacts and art objects, and to improve stabilization, conservation and restoration practices.

Selected applications of SR-FTIR methods are discussed with a special emphasis on the chemical and mineralogical characterization of ancient paintings, on the study of alteration and corrosion layers, and the separation and identification of pigments. New

DOI: 10.1180/EMU-notes.20.12

perspectives offered by existing facilities and new developments in IR imaging and advanced vibrational spectroscopy that may broaden the variety of archaeological and historical materials that may be studied are outlined.

1. Introduction

Infrared spectroscopy was developed at the beginning of the 20th century for analytical purposes. From early on, IR spectroscopy was revealed to be a powerful analytical tool for identifying and quantifying molecules and molecular structures in solids, liquids and gases. Coblentz (1905) released the very first database of IR spectra with a table of the characteristic absorption bands for a series of materials. Used extensively by organic chemists since the 1950s, it was recognized as a fundamental method in Mineralogy and Earth sciences (Keller and Pickett, 1949, 1950), although the first comprehensive compilation of IR spectra of minerals was published much later (Farmer, 1974). In the late 1960s/early 1970s, the Fourier-Transform Infrared (FTIR) spectroscopy technique was established thanks to the feasibility of new interferometers and the availability of personal computers suitable to run the Fast Fourier-Transform (FFT) algorithm on-line. Since then, several reviews have been published on the application of IR spectroscopy in mineralogy, and a complete treatment of the infrared theory and aspects of instrumentation and methods became available (Hawthorne, 1988; Beran and Libowitzky, 2004).

The present chapter is intended to highlight aspects of infrared microanalysis and imaging that have been developed recently for the characterization of works of art or other objects of relevance to the cultural heritage and that use SR. Although the interest of SR in this field is mainly in the use of X-rays, in particular X-ray microanalysis, proposals focussing on the use of IR radiation have increased steadily. Several applications in the IR domain have been pointed out previously by Salvadó *et al.* (2005), Cotte *et al.* (2009b) and Bertrand *et al.* (2012a). The type of cultural heritage objects studied ranges from paintings to historical buildings, and involves a very broad range of materials. Among the subjects investigated are research into alteration and corrosion processes, and also the most recent theories and technologies concerning conservation and restoration. The specific goal of this chapter is to give an up-to-date perspective on the use of SR-IR microspectroscopy, which is a mature method applied extensively in studies requiring high spectral quality and ultimate spatial resolution on extremely small samples, *i.e.* ~10 μm or less. Finally the chapter covers the latest development of SR-IR imaging techniques as applied to the investigation of materials of archaeological and cultural heritage interest.

At present, operational SR facilities around the world offer dedicated lines for IR microspectroscopy and imaging. They host an ever-increasing number of multi-disciplinary users such as mineralogists, archaeologists and researchers working in cultural heritage. The interest in SR-IR has increased over the last 20 years because of the unique properties and advantages of this non-thermal source, *i.e.* a spectral emission following the universal curve for accelerators: an asymptotic behaviour at long wavelengths for relativistic electrons with energy greater than 0.5 GeV. Specifically: (1) SR is among the most broadband of sources, spanning from the far

IR (mm wavelengths) to hard X-rays; (2) SR emission is a 'laser-like' small diverging beam, still due to the relativistic emission of electrons, originating from submicron 'wire of current' of circulating electrons, unique features that allow us to overcome the source size-limitations of conventional 'benchtop' instruments, characterized by mm-wide sources and macroscopic angle of emission; and (3) SR emitted light is intrinsically linearly polarized in the plane of the orbit of the accelerated particles. The current rapid development of IR lasers must be noted. Such lasers offer great IR microscopy capability, either at a fixed wavelength (quantum cascade lasers and free electron lasers), or over a limited spectral range (*e.g.* super-continuum laser). However, the image quality (*e.g.* visual artefacts due to the laser light coherence), signal reproducibility and full IR bandwidth, *i.e.* full spectroscopy capability, are still under development.

As detailed below, the main advantage of SR-IR over conventional/blackbody sources is its brilliance, defined as the number of photons emitted per second per unit area and solid angle (horizontal and vertical) at the source (photons/sec/mm^2/mrad^2). A fundamental theorem in optics states that brilliance cannot be increased by any optical system (*e.g.* by lenses and mirrors). In practice, most photons from a small and collimated source like SR can be focused onto a small IR spot. On the other hand, a large-diverging IR source, *e.g.* a conventional blackbody, is very ineffective for illuminating microsamples. The gain in brilliance of an IR-SR source with respect to a conventional broadband source is from 100 to 1000 times when going from the near-to-visible IR (wavelength >700 nm) to the far-from-visible IR (wavelength <cm). The greatest advantage of SR over a conventional source is obtained using mid-IR microanalysis, plus exploiting the broadband character especially towards the far-IR for spectroscopy experiments. For imaging applications, the best performances are given by confocal microscopy (Minsky, 1988), an optical layout in which the optical objective (imaging light out) and the condenser (focusing light in) share the same focal spot size on the sample plane. This can be achieved *via* apertures placed at the conjugated foci, which enables the illuminated and detected areas of the sample to be the same size. The use of small apertures such as pinholes enables images with the greatest spatial resolution to be obtained, but at the cost of a low photon throughput *i.e.* reduced signal. The figure of merit for spectral quality is the signal-to-noise ratio (SNR), which must be sufficient for the spectral recognition of small spatial features in the image. The drawback of the use of pinholes or small rectangular apertures (<20 μm^2) in IR microscopy is the dramatic degradation of the signal, hence the need for brilliant SR sources. It is worth remembering that IR light has longer wavelength than the visible, so the diffraction limit plays a bigger role in spatial resolution (theoretically $\delta x \geqslant \lambda/2$) for IR microscopy. In fact, the use of small apertures affects longer wavelength in the IR spectra to a greater extent. A practical rule of thumb for IR microscopy is that the slit size, *e.g.* $\delta x \approx 10$ μm, causes a cut-off in the spectrum at a comparable wavelength scale, *i.e.* $\lambda \approx 10$ μm.

The recent availability of mid-IR range multi-pixel detectors, equivalent to visible spectral range charge-coupled devices (CCDs) and known as Focal-Plane-Array (FPA), has revolutionized the use of FTIR imaging. These devices consist of a bidimensional array of small IR detectors (pixels) a few tens of microns in pitch. They allow the acquisition of thousands of IR spectra simultaneously, generating mid-IR images at high spatial sampling in a short time (~ minutes). The optical system of an IR microscope (Fig. 1) equipped with a FPA is an 'apertureless' imaging system, the ultimate spatial resolution of which is a factor two less than in confocal microscopes. Nevertheless, FPA sensitivity and read-out speed enable different imaging, namely oversampling specimens at the micron scale or providing a large field of view. Massive and fast data collection within several minutes is thus achieved (Colarusso *et al.*, 1998; Bhargava and Levin, 2001, 2007; Petibois *et al.*, 2010).

For Earth science materials, SR-IR has been applied successfully in studies which require either an improved spatial resolution, *e.g.* for fine-grained synthetic crystals, or high brilliance, such as in high-pressure (HP) studies using diamond-anvil cells (Scott *et al.*, 2007; Noguchi *et al.*, 2012). The new imaging capabilities of FTIR microscopes have also been used to address features such as zoning of volatile species across samples or possible conformational changes of structurally bound carbon molecular species during crystal growth, all features which are barely accessible using standard micro-analytical techniques. Recent studies have also explored the use of FTIR imaging to monitor processes in real time (Della Ventura *et al.*, 2010).

2. Basics of Fourier Transform InfraRed spectroscopy

Depending on convenience, electromagnetic (e.m.) radiation can be thought of either as a wave propagating and transporting energy and momentum, or as a flux of photons, *i.e.* massless particles travelling in space and absorbed or scattered by materials. The

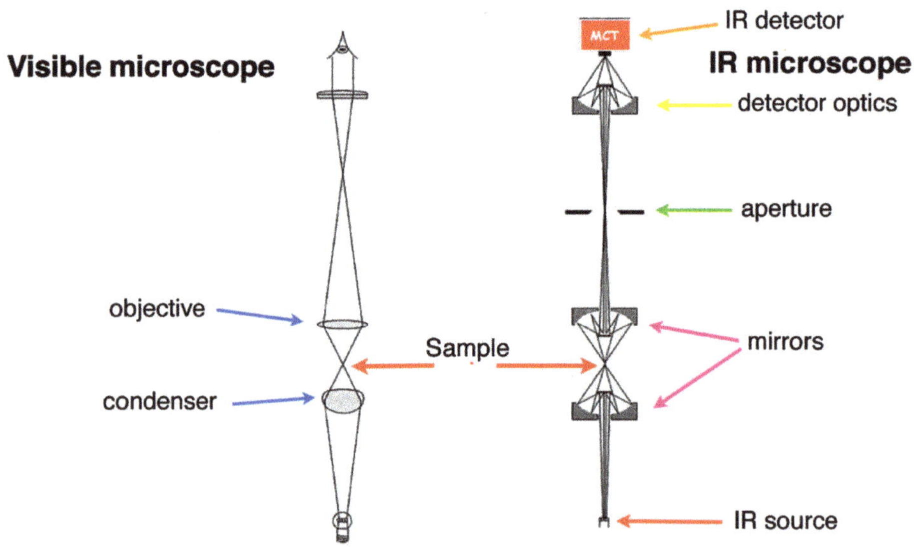

Figure 1. Comparison of the layout of an IR microscope (right) with that of a visible light microscope (left).

spectrum of e.m. radiation is the range of the wavelengths it covers, or the energy associated with photons; in practice, it extends from km- to cm-long wavelengths for radio-/microwaves, to mm- to μm-long for IR, or below 1 μm for the visible "rainbow" and the UV, up to Ångstrom-size wavelengths for X-rays (see Fig. 2a). Due to diffraction limit implications, this length of scale is crucial for microscopy purpose. The Abbe diffraction limit sets the greatest spatial resolution achievable at $\lambda/2$. Thus in the visible, sub-micron spatial resolution is optically possible, but such a resolution is not attained in the IR range.

On the other hand, the energy associated with photons dictates the interaction regime that the e.m. radiation can probe. X-rays and UV are ionizing radiations spanning energies above 10 eV, which means that they are capable of interacting with the atomic and electronic structure of matter. Together with visible light, the energy of which is a few eV, X-rays and UV radiation can induce sample luminescence and/or damaging/ bleaching. IR photons in contrast are in the sub-eV regime, *i.e.* so soft to be well below any damaging threshold, but capable of exciting molecular oscillations within solids. From the fundamental equation between photon energy and light wavelength $E = hc/\lambda$, a rule of thumb can be obtained: $\lambda E = 12400$ [μm meV] or equivalently $\lambda = 100$ μm ↔ E = 124 meV.

Specifically, the IR domain is generally divided into three parts named with reference to the 'spectral distance' from the visible: near-IR (wavelength 0.7 to 2.5 μm or from ~1800 to 500 meV), mid-IR (2.5 to 50 μm or from ~500 to 25 meV), and far-IR (>50 μm or < 25 meV). Historically, the way in which IR spectra were drawn was *vs.* wavenumber, the latter defined as the inverse of the wavelength in units of cm^{-1}. For an easier conversion, a practical rule correlates wavelength and frequency: $\lambda = 100$ μm ↔ $\nu = 100\ cm^{-1}$. Nowadays, it is common to refer to the THz region, which is the far IR with frequency from 0.3 to 3 THz. By rewriting the fundamental equation $c = \lambda * \omega$, where c is the speed of light, the equivalence 1 THz ↔ 300 μm (~ 4 meV) is obtained.

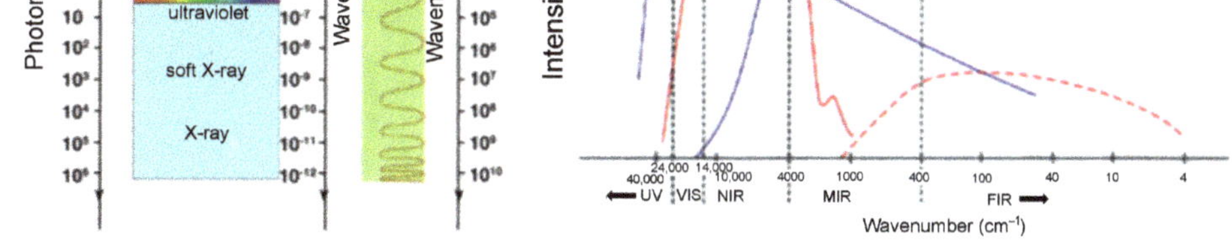

Figure 2. (a) The regions of the e.m. spectrum *vs.* energy, wavelength and wavenumber covered by a synchrotron radiation source; (b) the various conventional sources necessary to cover the IR domain from the THz to the NIR and visible to UV regions.

Lamps based on the incandescence of electrically heated filaments are still the most used conventional sources (Fig. 2b). They rely on the so-called e.m. emission of a blackbody: a perfect all-wavelength absorbing material which gives a universal spectrum as a function of its temperature only. A blackbody is a particularly powerful, isotropic and homogeneous source of visible-to-IR light useful for most purposes of spectroscopic analysis of macroscopic samples. In the IR field this is referred to as globar, *i.e.* a few mm element heated around or above 1000 K to emit a bell shape spectrum conveniently peaked at ~2.9 μm or below according to the Wien law, *i.e.* in the mid-IR region. As a variation, spectral emission from light bulbs ($T \approx 2000$ K) is used for near-IR, and Mg lamps for the gas emission in UV and far-IR due to light Doppler shift from N_2 buffer gas.

The basis of absorption spectroscopy lies in the idea of characterizing an unknown sample by measuring the light loss across it. The ratio of the light intensity measured after (I) and before (I_0) an illuminated sample is its transmission ($T = I/I_0$), and its transmission spectrum, $T(\lambda)$, is its plot as a function of wavelength. Lambert-Beer's law states that the intensity of an electromagnetic wave penetrating a material falls off exponentially with distance from the surface: $I/I_0 = e^{-\alpha(\lambda)L}$, where $\alpha(\lambda)$ (units of m^{-1} or submultiples) is the absorption coefficient of the specific material and L the sample thickness. The absorbance A, a dimensionless quantity sometimes called optical density is a property of the specific sample calculated as the inverse of transmittance in a log scale: $A = \log(1/T)$. Again, the absorbance spectrum $A(\lambda)$ is plotted *vs.* wavelength.

Comparison of these definitions shows that $A = \alpha(\lambda)\, L/2.303$. Also useful to note is that the penetration coefficient is the reciprocal of the absorption coefficient, that is the thickness of material for which the amount of light has reduced to $1/e$ ($T \approx 37\%$). An absorbance equal to one means that the amount of light has reduced by 10 times across the sample; an absorbance of 0.1 means ~80% transmission. Because the linearity range of most detectors used in the IR region is physically limited, reliable spectroscopy measurements are generally possible for samples with absorbance between 2 and 0.01 (or transmission between 1% and 98%). Above and below such limits, IR spectroscopic data are no longer quantitatively accurate.

At present, the most common spectrometer for IR spectroscopy is the Michelson interferometer (Fig. 3). In this system the IR radiation from a source goes through the instrument pupil and is split 50/50 into two beams by a semi-reflecting mirror called a beamsplitter. A series of mirrors in a 'cross' configuration makes the two beams recombine into one: this is

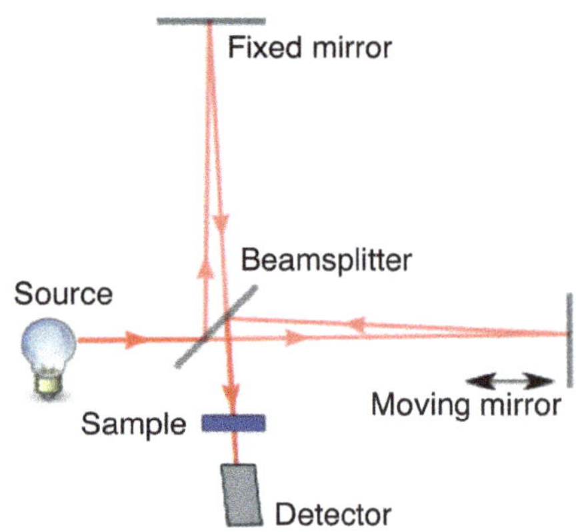

Figure 3. The optical layout with main optical elements of a Michelson interferometer.

brought onto a detector after having been through the sample. In one of the two arms of the 'cross' a moving mirror modulates continuously the radiation path for all the different wavelengths passing through the interferometer. The basic idea is that reflected and transmitted beams split by the beamsplitter are recombined to produce positive/negative interference: this happens for each wavelength travelling a distance equal to an even/odd multiple of $\lambda/2$.

All wavelengths interfere constructively for equal paths (centre burst), but their intensities modulate as a decreasing oscillation for longer path differences. Such a recorded signal is called the interferogram, and it represents the light output as a function of the mirror position, which is correlated to each wavelength IR intensity. It can be demonstrated that the interferogram is the Fourier Transform (FT) of the IR spectrum. By Fast Fourier Transform (FFT) algorithm, data are processed in real time and transformed in a spectrum where the absorbance of the sample is plotted *vs.* wavelength. As a consequence of the FT method, the relation between mirror path and spectral resolution (the minimal measurable separation between two close spectral peaks) means that the resolution expressed as wavenumber (cm^{-1}) is the inverse of the mirror path (cm). Compared to a dispersive instrument, the complexity of a FTIR spectrometer is largely compensated by a greater advantage, namely that the information in the entire frequency range is collected simultaneously, improving both acquisition speed and signal-to-noise ratio (SNR). This is the multiplex or Fellgett's advantage. The other merit of the interferometer is its high throughput, also called the Jacquinot's advantage. In fact, the whole beam from the source is used in an interferometer, while in any dispersive instrument slits are necessary to define the bandwidth, at the cost of less light being detected.

In condensed matter, any molecules can absorb e.m. radiation by reaching an excited vibrational state (transition in the IR), or an electronic state (transition in the UV or X-ray). Physically, the energy levels are quantized; therefore, the light absorption by a molecule takes place only for photons of energy corresponding to the gap from the ground to an excited state. Semi-classically, the criterion for IR absorption is the presence of, or change in, electric dipole moment in a molecule. Absorption occurs when there is a match in frequency between the radiation and the molecule harmonic oscillation. Molecules without a permanent/induced dipole moment cannot absorb IR radiation.

In polyatomic molecules and in a condensed phase, vibrational motions, distinguished by stretching (*i.e.* a change in the interatomic distances), and bending modes (*i.e.* a change in the angle between two bonds) provide the main IR evidence. A molecule with a number of atoms, N has $3N$ degrees of freedom in a system of Cartesian coordinates. For a generic molecule, 3 of these degrees of freedom are rotational and 3 translational; the number of fundamental vibrations or IR bands is thus $(3N - 6)$. In a linear molecule, there are only 2 rotational degrees of freedom: the fundamental modes are thus $(3N - 5)$. For example H_2O is a non-linear molecule composed of 3 atoms, thus it has 3 fundamental modes in the IR, namely asymmetric and symmetric stretching plus bending vibrations. CO_2 is linear molecule thus with 4 fundamental modes. Of the

2 stretching modes, the symmetric one is not IR active because there is no change in dipole moment. The other 2 bending modes are degenerate in energy so that they merge into one band only.

IR spectroscopy is actually the most routine technique employed in identifying molecular/chemical compounds based on the presence of different functional groups *via* their absorption bands. In general for complex molecules their recognition by vibrational spectroscopy is made *via* the so-called mid-IR fingerprint, *i.e.* using the part of the spectrum below 2000 cm^{-1}, mostly associated with the bending modes of the functional groups, where IR bands are affected by the surrounding atoms of the molecule structure. At a first approximation, the interpretation of stretching modes in an IR spectrum can be done with reference to the harmonic oscillator strength and reduced masses, so the C–C mode is $\sqrt{2}$ lower in frequency than the C=C one, while the C–H mode is $\sqrt{3}$ higher in frequency because of the reduced mass change carbon to hydrogen, *etc.*

In terms of the actual acquisition mode of the spectra, transmission analysis in the IR regime is done by first measuring the background, *e.g.* on the sample support to get the I_o, and then on the sample, to obtain I, at any wavelength. Then Transmittance (and Absorption) is calculated as explained above. IR opaque samples can sometimes be measured in specular reflectance if the specimen is optically polished: *via* the Kramer-Kronig relations linking refractive index and absorbance, the transmission spectrum can be calculated by most acquisition software, while the inverse is non-trivial. Thin samples can be measured if deposited on a reflective substrate, and the method is called transflectance. The Attenuated Total Reflection (ATR) method is used for semi-infinite samples. Here the IR light illuminates a sample through a high refractive index crystal such as Ge, Si or diamond, which is placed in contact with the material to be analyzed. The incoming beam is at a small angle with respect to the crystal–sample surface, so that it is below the critical angle of total reflection. At the interface some light leaks into the sample; if the material absorbs at some wavelengths, then the reflected light is modulated accordingly, and an IR spectrum is produced. This method can also be applied in IR microscopy using all-mirror based objectives and condenser in the Schwarzschild geometry, also known as Cassegrain optics (Fig. 4).

3. Synchrotron Radiation and IR microanalysis, principles and advantages

Synchrotron radiation exhibits several unique properties as a photon source: (1) high brilliance; (2) broadband and continuous intensity distribution over the entire e.m. spectrum; (3) a high degree of polarization; (4) ultrafast time structure; and (5) high stability. Historically, the IR SR emission first used was the classical bending magnet (BM) radiation, which is characterized by the broadest IR spectrum coverage extending from microwaves to the IR domain up to X-rays (Hofmann, 2004). More recently, e.m. radiation emitted by electrons travelling through the fringing magnetic field of a BM has been exploited; this takes the name ‘edge radiation’ at entrance/exit of a bending magnet. Such a source is limited to the lower energy of the spectrum, *i.e.* visible to IR,

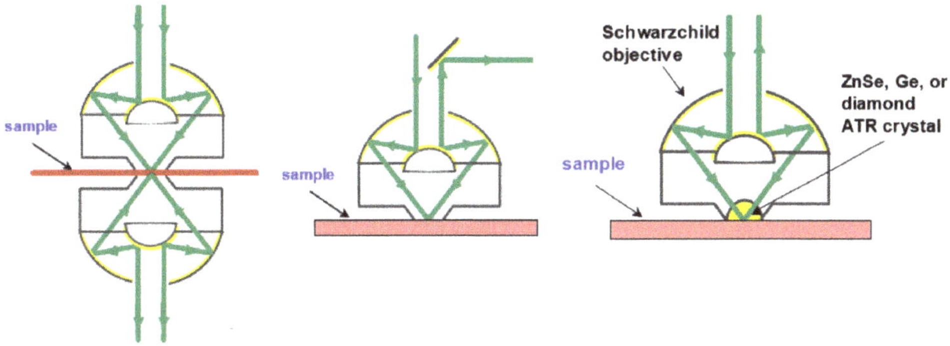

Figure 4. Optical layout of Cassegrain objective and condenser systems as used in transmission (left), reflection (centre) and micro Attenuated Total Reflection (right) geometry.

and in practical terms is comparable to the standard BM radiation for IR purposes, with a flux benefit towards the longer wavelengths spectrum.

The most exploited advantage of synchrotron radiation is its broadband brilliance. Rather large collection angles are required at a beamline 'front-end' to extract long wavelengths – such as the IR – out of bending magnet sources. This is because the intrinsic light divergence increases up to several orders of magnitude from the characteristic synchrotron divergence $\theta \sim 1/g$ (γ, relativistic factor for electrons) for X-rays at critical energy. On the contrary, long-wavelength radiation emitted as edge radiation has a significantly smaller opening angle than standard BM radiation.

The history of SR applications in the long wavelength region and in particular the IR is more recent than that for UV and X-rays (Marcelli and Cinque, 2011). In fact, SR sources are some orders of magnitude more brilliant than a conventional black body source with the same spectral range (Fig. 5)[1]. If a highly brilliant source is not a priority for standard spectroscopic measurement, it becomes mandatory in microscopy or microanalysis where a spot size down to the diffraction limit is highly desired in many applications. In fact, the signal and spectral quality decrease drastically if apertures are reduced to define a smaller beam for the sake of lateral resolution (for example, see Fig. 6 for organic matter).

The use of IR microscopes coupled to SR sources provides outstanding results because the high brilliance of this source allows the microscope apertures to be reduced to the diffraction limit, *i.e.* down to a few microns in the mid-IR region (Carr, 2001). The electrons in the storage ring do not have a continuous distribution around the orbit in a circular accelerator but are stored and travel in bunches, thus the resulting emission has a time structure. The SR is indeed emitted as light pulses with duration and separation characteristic of each ring, typically ranging from tens of psec to tens of nsec. This is another interesting characteristic of this source for users interested in time-resolved experiments. An additional and key feature of SR is its polarization. Owing to

[1] We are not discussing the use of IR laser sources because they are not yet standardized as routine instruments for FTIR.

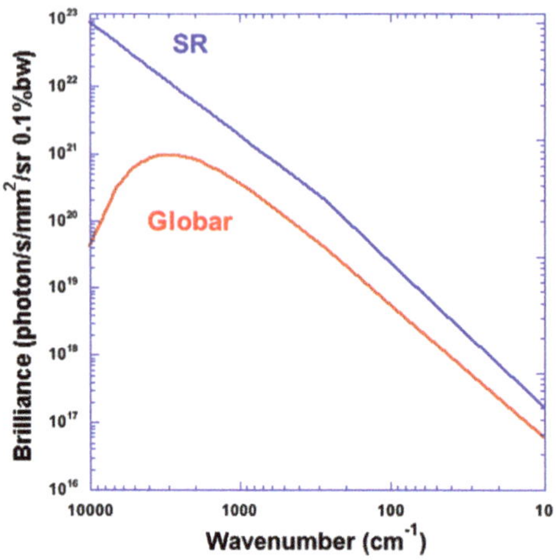

Figure 5. Comparison of the calculated brilliance of a 3 GeV synchrotron radiation bending magnet source at 1.4 T *vs.* a conventional source.

SR emission geometry, BM radiation is linearly polarized in the orbital plane. Since the early pioneering work in the visible region, studies performed both in the UV and X-ray regions demonstrated that when light is observed above or below the orbit plane, the radiation is circularly polarized. The polarization properties of the SR at long wavelengths are very interesting for many applications, in particular in the far-IR region. By placing a slit on the exit port, one can, in principle, select the desired degree of circular polarization and the flux of the emitted radiation. About polarization, we may define three polarization indices, of which two are more important: the linear and the circular polarization. Experimental observations indicate that values up to 80% of circularly polarized light can be obtained using a set of slit that selects ~50% (upper or lower part of SR fan) of the total flux available (Cestelli Guidi *et al.*, 2005).

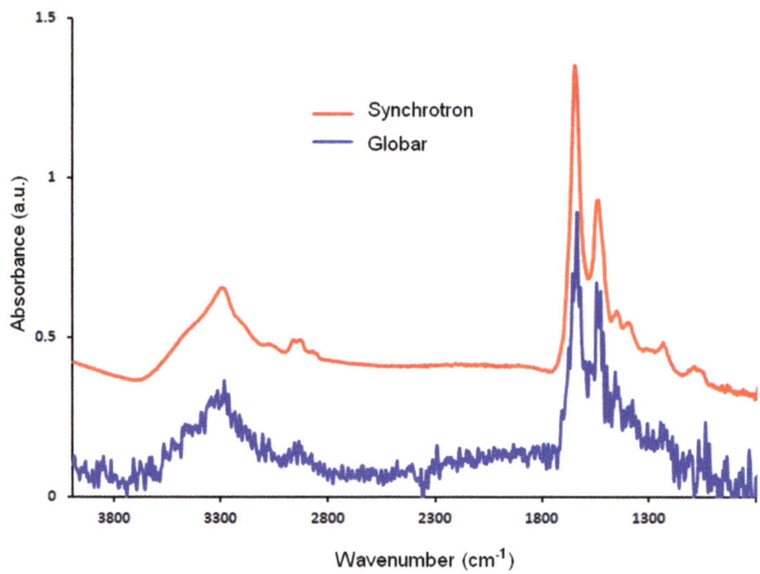

Figure 6. Comparison of the IR spectral quality obtained on an organic microsample by a conventional source (lower) and a SR-IR source (upper) both over the same area (slit size 5 μm × 5 μm) and in 30 s. The absorbance curves are in scale and shifted for clarity. The figure of merit SNR (rms) is <200 for the Globar (blue) and >5000 for the SR (red), respectively.

In SR-IR beamlines, commercial FTIR benches and microscopes are generally installed, and the SR replaces only the conventional source. The IR beam is focused in the upper part of the microscope (Fig. 1) and is projected onto the sample plane by a reflective objective; the slit dimensions determine the microbeam size, and the relations between spectral cut-off and resolution in these conditions have been discussed previously. More details can be found in the paper by Marcelli and Cinque (2011).

The goal of FTIR microscopy is to attain the best spatial resolution at the sample location, while exploiting the typical sensitivity of IR spectroscopy to molecular composition. The concept of spatial resolution in IR imaging is not an absolute parameter for non-monochromatic systems: it is limited by diffraction and affected by both the optical design, *e.g.* objectives and slits, and the final contrast achieved (Levin and Bhargava, 2005; Levenson *et al.*, 2006, Bhargava and Levin, 2007). According to Rayleigh's criterion, two points can be separated *via* a microscope when the central maximum of the diffraction pattern of the first (*e.g.* due to a plane e.m. wave encountering an infinitesimal pinhole in a screen) is at a distance greater than the first minimum of the Airy disk of the second. This is shown in the upper right part of Fig. 7.

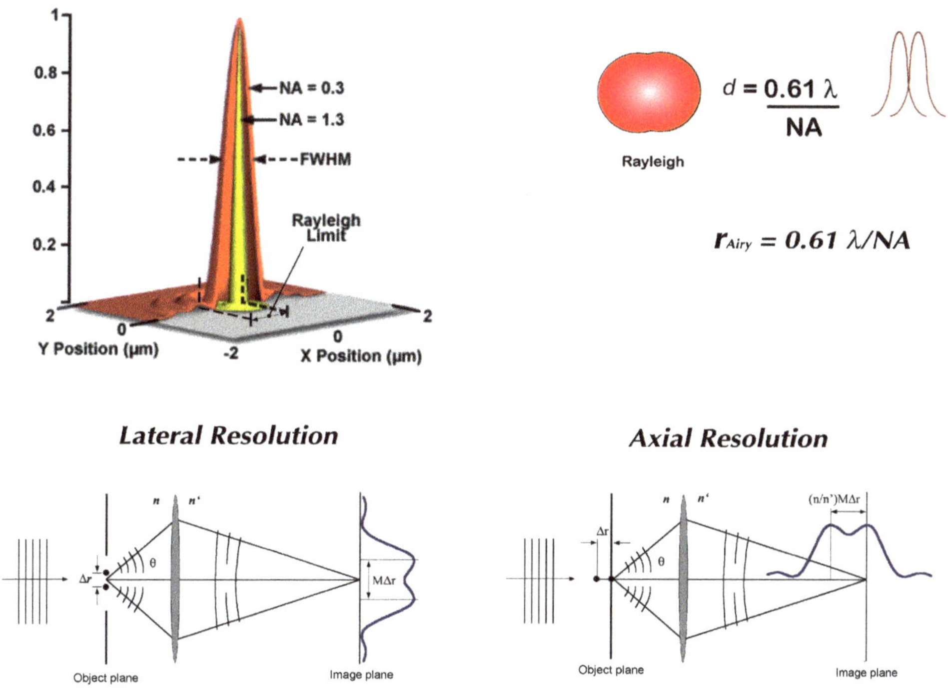

Figure. 7. (Upper, left) Comparison of the Point Spread Function (PSF) between two optical systems with different Numerical Apertures (*NA*). (Upper, right) The Rayleigh criterion where two points can be resolved spatially if the central maximum of the Airy disc of the first is at a distance of one radius (0.61 λ/*NA*) from the second. (Lower) Comparison of spatial lateral (left) and axial (right) resolution for confocal and widefield microscopy.

When a point source of monochromatic radiation goes through the microscope optics, the Airy pattern is observed at the beam focus, *e.g.* onto the sample. The central circular area of the Airy disc, known as the point spread function in 2D, in Fig. 7 is characterized by a radius $r = 0.61*(\lambda/NA)$, where λ is the wavelength and *NA* is the numerical aperture of the microscope. If we consider a standard optical microscope illuminating a large sample region, the ultimate resolution can be achieved by an 'optical system' made by the objective optics and the optical camera. Similarly, in an IR microscope using an IR area-detector, *i.e.* a focal-plane array (FPA) detector, the spatial resolution is determined by the magnification of the optical system, with the size of the individual pixel of the detector also being a relevant parameter (Miller and Smith, 2005).

In a microscope, the field of view (FOV) can be reduced by using a field stop, *e.g.* an aperture placed at an intermediate focal point; in this case, the spatial resolution is determined by the fraction of light from each point of the specimen that reaches the detector, *i.e.* its sensitivity pattern. As outlined above, if we reduce the illuminated region while maintaining an homogeneous illumination, then the spatial resolution is a function of the number of photons that illuminate the small selected region of the specimen. To probe this small region of a specimen, the microscope must illuminate it and simultaneously detect the light. A confocal microscope has identical apertures for both the illumination and detection systems. It achieves the ultimate spatial resolution by reducing to the same size both the illuminated region and the region to be detected. An image of the specimen under analysis can then be collected, although with much longer times, by rastering the sample through the focus of the confocal microscope. In this case, standard IR single-element detectors, typically with areas of 250 μm × 250 μm, are not suitable for work at high spatial resolution. In fact, the effective projection of a detector pixel size onto a sample by a standard IR objective (*e.g.* 15 ×, plus the microscope internal optics) can be one order of magnitude larger than the IR wavelength limit (*e.g.* ~170 μm × 170 μm), *i.e.* far beyond the minimum diffraction limit. Moreover, their intrinsic noise is proportional to the dimension of the detector (and the SNR proportional to $\sqrt{\text{area}}$), which is a real bottleneck for spectral quality. In practice, smaller IR detectors (commercially available <100 μm × 100 μm) are both less noisy and better match the SR beam size for best performances in IR microscopy.

In contrast to confocal microscopes with single-element detectors, the combination of a SR source with a two-dimensional FPA detector is an alternative way to take advantage of SR-IR brilliance for imaging and time-resolved experiments. Using IR area-detectors FTIR spectra can be collected simultaneously over large areas (generally less than hundreds of μm^2) depending on the magnification (generally, above 15 ×). IR snapshots containing hundreds of points are obtained within minutes, to be compared with acquisition times in the range of hours typically required by single-element detector mapping. Although in principle a sub-second timescale is necessary for a single FTIR spectrum, the raster scan method of IR mapping gives intrinsically the highest spatial resolution but is much slower in acquisition time. Ideally, the wide IR fan available from a high-current synchrotron radiation facility can be shaped to illuminate a limited number of pixels of a FPA detector, and the brilliance advantage per small pixel size can increase the SNR ratio of the IR image by a high contrast (Petibois *et al.*, 2010). For these reasons, SR-IR coupling with FPAs is being developed increasingly especially for biological studies; in contrast, very few studies of materials of interest to Earth Science have been done so far in spite of the enormous potential. Recently, a synchrotron-based multiple-beam FTIR chemical imaging beamline was used in the SR Madison facility (USA), providing new opportunities to examine samples of interest to the cultural heritage community (Unger *et al.*, 2013).

Focusing on resolution, great emphasis is usually placed on the lateral resolution. However, to interpret an image it is also important to consider the axial (or in-depth) resolution, *i.e.* the axial resolving power of the objective measured along the optical axis. Similar to lateral resolution, axial resolution (Fig. 7) depends on the numerical aperture (*NA*) of the objective (Inoué, 1995; Born and Wolf, 1997). Lateral and axial spatial resolution for a confocal microscope are better by ~50% and ~30%, respectively, compared to a corresponding microscope working with a wide-field illumination as shown in Fig. 7. A large *NA* improves both the lateral and axial spatial resolution of the microscope, although the *NA* of the microscope objective is much more effective for the axial resolution ($\sim\lambda/NA^2$) than the lateral one ($\sim\lambda/NA$). Knowledge of the axial resolution is fundamental for performing 'optical sectioning', a technique which enables a three-dimensional reconstruction of a sample *via* non-destructive depth profiling, *i.e.* collecting images as the focus is moved deeper into the sample. This method offers clear advantages for a non-destructive sample analysis such as that required for precious artefacts.

There are essentially three possible experimental set-ups in FTIR microscopy: (1) point analysis in a confocal layout; (2) FTIR mapping done by integrating the signal from successive locations of the specimen surface; and (3) FTIR imaging, performed with bi-dimensional arrays such as focal plane array (FPA) detectors. The main advantage of observing an entire field of view at the same time allows spatially-resolved spectroscopy of large and multi-phase samples.

Single-spot analysis can be done in both transmission and reflection modes. The spot size at the sample under the microscope illuminates images which are 30–100 μm across using a conventional source, which can be reduced to 3–20 μm across by using a SR source. To characterize the spatial distribution of an absorber across a sample, a standard practice is to isolate a small area of interest using apertures placed before or after (or both) the sample and then collect several spectra along traverses. Guilhaumou *et al.* (1998) presented the first application of this technique using SR-FTIR microscopy to monitor the evolution of the OH peak while using a small aperture.

In the FTIR-mapping mode the IR spectrum is measured at each in-plane point and then peak heights, peak areas, the integer performed in a defined spectral region or other criteria are used to visualize the distribution of the target molecule. This experimental set-up can be coupled to a confocal-like set-up to achieve good statistics and a high SNR. However, the method requires collection of a large number of single spectra and reliable movement of the sample on the stage; consequently, the acquisition time for collecting data from large areas, *e.g.* a few mm^2, is up to several hours. In the FPA-imaging mode, the whole image is obtained in a single data collection thanks to a multichannel detection, similar to recording images with CCDs in optical microscopy. The number of pixels (the detectors) and their effective size will depend on the FPA type used. Typical arrays, designed for conventional sources, range from the 64×64 channels, providing 4096 individual spectra, to 256×256 or 1024×1024 channels allowing imaging of much larger areas. However, these detectors are not optimized to match a SR source that, because of its brilliance, may illuminate effectively only a limited portion of a bi-dimensional device (Petibois *et al.*, 2010). Such arrays are coupled with $15 \times$ or $36 \times$ objectives, so that with a single image up to few mm^2 in size, a single pixel corresponds to a physical dimension in the range 2.6 to 1.1 μm. Although FTIR imaging has evolved rapidly as a tool in biological and biochemical sciences, very few applications have so far been published in Earth Sciences (Della Ventura *et al.*, 2010, 2014).

4. General applications of IR microspectroscopy to materials of historical interest

SR-IR applications cover the full field of the Earth Sciences at different scales, including crystals and minerals in natural rocks (rocks may contain amorphous phases but minerals are crystalline, although not always in crystal shape), assemblages of various solid phases, and even glasses. Minerals are often twinned or contain fluid inclusions. The analysis of inclusions such as gas and liquid molecules like H_2O/OH in solids has been applied to nominally anhydrous minerals, similar to the occurrence of H, C and N-based molecules in minerals and interplanetary dust particles, and in some cases also in high-pressure studies. Applications in cultural heritage have become more and more frequent in the last decade. Complementary to other techniques, vibrational spectroscopy – either IR or Raman – is used widely in the Earth Sciences to characterize thin sections of geological materials with a spatial resolution that can reach the micron scale. Both methods have high molecular specificity. IR has the advantage of geater sensitivity and is totally non-destructive; Raman can provide higher lateral and longitudinal resolution but requires significant laser power to be applied to the specimen to obtain a good spectral quality[2].

Guilhamou *et al.* (1998) and Bantignies *et al.* (1998) were among the very first to explore the potentiality of SR-FTIR microspectrometry for such studies. In particular, Guilhaumou *et al.* (1998) addressed the analysis of hydrocarbon-rich fluid inclusions in siliceous diagenetic materials, aimed at constraining petroleum formation and migration. SR-IR microspectroscopy has also been applied in archaeology,

[2] An excellent review paper on microRaman outlining key applications of the technique to Earth Sciences including a good reference catalogue of Raman spectra for representative minerals was published by Gillet (2006).

archaeometry and in the study of cultural heritage materials because it is a highly sensitive, reliable and non-destructive analytical tool. Most important are studies of ancient painting materials, which are typically multi-layered and structurally heterogeneous, and can contain various fine-grained mixtures of minerals, pigments and binding media. Salvadó *et al.* (2005) neatly outlined three experiments aimed at demonstrating the importance of synchrotron-based IR microanalysis and sample preparation in the cultural heritage field. More specifically, the characterization of mineral pigments present in polychromatic stucco used for the wall paintings in a Roman villa was made on a micrometric fragment spread in a diamond cell to obtain a thin section suitable for transmission. The same preparation method was used to reveal the evolution in medieval painting techniques, namely the change of the organic binding media from tempera to oil that occurred between the Gothic and the Renaissance periods. In this case, a further method was used, which required measurements in reflection mode on a polished surface after the fragment was embedded in a plastic material and cut into slices. Finally, the determination of the corrosion of a Corinthian-type helmet by direct IR measurement in reflection on the surface of this single bronze piece: this again revealed the presence of a thin protective organic coating on the metal (Pantos *et al.*, 2005).

In situ characterization of the various components of these materials is essential to extract information about the techniques used during manufacturing and the origin of the materials used, as well as to design appropriate restoration procedures. A combination of X-ray diffraction/absorption and SR-FTIR spectroscopy has provided considerable advancements in these studies. For an overview of such combination the reader may refer to Cotte *et al.* (2009a). Notable investigations involved the analysis of Romanesque wall paintings from Unha (Val d'Aran) (Salvadó *et al.*, 2008) and the study of the alteration of silver foils in medieval painting from the Museu Nacional d'Art de Catalunya, Barcelona (Salvadó *et al.*, 2011).

Another field of application is the characterization of historical findings, as well as prehistoric artefacts of interest. For example, lithic artefacts were frequently used in prehistory, and deciphering the origin of the raw material provides information on the trade and migration of the first human societies. Bernardini *et al.* (2011a,b) used SR-based microFTIR directly on specific points of polished thin sections of serpentinitic rocks and specimens of prehistoric axes in order to identify the different polytypes of serpentine minerals. Notably, IR microspectroscopy was performed in reflection mode, *i.e.* without a need to sample the objects. The use of a synchrotron radiation source was a prerequisite because it provides both the spatial resolution and the spectral quality needed for the discrimination of different occurrences of serpentine minerals intimately intergrown with other sheet silicates. In practice, spectra were measured on features <20 μm long, or on poorly reflecting material like most real samples, which gave diluted signals requiring SR microFTIR to achieve a signal-to-noise ratio higher by several orders of magnitude than those obtained with conventional systems.

A fundamental issue underlining the analysis of multi-layered or multi-phase samples by imaging or microscopy is sample preparation. Cotte *et al.* (2009b) reviewed different procedures in the particular context of synchrotron-based microanalysis concerning both IR and X-rays.

Besides the use of transmission and reflection modes in IR spectroscopy on thin samples or polished specimens respectively, micro Attenuated Total Reflection (ATR) is a valid alternative, which may be applied to many samples either non-reflective, or optically very dense. In the first two modes, spatial resolution is limited by optical diffraction to practically ~λ, which can hardly be achieved or useful in terms of the SNR ratio considerations unless a SR source is used. Micro-ATR imaging offers better spatial resolution (up to a factor of 2–4 times greater) because of the change in refractive index between the ZnSe or Ge-crystal element and the sample matrix, which, in practice, eliminates any sample preparation. Goodall *et al.* (2008) used a micro-ATR device to characterize Mayan wall paints from an ancient chamber with the objective of identifying the mineral phases in the paint and stucco. FTIR spectra identified successfully the various mineral components, and the greater spatial resolution given by the micro-ATR allowed identification, spatially, of the different particles in the paint: hematite and silicate particles with minor amounts of calcite, magnetite particles in a sub-micrometer hematite and a calcite matrix – also in the underlying stucco – *i.e.* a combination of calcite with fine carbon particles. The use of micro-ATR in combination with a FPA detector is of particular importance; in fact, IR images allowed the authors to decipher the orientation and morphology of individual grains.

5. Selected examples of IR imaging in archaeology and cultural heritage

The specialist use of synchrotron radiation to probe cultural heritage and archaeological specimens dates back ~20 years. In terms of the synchrotron techniques used, most of them were X-ray-based methods, including both spectroscopy (X-ray fluorescence/absorption), and diffraction (crystalline and small angle scattering) (Bertrand *et al.*, 2012a). In this context, synchrotron-based FTIR microanalysis started more recently and immediately for microscopy/imaging purposes. By exploiting the high sensitivity of IR to low-Z elements of interest in mineralogy, which are difficult to detect using X-ray analysis, IR high specificity to functional groups in organic molecules for chemical analysis is an invaluable asset.

The following choice of IR examples is not to be considered exhaustive, but rather a series of significant examples taken from recent works published in the field, with the purpose of highlighting the state of the art of the technique and the breadth of applications. The range of materials probed is very diverse and includes painting fragments, bones, pottery and ceramics, metal and stone artefacts, wooden materials, as well as organic-inorganic mixtures in ancient plaster and cosmetics, thus reflecting the large variety and complexity of archaeological and cultural heritage studies.

A first case study showing both the non-destructive character of SR-based FTIR microscopy and its high sensitivity concerns microanalysis of the patina deposited on Dogon statuettes (Mazel *et al.*, 2008). The Dogon are among the most studied ethnic groups in West Africa. Since the beginning of the 20th century, many ethnographical missions have been undertaken to study their beliefs and culture. These patinas were formed during religious and ritual ceremonies, and the anthropomorphic statuettes from the Dogon culture are well known examples of this kind of practice. For the purpose of better understanding details of the ceremonial practices throughout the centuries (the samples span the 15th to 17th centuries), the underlying idea was to extract unequivocal information about the original constituents and the method of deposition of these patinas. After preparation using ultra-microtomy of embedded samples, different chemical IR images led to the identification of minerals (clays, quartz and calcium carbonate) and organic products, as well as mapping their spatial distribution (*e.g.* proteins, starch, lipids as shown in Fig. 8). For some of these artefacts, the significant quantity of minerals and the lack of stratigraphy, as revealed by IR maps, suggested that the patina was deposited all at once, according to different recipes. The mineral portion consists mostly of quartz and alumina-silicates, although kaolinite was also identified in some samples. This composition corresponds to minerals that can be found readily in the Dogon area. However, differences suggest the existence of a variety of rituals more extensive than that described by the ethnologists.

A second example concerning the use of minerals in prehistory is the study of serpentinite stone, as the most frequently used raw material to produce polished lithic artefacts (Bernardini *et al.*, 2011a). Bernardini and coworkers characterized axes using a range of mineralogical and chemical techniques in order to locate the primary outcrops of their raw materials, and related areas of secondary exploitation. The evolution of mineral components in stone axes during the Neolithic to Copper Age transition in the so-called Caput Adriae (northeastern Italy, central and western Slovenia and northwestern Croatia) is of interest because it is linked to the development of early European metallurgy. Serpentinite shaft-hole axes in this regional archaeological context have a rounded, irregular shape, suggesting that the raw material came mainly from secondary deposits, thus the motivation for the identification of primary and secondary sources. Several conventional analytical techniques can identify the different serpentine minerals, but their application generally requires a powdered sample, which means damaging artefacts. The use of spatially resolved techniques is also a must in order to address sample non-homogeneity. SR-FTIR spectroscopy is ideal in many respects; it is a spatially

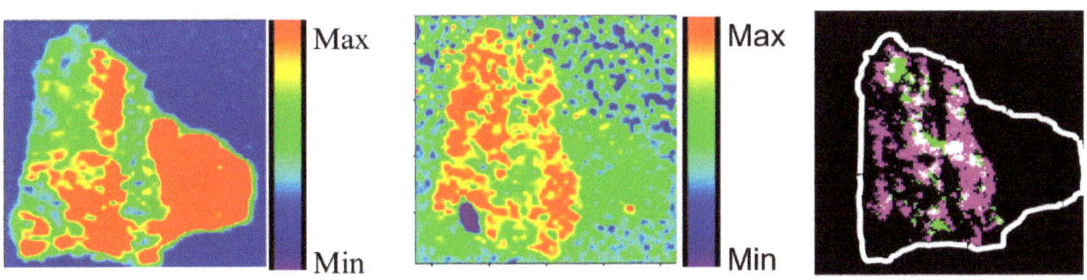

Figure 8. SR-IR images of the sample of a Dogon statuette (Mazel *et al.*, 2008): (left) distribution of proteins; (centre) distribution of carbohydrates; (right) spatial localization obtained by ToF-SIMS for starch and lipid ions.

resolved technique that can recognize the presence of and distinguish among serpentine polytypes directly in rocks. SR-IR high brilliance allows us to probe tiny areas, as small as 10 μm, as well as a good spectral quality able to distinguish the mineral composition reasonably quickly; finally, it is a non-destructive method. Several thin sections of serpentinitic rocks and prehistoric axes were analysed in reflection mode, and the spectra were acquired directly from specific points of polished stone artefacts. The key results are that all the artefacts analysed are derived from peridotites, ultramafic rocks metamorphosed under greenschist-facies conditions, and characterized by antigorite, diopside and magnetite, sometimes rimmed by penninite. Chemical data demonstrated also a selection of the most suitable raw material for axe production. Mineralogical and petrographic data exclude most Eastern Alps origins, limiting the options to the Hohe Tauern and related secondary deposits of the Drava River hydrographical system. This conclusion agrees with previous studies, as this region is rich in copper ore deposits, which have been exploited since prehistory. These results show the IR microspectroscopy technique to be a helpful tool for the characterization of archaeological lithic material, allowing easy discrimination of the different mineral polytypes, *e.g.* serpentine present in the stone, without reducing them into powder.

Archaeological bone materials record characteristic markers of life in prehistoric times (dating, climate, environment, diet, human migration). Their isotopic and chemical composition can be combined with palaeontological, archaeo-zoological, anthropological and palaeo-genetic information. However, archaeological bone materials are altered by different post-mortem processes. Although sometimes very challenging, it is key to estimating whether the archaeological information is still reliable, or has been modified during burial. Reiche *et al.* (2010) characterized for the first time the state of preservation of 5,000-year-old archaeological bone sections at the microscale. An archaeological mammal bone fragment ~8 cm in size from the Neolithic Chalain Lake site (station 19, 3,000 to 2,970 BC, Jura, France) was selected because of the very good preservation of its organic and inorganic fractions. Bones are nanocomposites composed mostly of two phases: an organic one consisting mainly of type I collagen, and a mineral one consisting of carbonate hydroxyapatite (carb-HAP), closely packed at the nanoscale. Using a method adapted from biomedical research, which avoids bias caused by variation in sample thickness, IR band intensity ratios were used to characterize the state of preservation of the organic and the mineral phases. Specifically the amide I/phosphate ratio (1,660/1,095 cm^{-1}) was used as a measure of the collagen content; carbonate/phosphate ratio (1,415/1,095 cm^{-1}) as a measure of the carbonate content and the (1,030/1,020 cm^{-1}) ratio to evaluate apatite crystallinity. Figure 9 shows the osteon structure of the archaeological bone; the collagen content is homogeneous around the Haversian canal, except in the upper left part, carbonates are clearly enriched in the Haversian canal, and apatite crystallinity is very heterogeneous in HAP. Most of the work done on archaeological bone materials established the state of preservation of the organic and the mineral part separately, without investigating the spatial distribution of the different phases. SR micro-FTIR imaging in transmission mode on archaeological bone thin sections allowed

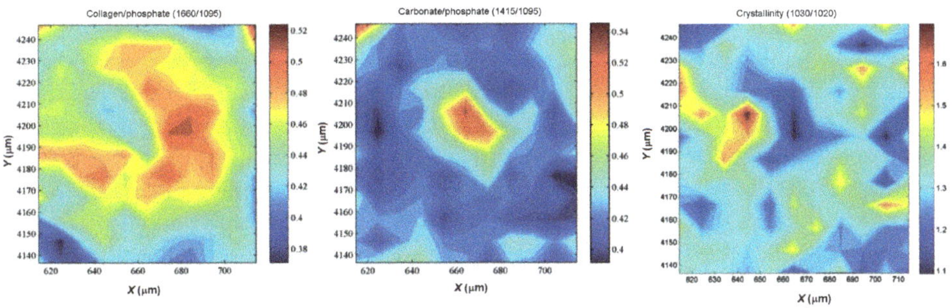

Figure 9. SR-FTIR images showing: (left) the amide I/phosphate ratio (1660/1095 cm^{-1}) corresponding to an osteon area of an archaeological bone sample from the Chalain Lake site (Reiche *et al.*, 2010); (centre) the carbonate/phosphate ratio (1415/1095 cm^{-1}) and (right) the apatite crystallinity (ratio of 1030 to 1020 cm^{-1} bands) for the same osteon. The colour scale goes from red (high intensity) to blue (low intensity) in all images.

investigation of both aspects simultaneously, and provided spatially resolved information on the state of preservation at the microscale. The fine changes in the bone ultrastructure revealed by IR were in agreement with complementary observations of the same specimen. SR micro-FTIR imaging is confirmed to be a powerful tool for obtaining relevant archaeological information and to localize and characterize the less altered zones in the archaeological bone. This information is very useful for the successive selective microsampling necessary for isotopic analysis or dating.

A similar line of research involving IR microspectroscopy concerns fossil bones, their spatial distribution and molecular structure (Lebon *et al.*, 2011). Diagenetic alteration modifying fossil bones over geological time can limit their use as archaeological and palaeontological proxies. The understanding of fossilization processes and the evaluation of the extent of diagenetic alterations of bones remains a major issue in much current research in this field. The complex hierarchical structure of bone tissue and the spatial heterogeneity of the diagenetic alterations, however, induce significant chemical variation at the microscale. To investigate these variations in bone composition at the histological scale, an SR-FTIR microprobe was applied to fossil bone samples of different age and distinct preservation states: bone from Magdalenian layers dated at 15 ka in the Bize-Tournal cave (Aude, France), and samples from the Song Terus site (Java island, Indonesia, 60 ka) were studied. These bones showed partial recrystallization and zones with locally preserved collagen and biogenic carbonates, whereas others showed a recrystallized mineral fraction without any remaining collagen.

Modern bones were used as a reference to determine crystallite size/crystallinity, in addition to the distribution of the collagen matrix and mineral properties. For this purpose, the collagen/phosphate ratio was used to monitor the distribution of the organic matrix. Other specific peak height ratios were used to monitor the composition and structure of the mineral phase, such as the carbonate/phosphate ratio (CO_3/PO_4), the acid phosphate/phosphate ratio (HPO_4/PO_4) content, as well as the mineral crystallinity after validation and calibration with modern and fossil bones (Fig. 10). Note that a preparation based on PMMA resin impregnation was developed to study heavily altered ancient bones, so that both large and thin sections of fossil bones could be obtained independently of their preservation state and maintaining their histological structure. The distribution of the different components of the mineral matter provided information about the state of preservation of the samples and about the underlying alteration processes. SR-FTIR micro-spectroscopy is indeed an almost unique technique for the identification of preserved and altered areas in bone samples. It may also be used to select samples/parts for palaeo-isotopic reconstructions, or for dating as well as for ancient DNA analysis.

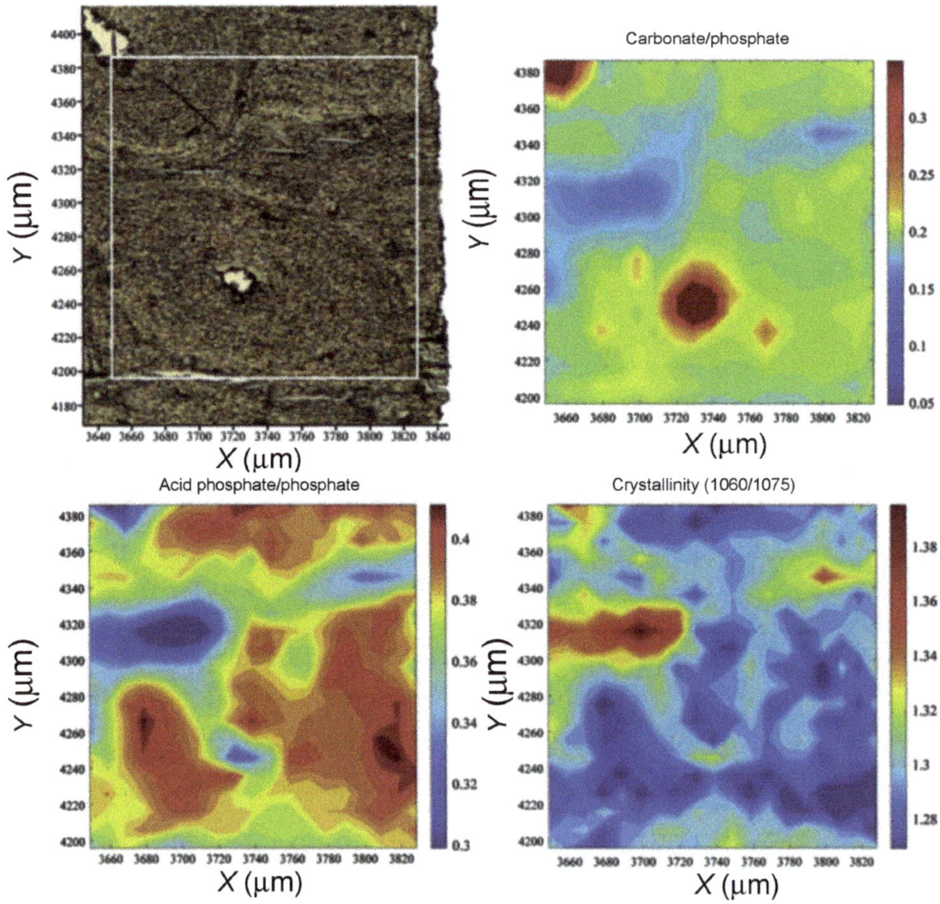

Figure 10. Comparison of (a) the optical view of the analysed area of a bone sample 60 ka old (Lebon *et al.*, 2011) and FTIR colour-coded images showing the spatial distribution of the bone composition; (b) the carbonate/phosphate (CO_3/PO_4) ratio; (c) the acid phosphate/phosphate (HPO_4/PO_4) ratio; and (d) the crystallinity (1060/1075 cm^{-1}).

Paints consist of different materials/layers: binders, pigments and fillers. Learning about their aging is important for the conservation and restoration of artworks. In an altarpiece painting from the 15th century, traditional egg tempera coexists with the previous drying oil technique. Also, they can exhibit sometimes golden but mostly silvery areas (lustre) obtained by the application of a very thin foil of metal, normally protected from the atmosphere by a varnish or resin, and glued by adhesive substances: egg yolk, drying oil or animal glue. The latter materials often produce reaction compounds which, with time, may result in silver alteration, blackening or even disappearance. In an archetype study (Salvadó *et al.*, 2011), synchrotron-based microFTIR was used to identify organic coatings used to protect silver, the metal foil gluing layers and the reaction compounds in 15th Century Catalunya altarpieces. For example, the Silver degradation area on the altarpiece of 'Santa Llúcia' (by B. Martorell 1400?–1452, Museu Nacional d'Art de Catalunya) is shown in Fig. 11. In the same figure, a series of IR spectra obtained from the area in red were obtained from freshly fractured fragments pressed in a diamond cell and in transmission mode *via* a 10 μm × 10 μm SR-IR microbeam. The various layers are identified easily by their IR fingerprint. The more general result was that the compounds formed were due to atmospheric corrosion and the apparent degree of alteration depends mainly on the state of conservation of the organic protecting coatings. The mechanism of formation of the silver compounds involves Ag chloride and Ag sulfide. The presence of cracks and pores helps the penetration of the atmosphere into the inner layers, and the growth and efflorescence of silver corrosion products induces tensions and flaking of the layers. The sub-millimetric layered structure, the high variability and small amount of compounds present as well as their organic-inorganic nature made the use of micron-sensitive high-resolution technique as SR micro-FTIR essential for this study.

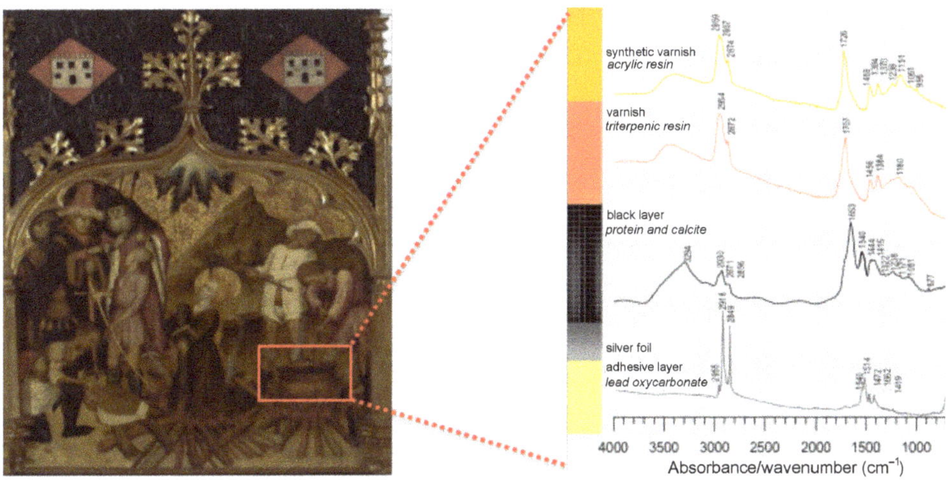

Figure 11. (left) Silver degradation area in red on the altarpiece painting of 'Santa Llúcia' (by B. Martorell 1400?–1452, Museu Nacional d'Art de Catalunya, Barcelona). (right) Series of IR spectra showing the various layers of the protected silver area from surface: acrylic varnish (modern restoration), original varnish, black paint over Ag metal foil and adhesive preparation base, respectively.

The high brilliance of SR allows IR spectra to be obtained routinely from the surfaces of reflecting samples, and clearly opens the possibility for non-destructive testing of museum objects with no sampling. The historical contribution (Salvadó *et al.*, 2005) made by SR-FTIR analysis of alteration and corrosion layers in bronze by means of direct IR reflection measurement is interesting. The sample was a 7th Century BC Corinthian-type bronze helmet from The Manchester Museum. This was a fine example of ancient Greek technology, made from a single piece of metal. The art of making such helmets survived until around 1500 AD, but clear information about its manufacture is still lacking. Although FTIR in reflection mode on unpolished surfaces gives poorer results than the ideal transmission mode, the use of SR helps to overcome most issues concerning spectral quality. In this specific example, the detection system had to be adapted to focus on the surface of a large 3D object (Fig. 12). Although only the compounds present on the surface could be determined, the approach allowed selecting different areas of the surface of the object, thus measuring a representative choice of sampling points without extracting fragments from the specimen. One of the outcomes of this study was the determination of the authenticity of the noseguard piece soldered onto the helmet. With FTIR a protein compound associated with animal glue was observed, which was distributed all over the helmet. This glue was very likely applied to prevent further corrosion well before the helmet was acquired by the Museum. Differences were observed between the corrosion products identified: in particular, hydrocerussite on the noseguard piece and malachite all over the rest of the helmet (Fig. 12). This indicates that the noseguard piece was made of a lead-containing copper alloy, whereas the helmet was made of a lead-free copper alloy. Consequently, it is highly probable that the noseguard had been replaced and is not an original element.

In conclusion, SR-IR reflectivity allows the 2D and 3D study of complete objects with a spatial resolution of 10 μm or better, a condition extremely useful for instance in the study of manuscripts (inks, paintings, alterations), and for monitoring the surface degradation and corrosion of large museum objects in a truly non-destructive way.

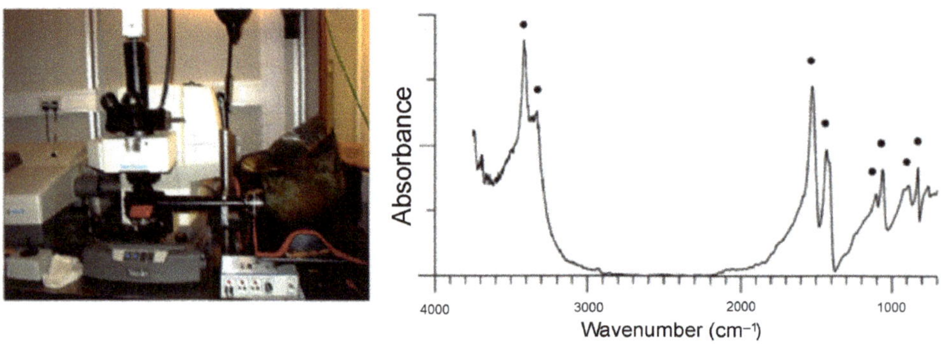

Figure 12. (left) IR microscope coupled to the SR source at SRS Daresbury (UK) and the accessories necessary to collect spectra from an integer bronze helmet in the reflection mode; (right) SR-FTIR spectrum (128 scans, 4 cm^{-1} resolution, spot size 10 μm × 10 μm) collected in reflection mode. The malachite, the $CuCO_3$ and $Cu(OH)_2$ bands marked with a dot, allow the alteration of the alloy probed to be monitored.

Ancient paintings are complex materials because of their mixed organic/mineral as well as amorphous/crystallized nature. Restorers and conservators are deeply interested in any useful insight into ancient artistic practices, *e.g.* through the identification of pigments and binders in the matrix. The example chosen concerns Buddhist mural paintings from the Bamiyan site, in the highlands of Afghanistan – middle of Bamiyan valley. Here Buddhism flourished between the 5th and 9th centuries, with the iconic outcome of two Giant Buddhas, magnificent Buddhist paintings in caves, temples, *etc.* Unfortunately famous for its partial demolition by the Taliban, the Bamiyan site was included in 2003 in the World Heritage Danger List, with a UNESCO project developed to decipher the painting techniques and deterioration mechanisms specifically for the Bamiyan Mural Paintings.

In this context, a synchrotron-based IR microprobe was a valuable analytical tool for revealing the two-dimensional molecular and structural compositions of such complex matrices at micrometric resolution (Cotte *et al.*, 2008). Figure 13 shows a painting stratigraphy with average FTIR spectra calculated from the different coloured layers with 1–3 = paint layers, 4 = a ground layer, 5 and 6 = sizing layers applied on the earthen plaster render to provide a better surface for depicting images. Similarities and differences can be observed in spectra 1 to 4, with C–H and C=O ester bands supporting the identification of an oily binder. Hydrocerussite ($Pb_3(CO_3)_2(OH)_2$), identified through the characteristic O–H stretching, is concentrated in layer 4. In layer 2, high spots of hydrocerussite are the white visible grains. Cerussite ($PbCO_3$) – associated mainly with lead white – is concentrated in the white layer 4. In layer 1, copper oxalates and copper hydroxyl-chlorides are observed *via* the large C=O band of metallic carboxylates. Based on shape and position, a mixture of both copper and lead soaps is then hypothesized. This mixture is present in the two green layers (1 and 2), and the upper layer with copper oxalates could be an alteration of layer 2 due to ageing. In layer 3, the red colour is partially due to goethite (FeOOH), identified by the doublet at 896–796 cm^{-1}, and the large O–H band. Layer 5 contains a mixture of proteins (see amide I band) and polysaccharides (C–O). This suggests the use of mucilage and/or protein-based material, such as hide glue or egg, for the surface preparation. Finally, in layer 6 the specific C=O acid mode is typical of a natural resin (*e.g.* mastic).

In this example, IR spectral quality obtainable by SR was essential to enable distinguishing the various compounds in the stratigraphy. By choosing appropriate regions of interest, it was possible to distinguish ester, acid, amide, carboxylate, carbonate and oxalate groups with the required micron spatial resolution across the specimen.

In the study of the beginning of civilization, the archaeological sites of the first towns in the Neolithic age are rich in information on human life, obtained *e.g.* from house relics. In a recent study, samples from two Neolithic towns were investigated (Anderson *et al.*, 2014): Çatalhöyük in central Turkey with the remains of a town occupied between 7400 and 6000 BC (renowned for elaborate wall paintings in many houses, with designs ranging from patterns/geometric shapes/abstract designs to hunting scenes) and Sheikh-e Abad in the Central Zagros Mountains of Iran which was occupied between 9800 and 7600 BC as a much smaller but earlier site.

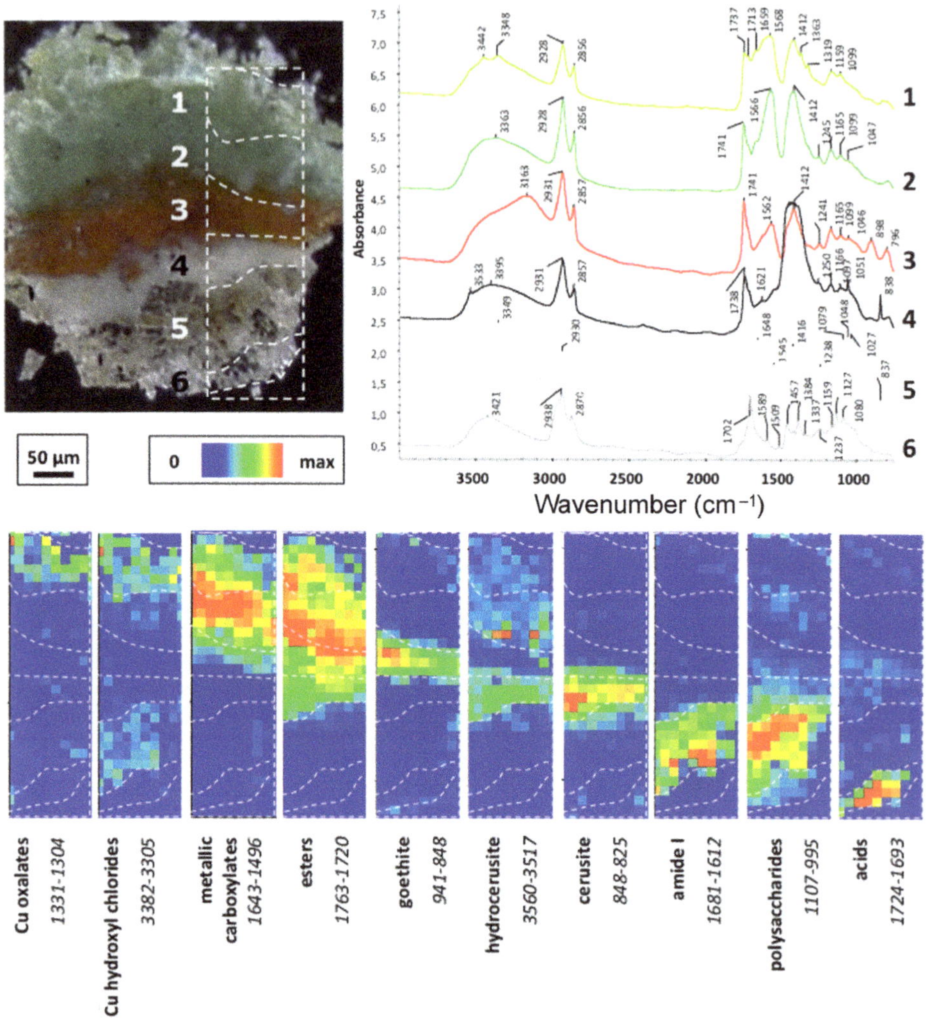

Figure 13. SR-IR stratigraphy of a Buddhist mural painting from Foladi. (Upper, left) Visible light picture; (upper, right) average FTIR spectra; (lower) IR chemical mapping over an area 144 μm × 384 μm, step and beam size 12 μm × 12 μm.

Interestingly, the internal walls of the mud brick houses in the first had been plastered very frequently by inhabitants. Paintings were applied occasionally to the surface of the plasters, but this artwork would only have been displayed for a short time before being covered over with further layers of plaster. The key sample from Sheikh-e Abad post-dating *ca.* 8000 BC is a roughly shaped ball of red sediment: ~5 cm in diameter; it was found in the debris between two pale coloured platform surfaces, which were part of a series of surfaces built in succession and covered with debris like ashes, lithics, burnt bones, shell and reed phytoliths.

Samples containing red pigment have been collected and analysed using a range of techniques, sub-samples were examined also by SR-based IR microscopy (Fig. 14). Thin layers of red paint in a wall painting from Çatalhöyük were found to contain ochre, hematite and clay, as well as an unexpected component: grains of red and colourless obsidian not identified in any previous studies. These small grains of obsidian may have improved the reflective properties of the paint and made the artwork more vivid in the

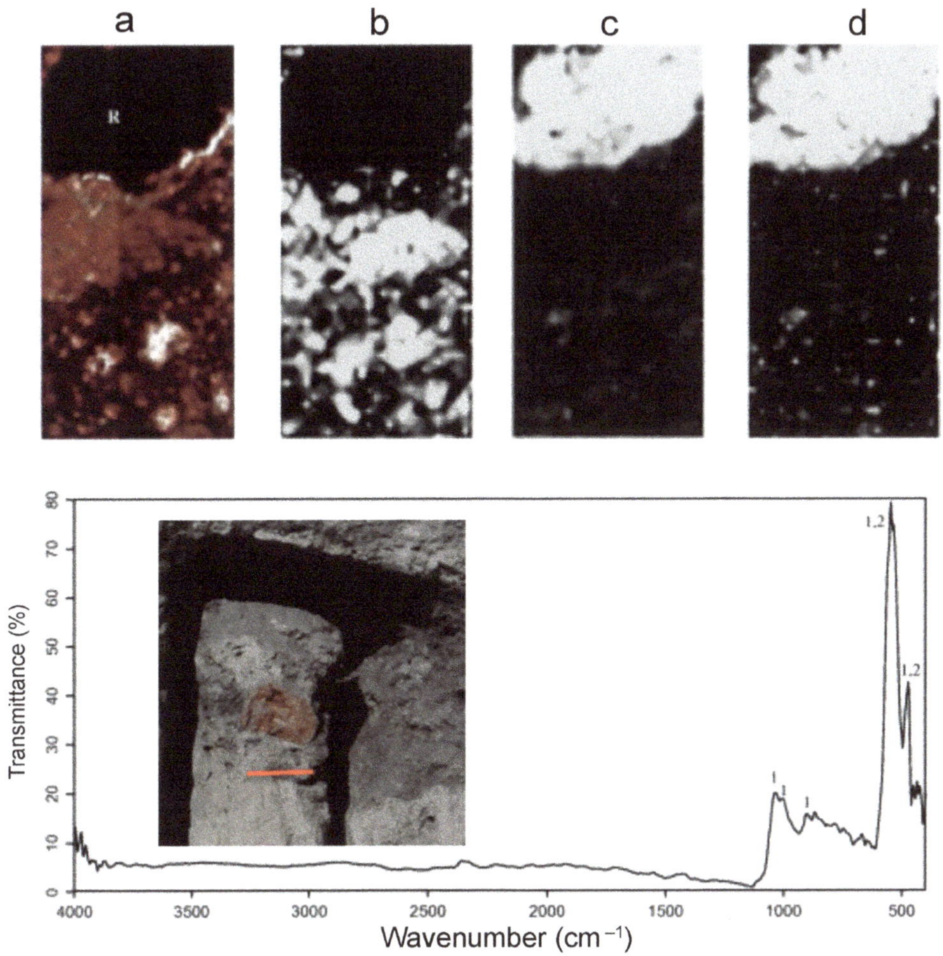

Figure 14. (Upper) SR-IR maps (horizontal size 600 μm) with distribution of minerals across the red sediment from Sheikh-e Abad (inset): (a) visible image of IR mapped area; (b) calcite (CO_3^{2-} asymmetric stretch at 1412 cm^{-1}); (c) kaolin group clay (Si–O–Si and Si–O–Al stretches at 1046 cm^{-1} and Al–O–H bend at 998 cm^{-1}); and (d) clay and hematite (Si–O–Al and Si–O bends and Fe–O lattice vibrations at 547 and 477 cm^{-1}). (Lower) IR spectrum identifying 1 = Clay and 2 = hematite. Inset, locations of block samples at the archaeological site of Tappeh Sheikh-e Abad, Neolithic town in the Central Zagros Mountains of Iran post-dating *ca.* 8000 BC. The red bar $\approx$5 cm.

darkness of the buildings. The obsidian was clearly distinguishable by IR microscopy using a synchrotron radiation source, which enabled appropriate spatial resolution. At Sheikh-e Abad the Neolithic people used a red pigment to paint animal skulls and human bones, but no natural source has yet been found in the area. Analysis of the ball of red sediment found on a possible working surface at Sheikh-e Abad revealed that the red colouring is due to the mineral hematite, which was probably obtained from a source of a *terrarossa* sediment in the local area that had not been modified or altered by the Neolithic people. It may have been quite commonly used selectively for specific objects or architectural materials, which is an exciting prospect for new excavations.

Recent papers (*e.g.* Echard *et al.*, 2008; Bertrand *et al.*, 2013) suggest that greater emphasis must be placed on the specificities of ancient materials, and their theoretical implications for the analytical process. These works point out the importance of the *a posteriori* framework of the study of heterogeneous materials considered in their historicity as structuring parameters of a study. Indeed, in many ancient artefacts, the materials under consideration, both of biological or mineral origin *e.g.* pigments, metals, earths, stone, *etc.* show an heterogeneity from the centimetre to the nanometre scale.

Transformation during the lifetime of these materials by various natural processes such as heating, hydration in wet/dry cycles, chemical reactions, or mechanical constraints (*e.g.* hammering) may introduce additional heterogeneities. Ageing and long-term reactivity with the environment may also contribute with non-negligible physical-chemical perturbations.

Examples combining high spatial resolution and spectroscopy had already shown the importance of this approach when looking at the presence of traces of crystalline mineral phases in copper-based green pigments in cross-sections of a painting of the Catalan Gothic painter Jaume Huguet (*ca.* 1415–1492), as markers of the technology used to prepare pigments and of their alteration state (Salvadó *et al.*, 2002). For this specific analysis only the use of a high-brilliance SR source working with a spot size down to a few microns will guarantee a careful investigation of heterogeneities down to the micrometre spatial dimension and mineral compositions of layers of a few micrometres thick.

An example of how spectral and spatial information through SR-FTIR micro-spectroscopy may permit identification of a pigment in the varnish layer of a wooden musical instrument is given in Fig. 15 (Echard *et a*l., 2008; Bertrand *et al.*, 2013). Quantitative information can be extracted by combining the trace chemical composition and the spatial distribution of data. The spatial distribution of data from ancient materials is, in fact, often informative even as a comparative measurement. This is especially true when the spatial resolution and the uncertainty of experimental methods, taking into account matrix and morphological corrections, makes it challenging if not impossible to attain quantitative information.

A typical case study is the observation of the traces of earth-based pigments in the stratigraphy of the varnish of a theorbo (Echard *et al.*, 2008). Figure 15 clarifies why the visualization of the layered build-up in paint cross sections is often more

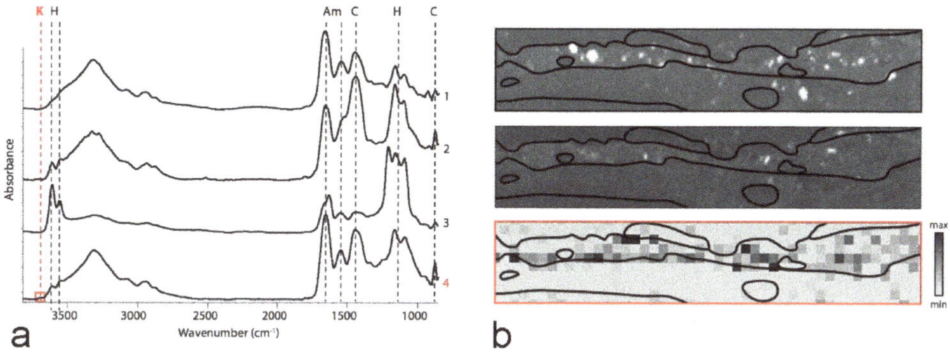

Figure 15. (Left) Comparison among spectra corresponding to a pixel of a first layer (L1) characteristic of a proteinaceous material (1), a grain of calcium carbonate (2), a grain of calcium sulfate hemihydrate (3). The spectrum (4) corresponds to a grain of the earth material located in the layer L2 where the red box is a low-intensity contribution due to the OH-stretching of kaolinite at 3698 cm^{-1}; (right) comparison of the maps of SR-FTIR signal integrated in the 3715–3679 cm^{-1} range (upper) and Al (centre) and Si (lower) SEM/EDX elemental maps.

informative regarding the artistic or technological gesture than the accurate local determination of the content in one of binder, pigment, and other components. Indeed, the FTIR spectral signature, although observed with the high brilliance of the SR source, is still weak and only the spatial distribution of this signal over a whole stratum confirms the presence of the compound in the varnish stratigraphy.

Another characteristic example is the identification and localization of varnishes on a musical instrument to reconstruct and understand the violin makers, finishing techniques (Bertrand *et al.*, 2012a). Varnishes from historical wooden instruments are series of layers deposited on the wood surface, each one being a few micrometres thick. Their composition is complex, usually a mixture of organic compounds, occasionally admixed with inorganic compounds, the interactions of which produce strong inhomogeneity from the macro- to the micro-scale. The stratigraphic reconstruction of these layers may provide unique information for conservation scientists, historians and curators. For old instruments written sources are scarce so that information obtained by these studies can be precious not only for recognizing the sources of material, but also for reconstructing differences and similarities among workshops. For these particular studies, sample preparation is a crucial step in accessing high-quality IR spectroscopy data in transmission mode.

Figure 16 shows the comparison of FTIR spectra of the surface stratum and the underlying stratum, both dominated by a very strong and broad carbonyl stretching absorption and by alkyl stretches. The broad band corresponding to the distribution of C=O stretching vibrations is centred at 1.709 cm^{-1}, a value intermediate between that of aged drying oils at higher wavenumbers and that of natural terpenic resins at lower wavenumbers. The discrimination between the two strata can be outlined clearly by mapping the main spectral signatures (Fig. 16 right). The selectivity and sensitivity at the micron-scale of SR-FTIR micro-spectroscopy is demonstrated clearly here, and

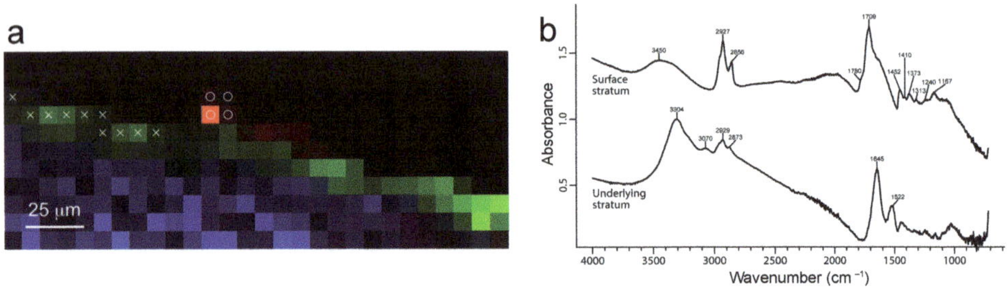

Figure 16. (Left) Mean spectra of SR-FTIR data collected on the underlying stratum (the 10 selected "x" pixels in the map on the right) and on the surface stratum (the 4 selected "o" pixels) of the cello timber; (right) RGB visualization of the FTIR map illustrating the major spectral variations of the stratigraphy where the C–O deformation from cellulose/ hemicellulose is in blue, the amide I band is in green and the C=O stretch from oil/resin compounds is in red.

complementary tests using other techniques did not lead to the identification of additional mineral compounds (driers, pigments) or dyes within the two strata. Although these results have been obtained from a restricted number of artefacts, they allowed the authors to decipher clear differences between major manufacturing techniques used at that time in Europe by famous violin-makers, such as Jacques Boquay and Antonio Stradivari from Cremona. In the studied Stradivari instruments, the underlying stratum was made mainly of a siccative oil (Echard *et al.*, 2010), whereas Boquay used a proteinaceous material on the cello tested.

6. Perspectives for IR and Synchrotron Radiation in conservation and restoration

Notwithstanding the restraints imposed by access to SR facilities, such as typically a six-month call for proposal or beamtime availability at imaging/spectroscopy beamlines, the number of cultural heritage and archaeological studies is growing steadily. At present, most heritage and archaeological studies using SR techniques are limited to materials which can be removed from museums or collections. These include pottery and decorations, glasses, metals, paints and biological remnants, or, very commonly, combinations of organic matter with mineral phases. Typically, the purpose of the investigation is to determine alteration and corrosion dynamics, and/or to understand manufacturing techniques, past-restoration procedures and the provenance of the raw/secondary materials. Due to the difficulty in achieving a statistical validation, in many cases these studies require a comparison with data from similar objects, or samples artificially aged, or contemporary reference materials. This is generally true for microscopy, and more particularly for these challenging specimens an overlooked aspect is the sample preparation (Beltran *et al.*, 2015), which is truly crucial when working on fragile/precious specimens such as mural painting fragments, or brittle bone samples. Many new front-line research proposals have started recently, such as the studies of old documents or ancient books sometimes decorated with precious bezels and gems. For example, the *Evangelarium Marcianum* was the subject of extensive characterization by various techniques (Sebastiani and Crisostomi, 2012). This fine work of Byzantine jewellery, made in the 13th century and then adapted to be a Missal on the altar of San Marco's Basilica in Venice, is a precious book containing ~100 gems on its front and back binding plates. The majority are coloured glass, polished either in the shape of a fancy cut or a cabochon; others are garnets, almost pure almandine in composition and typical of Czech Moldanubian rocks. Other gems, not well identified, are assumed to be amber. An IR chemical and spectroscopic micro-examination showed this supposed amber to have a silicate composition, probably originally obsidian, which was profoundly altered by the effect of secondary hydration. This volcanic glass is widely distributed throughout the source area, referring to the Mediterranean Sea, but seldom used in gemmology in the rare studies of various forms of its degradation (Mottana *et al.*, 2012).

This is a particularly suitable application of SR-based FTIR microspectroscopy because of its unique specificities: high SNR spectra from its micrometre size spot and imaging capabilities suitable for investigating with a high spatial resolution and in a non-destructive way complex and heterogeneous materials, where small amount of inorganic/organic components are present.

Since the original idea to use synchrotron radiation for historical research studies (Smith, 2003), examinations of cultural and archaeological materials have involved all IR spectral ranges from the near- (NIR), to the far-infrared (FIR) and the THz domains. In the last case, 3D imaging capabilities of small objects can be foreseen in the near future. Development of these measurements is indeed expected to benefit both from the large light flux density and from the long-wavelength IR domain coverage of SR sources.

Studies involving minute scale features or small molecular variations related to local chemical changes are those that may really benefit from the high sensitivity of SR-IR microprobe methods. Anomalies with respect to standard behaviour may occur in mineral phases, crystal defects, *etc.* such as in pigments. Trace minerals can be seen as markers of the technology used to prepare pigments and as the base of their alteration (Salvadó *et al.*, 2002).

In complex mixtures, spectral fingerprinting is essential in order to achieve selective identification. Detection of trace features buried in the intense signal from the matrix will require specific strategies, and, again, the availability of broadband and brilliant SR sources. Conventional laboratory IR microscopy gives useful average spectra over areas of ~50 μm or above. Synchrotron-based micro-FTIR easily allows high spectral quality on most matrices by a micro-spot as small as 5 μm × 5 μm. Improved spatial resolution may be achieved by the use of objectives incorporating attenuated total reflectance (ATR), because of its high refractive index and consequent higher optical magnification obtainable (see Section 2). The high brilliance of the synchrotron IR source has enormous advantages and allows collection of high-quality spectra from a few μm^2 area of *e.g.* paint cross-section sample in contact with the micro-ATR crystal (Slogget *et al.*, 2010). An intrinsic advantage of the micro-ATR method should be noted, which is the possibility of analysing thick samples in their integral form, or those specimens not suitable for any thinning method or microtome cutting because of mechanical fragility, *etc*. The counterpart is the contact needed by a 100–250 μm crystal tip with the sample, which may leave a micro-indentation on soft materials.

The combined use of non-destructive structural, spectroscopic and imaging methodologies using SR sources is probably one of the most effective ways of addressing both the spectral quality and spatial resolution requirements typical of cultural heritage and archaeological studies (Cotte *et al*., 2008; Lluveras-Tenorio *et al*., 2012). If IR spectroscopy is molecule-specific – thus valuable in providing chemical imaging – X-ray based techniques are relevant for elemental analysis and speciation (Cotte *et al*., 2010; Bertrand *et al*., 2012a). X-ray absorption spectroscopy (XAS) experiments can be performed on thin samples, and focusing the beam to sub-micron levels enables 2D selective mapping of many elements. When XAS spectra are acquired in X-ray fluorescence (XRF) mode, the sensitivity is high and can be applied to bulk samples, if not to entire objects. Synchrotron-radiation-based micro-XRF thus allows detection not only of major but also minor or trace elements; in particular, transition metals and rare earth elements which may have major impacts on artefact composition both in the bulk matrix or the decoration. Conversely, XRD can yield information about the atomic order and structure of material (Salvadó *et al*., 2009, 2013, 2014). For art and archaeology, a key application lies in the use of micro-XRD for the identification and quantification of crystalline phases in a mixed matrix, *e.g*. with reference to diffraction patterns from databases or reference pure standards. The X-ray Absorption Near Edge Spectroscopy and Extended X-ray Absorption Fine Structure (XANES and EXAFS) techniques are highly sensitive and provide precise information on the redox state and local structure of specific elements, respectively. This is particularly significant in artefacts based on amorphous phase materials, such as glasses, ceramics alloys, *etc*. Finally, this chemical sensitivity resulting from SR high brilliance and broadband energy coverage in the X-ray range, provides an unrivalled means of generating elemental maps, especially of higher-Z elements, which is an ideal match to SR micro-IR ability to deliver molecular maps of low-Z-based materials on the same specimens area by a comparably high spatial resolution at the micron scale.

7. Conclusions

Art and archaeological objects are precious, sometimes too large and/or heavy to allow entire objects to be moved to analytical laboratories. Rigorous conditions are applied to their sampling, so that non-invasive and non-damaging studies are a must in research dealing with cultural heritage and historical artefacts. This is the case for jewels as well as for tools, manuscripts and frescos, paintings and statues, *etc*. Most of the time chemical-physical analysis cannot be applied to these specimens, and few spectroscopic methods can be carried out successfully *via* portable instruments. Vibrational microspectroscopy *via* IR offers the possibility of a complete non-destructive analysis with high molecular sensitivity and specificity. The use of SR-IR is of added value because it secures both the highest spectral quality and true micron resolution (Osmond *et al.*, 2012). Among the complementary technique based on conventional sources, Raman microanalysis must be mentioned for its greater (submicron) spatial resolution. Although with a lower sensitivity, it is a very valuable technique for molecular fingerprinting and mapping, especially of inorganic samples, which are less affected by intense laser beams. On the other hand, organic matrices are more prone to interference issues due to sample auto-fluorescence, sometimes resolved by the use of long wavelength excitation at the cost of a weaker Raman signal.

Vibrational spectroscopy by FTIR *via* compact conventional sources provides valuable molecular specificity, and the use of synchrotron radiation overcomes the spatial limitation when the materials or the objects under analysis are particularly complex or heterogeneous. Spectral imaging of ancient material gives access to topological data that contains additional precious information. Sometimes the 'spatialization' of data allows the 'contextualization' of the data after recognizing spectral and spatial signatures, *e.g.* with reference to the historical period or geographical provenance. The spatial information of IR spectra can be further processed using a chemometric approach that may identify components or compounds inside the investigated object (Bertrand *et al.*, 2013).

In conclusion, the combination of various micro-analytical and multiple techniques based on X-ray and IR at Synchrotron Radiation facilities is now feasible, and provides a unique asset in such multi-material/multi-disciplinary studies (Salvadó *et al.*, 2010). Having positively passed the exploratory and demonstrative phase described in the previous section, SR micro-FTIR, micro-XRF and micro-XRD can now be considered as mature techniques and recognized methods. They are capable of shedding new light on scientific questions in cultural heritage, conservation and archaeological research that cannot be addressed using conventional laboratory methods. If portable devices support important preliminary characterization, SR brilliance and broadband attractiveness guarantees a full spectrum coverage down to the THz level and a large signal/noise ratio at the highest spatial resolution. Indeed, these unique SR parameters are essential for the compositional analysis of extremely small features on valuable specimens, often heterogeneous and with variable combination of molecular composition, which is a characteristic of most ancient materials (Bertrand *et al.*, 2012b).

Acknowledgements

The authors acknowledge all of the several people who helped and are still contributing to the improvement of our spectroscopic work, and in particular the technical and scientific staff at the DAΦNE-L (Frascati - Italy) and Diamond (Oxfordshire - UK) synchrotron radiation facilities and specifically at those IR beamlines. They also thank Ulrich Schade (BESSY) and Liliana Puskar (formerly Australian Synchrotron) for the useful material provided. Special thanks are due to Marine Cotte (ESRF), Loïc Bertrand (SOLEIL) and Nati Salvadó (Universitat Politècnica de Catalunya - Barcelona) for some of the figures used in this chapter and for many useful discussions. Finally, particular thanks are due to Manolis Pantos (formerly SRS at Daresbury - UK) for his careful and interested reading of the manuscript.

References

Anderson, E., Almond, M.J., Matthews, W., Cinque, G. and Frogley, M.D. (2014) Analysis of red pigments from the Neolithic sites of Çatalhöyük in Turkey and Sheikh-e Abad in Iran. *Spectrochimica Acta Part A: Molecular and Biomolecular Spectroscopy*, **131**, 373–383.

Bantignies, J.L., Carr, L., Dumas, P., Miller, L. and Williams, G.P. (1998) Applications of infrared microspectroscopy to geology, biology and cosmetics. *Synchrotron Radiation News*, **11**, 31–37.

Beltran, V., Salvadó, N., Butí, S. and Cinque, G. (2015) Micro infrared spectroscopy discrimination capability of compounds in complex matrices of thin layers in real sample coatings from artworks. *Microchemical Journal*, **118**, 115–123.

Beran, A. and Libowitzky, E. (editors) (2004) *Spectroscopic Methods in Mineralogy*. EMU Notes in Mineralogy, **6**, European Mineralogical Union.

Bernardini, F., De Min, A., Eichert, D., Alberti, A., Demarchi, G., Veluscek, A., Tuniz, C. and Montagnari Kokelj, E. (2011a) Shaft-hole axes from Caput Adriae: mineralogical and chemical constraints about the provenance of serpentinitic artefacts. *Archaeometry*, **53**, 261–284.

Bernardini, F., Eichert, D., Lenaz, D., De Min, A., Tuniz, C., Veluššek, A. and Montagnari Kokelj, E. (2011b) Synchrotron FTIR micro-spectroscopy applied to the study of polished serpentinite artefacts: a non-destructive analytical approach. *Archaeometry* , **53**, 753–764.

Bertrand, L., Robinet, L., Thoury, M., Janssens, K., Cohen, S.X. and Schöder, S. (2012a) Cultural heritage and archaeology materials studied by synchrotron spectroscopy and imaging. *Journal of Applied Physics A*, **106**, 377–396.

Bertrand, L., Cotte, M., Stampanoni, M., Thourya, M., Marone, F. and Schöder, S. (2012b) Development and trends in synchrotron studies of ancient and historical materials. *Physics Reports*, **519**, 51–96.

Bertrand, L., Thoury, M. and Anheim, E. (2013) Ancient materials specificities for their synchrotron examination and insights into their epistemological implications. *Journal of Cultural Heritage*, **14**, 277–289.

Bhargava, R. and Levin, I.W. (2001) Fourier Transform Infrared Imaging: theory and practice. *Analytical Chemistry*, **73**, 5157–5167.

Bhargava, R. and Levin, I.W. (2007) Fourier transform mid-infrared spectroscopic imaging. Pp. 1–24 in: *Spectrochemical Analysis Using Infrared Multichannel Detectors* (R. Bhargava and I.W. Levin, editors) Blackwell Publishing, Oxford, UK.

Born, M. and Wolf, E. (1997) *Principles of Optics*. Cambridge University Press, Cambridge, UK.

Carr, G.L. (2001) Resolution limits for infrared microspectroscopy explored with synchrotron radiation. *Review of Scientific Instruments*, **72**, 1613–1619.

Cestelli Guidi, M., Piccinini, M., Marcelli, A., Nucara, A., Calvani, P. and Burattini, E. (2005) Optical performance of SINBAD, the IR source at DAΦNE. *Journal of the Optical Society of America A*, **22**, 2810–2817.

Coblentz, W.W. (1905) Investigations of infra-red spectra. Carnegie Institution of Washington, Washington DC.

Colarusso, P., Kidder, L.H., Levin, I.W., Fraser, J.C., Arens, J.F. and Lewis, E.N. (1998) Infrared spectroscopic imaging: from planetary to cellular systems. *Applied Spectroscopy*, **52**, 106A–120A.

Cotte, M., Susini, J., Solé, V.A., Taniguchi, Y., Chillida, J., Checroune, E. and Walter, P. (2008) Applications of synchrotron-based micro-imaging techniques to the chemical analysis of ancient paintings. *Journal of Analytical Atomic Spectrometry*, **23**, 820–828.

Cotte, M., Dumas, P., Taniguchi, Y., Checroun, E., Walter, P. and Susini, J. (2009a) Recent applications and current trends in Cultural Heritage Science using synchrotron-based Fourier transform infrared micro-spectroscopy. *Comptes Rendus Physique.*, **10**, 590–600.

Cotte, M., Checroun, E., Mazel, V., Solè, V.A., Richardin, P., Taniguchi, Y., Walter, P. and Susini, J. (2009b) Combination of FTIR and X-rays synchrotron-based micro-imaging techniques for the study of ancient paintings. A practical point of view. *e-Preservation Science*, **6**, 1–9.

Cotte, M., Susini, J., Dirk, J. and Janssens, K. (2010) Synchrotron-based X-ray absorption spectroscopy for art conservation: Looking back and looking forward. *Accounts of Chemical Research*, **43**, 705–714.

Della Ventura, G., Bellatreccia, F., Marcelli, A., Cestelli Guidi, M., Piccinini, M., Cavallo, A. and Piochi, M. (2010) FTIR imaging in Earth Sciences. *Analytical and Bioanalytical Chemistry*, **397**, 2039–2049.

Della Ventura, G, Marcelli, A. and Bellatreccia, F. (2014) SR-FTIR Microscopy and FTIR Imaging in the Earth Sciences. Pp. 447–479 in: *Spectroscopic Methods in Mineralogy and Materials Science* (G.S. Henderson, D.R. Neuville and R.T. Downs, editors). Reviews in Mineralogy and Geochemistry, **78**, Mineralogical Society of America and Geochemical Society, Chantilly, Virginia, USA.

Échard, J.-P., Cotte, M., Dooryhée, É. and Bertrand, L. (2008) Insights into the varnishes of historical musical instruments using synchrotron microanalytical methods. *Applied Physics A*, **92**, 77–81.

Echard, J.P., Bertrand, L., von Bohlen, A., Le Hô, A.S., Paris, C., Bellot-Gurlet, L., Soulier, B., Lattuati-Derieux, A., Thao, S., Robinet, L., Lavédrine, B. and Vaiedelich, S. (2010) Nature of the extraordinary finish of Stradivari's instruments. *Angewandte Chemie International Edition*, **49**, 197–201.

Farmer, V.C. (1974) *The Infrared Spectra of Minerals.* Monograph, **4**, The Mineralogical Society of Great Britain and Ireland, London.

Gillet, P. (2006) Applications of vibrational spectroscopy to geology. Pp. 1–14 in: *Handbook of Vibrational Spectroscopy*, vol. 4 (J.M. Chalmers and P.R. Griffiths, editors). J. Wiley.

Goodall, R.A., Hall, J., Sharer, R.J., Traxler, L., Rintoul, L. and Fredericks, P.M. (2008) Micro-attenuated total reflection spectral imaging in archaeology: Application to Maya paint and plaster wall decorations. *Applied Spectroscopy*, **62**,10–16.

Guilhamou, N., Dumas, P., Carr, G.L. and Williams, G.P. (1998) Synchrotron infrared microspectroscopy applied to petrography in micrometer-scale range: fluid chemical analysis and mapping. *Applied Spectroscopy*, **52**,1029–1034.

Hawthorne, F.C. (editor) (1988) *Spectroscopic Methods in Mineralogy and Geology.* Reviews in Mineralogy, **18**, Mineralogical Society of America. Chantilly, Virginia, USA.

Hofmann, A. (2004) *The Physics of Synchrotron Radiation*, Cambridge University Press

Inoué, S. (1995) Foundations of confocal scanned imaging in light microscopy. Pp. 1–18 in: *Handbook of Biological Confocal Microscopy* (J.B. Pawley, editor). Plenum Press, New York.

Keller, W.D. and Pickett, E.E. (1949) The absorption of infrared radiation by powdered silica minerals. *American Mineralogist*, **34**, 855–868.

Keller, W.D. and Pickett, E.E. (1950) The absorption of infrared radiation by clay minerals. *American Journal of Science*, **248**, 264–273.

Lebon, M., Müller, K., Bahain, J.-J., Fröhlich, F., Falguères, C, Bertrand, L., Sandtd, C. and Reiche, I. (2011) Imaging fossil bone alterations at the microscale by SR-FTIR microspectroscopy. *Journal of Analytical Atomic Spectrometry*, **26**, 922–929.

Levenson, E., Lerch, P. and Martin, M.C. (2006) Infrared imaging: synchrotron *vs.* arrays, resolution *vs.* Speed. *Infrared Physics and Technology*, **49**, 45–52.

Levin, I.W. and Bhargava R. (2005) Fourier Transform Infrared Vibrational Spectroscopic Imaging: Integrating Microscopy and Molecular Recognition. *Annual Reviews in Physics and Chemistry*, **56**, 429–474.

Lluveras-Tenorio, A., Andreotti, A., Bonaduce, I., Boularand, S., Cotte, M., Roqué, J., Colombini, M.P. and Vendrell-Saz, M. (2012) Mass Spectrometric and Synchrotron Radiation based techniques for the identification and distribution of painting materials in samples from paints. *Chemistry Central Journal*, **6**, 45–62.

Marcelli, A. and Cinque, G. (2011) Infrared synchrotron radiation beamlines: high brilliance tools for IR spectromicroscopy. A practical guide to the characteristics of the broadband and brilliant non-thermal sources. Pp. 67–104 in: *Biomedical Applications of Synchrotron Infrared Microspectroscopy* (D. Moss, editor). Royal Society of Chemistry

Mazel, V., Richardin, P., Debois, D., Touboul, D., Cotte, M., Brunelle, A., Walter, P. and Laprevote, O. (2008) The patinas of the Dogon-Tellem statuary: A new vision through physico-chemical analyses. *Journal of Cultural Heritage*, **9**, 347–353.

Miller, L.M. and Smith, R.J. (2005) Synchrotrons versus globars, point-detectors versus focal plane arrays: Selecting the best source and detector for specific infrared microspectroscopy and imaging applications.

Vibrational Spectroscopy, **38**, 237–240.

Minsky, M. (1988) Memoir on inventing the confocal scanning microscope. *Scanning*, **10**, 128–138.

Mottana, A., Gorghinian, A., Cestelli Guidi, M. and Marcelli, A. (2012) Esame gemmologico quantitativo, Pp. 91–107 in: *Splendore Marciano* (M.L. Sebastiani and P. Crisostomi, editors). Nova Charta, Padova, Italy.

Noguchi, N., Moriwaki, T., Ikemoto, Y. and Shinoda, K. (2012) OH group behavior and pressure-induced amorphization of antigorite examined under high pressure and temperature using synchrotron infrared spectroscopy. *American Mineralogist*, **97**, 134–142.

Osmond, G., Boon, J.J., Puskar, L. and Drennan, J. (2012) Metal stearate distributions in modern artists' oil paints: Surface and cross-sectional investigation of reference paint films using conventional and synchrotron infrared microspectroscopy. *Applied Spectroscopy*, **66**, 1136–1144.

Pantos, E., Kockelmann, W., Chapon, L.C., Lutterotti, L., Bennet, S.L., Tobin, M.J., Mosselmans, J.F.W., Pradell, T., Salvadó, N., Buti, S., Garner, R. and Prag, A.J.N.W. (2005) Neutron and X-ray characterisation of the metallurgical properties of a 7th century Corinthian-type bronze helmet. *Nuclear Instruments and Methods in Physics Research. B*, **239**, 16–26.

Petibois, C., Cestelli Guidi, M., Piccinini, M., Moenner, M. and Marcelli, A. (2010) Synchrotron radiation FTIR imaging in minutes: a first step towards real time cell imaging. *Analytical and Bionalytical Chemistry* **397**, 2123–2129.

Reiche, I., Lebon, M., Chadefaux, C., Muller, K., Le Hô, A.-S., Gensch, M. and Schade, U. (2010) Microscale imaging of the preservation state of 5,000-year-old archaeological bones by synchrotron infrared microspectroscopy. *Analytical and Bionalytical Chemistry*, **397**, 2491–2499.

Salvadó, N., Pradell, T., Pantos, E., Papiz, M.Z., Molera, J., Seco, M. and Vendrell-Saz, M. (2002) Identification of copper-based green pigments in Jaume Huguet's Gothic altarpieces by Fourier transform infrared microspectroscopy and synchrotron radiation X-ray diffraction. *Journal of Synchrotron Radiation*, **9**, 215–222.

Salvadó, N., Butí, S., Tobin, M.J., Pantos, E., Prag, A.J.N.W. and Pradell, T. (2005) Advantages of the use of SR-FT-IR microspectroscopy: Applications to Cultural Heritage. *Analytical Chemistry*, **77**, 3444–3451.

Salvadó, N., Buti, S., Pantos, E., Bahrami, F., Labrador, A. and Pradell, T. (2008) The use of combined synchrotron radiation micro FT-IR and XRD for the characterization of Romanesque wall paintings. *Applied Physics A*, **90**, 67–73.

Salvadó, N., Butí, S., Nicholson, J., Emerich, H., Labrador, A. and Pradell, T. (2009) Identification of reaction compounds in micrometric layers from gothic paintings using combined SR-XRD and SR-FTIR. *Talanta*, **79**, 419–428.

Salvadó, N., Butí, S. and Pradell, T. (2010) The role of microspectroscopy techniques in the study of historic artworks. *Spectroscopy Europe*, **22**, 16–19.

Salvadó, N., Butí, S., Labrador, A., Cinque, G., Emerich, H. and Pradell, T. (2011) SR-XRD and SR-FTIR study of the alteration of silver foils in medieval paintings. *Analytical and Bioanalytical Chemistry*, **399**, 3041–3052.

Salvadó, N., Butí, S., Cotte, M., Cinque, G. and Pradell, T. (2013) Shades of green in 15th century paintings: combined microanalysis of the materials using synchrotron radiation XRD, FTIR and XRF. *Applied Physics A*, **111**, 47–57.

Salvadó, N., Butí, S., Aranda, M.A.G. and Pradell, T. (2014) New insights on blue pigments used in 15th century paintings by synchrotron radiation-based micro-FTIR and XRD. *Analytical Methods*, **6**, 3610–3621.

Scott, H.P., Liu, Z., Hemley, R.J. and Williams, Q. (2007) High-pressure infrared spectra of talc and lawsonite. *American Mineralogist*, **92**, 1814–1820.

Sebastiani, M.L. and Crisostomi, P. (2012) *Splendore Marciano*, Nova Charta, Padova, Italy.

Sloggett, R., Kyi, C., Tse, N., Tobin, M., Puskar, L. and Best, S. (2010) Microanalysis of artworks: IR microspectroscopy of paint cross-sections. *Vibrational Spectroscopy*, **53**, 77–82.

Smith, G.D. (2003) Infrared microspectroscopy using a synchrotron source for arts-science research. *Journal of the American Institute for Conservation*, **42**, 399–406.

Unger, M., Mattson, E., Schmidt Patterson, C., Alavi, Z., Carson, D. and Hirschmugl, C.J. (2013) Synchrotron-based multiple-beam FTIR chemical imaging of a multi-layered polymer in transmission and reflection: towards cultural heritage applications. *Applied Physics A*, **111**, 135–145.

Subject and Author index

Numbers refer to the page where a definition or explanation of, and/or a *figure* or a **table** for, a given subject is found. Author names are given with the number of the first page of the paper in which they are involved.

DOI: 10.1180/EMU-notes.20.index

www.ingramcontent.com/pod-product-compliance
Lightning Source LLC
LaVergne TN
LVHW050907210926
848482LV00012B/363

* 9 7 8 0 9 0 3 0 5 6 6 1 8 *